Perceptions of Knowledge Visualization:

Explaining Concepts through Meaningful Images

Anna Ursyn
University of Northern Colorado, USA

A volume in the Advances in Multimedia and Interactive Technologies (AMIT) Book Series

Managing Director: Lindsay Johnston
Editorial Director: Myla Merkel
Production Manager: Jennifer Yoder
Publishing Systems Analyst: Adrienne Freeland
Development Editor: Allyson Gard
Acquisitions Editor: Kayla Wolfe
Typesetter: Christina Henning
Cover Design: Jason Mull

Published in the United States of America by
Information Science Reference (an imprint of IGI Global)
701 E. Chocolate Avenue
Hershey PA 17033
Tel: 717-533-8845
Fax: 717-533-8661
E-mail: cust@igi-global.com
Web site: http://www.igi-global.com

Library of Congress Cataloging-in-Publication Data

Ursyn, Anna, 1955-
Perceptions of knowledge visualization : explaining concepts through meaningful images / by Anna Ursyn.
pages cm
Includes bibliographical references and index.
ISBN 978-1-4666-4703-9 (hardcover) -- ISBN 978-1-4666-4704-6 (ebook) -- ISBN 978-1-4666-4705-3 (print & perpetual access) 1. Information visualization. I. Title.
QA76.9.I52U77 2014
001.4'226--dc23
2013032108

This book is published in the IGI Global book series Advances in Multimedia and Interactive Technologies (AMIT) (ISSN: 2327-929X; eISSN: 2327-9303)

British Cataloguing in Publication Data
A Cataloguing in Publication record for this book is available from the British Library.

Advances in Multimedia and Interactive Technologies (AMIT) Book Series

ISSN: 2327-929X
EISSN: 2327-9303

Mission

Traditional forms of media communications are continuously being challenged. The emergence of user-friendly web-based applications such as social media and Web 2.0 has expanded into everyday society, providing an interactive structure to media content such as images, audio, video, and text.

The **Advances in Multimedia and Interactive Technologies (AMIT) Book Series** investigates the relationship between multimedia technology and the usability of web applications. This series aims to highlight evolving research on interactive communication systems, tools, applications, and techniques to provide researchers, practitioners, and students of information technology, communication science, media studies, and many more with a comprehensive examination of these multimedia technology trends.

Coverage

- Audio Signals
- Digital Games
- Digital Technology
- Digital Watermarking
- Gaming Media
- Internet Technologies
- Mobile Learning
- Multimedia Services
- Social Networking
- Web Technologies

IGI Global is currently accepting manuscripts for publication within this series. To submit a proposal for a volume in this series, please contact our Acquisition Editors at Acquisitions@igi-global.com or visit: http://www.igi-global.com/publish/.

The Advances in Multimedia and Interactive Technologies (AMIT) Book Series (ISSN 2327-929X) is published by IGI Global, 701 E. Chocolate Avenue, Hershey, PA 17033-1240, USA, www.igi-global.com. This series is composed of titles available for purchase individually; each title is edited to be contextually exclusive from any other title within the series. For pricing and ordering information please visit http://www.igi-global.com/book-series/advances-multimedia-interactive-technologies/73683. Postmaster: Send all address changes to above address.

Titles in this Series

For a list of additional titles in this series, please visit: www.igi-global.com

Computational Solutions for Knowledge, Art, and Entertainment Information Exchange Beyond Text
Anna Ursyn (University of Northern Colorado, USA)
Information Science Reference • copyright 2014 • 457pp • H/C (ISBN: 9781466646278) • US $180.00 (our price)

Perceptions of Knowledge Visualization Explaining Concepts through Meaningful Images
Anna Ursyn (University of Northern Colorado, USA)
Information Science Reference • copyright 2014 • 457pp • H/C (ISBN: 9781466647039) • US $180.00 (our price)

Exploring Multimodal Composition and Digital Writing
Richard E. Ferdig (Kent State University, USA) and Kristine E. Pytash (Kent State University, USA)
Information Science Reference • copyright 2014 • 352pp • H/C (ISBN: 9781466643451) • US $175.00 (our price)

Multimedia Information Hiding Technologies and Methodologies for Controlling Data
Kazuhiro Kondo (Yamagata University, Japan)
Information Science Reference • copyright 2013 • 497pp • H/C (ISBN: 9781466622173) • US $190.00 (our price)

Media in the Ubiquitous Era Ambient, Social and Gaming Media
Artur Lugmayr (Tampere University of Technology, Finland) Helja Franssila (University of Tampere, Finland) Pertti Näränen (TAMK University of Applied Sciences, Finland) Olli Sotamaa (University of Tampere, Finland) Jukka Vanhala (Tampere University of Technology, Finland) and Zhiwen Yu (Northwestern Polytechnical University, China)
Information Science Reference • copyright 2012 • 312pp • H/C (ISBN: 9781609607746) • US $195.00 (our price)

Multimedia Services and Streaming for Mobile Devices Challenges and Innovations
Alvaro Suarez Sarmiento (Universidad de las Palmas de Gran Canaria, Spain) and Elsa Macias Lopez (Universidad de las Palmas de Gran Canaria, Spain)
Information Science Reference • copyright 2012 • 350pp • H/C (ISBN: 9781613501443) • US $190.00 (our price)

Online Multimedia Advertising Techniques and Technologies
Xian-Sheng Hua (Microsoft Research Asia, China) Tao Mei (Microsoft Research Asia) and Alan Hanjalic (Delft University of Technology, The Netherlands)
Information Science Reference • copyright 2011 • 352pp • H/C (ISBN: 9781609601898) • US $180.00 (our price)

Table of Contents

Foreword ix

Preface xiii

Acknowledgment xxvi

Section 1
Perceiving

Chapter 1
Articulation and Translation of Meaning 1
Introduction 1
Articulation 2
Translation of Meaning 9
Conclusion 22

Chapter 2
Communication through Many Senses 25
Introduction 25
Modes of Gathering Information and Communication through the Senses 27
What Can Be Done with the Senses to Defend (And Attack)? 35
The Sense of Numbers: Communication Using Numerals 39
Expanding the Senses through the Use of Technology 42
Sensitivity 44
Spatial Perception and Cognition 48
Conclusion 55

Chapter 3
Essential Art Concepts 61
Introduction: Visual Literacy 61
Art Definitions: What is an Artwork and What is Not an Artwork? 64
Basic Art Concepts 65
Pastiche 68
Project: Drawing an Apple 70

Design, Craft, and Folk Art ... 74
Elements of Design in Art ... 77
Principles of Design in Art ... 87
Conclusion ... 94

Chapter 4
Creativity, Intuition, Insight, and Imagination ... 96
Introduction ... 96
The Meaning of the Creative Process ... 97
Approaches from Other Points of View ... 99
Some Earlier Studies on Creativity ... 101
Convergent vs. Divergent Thinking and Production ... 104
Critical Thinking and Creative Thinking ... 104
Creativity vs. Intelligence ... 105
Intuition ... 108
Insight ... 109
Curiosity ... 109
Imagery ... 111
Imagination ... 114
Initiatives Taken to Advance Digital Creativity ... 118
Technological Offerings Serving a Creative Process ... 118
The Role of Imagination in Learning and Teaching ... 120
Conclusion ... 122

Section 2
Visual Cognition

Chapter 5
Cognitive Processes Involved in Visual Thought ... 131
Introduction ... 131
Basic Information About Human Cognition and Cognitive Science ... 132
Cognition and Some of the Structures of the Brain ... 135
Methods and Tools for Studying Neural Structures and Functions ... 135
Some Ideas about Cognitive Thinking and Memory ... 137
Some Philosophical Approaches to Perceptual Thinking ... 149
Cognitive Informatics ... 150
Earlier Investigation About Cognitive Development ... 151
Visual Development ... 153
Mental Imagery ... 155
Gestalt Psychology ... 160
Intelligence and Visual Literacy ... 161
Conclusion ... 167

Chapter 6
Semiotic Content of Visuals and Communication ... 174
Introduction ... 174
Sign Systems, Codes, and Semiotics ... 177
The Semiotic Content of Product Design ... 187
Biosemiotics ... 188
Conclusion ... 190

Chapter 7
Pretenders and Misleaders in Product Design ... 193
Introduction ... 193
Canonical Objects ... 194
Product Semantics in Product Design and Marketing ... 196
Pretenders, Misleaders, Informers, Double-Duty Gadgets, and Multifunctional Tools ... 197
Pretenders vs. Camouflage ... 205
Conclusion ... 206

Chapter 8
Metaphorical Communication about Nature ... 208
Introduction ... 208
Translation of Meaning with Visual and Verbal Metaphors ... 209
Metaphors Using Iconic and Symbolic Images ... 212
Inspiration Coming from the Rules and Phenomena ... 213
The Use of Metaphors in Data, Information, and Knowledge Visualization ... 215
Nature Derived Metaphors Serving as the Enrichment of Interdisciplinary Models and Architecture ... 219
Metaphorical Way of Learning and Teaching Using Art and Graphic Metaphors for Knowledge Visualization ... 220
Conclusion ... 221

Section 3
Tools for Translating Data into Meaningful Visuals

Chapter 9
Visual Approach to Translating Data ... 226
Introduction ... 226
Computer vs. Pencil ... 228
Evolution of Imaging with the Use of Computing ... 230
Basic Ways of Graphical Display of the Data ... 234
Computer Simulation ... 242
Evolutionary Computing ... 244
Conclusion ... 245

Chapter 10
Digital and Traditional Illustration 247
Introduction 247
Background Information: Techniques 248
Illustration: Projects 254
Product Design: Making a Drawing for a Deck of Playing Cards 270
Audiences for Book Illustrations 274
Conclusion 275

Chapter 11
Making the Unseen Visible: The Art of Visualization 277
Introduction: From Data to Pictures and Then to Insight 277
Visualization 283
Tools and Techniques Used in Visualization 301
Music Visualization 311
The Meaning and the Role of Visualization in Various Kinds of Presentations 312
Cultural Heritage Knowledge Visualization 325
Conclusion 326

Chapter 12
The Intelligent Agents: Interactive and Virtual Environments 332
Introduction 332
Augmented and Virtual Reality 333
Information Systems and the Design Science 340
Social Networking and Collaborative Virtual Environments 341
Examples of Interactive Visualization Techniques 343
Ubiquitous Computing 346
Second Life 350
The Internet of Things 351
Startups 352
Conclusion 352

Conclusion 357

Further Reading 358

Compilation of References 359

About the Author 387

Index 388

Foreword

I attended a visualization conference some twenty years ago, with little idea of what I was doing there. I am a painter, and I use computer graphics. I must have been looking for connections, but this was a new field to me. I was aware of special effects in movies, and I knew that animators favoured ants because they were easy to render, but like most people, I took information displays for granted—traffic warnings, air traffic charts, weather systems. I did not think of these as being designed, but I could tell they had become "computerised."

Here we were, all gathered together: statisticians, academics, ecologists, physicists, chemists, medics, software engineers, artists; with presentations by librarians, armourers, car designers, market analysts. The visuals were eerily beautiful, rolling and rotating 3D meshes like models of galaxies or neurons in the brain. My attention wandered away from what they were supposed to communicate—the librarian needed to know how many readers borrowed simultaneously both Jane Austen and Hemingway. I was looking at them with delight, like a collector of gorgeous medieval maps who does not care whether they represented something real or imaginary. Then I was alerted to cases where it would be life or death: interactive 3D modelling let medical students explore the interior of the heart. Leonardo—anatomist, inventor, engineer—had made detailed drawings of those same ventricles some five hundred years before.

The past twenty years have seen some amazing science and art projects, wonderful art exhibitions in science museums, but it would be a mistake to blur the distinctions. Information design needs to be objective, clean, and articulate. It has to be tested. A "graphic," such as a desktop icon, logo, or a stop sign, has a job to do. An ambiguous message is no use. Recognition must be instant. An art piece can be vague and linger in the imagination. It may not "represent" anything, but some diagrams—information stripped down to the essentials—have the look of abstract art. The most often-cited visualization masterpiece must be Harry Beck's 1931 London Underground map, itself derived from electrical circuits. It was an inspired invention, seized on by commuters the moment it was tried out in 1933—despite it misrepresenting the actual positions of the subway lines. Mondrian, a London resident in the early forties, took up the theme in the late Boogie-Woogie paintings.

I should here mention something obvious. When stuck with a difficulty, such as being lost in a city, and then realising where we are, we naturally say, "I see." Seeing is synonymous with being aware, realising, recognizing. We draw maps in the sand, plumbing diagrams on envelopes. What we want is information we can act on. Apart from the occasional aesthete, we do not need to savour the image itself. More and more we just look, click, and go.

Sometime in the eighties, I began to draw on the computer. The flickering screen was more like a diagram than "art." I could get a fuzzy "print-out," but it was nowhere near as detailed and smooth as a photograph. You could play with shapes and colours. You could spin them about and rearrange them at will. I knew how squares and circles came to be "generated" by inputting simple directions. Therefore, I am of the generation who can think of putting information into the processor to get an image. Today, we take the information out. Compared with today's laptops, those computers were primitive, slow, bulky. It was a bit of a struggle. We did not have Facebook or Wikipedia. I now use a Cintiq, which means I draw directly on the screen, but one advantage back then was that you sensed the affinity with the logic of the program. You knew how to build forms in the right and simplest manner. I recall walking through a park after a session and looking at the trees and seeing the branching command—go up, divide in two, repeat. Computing gave you this after-image, this glimpse of nature's systems. In fact, representations of the tree were the default 3D modelling demos, and my observation was something of a cliché, but here and there a theoretical conundrum has been resolved in a moment of pattern recognition. One such event took place in 1951 at Kings College, London: Rosalind Franklin looked at a foggy X-ray slide and identified the double helix of DNA. Once "seen," the problem was solved.

Can we expect an interchange between researchers in labs, information modellers and artists, now that they all use Photoshop? Could we find these connections within the field of abstract art? I have long been an abstract artist—not exclusively. I admire Paul Klee's "Thinking Eye" method of improvising, of "taking a line for a walk." In the thirties, anthologies of modern art included photos of bacteria, but we should be wary of the term "abstract" because it implies the artist is attempting to represent something immaterial and distant, to "visualize" an idea perhaps, something less tangible than a bowl of fruit. This is not necessarily the case.

Traditionally, artists have always begun their studies with drawing. You can get an idea of how teaching methods have evolved over the past hundred years by looking at drawing manuals. Here you find, in embryonic form, some of the concepts that inform today's computer-aided visuals. It is doubly interesting, because much of the explaining is done with drawn illustrations – drawings of hands holding the pencil the correct way, vertiginous lines heading to the vanishing points of perspective. In other words, visualizations of how to visualize. There were fierce arguments about whether children should be disciplined or allowed to express themselves freely, and whether it really was possible to draw what was in front of you without first understanding a good deal about it. Before you drew the chair, you might need to know perspective. Before you drew the model sitting in it you might need to know how the skeleton and muscles fit together. Others argued that this was beside the point: all you needed to do was to look and see what was there, but they did agree on one principle: the trick was not so much in learning to draw as in learning to see.

Today teachers tend to encourage the student's individual creativity, rather than strive for uniformity and accuracy. Drawing, generally, is not seen as an examination subject – except in China. Ideas about drawing, about what else drawing might be, how it might include video and performance for example, are in flux. The precision drawing necessary in aeronautics and architectural fly-throughs is necessarily computer strict. Educators are left speculating how drawing should be taught. Forget about pencil and paper, the still life and the posed model. Have the cameraphone and iPad finger-painting taken their place? Have smart phones made us smarter? Because we speed dial and know all the icons? Whatever the answer, if we want to work in this field—as artists or as information designers—we should probably still be able to draw a table. That would not have been much of an attainment a hundred years ago. Art students then had to copy geometric figures pinned to the classroom wall, copy approved drawings by the masters, draw simple still life objects in correct perspective. They would study plants and animals. They would learn to draw from memory. There are samples of students' memory drawings—bell-towers, fire engines, clocks—whose virtuosity would stun today's tutors, but that previous generation would probably be as shocked by what we can draw with our gadgets.

It would be impossible to provide a comprehensive how to draw book today with the scope of those publications. "Nature" meant botany, animal studies, and landscape scenery. Much of what we know of our universe—from the very small to the very large—we know second or third hand through TV documentaries. Direct observation is not an option.

I recall another lucky encounter at a conference. This time an astronomy conference. I just gasp in awe when I look at a Hubble photo of a tiny fraction of the night sky, clouds of luminous gas, millions of galaxies. I assume I am looking at a photo, where what is "out there" more or less corresponds with what I am looking at. Not so. I am looking at a sophisticated simulation, a visualization. The data for these images is processed and edited; the colour is cosmetic. Asked what would I see were I suspended amidst these unimaginably distant galaxies, my astronomer colleague looked perplexed. Do not take these literally. It would be like looking at a paper map of the Atlantic Ocean and thinking that was the same as the miles of turbulent sea. It is just a way of representing something unknowable, massaged until it looks like something familiar.

Therefore, I would sidestep these questions of whether we need to draw, whether we have become so much better than previous generations at understanding the world around us, better at making serviceable information boards, and better at making art works that embrace the insights of science. For the most part, I am skeptical about progress and technology, but I feel we should do all we can to follow our curiosity and see where it leads. I like the ambition of this remarkable book, and following some of its hints, we can afford a little humility as a species. Yes, we navigate with GPS, we Skype, we shop on the Web, but what of the birds that migrate across continents? Do we dismiss that as blind instinct, or do we respect it as visual thinking? I was intrigued to learn that experiments have shown that pigeons—which do not migrate—have differentiated vision, in that they navigate with the left eye and find food to peck with the right eye. Pigeons "were found to be at least as good as humans at memorising and categorising visual images. At the time, this seemed bizarre because pigeons appeared to be so inept at other tests, but when it was later realised that pigeons rely on visual maps to navigate… it made perfect sense" (Birkhead, 2012, p. 184).

James Faure-Walker
University of the Arts, UK

James Faure-Walker *(born London 1948, St Martin's 1966-70, RCA 1970-72) has been incorporating computer graphics in his painting since 1988. In 1998, he won the "Golden Plotter" at Computerkunst, Gladbeck, Germany. He was one of five English artists commissioned to produce a print for the 2010 South African World Cup. He has eleven works in the collection of the Victoria and Albert Museum, and his work was featured in "Digital Pioneers" there. One-person exhibitions include Galerie Wolf Lieser, Berlin (2003); Galerie der Gegenwart, Wiesbaden, Germany (2000, 2001); Colville Place Gallery, London (1998, 2000); the Whitworth, Manchester (1985). Group exhibitions include "Imaging by Numbers," Block Museum, Illinois, USA (2008); Siggraph, USA (eight times 1995 -2007); John Moores, Liverpool (1982, 2002); DAM Gallery (2003, 2005, 2009), Bloomberg Space, London (2005); Digital Salon, New York (2001); Serpentine Summer Show (1982); Hayward Annual, London (1979). He co-founded Artscribe magazine in 1976, and edited it for eight years. His writings have appeared in Studio International, Modern Painters, Mute, Computer Generated Imaging, Wired, Garageland, and catalogues for the Tate (Patrick Heron), Barbican, Computerkunst, Siggraph. His 2006 book, Painting the Digital River: How an Artist Learned to Love the Computer, Prentice Hall (USA) won a New England Book Award. In 2013, he won the Royal Watercolour Society Award. He is Reader in Painting and the Computer at CCW Graduate School, Chelsea, University of the Arts.*

REFERENCE

Birkhead, T. (2012). *Bird sense: What it's like to be a bird*. London: Bloomsbury. ISBN 978408820131

Preface

A SUMMARY

The focal premise of this book is a conviction that multisensory perception is becoming an important factor in shaping the current lifestyle, technology, and reasoning. This is because a growing number of biologically inspired technological solutions are based on our knowledge about living organisms that communicate in ways not resembling the traits of human senses. This is also because social communication is becoming multisensory, interactive, interdisciplinary, and technology-augmented. Researchers are often solving problems according to social behaviors and the heuristic ways the social societies of insects, fish, or birds solve their difficult situations.

With projects involving readers' cooperation, this book discusses background material that might be useful for computational solutions for knowledge, art, and entertainment. The book offers a discussion of issues related to visualization of scientific concepts, picturing processes and products, as well as the role of computing in advancing visual literacy skills.

The topics introduced above are spread between the two books titled: *Perceptions of Knowledge Visualization: Explaining Concepts through Meaningful Images* and *Computational Solutions for Knowledge, Art, and Entertainment: Information Exchange beyond Text.*

NEEDS AND ISSUES THAT AMOUNTED TO SHAPE THIS BOOK

Multisensory Perception of Science and Physical Manifestations

This book emphasizes a need for increasing interest in multisensory, especially visual, ways of thinking and presenting knowledge. Computer scientists, cognitive scientists, science, and technology-oriented professionals are often communicating this need. The world is enthralled by the multisensory solutions because interactive application software such as apps, installations, and multimedia presentations are pervasive in technology, education, and everyday life. Computer scientists, engineers, and technology experts see and acknowledge the power existing beyond visual explanations with comparative power. However, many are not ready to approach this subject in practical terms. Other people take an attitude that an active approach toward visual thinking and presentation of scientific and computational concepts may be marginally interesting to computer scientists or cognitive scientists. The goal of the book is to connect theory with practice, processes with products, and to give the reader an active, engaging experience, which would enhance perception of the role of computer graphics.

Suggestions on How to Read This Book

The chapters of this book include mini-topics that encourage the reader to explain concepts in a visual and verbal way. They serve as a link between theory and the reader's own practice, and encourage the reader to explain and reveal a visual aspect of a theme under discussion. Reader's visual solutions will link and connect the conceptual with the depicted and include the reader in an active, visual style of processing and outputting information. In contrast to a textbook-like style, this book offers information about basic concepts and facts as inspiration for creating visual solutions coming from examples of applications of knowledge, as well as trends resulting from developments in technologies.

Many figures in this book have QR codes (Quick Response codes) for the URL of the Website containing color pictures. Many figures containing art works comprise the QR codes in order to bridge the offline text with online presentation of art by enabling the reader to access the Webpage and look at art works. "Digital and Traditional Illustration" provides information and a picture about a structure of QR code matrices designed to be detected as a 2-dimensional digital image by a semiconductor image sensor and then digitally analyzed by a programmed processor. A QR code is a matrix barcode consisting of black modules (square dots) arranged in a square pattern on a white background that records information about an item.

Interdisciplinary Way of Presenting Topics

This book provides a selection of concepts, data, and information belonging to a number of disciplines. This is because most of recent advances in knowledge result from cooperation of specialists in seemingly unrelated domains. Moreover, the progress often moves forward through networking, chatting, using Skype, or simply updating the school-based knowledge. Fields of research become interdisciplinary, interactive, and often integrated. Many themes discussed in this book have been annotated with explanatory notes, some of them being obvious for readers focused on the issues under discussion, and many appearing to be unrelated for those concentrated on other fields of interest.

A question arises about the ways the teaching about art and design could be combined with programming and computing. Both are aimed at enhancing higher-level thinking skills, abstract thinking, creativity, and novelty. Many artists apply programming to create art works or visualizations, and many computing scientists and programmers do the same. The content of these programs becomes a question belonging to the art domain, while inquiries about what can be done to make these programs aesthetic becomes the problem of the usability territory. After pursuing a study of the arts, a programmer may gain a viewpoint about the purpose for programming the individual projects and making sense of it in further phases, and thus achieve a more ontological attitude relative to the essence of being.

While constantly immersed in the mind puzzling natural phenomena, objects, and processes explored by sciences, we gain knowledge and experience, a good deal of it ensuing from our school education. However, educational assessment involves multiple-choice tests as a typical form of testing. In order to prepare ourselves to tests, we have often memorized particular facts, laws, and formulas, each and every one with the test questions in mind. This kind of knowledge interweaves with a whole landscape of knowledge we acquire later. Our knowledge constantly changes along with the developments in technology. At the same time, the school tests are the same for all students, disregarding the diversity of the intelligence types described by Howard Gardner (1993/2011, 1993/2006): visual/spatial intelligence, verbal/linguistic, logical/mathematical, bodily/kinesthetic, musical/rhythmic, interpersonal, intrapersonal, naturalistic, and existential intelligence. We may feel our own visual or verbal preferences in dealing with our tasks. The projects presented in this book are designed to inspect selected themes from a totally different perspective.

A Place for the Arts in the Multimedia-Oriented Social Environment

One may say art is an interpretation of human perception saved accordingly. This book focuses on a visual approach to natural events rather than on their detailed analyses. It encourages the readers to perform some mental activities in a visual way. Many agree that our ways of communication are drifting toward visual media; our efficiency in sharing knowledge and emotions may depend on our adaptability and ability to convey them in an up-to-date way. It may have something in common with Barbara Smaller's wish that was pictured in a June 4, 2012 issue of *The New Yorker* (p. 114): "I'm looking for a career that won't be obsolete before my student loan is paid off." This book attempts to respond to the changing role of art and promotes including the learning of art into the technology-oriented world.

Viewers used to appreciate art they considered beautiful, which often was meant as the lifelike art works that resembled real-life objects. At the present time, due to the pervasive presence of social networking sites, groups of interconnected people exchange information and cooperate applying computing. Their creative activities involve higher-level thinking processes aimed at approaching multisensory, interactive actions. We may notice art-related schools, which were traditionally named the Art and Design departments, now introduce themselves as the Art and Media or the Art and Technology schools, with computing and programming described as a requisite both for the studies and future work.

This Book as a Form of Entertainment

Many agree that mental exercises make the best entertainment. Japanese prize-winning writer, Haruki Murakami (2011, p. 175), assumes that what may be called intellectual curiosity, a desire to obtain knowledge at the universal level, is a natural urge in people. Jean-Baptiste Dubos (1670-1742) wrote that man does nothing but what fulfills his needs; one of them is a need for keeping his own mind busy; otherwise, he becomes bored and unhappy:

The soul hath its wants no less than the body; and one of the greatest wants of man is to have his mind incessantly occupied. The heaviness which quickly attends the inactivity of the mind, is a situation so very disagreeable to man, that he frequently chuses to expose himself to the most painful exercises, rather than be troubled with it. (Dubos, 1717)

With Facebook becoming the most popular social networking site involving about a billion active users, Google being probably the Internet's most visited Website, console gaming becoming a widely used instructional tool, and cinematic effects in motion pictures and games valued as motivational tools, we often consider play as a tool for learning, sharing, and entertainment. Within this template, learning can provide entertainment and amusement.

Dean Simonton (2003, 2004) points out that creativity of scientists is a constrained, stochastic, randomly determined behavior, as the new theories in all sciences are. When we realize that the results of our research are characterized by conjecture and accidental or unpredictable events (Simonton, 2004, p. 41), our curiosity may be enhanced. We cannot predict the results we can only know after computing them. Simonton reminds us of the Albert Einstein's remark:

It is, in fact, nothing short of a miracle that the modern methods of instruction have not yet entirely strangled the holy curiosity of inquiry; for this delicate little plant, aside from stimulation, stands mostly in the need of freedom; without this it goes to wreck and ruin without fall. (Schlipp, 1951, p. 17)

Many would agree our thinking often depends on the tools we use; more areas are available for thought experiments by reason of the developments in technology. Tools may enhance our imagination. Thought activities are often shared, and thus become more entertaining, because almost everything we examine can be visualized. This allows creating technology-based entertainment such as films based on scientific books, worlds populated by avatars and beings existing in the past, the future, or in fictitious environments. Charles Jencks, an architect and writer questioning postulates expressed by the Modern architecture and describing its successors – the Late, Neo, and Post-Modern architecture, wrote:

Whatever the reasons, contemporary science has not yet transformed the cultural landscape not led to a renaissance in thought… In any case, I believe that the ideas of contemporary science do provide the basis for a cultural reawakening and that a new iconography must be made more tangible through art if it is to be assimilated. (Jencks, 2003, p. 20)

Nathan Yau (Lima, 2011, p. 248) describes the citizen science that is based on social data collection, "Although not everyone who 'analyzes' this data will have a background in the proper techniques, a certain level of data literacy must be developed. Visualization will be essential in making the data more accessible." Yau (2011) emphasizes the engaging quality of interactive, flying data that he finds not only explanatory but also compelling and entertaining. Non-professionals become involved in visualization and analysis when they take on microblogging and engage with social applications like Twitter and Facebook. The task is to add structure and tools that take advantage of these open applications, to see the undiscovered relationships, and to interact with our surrounding.

The Power of Visualization and Visualizing Thoughts

According to the pioneer in the field of data visualization Edward Tufte (1983/2001), vision is the only universal language. Gyorgy Kepes (1906-2001), who published an influential book about design and design education *Language of Vision* (1944/1995, p. 13) wrote, "Visual communication is universal and international; it knows no limits of tongue, vocabulary, or grammar, and it can be perceived by the illiterate as well as by the literate."

The Voyager, which is conveying the data about the heliosphere and the interstellar space, had sent into the deep space a gold-plated copper disk containing visual descriptions as a record of our civilization (Figure 1). A committee chaired by Carl Sagan of Cornell University selected the content of the record for NASA. The spacecraft may approach another planetary system in at least 40,000 years (NASA Jet Propulsion Laboratory, 2012).

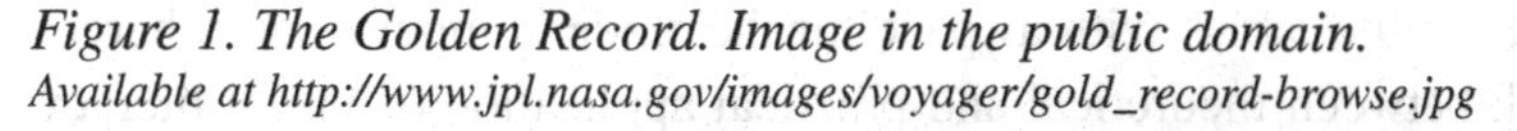

Figure 1. The Golden Record. Image in the public domain.
Available at http://www.jpl.nasa.gov/images/voyager/gold_record-browse.jpg

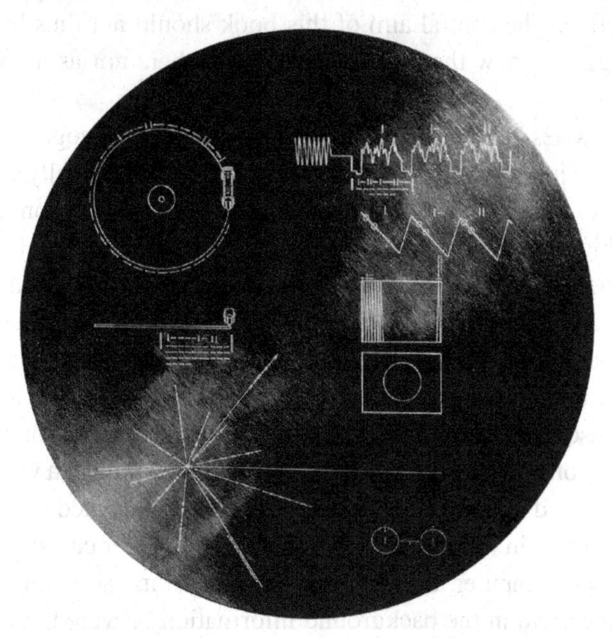

Projects for the Reader

Themes for particular chapters in this book have been selected with several objectives in mind. First of all, descriptions of natural and technological processes are focused on visual ways of explanation. Second, the readers are invited to look at the underlying physical and natural laws and actively react to the nature- and science-derived facts and processes. Translation into the art and visualization oriented frames of thinking supports current methods of communication going through networked, linked, and shared media. Developments in science, technology, and art created with the use of a computer come in a great part from biology-inspired sources. As Gérard Battail (2009, p. 323) wrote, "Life is an outstanding expert in solving engineering problems." The selected nature- and science-inspired themes are intended to encourage the reader to respond in one's own way, by creating, designing, writing, and programming individual reactions to these themes.

Many of us scan and copy items or use Internet resources, and then apply filters to transform them into line drawings; many do not draw at all. We can see this trend in animations and feature films. Perhaps the meaning of drawing is different in electronic media, where images are interactive, linked, and open-ended. Paul Fishwick (2008, p. 4) defines aesthetic computing as "the application of the theory and practice of art to the field of computing." For him, it is the study of artistic, personalized, formal model structures in computing that go beyond representation and events in technology.

The book tells about a number of nature-inspired projects, applications, and technologies, selected with a focus on visual way of communicating solutions. Readers will find framed spaces for their visual and verbal responses. The spaces left for reader's thought and action are the decisive parts of this book; it is a place for novel, personal interaction in the form of drawing and writing. Combining selected fields of knowledge with practical applications in terms of the visual and verbal expression serves as a tool used to show the way of applying one's visual way of solving particular tasks and to work on one's ability to do this. For these reasons, I see the tables with a space for the reader's input as an inherent part of the book. This serves as a link between theoretical and practical application of visual literacy seen from a new perspective. The central aim of this book should not thus be misunderstood, neither as a research source suggesting new themes for other researchers nor as a collection of exercises for particular groups of people.

The goal of art therapists lies in helping people with problems at the cognitive, motor, emotional, and psychomotor levels, to name just a few. It is not necessary that "patients" fully comprehend this material, with its scope spanning from science, engineering, and computing to art concepts. This book is meant for those thinking at the higher, abstract thinking level, who grew to the point of opening themselves to current venues and experiences. Thus, the main thesis of the book is in proclaiming a need for shifting the readers' thinking and acting towards creating visual explanations and solutions based on a selected knowledge base. Filling out by the reader each framed space adheres to the book's intention.

Projects suggested in this book are meant to support visual way of thinking and developing visual communication with the use of visual semiotics by constructing signs, symbols, icons, iconic objects, analogies, and metaphoric connotations, thus conveying some meaning in a visual way. The text is interlaced with projects to be solved by the reader within the boxes designated to their visual/verbal answer to the project. The empty boxes in the text are for sketches; the reader can sketch or can choose to continue working further on the computer. Projects are open-ended in nature and integrative. The sources for inspiration are contained within the background information provided, rather than in a description of an expected outcome. The reader may go any direction one would choose, look for answers on the Internet, or try to create something totally new. My students' solutions accompany the text, along with the author's visual solutions, which are printed black-and-white here.

Each project challenges you to react to a theme under discussion, add your input or modify the content, visualize the concept, and then complete your visual/verbal answer. Each empty space is a place intended for your planned idea for a project. First, you may want to describe it, to sketch, draw, design a concept map, or draw some key frames. Then you may feel ready for writing a program, designing a software application or an app for mobile devices, use graphic software, and create a picture or a sculpture (for example from wooden blocks or the found objects). Finally, you may want to make a photo or a short video of your project, post it online, thus adding your active, creative, independent solution or interpretation and explaining it to others. Your projects may take form of an artwork, a verse, a story, a concept map, animation, comics or manga, or a smart phone app.

The purpose behind these activities lies in their explanative and motivational power to enhance one's visual, graphical, and visualization literacy (both of the readers and of those who would look at their projects). Our environment and its changes influence our thinking and our acting, which we do mostly with the use of computing. For this reason, the following text tells about the connectivity between our daily life, knowledge, art, and entertainment. In a quest for things that last longer, people work on making materials indestructible and designing intelligent applications. This connectivity becomes even stronger because of the changes we experience when our knowledge about the world we live in becomes bio-inspired, nano-oriented, and progressively shared because:

1. The impact of biology-inspired knowledge, technology, and art is growing.
2. The focus on nanotechnology drives the advances in many domains and brings changes in materials, technologies, and applications, influencing each other.

3. The Web-based networking results in changing the way we now solve our problems (with the immediate help coming from often unidentified sources), entertain (we can enjoy gaming with people from far away), and develop in social media, new media art, or networked art (existing in real time and/or in virtual spaces).
4. Programming became accessible and easier due to the visual way of instruction such as processing, with free online instruction (such as HTML, free courses, Apple Developers' kits, SDK – Software Development Kits, etc.), so the art creating often fuses with the manufacture, while the designing of games becomes an art medium.

In a quest for objects that would last longer, people work on making materials indestructible and designing intelligent applications. Projects interweaving the text are intended to associate knowledge with practical applications, facilitate the integration of particular facets of science that have been routinely segregated into special fields, and to follow the current advances in various areas. Our thinking may probably change not only with the technical progress but also with the experienced reality changing along with the advancements in technology and everyday life. Projects are aimed to hopefully engage the readers in practicing visual communication and visual organization of data and knowledge, with a focus on the meaning, not exclusively on data or numbers. When working on these projects, you may hesitate to look at or copy the ready examples, because copying may influence a person who copies and may have an impact on one's personal visual statement. As a summary, the projects offered in this book will most likely prompt inspiration to find progressive solutions based on the informed way of thinking.

As a conclusive remark, with the advent of pervasive computing, with computer-mediated way of thinking and living at many fronts, one might ponder about a need for a talent search and support for all talents that could further advance our ways of living. Three issues come to mind:

1. A need for a free access to the Internet for everyone, disregarding all differences and levels, so every idea-driven and motivated individual could explore, learn, produce, and share knowledge and achievements. This issue seems to face similar obstacles as a free access to water.
2. A need for solving the image- and video-related copyright problems, so every author could freely illustrate one's writings with visual examples, rather than provide complex, lengthy, and often short-living links. For that, an international agreement would be needed to address profit-based issues.
3. In regard to mining and supporting talents, training and education of children should be focused on recognizing and supporting the innate abilities of children. This would allow starting a holistic training of young minds by providing knowledge visualization early, that means from kindergarten (Figure 2). Knowledge visualization has a power to introduce an outline of major ideas and connections between science, mathematics, and programming.

Figure 2 conveys an opinion that before a child learns typical attitudes, misconceptions, and classifications, we can introduce a big picture as inspiration to finding their own interest, focus, and future path. Ongoing developments in computer graphics and visualization techniques may make us to reconsider the needs of education. With a shortage of programmers combined with usual fear of mathematics, programming, and science, one may consider knowledge visualization as a tool for showing the world at the time when attitudes are open and children' brains are curious.

Figure 2. Anna Ursyn, "Visualize Knowledge in Kindergarten" (© 2013, A. Ursyn. Used with permission)

With the use of visualization techniques, themes related to science, nature, math, art, and how they mutually influence each other might be presented to young children as a big holistic spectrum of knowledge. We may instill abstract thinking in young children by supporting an understanding of the surrounding world, which would allow making connections. On the basis of openness to a wider picture, they may have a chance to shape their own, individual focus on what stirs their curiosity, in relation to other levels of knowledge. To enhance instruction with knowledge visualization component, early childhood specialists and departments would need to welcome knowledge visualization specialists on the board.

THE BOOK CONTENT

Section 1: Perceiving

Chapter 1: "Articulation and Translation of Meaning"

This chapter is about concepts of articulation and translation as the ways of exploring meaning. Articulation is discussed as units combined into complete structures and thus meaningfully expressed. The text includes examples of double and triple articulation of signs in languages, programs, and several other fields. Translation—another common thread interweaving distinctive processes and events—may include translation from nature to art (with the use of technology), as well as many forms of visual, verbal, and numeral translation. Two-way translation is discussed, from nature to idea and production (technical solutions) and from products to human perception and creation.

Chapter 2: "Communication through Many Senses"

Sensory messages are examined as electromagnetic waves clearly identified by our senses, consisting of interacting electric and magnetic currents or fields and having distinctive wavelengths, energy, and frequency. Further text discusses modes of gathering information and communication that include sensory responses to electromagnetic waves, visible vibrations exemplified by cymatics, the pitch response, the senses of vision, smell, touch, and taste, all of them further expanded by the developments in current technologies. The sense of numbers is examined next, involving numerical and verbal cognition and communication with the use of numerals. Sensitivity, spatial abilities, and the threshold of sensory information make a part of the issues about biology-inspired computational solutions for enhancing our particular or synesthetic abilities, and the role of imagination in biology-inspired research and technology, learning, and teaching. The role of the sensory input in art, which pertains in some extent to individual curiosity and sensibility, concludes the chapter.

Chapter 3: "Essential Art Concepts"

This chapter comprises a basic overview of visual literacy, computer art graphics, and visual communication design, basic art concepts, elements and principles of design in art, design, craft, and folk art, technical issues related to art and design, and the quality of display, among other topics. A short introduction of this kind may be useful for those focused on domains other than visual arts, and may be helpful for those readers who would like to be reactive to further themes. The advantage of visual displays of information over speech or writing is in its nonlinear and flexible time of viewing, multiple dimensions, and possibility of restructuring its content. The projects resulting from reading this book will hopefully display information aesthetically, with visible traces of reasoning about concepts.

Chapter 4: "Creativity, Intuition, Insight, and Imagination"

Notions such as art creation, creativity, and the creative process are changing in response to the developments in information technologies. Countless options of social networking provide fuel for many forms of online creative works. Comprehension of the role of creativity in new media art involving concepts beyond 2D and 3D graphics, such as interactive and time-based art, networking, online and Second Life presence, evoke initiatives taken in journals, books, college curricular programs, conferences, and the new options taken by artists and designers. This results in the quest of the new role of digital creativity and an emerging need for boosting digital creativity in schools. The text looks at the role of creativity in a process of digital art image creation.

Section 2: Visual Cognition

Chapter 5: "Cognitive Processes Involved in Visual Thought"

Cognitive thinking is discussed here in terms of processes involved in visual thought and visual problem solving. This chapter recapitulates basic information about human cognition, cognitive structures, and perceptual learning in relation to visual thought. It tells about some ideas in cognitive science, cognitive functions in specific parts of the brain, reviews ideas about thinking visually and verbally, critical versus creative thinking, components of creative performance, mental imagery, visual reasoning, and mental images. Imagery and memory, visual intelligence, visual intelligence tests, and multiple intelligences theory make further parts of the chapter. This is followed by some comments on cognitive development, higher order thinking skills, visual development of a child, the meaning of student art in the course of visual development, and the role of computer graphics in visual development.

Chapter 6: "Semiotic Content of Visuals and Communication"

The semiotic content of visual design makes a foundation for non-verbal communication applied to practice, especially for visualizing knowledge. The ways signs convey meaning define the notion of semiotics. After inspection of the notions of sign systems, codes, icons, and symbols further text examines how to tie a sign or symbol to that for which it stands, combine images, and think figuratively or metaphorically. Further text introduces basic information about communication through metaphors, analogies, and about the scientific study of biosemiotics, which examines communication in living organisms aimed at conveying meaning, communicating knowledge about natural processes, and designing the biological data visualization tools.

Chapter 7: "Pretenders and Misleaders in Product Design"

This part of the book is about the meaningful message in product design and the use of pretenders in product design as the carriers of hidden messages that refer to visual practices in design and visualization. The notion of iconic objects, or iconcity of an object, makes a basis of product semantics. Proper design versus pretenders, misleaders, informers, double-duty gadgets, and multitasking tools are discussed and then contrasted with the notion of camouflage.

Chapter 8: "Metaphorical Communication about Nature"

Metaphors are present in our thoughts and make invisible concepts perceivable. The metaphorical way of perceptual imaging is discussed in this chapter, particularly the use of art and graphic metaphors for concept visualization. We may describe with metaphors the structure and the relations among several kinds of data. Metaphors may represent mathematical equations or geometrical curves and thus make abstract ideas visible. Most metaphors originate from biology-inspired thinking. Nature-derived metaphors support data visualization, information and knowledge visualization, data mining, Semantic Web, swarm computing, cloud computing, and serve as the enrichment of interdisciplinary models. This chapter examines examples of combining metaphorical visualization with artistic principles, and then describes the metaphorical way of learning and teaching with art and graphic metaphors aimed at improving one's power of conveying meaning, integrating art and science, and visualizing knowledge.

Section 3: Tools for Translating Data into Meaningful Visuals

Chapter 9: "Visual Approach to Translating Data"

This chapter examines some of the tools that enable a visual approach to translating data, beginning with a comparison of the use of a computer versus pencil in visual communication. A short note follows, discussing the evolution of imaging with the use of computing: the history of computers and then some examples of graphic display and early computer-generated art works. This is followed by a discussion of the basic ways of graphical display of data and strategies for visual problem solving in the context of art and design. Thoughts on visual translation of data include an introduction to computer simulation. Examples of computer simulation and evolutionary computing conclude the chapter.

Chapter 10: "Digital and Traditional Illustration"

Traditional and computing-based illustrations make a great part of our everyday experience. This part of the book examines how traditional illustration types have found their continuation in computing-based media, even when the products mimic the old appearance. The next part includes several projects addressed to the reader and illustrated by student solutions, which refer to various fields of interests or areas of activities and apply selected illustration techniques.

Chapter 11: "Making the Unseen Visible: The Art of Visualization"

Themes and examples examined in this chapter discuss the fast growing field of visualization. First, basic terms: data, information, knowledge, dimensions, and variables are discussed before going into the visualization issues. The next part of the text overviews some of the basics in visualization techniques: data-, information-, and knowledge-visualization, and tells about tools and techniques used in visualization such as data mining, clusters and biclustering, concept mapping, knowledge maps, network visualization, Web-search result visualization, open source intelligence, visualization of the Semantic Web, visual analytics, and tag cloud visualization. This is followed by some remarks on music visualization. The next part of the chapter is about the meaning and the role of visualization in various kinds of presentations. Discussion relates to concept visualization in visual learning, visualization in education, collaborative visualization, professions that employ visualization skills, and well-known examples of visualization that progress science. Comments on cultural heritage knowledge visualization conclude the chapter.

Chapter 12: "Intelligent Agents: Interactive and Virtual Encounters"

Tools available for enhancing and sharing knowledge include intelligent agents, Augmented Reality (AR), and Virtual Reality (VR), among other solutions and paradigms. Collaborative computing became possible due to the advances in social networking, collaborative virtual environments, multi-touch screen-based technologies, as well as ambient, ubiquitous, and wearable computing. Examples of simulations in various domains include virtual computing machines, transient public displays of the data, mining for patterns in data, and visualizations of past events with the use of immersive technologies, virtual reality, and augmented reality. Further discussion relates to the tools for creating and publishing interactive 3D media and the Second Life culture.

Anna Ursyn
University of Northern Colorado, USA

REFERENCES

Battail, G. (2009). Living versus inanimate: The information border. *Biosemiotics, 2*, 321–341. DOI 10.1007/s12304-009-9059-z. Retrieved October 19, 2012, from http://www.springerlink.com/content/r376x87u5mk68732/fulltext.pdf

Dubos, J. B. (1748). *Critical reflections on poetry, painting, and music*. Retrieved October 19, 2012, from http://archive.org/details/criticalreflecti01dubouoft

Fishwick, P. A. (2008). *Aesthetic computing*. The MIT Press.

Jencks, C. (2003). *The garden of cosmic speculations*. Frances Lincoln.

Kepes, G. (1995). *Language of vision*. Dover Publications.

Lima, M. (2011). *Visual complexity: Mapping patterns of information*. New York: Princeton Architectural Press.

Murakami, H. (2011). *1Q84*. New York: Knopf.

NASA Jet Propulsion Laboratory. (2012). Retrieved October 19, 2012, from http://voyager.jpl.nasa.gov/spacecraft/goldenrec.html

Schlipp, P. A. (Ed.). (1959). *Albert Einstein: Philosopher-scientist*. New York: Harper & Row.

Simonton, D. K. (2003). Scientific activity as constrained stochastic behavior: The integration of product, process, and person perspectives. *Psychological Bulletin, 129*, 475–494. doi:10.1037/0033-2909.129.4.475 PMID:12848217.

Simonton, D. K. (2004). *Creativity in science: Chance, logic, genius, and zeitgeist*. Cambridge, UK: Cambridge University Press. doi:10.1017/CBO9781139165358.

Tufte, E. R. (1983/2001). *The visual display of quantitative information*. Cheshire, CT: Graphics Press.

Yau, N. (2011). *Visualize this: The flowing data guide to design visualization and statistics*. New York: Wiley.

APPENDIX: SELECTED WEBSITES WITH ART WORKS SUPPORTING THE TEXT

Listed are some art related Web addresses that may support the text. While links become perishable, there are still Google Images available.

Figures in color and time-based works can be viewed at http://ursyn.com/student%20gallery/index.html; color figures have the QR codes (quick response codes) for the URL of the author's Website.

- **Art History Resources on the Web:** http://arthistoryresources.net/ARTHLinks.html
- **Contemporary Art:** http://arthistoryresources.net/ARTHcontemporary.html
- **WEB Museum:** (Nicolas Pioch) http://www.ibiblio.org/wm/
- **Guggenheim Museum:** New York, Venice, Bilbao, Berlin, Abu Dabi, International Exhibitions: http://www.guggenheim.org
- **SFMOMA ArtScope:** Established for exploring the Museum collection, http://www.sfmoma.org/projects/artscope/index.html#artwork=48370&r=73&zoom=4 shows art works from the San Francisco Museum of Modern Art; artwork images and descriptions pop-up
- **Whitney Museum:** http://www.whitney.org/
- **Whitney Artport:** http://artport.whitney.org - Artport is the Whitney Museum's portal to net art and digital arts, and an online gallery space for commissioned net art projects.
- **Emerging Artistic Practices:** http://www.rhizome.org
- **National Gallery of Art, Washington, DC:** http://www.nga.gov/copyright/toc.htm
- **New Museum, New York:** www.newmuseum.org/
- **Book Sources:** http://en.wikipedia.org/wiki/Special:BookSources/9780520204782. This page allows users to search for multiple sources for a book given the 10- or 13-digit ISBN number. Spaces and dashes in the ISBN number do not matter. In Wikipedia, numbers preceded by "ISBN" link directly to this page.
- **Published Reproductions of Art:**
  - *The Art Book*. Phaidon Press. ISBN 0714864676. Available in a big format or a pocketsize edition (4.8 x 1.1 x 6.4 inches) for $ 9.34, ISBN 071484487X / 9780714844879.
  - *The American Art Book*. Phaidon Press. ISBN 9780714838454.
  - *The 20th Century Art Book*. Phaidon Press. ISBN 0714847984. Works of selected artists are presented in alphabetical order with short descriptions; a glossary of artistic movements complements the collection.
- **Photoshop Resources:** http://sixrevisions.com/photoshop/70-excellent-photoshop-resources/

Acknowledgment

First, I would like to thank the members of the IGI Global publishing team: Erika Carter, Lindsay Johnston, Jan Travers, Monica Speca, and Allyson Gard, as well as all team members for their cheerful and personal assistance with this project.

My thanks go to James Faure-Walker for writing a Foreword to this book.

I would like to thank my students, who contributed to this book by providing images and texts illustrating their projects.

Many thanks go to the reviewers who diligently carried out a double blind review of this book, provided supportive suggestions and critiques, and thus helped to make each part of the book better.

My warmest thanks to my family, friends, and colleagues.

Anna Ursyn
University of Northern Colorado, USA

Section 1
Perceiving

Chapter 1
Articulation and Translation of Meaning

ABSTRACT

Notions of articulation and translation pertain to a great deal of concepts and events described in this book such as communication, cognition, and computing, so they will return as themes for discussion in chapters that follow. It seems particular areas of interest associated with ostensibly unrelated disciplines may have some common features. Both the articulation of units and translation of a meaning or a structure may hold common traits. Inquiring into concepts of articulation and translation may be considered the way of exploring the meaning. The articulation is discussed as units combined into complete structures and thus meaningfully formulated. The further text includes examples of double and triple articulation of signs in languages, programs, and several other fields. The concept of translation—another common thread interweaving distinctive processes and events—may include translation from nature to art (with the use of technology), as well as many forms of visual, verbal, and numeral translation. Two-way translation is discussed, from nature to idea and production and from products to human perception and creation.

INTRODUCTION

In this book, the meaning of things, concepts, and experiences is examined in several ways, for example in terms of perception, cognition, aesthetics, technology, or art production. This chapter is about the concepts of articulation and translation as the ways of exploring meanings. The further discussion will return to both concepts in following chapters.

It seems areas of interest belonging to apparently unrelated disciplines might have some common features. It can be assumed both the articulation of units combined into complete structures, as well as translation of a meaning or

DOI: 10.4018/978-1-4666-4703-9.ch001

a structure may hold common traits. However it is difficult to assign one common definition to the notion of articulation when it refers to mathematics, geometry, and fractal geometry, computer languages, computer graphics, linguistics, music, telecommunication, architecture, material sciences, medicine, and biology, among other options. Several structures existing in various systems, such as natural or programming languages or biological forms are combined in a set of signs that makes a code. They are thus meaningfully articulated. In a similar way, the concept of translation has various meanings in individual disciplines such as mathematics, geometry, and fractal geometry, computer languages, computer graphics, verbal communication, language arts and linguistics, literature, fine arts, art history, philosophy, phenomenology, molecular biology and genetics, material science, and many technology systems.

ARTICULATION

Articulation Applied in Many Domains

The notion of articulation may assume disparate meanings, so it may lack a common definition. We may spot the concept of articulation in various disciplines and areas of life. We may say that articulation happens when we assign a form (in words, notes, or algorithms) to an idea, information, or a feeling. Many of the meanings relate to linguistics where articulation may refer to putting an idea into words by the marking of information in a clause. In human speech articulation may mean the clarity, sharpness, and expressiveness of one's pronunciation and a way of uttering sounds, but also the way individuals adopt cultural forms and practices that are characteristic of their social status. As linguistics, the science of language concerns with several domains such as phonetics, morphology, syntax, semantics, phonology, pragmatics, and historical study, articulation may take a specific meaning for a particular context. Musicians articulate multiple notes or sounds supporting their continuous transition. When it comes to telecommunication, specialists measure the intelligibility of a voice system as the articulation score. Instructional designers and educators articulate curricula by comparing the content of courses between colleges and universities. Architects use the notion of articulation to describe the styling of the joints as the formal elements of design. The term articulation is also used to denote mechanical properties of parts joined into a whole. For example, the articulation in anatomy tells about the ability of flexing limbs where two or more bones make contact; in a similar way, axle articulation in engineering means a car's ability to flex its suspension: pivoted joints are articulated or hinged. In botany as a study of plant life, types of joints between separable parts, as a leaf and a stem, are described as articulation. In a graph theory, the articulation points of a graph are vertices in this graph.

Articulation of Signs in Natural and Programming Languages

A collection of signs used to communicate or to express something can be combined in a system and make a code, simple or double articulated one. Semioticians use the term articulation with reference to a code structure applied in structural linguistics (Martinet & Palmer, 1982); they are studying signs, relations between signs and meaning, codes, and communication related cultural events. Roland Posner (1992, 1993), a Czech-German semiotician and linguist called the string codes those sign systems that have all complex signs reducible to strings; as examples of sign systems, he included natural languages, writing systems, musical notations, codes related to clothing, culinary codes, and traffic signs. A formal language indicates in linguistics, mathematics, and computer science a set of strings of symbols defined by their structural patterns. Communication among people is possible due to articulation of the signs used.

Double Articulation of Signs in Natural and Programming Languages

In a double articulated code an infinite number of possible combinations may be done from a finite number of signs. This is possible because the rules controlling communication are acting on two structural levels, in a similar way as in speech. A study of syntax tells about principles and processes decisive in the construction of phrases and sentences. It has been generally accepted that a possibility of making a great variety of sentences results from double articulation on a syntactic level. Two types of elements interact in creating the meaning in language. Phonemes are the single units – speech sounds without any meaning. Morphemes are the smallest grammatical elements of a sentence: words or parts of words. In order to convey meaning, morphemes are developed through the combining of phonemes. A finite number of morphemes may be combined into a sentence according to the rules of syntax. An infinite number of correct sentences may be constructed this way. German linguist and semiotician Winfried Nöth (2000, 1990) finds the double articulated codes in the systematic codes used in library or warehouse catalogues, and the codes of data processing.

A study on the articulation of sign systems may relate also to formal languages that serve as a basis for creating programming languages. Computational semiotics, for instance algebraic semiotics that also supports social semiotics, applies theories of natural languages, cognitive study, logic, mathematics, and computing to the domain of programming in order to expand the theory of signs and apply it to methods of knowledge representation, studies on artificial intelligence, and computer-human interaction. A question has been around whether computer codes have double articulation. Some (for example, thbz, 2000) hold that computer language designers create double articulated languages, such as Java or C++ by using compilers that perform lexical analysis of vocabulary; tokens that have well-defined meaning are formatted from letters belonging to a fixed set (ASCII). At the second level of articulation, numbers, letters, and symbols are combined to form words. Then, in the first level of articulation, those words are ordered into statements. This enables computer programmers to create a huge variety of programs from a very small number of symbols (Wisegeek, 2012). Machines can process the web of data and sign systems directly and indirectly. Semantic Web, a collaborative movement led by the World Wide Web Consortium develops a framework for sharing and reusing the data across applications.

Double Articulation of Signs in Various Domains

Possibly, double articulation is not unique for verbal messages, so one may look for some analogy of double articulation of speech in other domains, for example, in organization of the Japanese calendar, in formation of protein molecules and a genetic code as well. Maybe organization of the Japanese calendar could be discussed as an example of the double articulation of symbols. In the Japanese calendar one may see the pairing of symbols as a combination of 12 zodiac signs and the stems of wood, fire, earth, metal, and water. The seven-day week names for the days correspond to those used in the Western world but they are also paired with symbols. Combination comprises the 12 zodiac signs, traditional names, and the names that come from the five visible planets (with names of planets given according to the Chinese elements – wood, fire, earth, metal, and water), and the moon and sun (Schumacher, 2011).

According to Manuel DeLanda (2000), double articulation is the process that makes up layers or strata. In the first articulation, substances (types of matter) are combined into forms. The second articulation creates stable forms in which these processes of becoming are actualized. Each articulation has a form component and a substance component; it is the sum of the two articulations that produces structure, or strata. De Landa asserted that Deleuze introduced the concept of a process of double articulation through which geological, biological, and even social strata are formed. The first articulation concerns the materiality of a stratum: the selection of raw materials out of which it will be synthesized (such as carbon, hydrogen, nitrogen, oxygen, and sulfur for biological strata) as well as the process of giving populations of these selected materials some statistical ordering. The second articulation concerns the expressivity of a stratum. The content has both a form and a substance: for example, the form is a hospital and the substance consists of patients hospitalized there and receiving medical treatment.

What really matters is not to confuse the two articulations with the distinction between form and substance, since each articulation operates through form and substance: the first selects only some materials, out of a wider set of possibilities, and gives them a statistical form; the second gives these loosely ordered materials a more stable form and produces a new, larger scale material entity… the first articulation is called "territorialization" and concerns formed materiality, the second one "coding" and deals with a material expressivity. (DeLanda 2010, pp. 32- 33)

Deleuze, Guattari, & Massumi (1987) and then Deleuze & Guattari (2009) apply the concept of double articulation to their divagations related to philosophy, psychology, politics, and semiotics (among other themes). They pose that the strata, which are formed by double articulation, make things out of the formless masses. The existence of one stratum causes that there must be another to border it. Deleuze and Guattari conclude that strata come in at least pairs, one serving as the sub-stratum for the other:

Double articulation is so extremely variable that we cannot begin with a general model, only a relatively simple case. The first articulation chooses or deducts, from unstable particle-flows, metastable molecular or quasi-molecular units (substances) upon which it imposes a statistical order of connections and successions (forms). The second articulation establishes functional, compact, stable structures (forms), and constructs molar compounds in which these structures are simultaneously actualised (substances). In a geological stratum, for example, the first articulation is the process of 'sedimentation,' which deposits units of cyclic sediment according to a statistical order: flysch, with its succession of sandstone and schist. The second articulation is the 'folding' that sets up a stable functional structure and effects the passage from sediment to sedimentary rock. (Deleuze, Guattari, & Massumi, 1987, pp. 40-41)

Possibly, double articulation may be also seen in a formation of protein molecules. Proteins are large biological molecules consisting of a chain (or more than one) chain of amino acids. A unique amino acid sequence specific for each protein depends on the genetic code – a three-nucleotide combination called a codon, which determines an amino acid according to the four-sign code created by four nucleotides contained in a DNA molecule. Nucleic acids contain four kinds of basic amino acids joined in triplets; they code for specific amino acids, which are the structural units that make up a molecule of a protein. Four bases assembled this way in groups of three in each triplet make possible 64 different codings; however, only 20 from about 500 known amino acids (called standard amino acids) are encoded this way by triplet codons into the genetic code. A wide margin of coding possibilities makes the code open, ambiguous, so more than one coding is possible for the most of amino acids.

Maybe also molecular information processing occurring in a living cell by means of transport of biologically active molecules may be interpreted as another example of a double articulation. Cells communicate by transmitting the first messenger signaling molecules from a cell membrane to membrane receptors located on other cells. Receptors activate the signaling proteins, called secondary messengers, located inside of these cells.

The traditional way of presenting a double helix of amino acids is one of the most popular educational applications of scientific visualization. Victorri (2007) adopts a functional approach to the classical comparison between language and biology. He parallels events with a functional signification in each domain, by matching the utterance of a sentence with the release of a protein. Proteins and sentences are both characterized by a complex hierarchical structure. The author makes a distinction between an I-language (the idiolect – individual speech pattern) and an E-language (a language in the common sense).

He argues that the proteome of an organism corresponds to an I-language and the proteome of a species is equivalent to an E-language. Victorri found the same intimate relation between structure and meaning in sentences and proteins (syntactic structure for sentences and three-dimensional conformation for proteins); however, the combinatorial power of language is not shared by the proteome – the entire set of proteins in an organism (but the immune system possesses interesting properties in this respect, and the analogy works to a certain extent regarding the evolutionary aspects). According to Paulo Correa & Alexandra Correa (2004, p. 140), "There is no way to compress time and reproduce in the laboratory all the conditions required to form all the elements of double articulation, the evolutionary advanced nucleic base combinations and the proteins they encode." The authors state, "The emergence of the double articulation between very different classes of molecules underlies the emergence of life" (p. 144). Correas examine whether the molecular event of the biological double articulation evolved in as a cellular or pre-cellular system. Possibly, "the system of double articulation was selected prior to cellularization" (p. 275), in a simple proto-system of double articulation in geological time, a part of a proto-complex of RNA and proteins, and then the emergence of cellular structures and genomic complexes with double articulation system.

Triple Articulation

According to Chandler (2013), semiotic codes have single articulation, double articulation or no articulation. However, media studies specialists focused on new technology domestication and appropriation by users turned their attention to the double and triple articulation of media technologies. Courtois, Verdegem, & De Marez (2012) incorporate triple articulation in the field of convergent audiovisual media consumption. They argue that audiovisual media technologies can be meaningfully articulated as objects, texts, and contexts, which would be useful in uncovering of articulation interactions. Berker, Hartmann, Punie, & Ward (2006) argue that there is a need for research on both double and triple dimensions of message articulation of information and communication technologies in media, examined as technological objects, symbolic environments, and individual texts. According to Robert Stam (2000, p. 114), Umberto Eco argued that cinema has a triple articulation combined from iconic figures, semes (combinations of iconic figures), and kinemorphs (combinations of semes).

One may ponder whether it is possible to have fourfold articulation or multiple articulation, and are there existing examples of it in art, science, or nature. One may also ask whether the fractal geometry design can be seen as a form of articulation, as it represent self-similar or scale symmetric objects, which appear when an object magnified or reduced in size has the same properties.

Figure 1. Anna Ursyn, "Two Skies" (© 1989, A. Ursyn. Used with permission.)

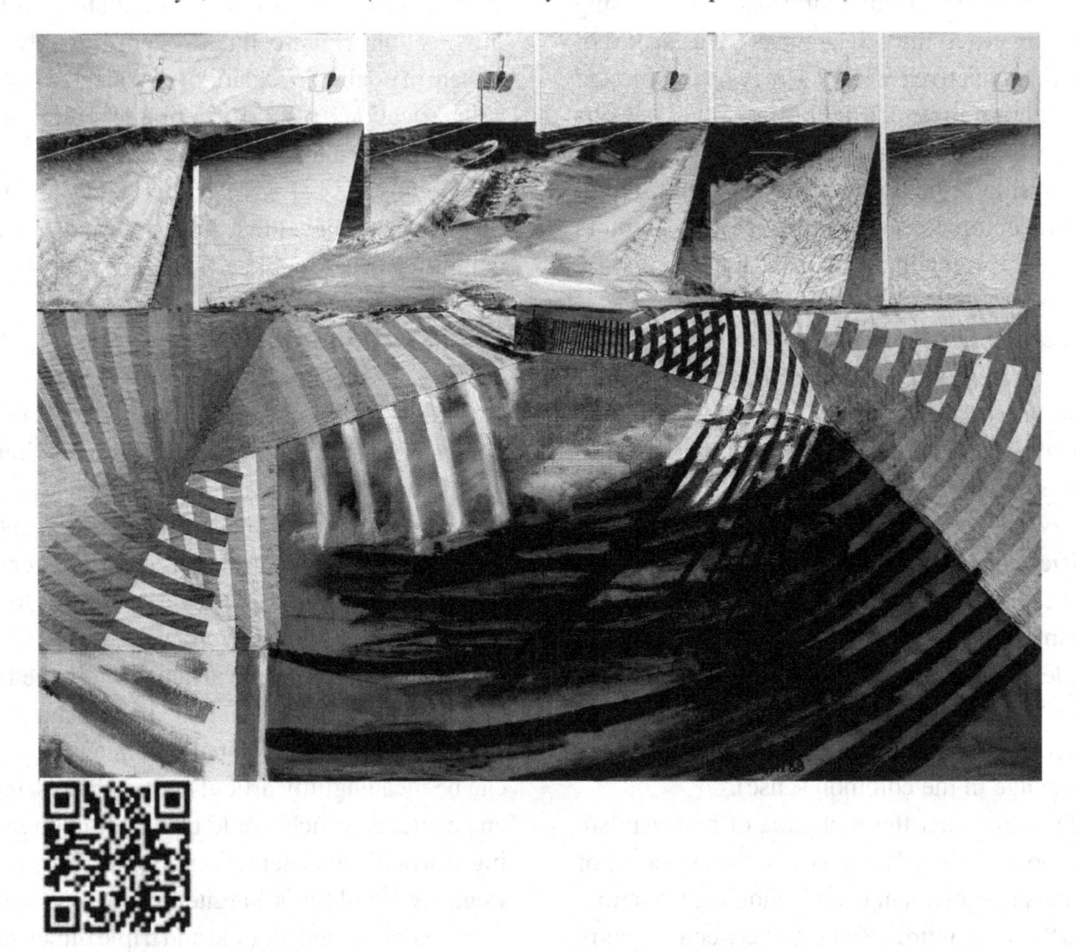

Articulation, Double- and Triple Articulation of Messages in the Arts

We may find an analogy of double articulation in art as well. Like in verbal language, a painting is supposed to articulate elements endowed with meaning; these elements, composed from colors and forms, may be considered equivalent to morphemes. An artist may use masses, rhythms, and relations of colors as the signs of second articulation. Particular colors and simple signs (e.g., in form of angles or curves) may be meaningless by themselves, but they stimulate imagination and evoke emotions. Colors and signs may be combined by the artist according to the artist's own principles of aesthetics, composition, or other concepts. Thus they become meaningful in a particular work of art.

Figures 1 and 2 present the results of this type of thinking. The viewers are invited to devise their individual explanations of the meaning contained in the pictures. Captions below figures may suggest one of possible interpretations of their visual content. One can also find a quick response code for the art works' URLs on the author's website.

The "Two Skies" (Figure 1) implies that the significant characteristic of the Western range is its legend of severe nature and austere cowboy life contrasted with the beauty of red rocks and changing sky. Snow fences and corrals, all made from planks, are of importance for the landscape character: a traveler may perceive them as moiré

Figure 2. Anna Ursyn, Fragile Balance (© 2004, A. Ursyn. Used with permission)

patterns. Algorithmically determined artwork fits in the very essence of these images; it indicates technological impact on our surroundings and links the beauty of man-made technical products and the aesthetics of painting.

Purposefully distorted objects and simplified shapes may be projected as visual symbols, acting as both objects' features and signs. One has an option of choosing artistic means of expression, such as colors and signs, from infinite possible ways of representation, and thus causing that the artwork becomes an ambiguous, open work with a loose relation between its signs and their meanings. The artwork is interpreted by the viewers according to knowledge they share with the artist as the sender of the message. The semiotician and writer Umberto Eco (1979) argues that traditional or 'classical' art was in an essential sense unambiguous, while the ambiguity of modern art as an open work is associated with its high degree of formal innovation. In most cases art education brings students to create conventional and unambiguous images; this was intended to happen to my student Lena who one day in kindergarten received an F for coloring a blank apple blue. Now she writes, "The blue apple became my inner trademark of someone who is proud to be an outcast; I look at the world with a unique sense of knowing. It is good to be different and I take pride that I'm not like everyone else."

In the domain of music, double articulation of multiple notes or sounds may support their continuous transition. For example, Robert Samuels

(2010) examines in these terms a song composed by Franz Schubert, Der Doppelgänger. An audio CD album "Double articulation" (2000) comprises songs of various artists.

In a Figure 2, "Fragile Balance" abstract, programmed elements merge in a repetitive manner with natural forms, thus providing general structure for the artwork. We strive for maintaining balance between our progressive actions and natural habitat where volcanic ridges catch rainwater on the windward slopes, rainforests catch dew, and deities of the clouds put forth their powers over water to sustain fishponds. At the same time, we attempt to compose our thoughts and keep going in good spirits.

"Still Life" is a verse written in a spirit of the double articulation code. The verse comprises a main text and three short parts combining into a whole displayed as a hypertext on the website, where the shorter parts, linked to the main one, can be accessed by a point-and-click action. In this book, places for the short parts have been marked with star symbols (*).

Still Life

Favorite color?
No color exists alone, like a note in a music composition
*or a frame in a movie**
I may say, a blue pen. You will say its cobalt one,
placed on a green book, actually a sap green,
*with one corner covered by a red apple, a symbol***
Well, maroon with salmon speckles;
Supported by a piece of yellow cheese, the kind with holes.
*Swiss cheese, we would say to simplify****
to communicate better, or faster.
Precise description of a color may confuse,
make conversation longer, less interesting,
or even destruct it; change the meaning of what we are saying,
like shouting or whispering.
Leave it to a painter.

*A Frame in a Movie

An artist works a lot to be able to see progress made
or not made
or visible after three weeks would pass.
Hard to see, judge, or value it
right after the task is completed.
Like in a key frame in animation:
two frames are looking almost the same.
Much is needed to make or see the progress.

** Are you an Artist because You Make Art, or You Make Art Because You are an Artist?

Symbols for a meaning shield a musician, a poet, or a mathematician.
A drawing is open for comments, suggestions, and friendly remarks
during the process of making.
No reason is left for rephrasing its visual sentence.
No timeframe is set for viewing it to finish its perception,
organize what's left of thoughts, connotations, and memory of order.

***Stereotypes

A small change makes a big difference.
We all know about the origins, uses, abilities, or problems.
It's faster and more efficient to classify, and find order and importance
knowing what to expect from a person
or from an object for each situation.
We have a saying, a sentence, a word or, much simpler, a gesture.
A vocabulary of symbols makes us well rounded only if we know how to apply our handy simplifications.

See Table 1 for Your Visual Response.

Table 1.

Your Visual Response: Articulation
A concept of articulation and double articulation may be presented in many visual ways. You are invited to make your own interpretation, or maybe you would prefer to enjoy constructing a project presented below. First, find a free and fitting major operating systems 3D modeling program Trimble SketchUp (www.sketchup.com). It is software for design and engineering offering 3D constructions through lines, shapes, and a content of its libraries. This package contains a set of libraries with people, trees, benches, plants, furniture, cars, and more. When you'd download SketchUp 8, you might build a house, furnish it, and then enter it to see how elements set together create a new, whole structure, which may be seen as an effect of a double articulation code where an infinite number of possible combinations may be done from a finite number of elements.   Create a house that blends up with a particular landscape. If this is a forest, make your house shaped after a tree; if you choose to build your house in the mountains, make it look like a rock. A house built on a place by the sea would have a reflective roof and transparent walls. Make the materials for the walls and roof matching the environment; for example, you may choose wooden shingles for the house in a forest, slate tiles for the mountain region, and aluminum or glass for the seaside house. You may also add some trees, shrubs, and flowers that would match the surroundings. Below is a space for making preliminary sketches and planning your solutions.

TRANSLATION OF MEANING

Many Meanings of a Concept of Translation (in Various Disciplines)

In a similar way as with many words, "translation" has acquired varied meanings in the particular disciplines, be it mathematics, geometry, and fractal geometry, computer languages, computer graphics, verbal communication, language arts and linguistics, literature, fine arts, art history, philosophy, phenomenology, molecular biology and genetics, material science, and many technology systems. The concept of translation includes translation from nature to art (with the use of technology), as well as many forms of visual, verbal, and numeral translation.

In *mathematical* terms, translation describes movement of an object in space, with every point of the object moving in the same direction over the same distance, without any rotation, reflection, or change in size. Transformations of drawings, such as reflection, translation, and rotation may help extend our knowledge about geometric relationship. In *geometry*, figures can display translational symmetry among other types of geometrical symmetry. Translation moves geometric figure by sliding, so each of its points moves the same distance in the same direction. *Fractal geometry* describes self-similar or scale symmetric objects called fractals, which are ragged at every scale, not as smooth as Euclidean lines, planes, and spheres, and they do not display translational symmetry. The mathematician Benoit B. Mandelbrot coined the name of fractals from the Latin verb frangere, "to break into pieces," as well as the related adjective fractious, "irregular and fragmented."

The concept of translation was applied to the construction of *computer languages*. At first instructions and data were entered through the medium of punched cards, coded in a symbolic language, which used mnemonic symbols to represent instructions. They must be translated into machine language before being executed by the computer, so the computer may understand high-level languages that are closer to natural languages. The first programmer was the Lord Byron's daughter, Augusta Ada Byron, the countess of Lovelace (1815-1852); a programming language Ada was named in her honor. Navy Commodore Grace Murray Hopper (1996-1992) developed one of the first translation programs for the Mark I computer in 1944. The machine code was recorded on a magnetic drum. Computing programming languages allow achieving translations and adaptations of artificial intelligence (AI) systems and genetic algorithms to digital media (The Carnegie Library of Pittsburg, 2011). However, as noted by Margaret Boden (2006), a perfect translation is not a simple matter, both in case of translating one computer language into another for the sake of artificial intelligence research and in human languages translations: "Even *Please give me six cans of baked beans* will cause problems, if one of the languages codes the participants' social status by the particular word chosen for *Please*" (Boden, 2006, p. 4). There are two types of translators. The compiler converts the program to low-level languages (machine- and assembly languages) to be executed later. The interpreter converts and executes each statement (Ebrahimi, 2003).

In *computer graphics*, construction methods serve to build the solids from simple shapes. In the context of the advanced modeling of complex surfaces, sweeping a two-dimensional figure through a particular area produces solids with translational or rotational symmetry. Solid geometry methods involve operations that joint objects to produce a single solid. Spline is a flexible strip used to produce a smooth curve through a set of plotted control points, and spline curves are drawn in this manner.

In the domain of *verbal communication*, one may notice several ways of translating and mediating a meaning to a verbal statement. Manuel Lima (2011), a designer, researcher and founder of the resource space for anyone interested in the visualization of complex networks, describes the syntax of a language as a new network visualization lexicon, the vocabulary of the particular language. The information-driven network culture unifies the two rising disciplines – network science (which examines interconnections of natural or artificial systems) and information visualization (which translates data into meaningful information thus bridging data and knowledge). A collection of the interconnectedness examples, arranged by Lima in alphabetical order (2011, pp. 97-158) includes:

- **Blogosphere:** Which changes the flow of online information across online social communities; blogosphere maps distinct aspects, from charting the link exchange to the dynamic blog space of the entire country.

- **Citations:** Serving as measure of popularity and credibility, relationships of similarity between subjects, and highlighting proximity across domains.
- **Social-Bookmarking Systems:** Such as Delicious, using non-hierarchical classification systems for posting publicly viewable bookmarks, e.g., easily accessible Delicious tag clouds stored at the servers at Delicious, and connections between tags.
- Donations, Scrutinizing business and political practices, connections between politicians and donors.
- **Email Depictions:** Representing rich containers such as in-boxes, looking by social structures. According to the About.com evaluation based on a Radicati Group's estimate of 2010, "294 billion messages per day, that means more than 2.8 million emails are sent every second and some 90 trillion emails are sent per year. Around 90% of these millions and trillions of messages are but spam and viruses" (About.com, 2012).
- **Internet:** A collection of servers and routers linked by copper wires and fiber-optic cables. Visual depiction of Internet is a first step in awareness of its inherent structure.
- **Literature:** Relationship between lines and words, e.g., in books.
- **Music:** Metaphors for the notes or artists; for example, music affinities in Last.fm that allows keep a record of what you listen to from any player.
- **News:** Alternative methods of dealing with the daily news, visualizations made by the media sources or the authors.
- Protein interaction networks and their intricate structures.
- **Terrorism:** With ties between the groups, people, and decentralized organizations.
- **Trajectories:** with GPS and video tracking devices, creating a lattice of individual networks, adding also other inputs such as GSR – galvanic skin response related to conducting electricity by skin.
- **Twitter:** Great to investigate the behavioral traits of social groups and a trend analysis tool (up to 140 characters). For example, John Maeda used his tweets for his book (Maeda & Bermont, 2011).
- **Wikipedia:** There are currently almost 4 million articles in the English-language Wikipedia, with all Wikipedias as total counting about 21 millions articles in 283 languages (English Wikipedia, 2012).

The notion of translation used in *linguistics*, as a process of translating words or text from one language to another, has been expanding into many domains. It may mean rendering of the meaning assigned to words, images, formulas or other forms of communication; it may also denote the conversion of one form of conveying information into another one, such as translation of an idea into material form, or translation from theoretical findings to a workbench stage.

In *literature* we transfer meaning (we are mediating) from the visual to the verbal, and beyond the verbal. Some say that the 21st century brings about the end of reading. Others counter with the opinion: study results show that because of the growth of social networking people read and write much more. For example, in urban environments such as Manhattan, people are constantly colliding with each other on the streets because they focus on their smart phones, read emails and postings, text their messages, read news on online publications, or watch and share videos at Vimeo, rather than watching their steps. (Vimeo is a video-sharing website that has over 14M registered members who can create, share, and discover videos there; it reaches a global audience of more than 85M each month). One may think that the interactivity factor existing in the network based environment makes the users of the wireless networks the co-authors of the web content, style, design, and structure, which may be conducive to the growing online readership, along with the readers' interest and motivation. One may notice some changes of meaning behind words such as technology, literacy, readership, and readability, which may cause misunderstandings of that kind. Studies on readership, usually designated as readers of certain publications considered as a group, indicate the existence of growing forces outside of newspapers' control, mostly caused by the growing number of website addresses and links providing information. Readership has been also linked with readability – the ease in which text can be read and understood, be it a large-circulation newspaper or a computer program. Also readability of graphics is of concern to visualization specialists. Rusu et al. (2011) incorporated Gestalt principles and introduced the concept of breaks in edges at edge crossings in graphs to improve graph aesthetics and readability. Text readability depends on the length of words and the difficulty of particular sentences, both the semantic difficulty and the syntactic complexity, among many other factors. Several formulas have been developed to evaluate the people' reading skills and increase readership of newspapers but also literary, research, government, teaching, law, and business publications. To enhance readership, both the majority of blockbuster writers and many government agencies write at the 7th grade level.

Figure 3 presents visually everyday experience of commuting, which provides familiar views as an input supporting thoughts in a not always conscious way. The commuters translate them to build individual perception of this regular journey to and from their place of work.

Figure 3. Anna Ursyn, "Commuter's Tunes" (© 2001, A. Ursyn. Used with permission)

And again you commute
Stable but roaming
Sitting quietly while driving in a haste,
Attentive, yet unobservant.
So distinct in your glass case
Yet immersed in milieus
Urban and rural anew,
Too familiar to disturb.
Composing tunes you whistle
Listening to yourself
Learning what you want for sure
Enjoying the company of you.

In a field of *fine arts*, artists write statements and manifestos to complement their art works and then convey to viewers, critics, and jurors the translation of visual messages into literary forms. Examination of nature-derived events and laws may result in developing biologically inspired computer techniques (evolutionary computation, artificial life, artificial neural networks, swarm intelligence, and other artificial intelligence techniques) for the creation of artistic systems (e.g., genetic or evolutionary art) and new media art, music, design, architecture and other artistic fields (Evomusart, 2012). Whichever source serves for choosing the imagery, artists examine mathematical, physical, chemical, biological, and other laws that are ruling the life on our planet and search for visual language to give an account of what they find out.

We may see a continuation of this approach in the domain of *art history* as well as in the work of curators who interpret works of art selected for an exhibition to supply information on labels and in catalog essays. For example, they explain a content of the paintings from the Baroque period or objects of oriental art, thus providing a translation of images into their symbolic meaning. As stated by Hover Collin (2012), language is a translation between logic and natural systems, while biological translation is a crossover between science and art through a biologically inspired computing.

Philosophers working in the field of *phenomenology* describe semiotic phenomenology as a method to advance translation studies (e.g., Kozin, 2008, Muralikrishnan, 2010); they refer to the works of Ferdinand de Saussure and Edmund Husserl and then later theories of translation, such as those formulated by a linguist Roman Jakobson (1995) and a philosopher Jacques Derrida. Exploration of possible meanings contained in particular works of art led Jacques Derrida (1991) to creating "The Truth in Painting" – a study of an idiom in painting, where possible meanings, translations, examples, etymologies, and supplementary parergas were composed into a piece of literary art. Translation of meaning can be augmented by the use of visual rhetoric, for instance metaphors acting as explanatory analogies, by making one-way or two-way connections, and performing comparisons aimed at identifying similarities and oppositions (Lengler, & Vande Moere, 2009). Deborah Harty (2010) explored the translation of experiences of water into drawing on iPad as the phenomenological experiences. She examined the physical elements of the water, its visual and tactile qualities, alongside with the psychological affect the water has on the state of consciousness; what it feels like to experience water.

The act of swimming and drawing about this experience can be examined in the context of artists' works, for example, the works of a British artist David Hockney. It can be presumed that a closer look at this artist's approach to the phenomenological experience of water in an act of swimming, presented in acrylic paintings of swimming pools (with vibrant colors, the light effects, and the visual side of the swimming pool culture in California of the sixties) would create an occasion to view it in wider context. One might also want to make reference to his studies on imaging techniques (Hockney 2006/2001), drawings from the eighties with the Quantel Paintbox (a computer program that allowed the artist to sketch directly onto the screen), and then, since 2009, his paintings of portraits, still lifes, and landscapes with the use of the Brushes iPhone and the iPad application (presented at his show Fleurs fraîches at La Fondation Pierre Bergé in Paris). Hockney is considered the first acknowledged artist using the networkable new media with portable applications running independently from the computer's operating system.

These days a growing number of people, including computer art students, use the iPad in several ways and create their work using an iPad (Fuglestad & Tiedemann, 2013; Apple in Education, 2013; Artlyst London Art Network, 2012; Makeuseof, 2013; The Teaching Palette, 2013).

In *molecular biology* and *genetics* translation refers to processes in macromolecules such as a proteins, nucleic acids, or synthetic polymers, which contain a very large number of atoms. Translation denotes a stage in protein biosynthesis in gene expression. To synthesize a protein or a polypeptide (a polymer consisting of amino-acid residues connected by peptide bonds to form a protein molecule), molecules of the messenger RNA transfer, in a translation stage, the triplets of nucleotides to amino acids. Following the solution of the DNA structure and the deciphering of the genetic code instructing the translation of RNA transcripts, Francis Crick introduced the term 'Central Dogma' (Crick, 1958 and then Crick, 1970). The central dogma of molecular biology depicted the flow of genetic information between macromolecules as proceeding from DNA to RNA to protein. Many animations on the web present the essence of the translation process.

In technological systems, translation of the data serves many managerial, security, or military purposes. For example, translation and adjustment of the data from the street webcam images served to generate a response to civic space. Translation of the viewer/object relationship into material form was achieved with the use of an array of open-source digital software (Matthews & Perin, 2011).

Translation from Nature to Idea and Production: Biology-Based Strategies

Images enhance connections between biology, engineering, and material sciences resulting in growing partnership among academia, laboratories, and industry. Scientists focus on biology-inspired research to understand how biological systems work, and then create systems and materials that would have efficiency and precision of living structures. The use of bio-inspired ways for developing new solutions applies to a number of venues. Translation of form to function is one of the ways of designing new applications and devices. It is often discussed in terms of the 'form follows function' approach; for example, by examining a symmetrical makeup of the butterfly's wings that allows flying. Imitating living systems to create materials that function in a similar way as living beings and respond to external stimuli with a response is another way of translation of biological data to materials science. Inventors develop systems inspired by structures that can be seen in nature, such as optical fibers, liquid crystals, or structures that scatter light. Many times designers combine biomaterials with artificial ones to create hybrid materials and technologies. Numerous authors describe how bio-inspired technologies change the way people think in the fields of computing, software management, material science and material design, resource management, developments in computer technologies, and many other fields.

According to the National Research Council of the National Academies (2008), strategies for creation of new materials and systems may be characterized as (1) bio-mimicry, (2) bio-inspiration, and (3) bio-derivation. By applying bio-mimicry people design structures that function in just the same way as living systems and create synthetic materials that respond to external stimuli. With bio-inspiration people strive to create structures acting in different scheme than the living systems to perform the same functions. Bio-derivation means incorporation of biomaterials into human-made structures. The National Research Council provided examples of these three strategies:

1. **Bio-Mimicry:** Learning the principles used by a living system to achieve similar function in synthetic material and also create materials that mimic cells in their response to external stimuli. For example, certain cells such as T-lymphocytes can sense particular external stimuli, and then deal with pathogens. The challenge is to design bio-inspired systems and devices for detecting hazardous biological and chemical agents and strengthen national security systems.
2. **Bio-Inspiration:** Developing a system that performs the same function, even with different scheme. For example, the adhesive gecko foot, the self-cleaning lotus leaf, and the fracture-resistant mollusk shell are examples of inspiring structures. The cutting-edge optical technology solutions can be found in animals as well: multilayer reflectors, diffraction gratings that spread a beam of light waves, optical fibers, liquid crystals, and structures that scatter light. For instance, Morpho butterfly has iridescence sparkle and blue color visible from hundreds of meters due to periodic photonic nano-structure in scales on wings that responds to specific wavelengths of light without any dye involved.
3. **Bio-Derivation:** Using existing biomaterial to create a hybrid with artificial material. For example, it may be incorporation of biologically derived protein into polymeric (having many similar units bonded together) assemblies for targeted drug delivery. The eyes of higher organisms and the photosynthesis mechanism (in green plants and other organisms which use sunlight to synthesize carbohydrates as food from carbon dioxide and water) are examples of biological structures and processes that can support harvesting light and also fuels (by converting cellulose polymer to ethanol). Deciphering force and motions in proteins driven by sub-cellular, molecular motors can advance clinical diagnostics, prosthetics, and drug delivery. Molecular motors convert chemical energy (usually in form of ATP – adenosine triphosphate) into mechanical energy. Contrary to Brownian movements, they are not driven solely by thermal effects. Scientists strive to create self-evolving, self-healing, self-cleaning, and self-replicating super-materials that could mimic the ability to evolve and adapt. The challenge is not easy to meet: for example, the gecko's adhesive works like a glue in vacuum and under water, leaves no residue, and is self-cleaning; adhesion is reversible, so geckoes alternatively stick and unstuck themselves 15 times per second as they run up walls. As for now, "all attempts to mimic their design or to synthesize artificial polymers that are analogous to the bioadhesives in structure or function have been largely unsuccessful … and the magic of a gecko's "dry" glue with its reversible attachments remains unsolved, unmatched, and more challenging than ever" (National Research, 2008, pp. 63-64).

In a process of nature-inspired inquiry about computing for art creation artists and scientists examine those rules and formulas in science, which define natural processes by abstracting the essentials from specific events or objects, such as elements (e.g., carbon or oxygen) or molecules (e.g., water). One may see organic chemistry as a study of structure, properties, and reactions carbon-based compounds, such as hydrocarbons consisting of hydrogen and carbon, carbohydrates consisting of carbon, hydrogen and oxygen (for example, with a ratio 2:1 as in water molecule), and other carbon-based compounds. We assume biological chemistry examines chemical processes in living matter, information transfer and flow of energy through metabolism, mostly in cellular components such as proteins, carbohydrates, lipids, and nucleic acids including DNA and RNA. At the atomic level, scientists examine soft condensed materials in states of matter neither liquid nor crystalline solid. Soft matter builds membranes and cytoplasm within our cells, and it is so omnipresent in biological systems that we may say we are soft matter examples. Scientists apply abstract concepts, for example permeability (an ability of a material to pass molecules and ions or transmit magnetic flux) or electromagnetism (interaction of the electric and magnetic currents or fields) to find rules and patterns that govern these materials. Explorations on structure and functions occurring in living and artificial matter involve intensive use of visualization techniques providing visual representation of information, data, and knowledge through pictures, information graphics, and also artistic display.

In further parts of this book we will focus on visualization – translation of mental, abstract, formal concepts into images by looking and seeing objects and processes, an "ability to perceive objects and events that have no immediate material existence made possible the visualization and creation of tools" (Ittelson, 2007, p. 279). It has been said that visualization means making the unseen visible; in our projects we will also build a meaningful net of associations – mental connections between the ideas, and connotations – additional, secondary meanings. A large database of visualization projects can be found at an online gallery "Visual Complexity" (Lima, 2012). Visitors can choose a subject from a selection of domains such as art, biology, business networks, computer systems, food webs, Internet, knowledge networks, multi-domain representation, music, pattern recognition, political, semantic, social, and transportation networks, World Wide Web, and others.

Two-Way Translation: From Nature to Technical Solutions and from Products to Human Perception

Many of us would agree with propositions by Mitchell (2005), first that perception changes thought, because what a person sees (and hears, and feels) influences what a person thinks about it; second, that thought changes perception, as how a person sees depends on experiences and memories built up over a lifetime; and third, that the body shapes thought and perception.

Figure 4 presents a visual complement to this thought. Philosophers used to believe nothing comes *ex nihilo* but everything comes out of Nature. We translate our perceptions into the content of our inner thoughts, but they influence what we see.

One may suppose that translation of nature-, science-, or technology-derived concepts into art production and appreciation may go in two directions. First, artists' inspiration from nature (water, sky, earth, animals), science (physical events and processes, mathematical order), and human generated environments (such as rural or urban surroundings) may result in patterns, shapes, and colors that translate the regularities and laws found in nature or science into visual language. Second, it has been generally accepted that advances in science, technology and the emergence of new media art change our perception of reality shifting our frames of reference in criticism, aesthetics, and philosophy of art.

On the other hand, biology-inspired technologies along with biology-inspired art increase the opportunities to advance our way of thinking about materials, resources, their management, and to apply new solutions for the developments in computer science and applications. We may see nature itself and natural processes in a different way due to information coming from the current research results and evolving technologies. As consequence, our ideas and concepts about nature change along with our knowledge base. Redström,

Figure 4. Measure your hopes, size them flat, but do not give up. Anna Ursyn, "Serenity" (© 2001, A. Ursyn. Used with permission)

Skog and Hallnäs (2000) define Informative Art as "computer augmented, or amplified, works of art that not only are aesthetical objects but also information displays, in as much as they dynamically reflect information about their environment. Informative art can be seen as a kind of slow technology, i.e. a technology that promotes moments of concentration and reflection" (p. 125). In many instances info art takes form of ambient display that informs viewers through the periphery of their attention.

Effects, changes, and impacts flow in both directions between nature-inspired technology and nature-inspired art. We may see natural processes differently due to information coming from the current research results and evolving technologies.

Figure 5. Anna Ursyn, "Man" program (© 1987, A. Ursyn. Used with permission.)

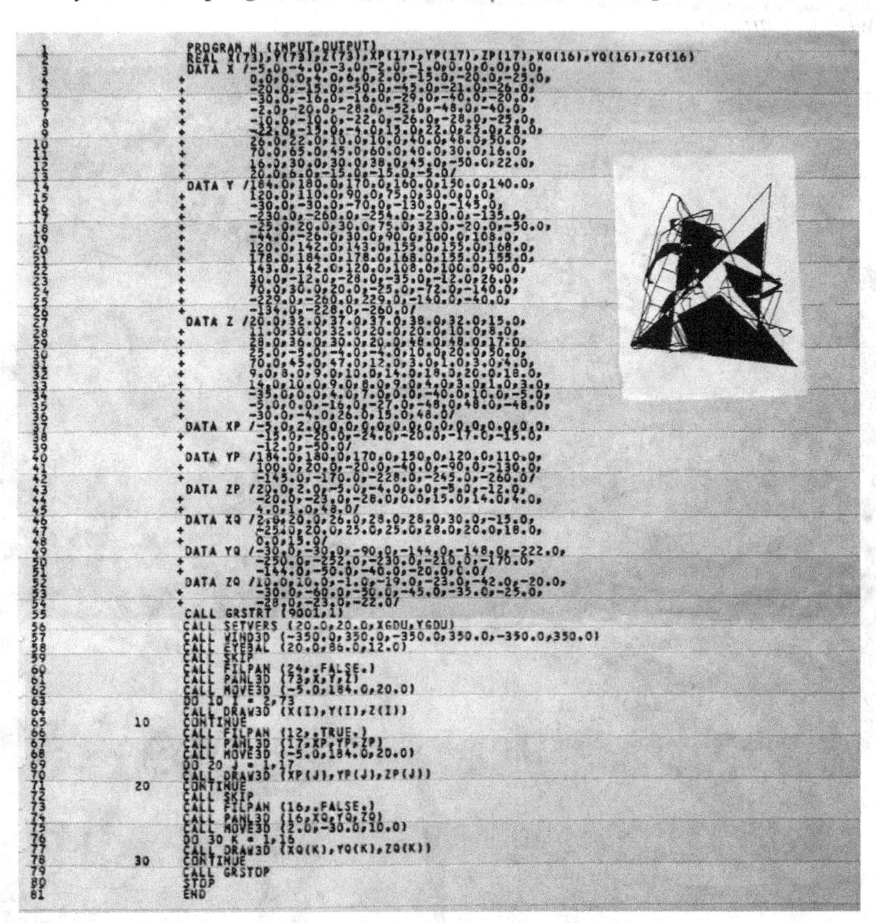

```
 1          PROGRAM M (INPUT,OUTPUT)
 2          REAL X(73),Y(73),Z(73),XP(17),YP(17),ZP(17),XQ(16),YQ(16),ZQ(16)
 3          DATA X /-5.0,-4.0,-3.0,-2.0,-1.0,0.0,0.0,0.0,
 4         +        0.0,0.0,4.0,6.0,2.0,-15.0,-20.0,-25.0,
 5         +        -20.0,-15.0,-50.0,-45.0,-21.0,-26.0,
 6         +        -30.0,-16.0,-16.0,-29.0,-40.0,-20.0,
 7         +        -2.0,-20.0,-28.0,-52.0,-48.0,-40.0,
 8         +        -10.0,-10.0,-22.0,-26.0,-28.0,-25.0,
 9         +        -22.0,-15.0,-4.0,15.0,22.0,25.0,28.0,
10         +        26.0,22.0,10.0,10.0,40.0,48.0,50.0,
11         +        70.0,65.0,45.0,60.0,40.0,30.0,16.0,
12         +        16.0,30.0,30.0,38.0,45.0,-50.0,22.0,
13         +        20.0,6.0,-15.0,-15.0,-5.0/
14          DATA Y /184.0,180.0,170.0,160.0,150.0,140.0,
15         +        120.0,110.0,90.0,75.0,30.0,0.0,
16         +        -30.0,-30.0,-70.0,-130.0,-145.0,
17         +        -230.0,-260.0,-254.0,-230.0,-135.0,
18         +        -25.0,20.0,30.0,75.0,32.0,-20.0,-50.0,
19         +        -44.0,-26.0,30.0,90.0,100.0,108.0,
20         +        120.0,142.0,143.0,155.0,155.0,168.0,
21         +        178.0,184.0,178.0,168.0,155.0,155.0,
22         +        143.0,142.0,120.0,108.0,100.0,90.0,
23         +        30.0,-12.0,-28.0,-35.0,-12.0,26.0,
24         +        70.0,30.0,20.0,-25.0,-72.0,-140.0,
25         +        -229.0,-260.0,229.0,-140.0,-40.0,
26         +        -134.0,-228.0,-260.0/
27          DATA Z /20.0,32.0,37.0,37.0,38.0,32.0,15.0,
28         +        11.0,30.0,32.0,20.0,20.0,10.0,8.0,
29         +        28.0,36.0,30.0,20.0,48.0,48.0,17.0,
30         +        25.0,-5.0,-4.0,-4.0,10.0,20.0,50.0,
31         +        70.0,45.0,47.0,12.0,3.0,1.0,3.0,4.0,
32         +        9.0,8.0,9.0,10.0,14.0,18.0,20.0,18.0,
33         +        14.0,10.0,9.0,8.0,9.0,4.0,3.0,1.0,3.0,
34         +        -35.0,0.0,4.0,7.0,0.0,-40.0,10.0,-5.0,
35         +        -5.0,0.0,-16.0,-27.0,-48.0,48.0,-48.0,
36         +        -30.0,-4.0,26.0,15.0,48.0/
37          DATA XP /-5.0,2.0,0.0,0.0,0.0,0.0,0.0,0.0,0.0,
38         +        -15.0,-20.0,-24.0,-20.0,-17.0,-15.0,
39         +        -12.0,-50.0/
40          DATA YP /184.0,180.0,170.0,150.0,120.0,110.0,
41         +        100.0,20.0,-20.0,-40.0,-90.0,-130.0,
42         +        -145.0,-170.0,-228.0,-245.0,-260.0/
43          DATA ZP /20.0,2.0,-5.0,-4.0,0.0,-5.0,-12.0,
44         +        -20.0,-23.0,-28.0,0.0,15.0,14.0,4.0,
45         +        4.0,1.0,48.0/
46          DATA XQ /2.0,20.0,26.0,28.0,28.0,30.0,-15.0,
47         +        -25.0,20.0,25.0,25.0,28.0,20.0,18.0,
48         +        0.0,15.0/
49          DATA YQ /-30.0,-30.0,-90.0,-144.0,-148.0,-222.0,
50         +        -250.0,-252.0,-230.0,-210.0,-170.0,
51         +        -144.0,-50.0,-40.0,-20.0,0.0/
52          DATA ZQ /10.0,10.0,-1.0,-19.0,-23.0,-42.0,-20.0,
53         +        -30.0,-60.0,-50.0,-45.0,-35.0,-25.0,
54         +        -28.0,-23.0,-22.0/
55          CALL GRSTRT (9001,1)
56          CALL SETVERS (20.0,20.0,XGDU,YGDU)
57          CALL WIND3D (-350.0,350.0,-350.0,350.0,-350.0,350.0)
58          CALL EYEBAL (20.0,86.0,12.0)
59          CALL SKIP
60          CALL FILPAN (24,.FALSE.)
61          CALL PANL3D (73,X,Y,Z)
62          CALL MOVE3D (-5.0,184.0,20.0)
63          DO 10 I = 2,73
64          CALL DRAW3D (X(I),Y(I),Z(I))
65    10    CONTINUE
66          CALL FILPAN (12,.TRUE.)
67          CALL PANL3D (17,XP,YP,ZP)
68          CALL MOVE3D (-5.0,184.0,20.0)
69          DO 20 J = 1,17
70          CALL DRAW3D (XP(J),YP(J),ZP(J))
71    20    CONTINUE
72          CALL SKIP
73          CALL FILPAN (16,.FALSE.)
74          CALL PANL3D (16,XQ,YQ,ZQ)
75          CALL MOVE3D (2.0,-30.0,10.0)
76          DO 30 K = 1,16
77          CALL DRAW3D (XQ(K),YQ(K),ZQ(K))
78    30    CONTINUE
79          CALL GRSTOP
80          STOP
81          END
```

For example natural logic defines the mechanics of biophysics – application of physical laws to biological systems; models of computation draw from the analogs of action potentials, biophysical cells, and membranes consisting of many cells (IASE, 2012.) As a consequence, our ideas and concepts about nature, and inspiration coming from nature change along with our knowledge base, and so our art works are changing. Art production may evolve according to the emergence of new scientific theories. I hope to demonstrate, using projects described in this book and other examples, the reasons for exploring the increasing crossover between science and art through a biologically inspired, creative computing process I refer to as biological translation. Particular chapters of this book provide parallel descriptions of the art works (on the city and rural themes) and of the selected concepts and processes coming from science (such as physics or biology). This is aimed at making translation of common patterns and structures into visual language.

The semiotic approach to visual rhetoric plays a crucial role in visual arts (metaphors, open messages contained in and conveyed through the art works); design (scissorness of the scissors, pretenders, misleaders, and informers in product design); signage production (the design or use of signs, symbols, and words, which may include billboards, posters, placards, etc.), and marketing (conveying messages through metaphors, for example, a feeling of power through images

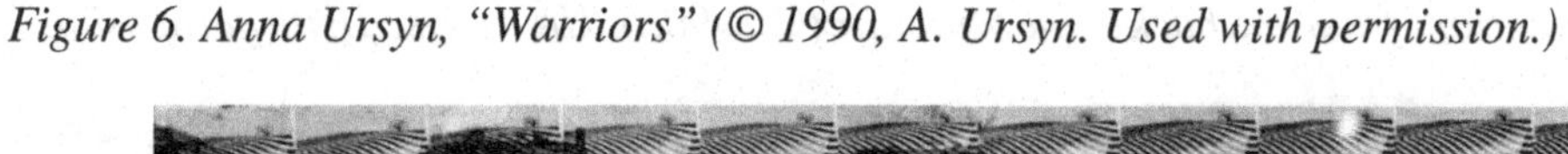

Figure 6. Anna Ursyn, "Warriors" (© 1990, A. Ursyn. Used with permission.)

of the prey animals). Visual power in business communication, training, and marketing is well understood by those who advertise their markets visually and name their companies accordingly, for example, through visual translations (http://visualtranslations.com/). Science Translational Medicine is a weekly journal from AAAS (American Association for the Advancement of Science) integrating science with clinical medicine, focused on applications of basic research knowledge to improve human health: to translate biological discoveries into medical advances.

Figures 5 - 7 present making translation of patterns and structures into visual modes in a series of sequential art media. A program "Man" (Figure 5) is an example of a code used for a three dimensional form of a transformed and abstracted warrior. Since this image involves a z-axis, it could be freely rotated, and altered. It was embedded into different types of art. A synthetic image of a man served as an inception for further transformations into a painting and a sculpture. First, the image was repeated and transformed into a two dimensional composition "Warriors" (Figure 6).

Image of a transformed man based on a program was used in a computer graphics entitled "Warriors" (Figure 6). The repetition of figures of warriors depersonificated for the purpose of fulfilling the common goal has been put into the endless landscape. I unified the meaning of men and a landscape using the same approach: rigid order created with a computer.

Figure 7. Anna Ursyn, "Warrior" front and back of the sculpture (© 1990, A. Ursyn. Used with permission.)

Then, programs for a figurative three-dimensional design were used as a guide for a sculptural form titled "Warrior" (Figure 7) created as a series of horizontally hanged planks. Their overall shape follows the outline of the image. The planks have an image photosilkscreened on their both sides. One side possesses color, while the other side of the sculpture is black and white.

See Table 2 for Your Visual Response.

Figure 8, 'Frame of Reference' is an almost abstract presentation of a complex combination of factors and variables that shape our frame of mind. We shape the image how we see the world and these constructs direct us in our decisions. "Translation" comments issues under discussion.

Translation

A square makes a cube,
a line can build a spiral,
a circle rotates into a sphere,
a triangle into a cone.
We study elements and principles of design,
in order to learn
about shape and form,
then we furnish our home, office, and
our mind's constructs accordingly.

Table 2.

Your Visual Response: Translation
Make a robot with long legs; your robot will dance. What function/advantage the legs provide to the robot in respect to its environment and coexisting creatures? Make your robot bio-inspired without copying nature. You may want to inspire yourself with the works of a Dutch artist Theo Jansen, who builds large mechanisms called Strandbeest (http://www.strandbeest.com/) that can move on their own. Your robot will dance having every point of its body moving in the same direction over the same distance, without any rotation, reflection, or change in size. This means your robot will carry out translation, which describes movement of an object in space. Let your robot move forward, backward, right or to the left. Now you may want to compose music for your robot. What kind of music will you compose to make this special kind of dance with restrictive movements the most attractive? Maybe you can also write a score?

Figure 8. Anna Ursyn, "Frame of Reference" (© 2006, A. Ursyn. Used with permission)

CONCLUSION

Both articulation and translation are common treads interweaving distinctive processes and events. These concepts are present in a great amount of fields related to nature, science, art, and technology, which are further discussed in this book. Common traits that may be revealed both in the articulation of units and translation of a meaning may pertain the structure of things and processes under discussion. Concepts of articulation and translation have been discussed as the means to explore the meaning: articulation as units combined into complete structures and thus meaningfully formulated, while translation may go from nature to art or may assume many forms of visual, verbal, and numeral translation. Examples of double and triple articulation of signs in languages, programs, and several other fields complement the conceptual landscape. Effects, changes, and impacts flow in both directions between nature-inspired technology and nature-inspired art. Two-way translation was discussed, from nature to idea and production and from products to human perception and creation.

REFERENCES

About.com. (2012). *How many emails are sent every day?* Retrieved May 30, 2012, from http://email.about.com/od/emailtrivia/f/emails_per_day.htm

Apple in Education. (2013). Retrieved March 30, 2013, from http://www.apple.com/education/apps/

Artlyst, London Art Network. (2012). Retrieved March 30, 2013, from http://www.artlyst.com/articles/art-students-to-be-given-free-ipads

Berker, T., Hartmann, M., Punie, Y., & Ward, K. J. (2006). *Domestication of media and technology*. McGraw-Hill International.

Boden, M. A. (2006). *Mind as machine: A history of cognitive science*. Oxford, UK: Clarendon Press.

Chandler, D. (2013). *Semiotics for beginners: Articulation*. Retrieved March 3, 2013, from http://users.aber.ac.uk/dgc/Documents/S4B/sem08a.html

Collin, H. (2012). Biological translation: Virtual code, form, and interactivity. In A. Ursyn (Ed.), *Biologically-Inspired Computing for the Arts: Scientific Data through Graphics*. Hershey, PA: IGI Global Publishing.

Correa, P. N., & Correa, A. N. (2004). *Nanometric functions of bioenergy*. Akronos Publishing.

Courtois, C., Verdegem, P., & De Marez, L. (2012). The triple articulation of media technologies in audiovisual media consumption. *Television & New Media, 13*(2). Retrieved May 31, 2012, from http://tvn.sagepub.com/content/early/2012/04/11/1527476412439106.abstract

Crick, F. H. C. (1958). On protein synthesis. *Symposia of the Society for Experimental Biology, 12*, 139–163. PMID:13580867.

Crick, F. H. C. (1970). Central dogma of molecular biology. *Nature, 227*(5258), 561–563. doi:10.1038/227561a0 PMID:4913914.

DeLanda, M. (2000). *A thousand years of non-linear history*. The MIT Press.

DeLanda, M. (2010). *Deleuze: History and Science*. Atropos Press.

Deleuze, G., & Guattari, F. (2009). *Anti-oedipus: Capitalism and schizophrenia*. Penguin Classics.

Deleuze, G., Guattari, F., & Massumi, B. (1987). *A thousand plateaus: Capitalism and schizophrenia*. University of Minnesota Press.

Derrida, J. (1991). *The truth in painting*. Chicago: The University of Chicago Press.

Ebrahimi, A. (2003). *C++ programming easy ways*. Boston: American Press.

Eco, U. (1979). *The open work*. Cambridge, MA: Harvard University Press.

English Wikipedia. (2012). Retrieved January 19, 2012, from http://en.wikipedia.org/wiki/English_Wikipedia

Evomusart. (2012). *Proceedings of the 1st international conference and 10th European event on evolutionary and biologically inspired music, sound, art and design*. Retrieved May 31, 2012, from http://evostar.dei.uc.pt/2012/call-for-contributions/evomusart/

Fuglestad, T., & Tiedemann, S. (2013). *iPads in art education*. Retrieved March 30, 2013, from http://ipadsinart.weebly.com/

Harty, D. (2010). *Drawing//experience: A process of translation*. (Ph.D. Thesis). Loughborough University, Leicestershire, UK.

Hockney, D. (2006). *Secret knowledge: Rediscovering the lost techniques of the old masters*. Studio, ISBN-10, 0142005126.

IASE. (2012). *Institute for advanced science & engineering*. Retrieved May 31, 2012, from http://iase.info/

Ittelson, W. H. (2007). The perception of nonmaterial objects and events. *Leonardo*, *40*(3), 279–283. doi:10.1162/leon.2007.40.3.279.

Jakobson, R. (1995). *On language*. Boston: Harvard University Press.

Kozin, A. (2008). Translation and semiotic phenomenology: The case of Gilles Deleuze. *Across Language and Culture*, *9*(2), 161-175.

Lengler, R., & Vande Moere, A. (2009). Guiding the viewer's imagination: How visual rhetorical figures create meaning in animated infographics. In *Proceedings of the 13th International Conference on Information Visualization*. DOI 10.1109/IV.2009.102

Lima, M. (2011). *Visual complexity: Mapping patterns of information*. New York: Princeton Architectural Press. ISBN 978 1 56898 936 5

Lima, M. (2013). *Visual complexity*. Retrieved February 20, 2013, from http://www.visualcomplexity.com/vc/

Maeda, J., & Bermont, R. J. (2011). *Redesigning leadership (simplicity: design, technology, business, life)*. Boston: The MIT Press.

Makeuseof. (2013). Retrieved March 30, 2013, from http://www.makeuseof.com/tag/7-ways-ipad-students-excel-school/

Martinet, A., & Palmer, E. (1982). *Elements of general linguistics*. Chicago: University of Chicago Press.

Matthews, L., & Perin, G. (2011). Digital images: Interaction and production. *International Journal of Creative Interfaces and Computer Graphics*, *2*(1), 27–41. doi:10.4018/jcicg.2011010103.

Mitchell, L. (2005). *A posthuman methodology for nondual world*. (Unpublished Thesis). York University, Toronto, Canada.

Muralikrishnan, T. R. (2010). Translator as reader: Phenomenology and text reception: An investigation of indulekha. *Language in India*, *10*(1), 255–266.

National Research Council of the National Academies. (2008). *Inspired by biology: From molecules to materials to machines*. Washington, DC: The National Academies Press.

Nöth, W. (2000). *Handbook of semiotics*. Stuttgard, Germany: Metzler.

Posner, M. I. (1993). *Foundations of cognitive science*. Cambridge, MA: The MIT Press.

Posner, R. (1992). Origins and development of contemporary syntactics. *Languages of Design*, *1*(1), 37–54.

Redström, J., Skog, T., & Hallnäs, L. (2000). Informative art: Using amplified artworks as information displays. In *Proceedings of DARE 2000, On Designing Augmented Reality Environments*. Elsinore, Denmark: ACM. Retrieved January 20, 2012, from http://www.johan.redstrom.se/thesis/pdf/infoart.pdf

Rusu, A., Fabian, A. J., Jianu, R., & Rusu, A. (2011). Using the gestalt principle of closure to alleviate the edge crossing problem in graph drawings. In *Proceedings of the Information Visualisation 15th International Conference*, (pp. 329-336). ISBN 978-1-4577-0868-8

Samuels, R. (2010). The double articulation of Schubert: Reflections on der doppelgänger. *The Musical Quarterly*, *93*(2), 192–233. doi:10.1093/musqtl/gdq008.

Schumacher, M. (2011). *Zodiac calendar & lore*. Retrieved November 1, 2012, from http://www.onmarkproductions.com/html/12-zodiac.shtml

Stam, R. (2000). *Film theory*. Oxford, UK: Blackwell.

Thbz. (2000). The double articulation of language: Computer languages. *Everything2*. Retrieved June 2, 2012, from http://everything2.com/title/The+double+articulation+of+language

The Carnegie Library of Pittsburg. (2011). *The handy science answer book*. Visible Ink Press.

The Teaching Palette. (2013). Retrieved March 30, 2013, from http://theteachingpalette.com/2012/02/24/theres-an-app-for-that-ipads-in-the-art-room/

Various Artists. (2000). *Double articulation*. Audio CD album. ASIN B000024ENV

Victorri, B. (2007). Analogy between language and biology: A functional approach. *Cognitive Process, 8*(1), 11-9. Retrieved July 9, 2012, from http://www.ncbi.nlm.nih.gov/pubmed/17171371

Wisegeek. (2012). *What is double articulation*? Retrieved June 2, 2012, from http://www.wisegeek.com/what-is-double-articulation.htm

Chapter 2
Communication through Many Senses

ABSTRACT

Sensory messages are examined as electromagnetic waves clearly identified by our senses, consisting of interacting electric and magnetic currents or fields and having distinctive wavelengths, energy, and frequency. Further text discusses modes of gathering information and communication that include sensory responses to electromagnetic waves, visible vibrations exemplified by cymatics, the pitch response, the senses of vision, smell, touch, and taste, all of them further expanded by the developments in current technologies. The sense of numbers is examined next, involving numerical and verbal cognition and communication with the use of numerals. Sensitivity, spatial abilities, and the threshold of sensory information make a part of the issues about biology-inspired computational solutions for enhancing our particular or synesthetic abilities, and the role of imagination in biology-inspired research and technology, learning, and teaching. The role of the sensory input in art, which pertains in some extent to individual curiosity and sensibility, concludes the chapter.

INTRODUCTION

It may be useful to keep in mind that communication with others, as well as exchange of knowledge, insight, and information can be done both in verbal and non-verbal way. Many agree that a written text alone may not be the most effective way to communicate ideas and information, even if it goes in a language shared by both parts. Moreover, our communication with the world and people, conscious and unaware of, goes through our senses in much more ways than only by sight, hearing, touch and haptic experience, smell, or taste. For these reasons, the issues discussed in this chapter

DOI: 10.4018/978-1-4666-4703-9.ch002

will return repeatedly in the chapters that follow, assuming diverse frames of reference, from physiological, physical, technological, to aesthetical.

Other faculties that are usually described as senses include many internal and external senses: a sense of temperature, kinesthetic sense that gives us balance, a sense of motion, a sense of acceleration and velocity changes (e.g., pressure caused by the wind), proprioception that allows sensing the relative position and movement of parts of the body, a feel of direction, responsiveness to pheromones, and sensitivity to pain. We may sense someone's feelings or mood through the tone of their voice, body language, even from the look in their eyes; it may happen also in one's communication with animals. We are constantly processing sensory information coming from our external and internal receptors that respond to and transmit signals about our body. Our own feelings, for example a feeling of being tired and exhausted, hungry, or just thirsty and dehydrated after a vigorous physical exercise or after a long discussion, can add or subtract the intensity of the sensory input.

Animal senses are seen analogous, and comparable to human ones but they often act differently, as it for instance happens with worms, butterflies, or birds. Some animals have more acute sense of smell, some have better balance, and other have the wider or more narrow ranges of frequencies used for vision and hearing; some may receive ultrasound signals. Many kinds of animals have also other kinds of senses, such as echolocation and different kinds of receptors such as electrically sensitive electroreceptors found in sharks, electric eels, catfish, and other fish (Wueringer, Squire, Kajiura, Hart, & Collin, 2012), and thus different ways of sensing and interpreting data from the environment. Every now and then scientists give a new account of animal facilities for searching through their environment. Some aquatic animals generate the bioelectric dipole fields created by the opposite electromagnetic charges separated by a small distance. Sharks and rays can detect such fields and attack their prey. Wueringer et al. (2012) analyzed the predatory behavior of sawfish. In a saw of a sawfish (all species of sawfish are critically endangered), an elongated cranial cartilage with teeth is covered in a dense array of electroreceptors. The sawfish's saw is unique in its use for both detecting and manipulating prey.

In general terms, senses provide input to an organism due to their physiological capacities. We examine these capacities and use this information for theoretical, practical, and computational solutions within the domains of physiology, neuroscience, cognitive science, cognitive psychology, sociology, anthropology, medicine, computer science, but also human perception, philosophy, and art. One may say that art media of the 21st century, including music, theater, new media art, and design, is the art inspired by the input from our senses, and incorporating the viewer's senses. It is often aimed to visualize the unseen and give the viewer the phenomenal, immediate experience.

Our sensory receptors receive signals from our surroundings or our internal environment, such as temperature, velocity changes (for instance, caused by air changes in the pressure – wind), touch, haptic experiences, and proprioception. Sometimes our senses are below par with the animal senses, for example we cannot sense from a distance the body heat of people or animals, and cannot perceive the signals related to the degree of water pressure or water current. Signals become stimuli that cause physical or physiological reflex responses, which may be performed without our consciousness or as intentional reactions; they also cause psychological reactions. Some hold that humans, and maybe some animals are experts (without training) of picking up emotional expression in faces (Loizides, 2012). The answers depend on the sensitivity of our senses or our organism. However we lack senses or not have enough sensitivity to detect many physical or chemical qualities. We seem to lack the sense of ultrasounds, we are unsure (in variable degree) about our circadian rhythms (biological processes recurring on a 24 hour cycle and influenced by the environment), our sense of time, our body temperature, or physiological responses to our emotions. Some animals have strong geospatial sense of the direction; we do not navigate like earthworms, fish, bees, or birds do. We cannot measure with our senses the strength of electric and magnetic fields, and also week electromagnetic fields, in spite of the warnings that electromagnetic fields produced by such man-made devices as mobile phones, computers, power lines, and domestic wiring might have harmful effects on living organisms: cell membranes, DNA, metabolism, and also neuronal activity (Goldsworthy, 2007). Also, we do not estimate without tools the air pressure or our blood pressure, to name a few. Therefore, we have to rely on man-made sensors and devices.

MODES OF GATHERING INFORMATION AND COMMUNICATION THROUGH THE SENSES

We may wonder what is the relation between information and communication: parents receive information about their infant's discomfort with their smell sense, and then a baby communicates by voice (so the parents can communicate also with the use of hearing). Thus information may be seen as a fact we are aware of, but not a signal. Also, the data is a fact that does not communicates anything if not received and interpreted. Experiments on sensory deprivation include the deliberate cutting out all sensory stimuli by putting a person in a dark, acoustically isolated space with lukewarm isotonic water with the water density equal to the density of one's body, where a person doesn't feel gravity, cannot sink nor go up to the surface. Sensory deprivation has been used in psychology to achieve therapeutic desensitization, to alleviate and mitigate phobias and other disorders, in alternative medicine as a factor conducive to meditation and altered states of mind, as a tool for training prospective pilots and astronauts, and in thriller or science fiction films; it has been also used for interrogations, causing harmful, negative effects. Bales & Kitzmann (in Yang, 2011) lists modes of communication (signal modalities) as visual, acoustic, chemical (pheromones produced and released by animals), tactile, electrical, and seismic signals resulting in vibration of earth. For example, a blind mole rat's thumping on burrow walls transmits over very long distances. The same message can be conveyed and received in many ways: visually with the semaphore flags (signals made with hand-held flags, rods, disks, paddles, or just hands), sonically (3 times 3 long versus short sounds: "· · · — — — · · ·" for the SOS, or through the international Morse code, for example as a distress signal.

First of all, describing senses in separate groups may be seen improper because senses are interconnected in many ways. Communication may be here seen as an exchange of sensory information in the form of different kinds of perception through the senses. Signals coming from the senses are often combined to convey a clear message. For example, we can receive information about numbers from various senses, looking at patterns, listening to sounds, feeling vibrations, or reading numbers.

Signals and Spectra

Generally speaking, we receive sensory messages as electromagnetic waves having the distinctive wavelengths, energy, and frequency that are clearly identified by our senses. Some could imagine that the electromagnetic spectrum (Figures 1 and 2), as a common link behind most of the sensory signals, might best characterize our world. The electromagnetic spectrum includes, from longest wavelength to shortest: radio waves, microwaves, infrared, visible light, ultraviolet, X-rays, and gamma rays (Table 1). Only part of the spectrum can be detected by our senses, for instance as sounds, heat, or light.

This can be also shown with the use of symbols (Figure 2).

Figure 1. Complete spectrum of electromagnetic radiation with the visible portion highlighted (http://en.wikipedia.org/wiki/Frequency. Text is available under the Creative Commons Attribution-ShareAlike License)

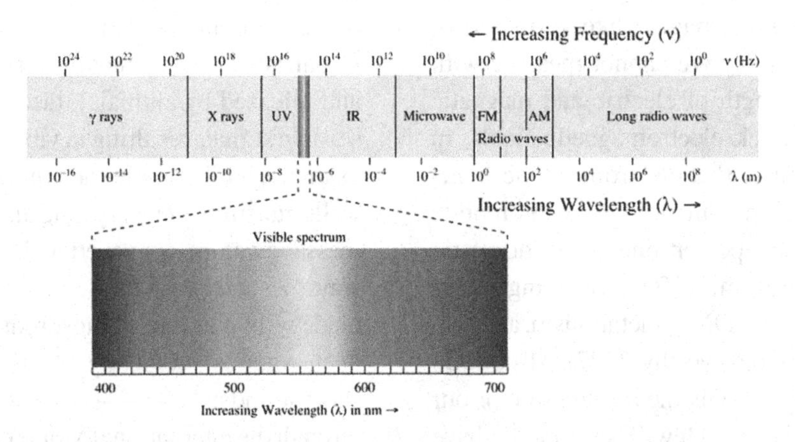

Figure 2. Wavelengths (in meters) and communication (© 2012, Matthew Rodriguez. Used with permission)

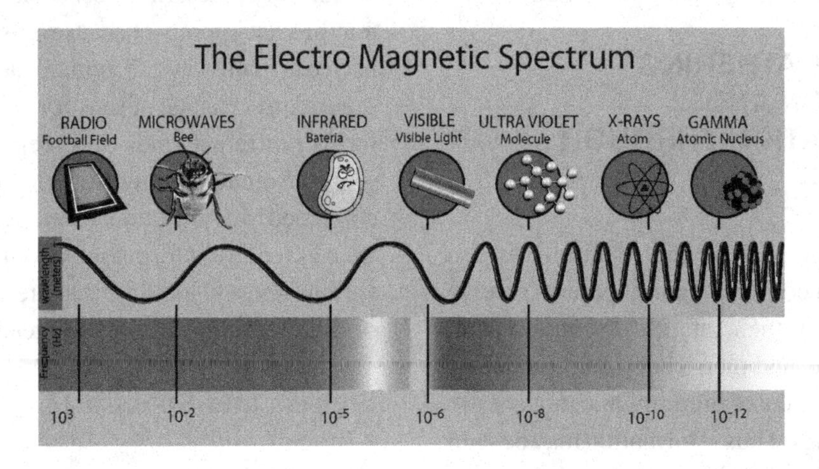

Table 1. Terms about electromagnetic spectrum

10^{-16}	Gamma rays	Information about Cosmos
10^{-12}	X rays	Instruments for gathering and communicating information
10^{-9}	m Nanometers	Information about organism or their parts, and inanimate structures
10^{-8}	10 nanometers	Ultraviolet; seen by some animals, e.g., birds, bats, dolphins, sharks, many insects, e.g., bees
10^{-6}	Visible	Light, colors, visual communication
10^{-5}	Infrared	Light invisible for humans but felt like heat; visible for some animals
10^{-4}	Microns	Recorded by humans with sensors
10^{-3}	Millimeters	Ultrasound has a wavelength (below 20 Hz) from 0.3 micrometers to nanometer. Ultrasounds (20 kHz to over 200 MHz) are received by animals, such as bats, insects, dogs, dolphins, whales, and fish
10^{0}	Meters	Radio; Sound and verbal communication; radio; drums as a code; music and visual music

Sounds as Electromagnetic Waves

We are immersed in a variety of sounds coming from our surroundings, music and various types of communications. Electromagnetic waves, seen as energy and matter that conducts the waves, oscillate. That means all parts of the system display sinusoidal motion with a given frequency and an uninterrupted pattern. Modal phenomena display a pattern of sinusoidal motion of a physical object (having the form of a sine wave of smooth, repetitive oscillations). We call a normal mode such pattern of vibrations when all parts of the system (for example, molecules, drums, pipes, bridges, or houses) vibrate with the same frequency on the whole surface and have a fixed phase relation. Such objects display their sets of normal modes that depend on material, structure, and surroundings of the object. A frequency of a normal mode of vibrating object is called its natural or resonant frequency. When more frequencies are involved, an object oscillates at resonant frequencies with greater amplitude, the maximum extent of oscillations. Sound waves cause that the ear structures pass and translate vibrations into neural messages sent through the auditory nerve from about 25,000 receptors to the brain. The ears contain also semi circular structures that provide the feeling of balance. Musical sound is produced by continuous, regular vibrations, while noise is not regular. Music, speech, and sound effects go together in a sound studio for a broadcast production, film, or video. Artists translate musical compositions into visual imagery and thus create visual music. For example, many artists presented their works at festivals, as it happened at the N. Y. Digital Salons, the 2009 Visual Music Marathon at the Northeastern University, and also published their pictures and still frames in journals such as the Leonardo, Journal of the International Society for the Arts, Sciences and Technology. A growing number of competitions (such as Eurographics – European Association for Computer Graphics, Imagina, or Ars Electronica) incorporate varied kinds of media: performances, live action, interactive installations, visual music, and movies into their events.

Cymatics

Cymatics is a study of vibration of matter that makes sound visible. When we cover a surface of a plate or a membrane – a thin, pliable sheet of material with a thin layer of particles or liquid and then produce a musical sound using for example, strings, drums, or air pipes, the surface will vibrate according to the music and we will see vibrations as patterns, thus making visible what the sound waves generate.

There is a long history of the study about patterns produced by vibrating systems. Galileo Galilei described the regular pattern of oscillating object in his work "Dialogue Concerning the Two Chief World Systems"

As I was scraping a brass plate with a sharp iron chisel in order to remove some spots from it and was running the chisel rather rapidly over it, I once or twice, during many strokes, heard the plate emit a rather strong and clear whistling sound: on looking at the plate more carefully, I noticed a long row of fine streaks parallel and equidistant from one another. Scraping with the chisel over and over again, I noticed that it was only when the plate emitted this hissing noise that any marks were left upon it; when the scraping was not accompanied by this sibilant note there was not the least trace of such marks. (1632, after McLaughlin, 1998)

Robert Hooke examined in 1680 the nodal patterns he obtained by running a violin bow across a glass plate covered with flour. Ernst Chladni described in 1787 such experiment, this time placing sand on metal plates and stroking the edge of the plate with a bow, in his book "Entdeckungen über die Theorie des Klanges" ("Discoveries in the Theory of Sound"). One can easily form and see symmetrical Chladni patterns following a demonstration manual "Chladni Plate" provided by the University of California, Los Angeles (UCLA, 2011). A Swiss physician Hans Jenny (1967/2001) studied the modal phenomena (called by him cymatics) that could be set into motion by sound waves. Gyorgy Kepes (1944/1995), the author of an influential book "Language of Vision" created a visual display of thermodynamic patterns resulting from conversion to heat of other forms of energy in an acoustically vibrated sheet of metal; he displayed small holes heated by the burning gas.

Communication by vibrations was a theme of an artistic research and audiovisual installation about the social organization in ants. Ants communicate mainly by pheromones, but they also produce the vibratory sounds. The life of the ant colony have been presented as a soundscape of scratching effects by constructing a sound-reactive installation comprising of contact microphones and video surveillance interfaced with the computer that feeds this bio-data to two turntables (Auson, 2012).

Figure 3 presents a work "Ecosphere." We communicate with our surroundings to receive information we need but we are feeling free of any constrains within the boundaries of our ecosphere.

Pitch

Absolute (perfect) pitch is the ability to recognize and remember a tone without a reference. It's a great asset for musicians who just know the exact tone they hear and name exact notes and chords by ear. Many musicians have no pitch recognition – they have relative pitch – an ability unique to people to identify pitches in relation to other pitches. At the same time, many researchers hold that a great deal of non-musicians and also newborn babies have absolute pitch: they prefer to respond to absolute rather than relative pitch differences (Saffran, 2003). This ability may possibly cease to exist in early maturation when absolute pitch would interfere with language perception (Bossomaier & Snyder, 2004). Infants learn to understand what their parents say; they are focusing on the content – the language perception. However, the pitch and the timbre (a tone quality of a sound that helps distinguish it from other sounds of the same pitch and volume) of the female or male voices are of no importance and of no use in this difficult task. Absolute pitch is prevalent among those who speak tonal languages where high/low pitch combinations are important (such as happens in Mandarin or Cantonese languages). The ability to acquire an accent-free second language, lessened after puberty, may possibly result from the loss of absolute pitch and hence ability to hear the raw sensory data: to accurately hear the phonemes, distinct units of sound in a foreign language and thus develop a good accent (Bossomaier & Snyder, 2004).

Some animal species display ability to learn songs. Vocal learning is considered a substrate for human language. This trait has been found in three groups of mammals (humans, bats, and cetaceans – dolphins) and three groups of birds (parrots, hummingbirds, and songbirds). Vocal learners gain vocalization through imitation, while auditory learners do that by making associations (Jarvis, 2004). Perfect pitch perception is present in many members of animal kingdom.

Figure 4 presents a work "Pitch and Volume." We may categorize sounds we receive through our sense of hearing depending on our actual mood, finding them interesting, pleasurable, or ear splitting and noisy.

Let's listen to the city music
Pitch our voices in key with others
Without confusion of languages and meanings
Let's hum a tune and rejoice in laugh
When we stay set in a traffic jam.

Figure 3. Anna Ursyn, "Ecosphere" (© 2007, A. Ursyn. Used with permission)

Figure 4. Anna Ursyn, "Pitch and Volume" (© 2004, A. Ursyn. Used with permission)

Vision

The part of the electromagnetic spectrum that can be seen by human eye is in the range of 390-750 nm.

A physician and physicist Hermann von Helmholtz (1821-1894), followed by researchers in several disciplines, explored the connection between the visible spectrum and color vision and developed theories about visual perception. From about 150 million light-sensitive cells, rods that identify shapes, especially in dim light, and cones that that identify colors (if light is bright enough) images are sent to the brain's visual cortex, the outer layer of the cerebrum composed of grey matter that relates to vision. Human vision is imperfect in comparison to many animals, such as owls, which can see in a dim light a mouse moving in a distance over 150 feet away. Contrary to us many animal species such as a horse have night vision. Eyes positioned on the sides of the horses' heads allow them seeing in a range of about 350°, while humans have about 180° field of vision. Part of vision that is outside of the center of our gaze is called peripheral vision; it is weaker in humans than in many animal species.

Figure 5 presents a work "Peripheral Vision." Both precise and abstract impressions that reach our vision combine into a whole reception of the landscape.

From a glass case of the car,
The outlook is different from the left,
a windshield, or the right window.
From behind, a big truck sends the harsh glare of the headlights.

Figure 5. Anna Ursyn, "Peripheral Vision" (© 2007, A. Ursyn. Used with permission)

Communication through Images

In a similar way as some musicians (and not only musicians) posses perfect pitch, some painters (and not only painters) have perfect color feel and color memory: they have an ability to recognize and remember color without an external reference for making a comparison, so many of them do not need to bring a color sample to a store in order to choose the right color of a paint. Visual communication that goes with the use of signals perceived by our sense of vision would thus involve images as well as written texts. Communication through images may take various forms: two-dimensional – such as drawings, art works, graphs, graphics, or typographic prints, among others, three-dimensional forms – such as architectural or sculptural works, 4-dimensional time-based media – such as moving images, and also can become interactive – going both ways, and virtual. Visual semiotics, a part of the domain of semiotics supports analysis of the visual way of communication and examines interaction between pictures of products or events and the audience as recipients. Images serve in such cases as iconic, indexical, or symbolic signs, according to the works of Peirce on formal semiotics from the 1860s. Communication through images uses visual metaphors, visualization techniques, symbols, analogies, icons, and time-based images; also, it often links haptic (related to the sense of touch) and visual modes using gesture, body language, dance, mime (a performance art involving body motions without the use of speech) and pantomime (a musical theatrical production using gestures and movements but not words). Visualization techniques often include small multiple drawings that represent the sets of data with miniature pictures, to reveal repetition, change, pattern, and facilitate comparisons. Since their introduction, Chernoff (1971) cartoon faces evolved into a tool for presenting data as empathic facial expressions (Loizides, 2012; Loizides and Slater, 2001, 2002). Thus a positive condition of the visualized data should be reflected as a smiling face and vice versa.

Smell

The sense of smell, called olfaction has chemosensory property as it can convert chemical signals into stimuli for our perception. Separate systems detect airborne substances and fluid stimuli. For example, pheromones – chemical substances produced by animals (mostly mammals and insects) and released into the environment trigger specific social responses. Depending on the type of a pheromone (the alarm, food trail, or sex pheromones) the response is in the form of aggregation of individuals, flight, aggression, but also attraction of mates or babies, and changes their activity, behavior, or physiology. A single receptor recognizes many odors. Odors are translated into patterns of neural activity and the olfactory bulb, olfactory nerve, and then the olfactory cortex interpret the patterns. Olfactory sensory neurons are replaced throughout life. Richard Axel and Linda B. Buck were awarded jointly the Nobel Prize in Physiology or Medicine 2004 "for their discoveries of odorant receptors and the organization of the olfactory system." The smell cells in dogs are 100 times larger than human cells and much more numerous: dogs have about 1 million smell cells per nostril and therefore they are so keen to go for a walk to 'read' (perceive) and 'write' (mark) their messages. Emoticons of smiley-face in a pictorial [☺] or digital [:)] form using punctuation marks seem to fit the description of a dog's smell perception because of a strong emotional component of the dogs' sharp perception of smells. Herz, Schankler, & Beland (2004) have shown that the senses of smell and taste are uniquely sentimental, prompting the feelings because they "are the only senses that connect directly to the hippocampus, the center of the brain's long-term memory (while) all our other senses (sight, touch, and hearing) are first processed by the thalamus, the source of language and the front door to consciousness."

The sense of smell is well connected to our memory. As often quoted in numerous books, French writer Marcel Proust (1871-1922) described the power of his early sensory experiences when he translated the taste of the madeleine pastry and the smell of the tea he remembered from his childhood into its psychological elements, and examined how he felt about the dessert and what it meant for him. In a book "Proust was a neuroscientist" Jonah Lehrer (2007) wrote that Proust intuited a lot about the structure of our brain. The insight of the writer was that senses of smell and taste bear a unique burden of memory. Below is a fragment of a book by Marcel Proust (2011), "Remembrance of Things Past: Swann Way:"

Many years had elapsed during which nothing of Combray, save what was comprised in the theatre and the drama of my going to bed there, had any existence for me, when one day in winter, as I came home, my mother, seeing that I was cold, offered me some tea, a thing I did not ordinarily take. I declined at first, and then, for no particular reason, changed my mind. She sent out for one of those short, plump little cakes called 'petites madeleines,' which look as though they had been moulded in the fluted scallop of a pilgrim's shell. And soon, mechanically, weary after a dull day with the prospect of a depressing morrow, I raised to my lips a spoonful of the tea in which I had soaked a morsel of the cake. No sooner had the warm liquid, and the crumbs with it, touched my palate than a shudder ran through my whole body, and I stopped, intent upon the extraordinary changes that were taking place. An exquisite pleasure had invaded my senses, but individual, detached, with no suggestion of its origin. And at once the vicissitudes of life had become indifferent to me, its disasters innocuous, its brevity illusory–this new sensation having had on me the effect which love has of filling me with a precious essence; or rather this essence was not in me, it was myself. I had ceased now to feel mediocre, accidental, mortal. Whence could it have come to me, this all-powerful joy? I was conscious that it was connected with the taste of tea and cake, but that it infinitely transcended those savours, could not, indeed, be of the same nature as theirs. Whence did it come? What did it signify? How could I seize upon and define it? ... Undoubtedly what is thus palpitating in the depths of my being must be the image, the visual memory which, being linked to that taste, has tried to follow it into my conscious mind.

One may ponder how much the beliefs and expectations concerning sensual qualities have the power to build perceived experiences and the resulting physiological and psychological effects. Rachel Herz (2009) explored the sentimental ways of perception and reaction to the sensory stimuli and examined the olfactory effects on mood, physiology, and behavior. She outlined an explanatory model for how odors produce emotional, cognitive, behavioral, and physiological responses, and concluded that of the two theoretical mechanisms that have been proposed to explain these effects: pharmacology or psychology, a psychological explanation could best account for the data obtained in the experimental studies. Aromatherapy is a folkloric tradition asserting the beneficial properties of various plant-based aromas on mood, behavior, and "wellness." It is true that many plants have therapeutic properties, for example, lavender may decrease heart rate, improve subjective mood, and reduce stress, anxiety, depression, and insomnia. However, the experimental results obtained by Herz illustrate that it is the perceived quality and the meaning of the aroma that induces psychological and/or physiological responses, while the chemical nature of the odorant itself plays a secondary role in the emotional and subjective changes that occur in the presence of an odor. For example, lavender, which is a culturally denoted "relaxing" odor, when prescribed as a stimulant, was able to elicit stimulatory effects in heart rate, skin conductance, and a self-reported mood. According to the psychological hypothesis of the author, responses to odors are learned through association with emotional experiences. Associative learning, perceptual experience, and expectation can account for the emotional, behavioral, and physiological effects produced by odor inhalation.

Touch

The sense of touch is of special importance for some people, which can be illustrated by the bringing the virtuoso musician's hand in contact with an instrument, the distinctive manner of performing a surgical operation, or the faculty of reading by touch (for example, Braille), and perception through physical contact. A celebrated ophtalmologist and an eye surgeon Vladimir Petrovich Filatov was famous for his ability to cut a stack of ultra-thin tissues, which were then used for rolling tobacco cigarettes, for a particular depth (for example, to cut 100 layers) as requested by an audience, so everybody could count and check the number of the cut tissues. The sense of touch is located in the dermis – the bottom layer of skin. About twenty different types of nerve endings in the dermis carry to the spinal cord and further to the brain information coming from the heat, cold, pressure, touch, pain, and other receptors, especially from the most sensitive body areas: hands, lips, face, neck, tongue, fingertips and feet. One may say our sense of touch may impose our need for creating the palpable, three-dimensional art forms; we can go around, and touch them.

See Table 2 for A Visual Response.

Taste

Taste – a physical ability to discern flavors is one of the senses; however, gustatory perception securing us a sense of taste can hardly be considered a tool for building communication. Computational solutions for testing taste took form of genetic programming where mathematical models of taste evaluations compete with each other to fit the available data, and a bio-inspired method of cross-pollinating to produce more accurate models (Hardesty, 2012).

WHAT CAN BE DONE WITH THE SENSES TO DEFEND (AND ATTACK)?

Senses may be considered the important part of the communication process, not only in about food, reproduction, and family matters but also

Table 2.

A Visual Response: Playing a Kinesthetic Game
You may now want to play a kinesthetic game based on the awareness of the position and movement gained by means of receptors, the sensory organs. Think of a symbol, a letter, or a simple image, draw it with your finger on somebody's back, and then ask whether this person can write down what it is. You may then want to change roles and recognize the shapes 'drawn' by another person on your back. It is amazing how human back can sense the touch, hot, cold, soft, sharp, and pressure signals, using different kinds of receptors.

about the self defense, group defense, and finding a prey or a competitor. Communication is an important factor in the defense systems for all populations of living things. One may say living organisms use their senses to become aware of the danger, defend their integrity to survive harsh environmental conditions, or attack other creatures in attempt to forage on them. Computing based approaches and technologies allow learning more about varied mechanisms of defense and communication between living organisms.

When we mention defense mechanisms in plants, spines may come to mind first. Most cacti are spiny plants. Spines are modified leaves, while thorn present in another kinds of plants are modified branches. Most of the species of a stinging nettle have stinging hairs on the stems and leaves. Venus flytrap catches (by a complex interaction between elasticity, turgor – pressure in a cell plasma, and growth and then digests (with the use of catalyzing enzymes) animal prey, mostly insects and spiders (arachnids).

Animals use many of their senses at the same time to develop their defense systems that fit their living conditions and their needs. The sense of smell is often more acute in animals than in humans, so most of animals manifest greater sensitivity to smells than people. A skunk secretes a foul-smelling liquid as a defense against predators. However, many keep a skunk as a pet and playful companion because skunks are intelligent, curious, and friendly.

The sense of touch allows the animals to feel the harmful things but also to develop the mechanical ways of defense and aggression. Porcupines (and also other rodents such as capybaras and agoutis belong here) use their sharp spines or quills when necessary; they defend and camouflage them from predators. The same can be told about hedgehogs (which are also often kept as pets). Swordfish, which usually reach ten feet in length, use their long, flat bill to slash (but not to spear) its prey and then catch it; they are among the fastest fish. Moreover, they efficiently use their sense of vision because they have excellent sight. Sea urchins, small globular animals with radial symmetry, have long protective spines. Starfish also have spines covering their upper surface and a soft bottom side.

A beak, bill, or rostrum helps a bird to eat, groom, feed young, courtship, or manipulate objects but also fight. A shape of a beak depends mostly on a bird's feeding method. Mammals use teeth and claws both for defense and aggression. Many mammals, for example cattle, goats, antelopes, use horns to defend and attack. A horn is a projection of the skin with a bone inside. However, there are many kinds of hornlike growths. Giraffes have bony bumps, deer have antlers, and rhinoceros have keratin horns. Also chameleons, horned lizards, some insects, and even some jackals display horny growth on their heads. Many mammals, such as musk deer, wild boars, elephants, narwhals, and walruses have tusks that are oversized teeth but serve the same function as horns.

Chemical defense and attack include the ability of sensing smells but also secreting, excreting, and in some cases injecting harmful substances. Jellyfish can sting their prey and inject venom; some of them may even kill a grown man. Puffer fish not only uses its external, spiky fins combining it with sudden speed burst, but it also puffs up, and many of them contain neurotoxins (tetrodotoxin) in their stomachs, ovaries, and livers that have lethal effect.

Spiders use chemical sensors providing information about taste and smell and touch sensors called setae located on their bristles. They also use their four pairs of eyes, some of them very acute, to detect direction of their prey movement. Spiders defense themselves against birds and parasitic wasps by their camouflage (a method of concealment) coloration. Venomous spiders have the warning coloration. Frogs defense themselves in many ways, by camouflage, making long leaps, or secreting mucus with diverse toxic substances (bufotoxins) from their parotoid glands behind the eyes. Poison dart frogs use their toxins for hunting. Some Australian frogs synthesize the alkaloids that are irritants, hallucinogenic, convulsants, nerve poisoning and vasoconstrictors, so they are severely affecting humans. Poisonous mushrooms, which contain many kinds of toxins, are often looking like the edible ones and taste good. For example, an edible chanterelle looks like a poisonous Jack-o-lantern.

Senses Related to Electricity and Magnetism

An electric eel that is often over 6 feet long is capable of generating electric shocks up to 600 volts and use them for self-defense and hunting. They can stun the prey by producing either a low voltage or high voltage electric charges in two pairs of electric organs made of electrocytes – muscle-like cells. Electric eel can also use electrolocation by emitting electrical signals.

Many kinds of animals can detect the direction of the Earth's magnetic field and use this information for orientation and navigation. Those are turtles, spiny lobsters, newts (including salamanders), and birds (Lohman, Lohmann, & Putman, 2007). Also earthworms can feel magnetic field. The directional 'compass' information extracted from the Earth's field provides the animals with positional 'map' information. It helps navigate; for example, small birds pied flycatchers navigate along the migratory pathways possibly using several magnetic navigational strategies in different parts of their journey. The tied to vision magnetic sense in the birds such as European robins, Australian silvereyes, homing pigeons, and domestic chickens allows them to sense the direction of the Earth's magnetic field and navigate when other landmarks are obscured. The magnetic compass orientation of birds is light dependent, in particular on blue-green wavelengths. The compass lies in the right eye of the robin. For example, if their right eye was blocked, they headed in random directions. Thanks to special molecules in their retinas, birds like robins can 'see' magnetic fields. Magnetic fields act upon the unpaired electrons and affect the sensitivity of a bird's retina (Stapput, Güntürkün, Hoffmann, Wiltschko, & Wiltschko, 2010).

Sight and Camouflage

Some fish show their colorful parts of fins only to their mates and hide them in other times while trying to stay invisible. Many mammals such as polar bears or foxes change their color according to the season so they are almost invisible from a distance. Many birds have colorful feathers on their sides only, so the predators cannot see them from above.

Chameleons, which have stereoscopic vision and depth perception can see in both visible and ultraviolet light. They can change their skin color; in three layers below their transparent outer skin they have the chromatophores – cells containing pigments (yellow, red, blue, or white) in their cytoplasm, and melanophores with a pigment melanin controlling how much light is reflected, which sets the intensity of each color. Prairie dogs use a dichromatic color vision. The common toad has golden irises and horizontal slit-like pupils, the red-eye tree frog has vertical slit pupils, the poison dart frog has dark irises, the fire-bellied toad has triangular pupils and the tomato frog has circular ones. The irises of the southern frog are patterned so as to blend in with the surrounding camouflaged skin (Beltz, 2009; Mattison, C., 2007).

Cuttlefish – the mollusk of the order Sepia, one of the most intelligent invertebrates, release brown pigment from its siphon when it is alarmed. The pupils of their eyes have a W shape. They do not see colors; they can perceive light polarization and contrast. The sensor cells (foveae) in their retina look forward and backward. According to Mäthger, Barbosa, Miner, & Hanlon (2005), the eyes focus by shifting the position of the entire lens with respect to the retina, instead of reshaping the lens as in mammals. Unlike the vertebrate eye, there is no blind spot because the optic nerve is positioned behind the retina. Cuttlefish can camouflage by intensity of their color. They have leucophores (the light reflectors) that help match the blue-green spectrum of their deep-sea environment.

Discussion of camouflage as an ability of pretending to be something else can be found in Chapter 7, Pretenders and Misleaders in Product Design.

Communication through Pitch

Birds communicate by sounds but they try to avoid being heard by predators. For this reason, a bird may emit a very high-pitched and a very short sound, which its mate but not its predator can hear. Some species display ability to learn songs. Vocal learning is considered a substrate for human language. This trait has been found in three groups of mammals (humans, bats, and cetaceans – dolphins) and three groups of birds (parrots, hummingbirds, and songbirds). Vocal learners gain vocalization through imitation, while auditory learners do that by making associations (Jarvis, 2004). Perfect pitch perception is present in many members of animal kingdom. According to Slobodchikoff (2002), prairie dogs have highly developed cognitive abilities; they use vocal communication to describe any potential threat, send information about what the predator is, how big it is, and how fast it is approaching. As described by the author, alarm response behavior varies according to the type of predator announced. If the alarm indicates a hawk diving toward the colony, all the prairie dogs dive into their holes, while those outside the flight path stand and watch. If the alarm is for a human, all members rush inside the burrows. For coyotes, the prairie dogs move to the entrance of a burrow and stand outside the entrance, observing the coyote, while those prairie dogs that were inside the burrows will come out to stand and watch also. For domestic dogs, the prairie dogs stand where they were when the alarm was sounded, again with the underground prairie dogs emerging to watch (Slobodchikoff, 2002).

Humans hear sounds of about 12 Hz to 20 kHz. With their hearing superior to human hearing, with a range 40 to 60 kHz (dog whistles are set at about 44 kHz), dogs can convey a lot of information and emotions through their barking. Also bats have sensitive hearing, between 20 and 150 kHz; they use it for navigation and echolocation for locating and tracking their prey. Mice hear sounds of about 1 kHz to 70 kHz. They communicate using high frequency noises partially inaudible by humans. Marmots and ground squirrels, groundhogs, squirrels, and prairie dogs all are highly social; they whistle when alarmed to warn others. Chameleons can sound frequencies in the range 200 to 600 Hz.

Other Defense Systems

Leukocytes, white blood cells are involved in defending an organism against infection. They evolve from a hematopoietic stem cell produced in bone marrow. The number of leukocytes in blood is an indicator of a disease. Immunology – a branch of biomedical science study the immune system that protects an organism. It is present in all organisms: phagocytosis is present in single-celled organisms, production of the antimicrobial peptides in arthropods, and the lymphatic system has developed of in vertebrates; it uses both ways of protecting an organism, phagocytosis and production of immunoactive substances.

The social grouping animals develop many kinds of the collective defense systems, for example by keeping their females and youngsters inside a herd. Animal communication in associations of animals goes often through pheromones – secreted or excreted chemical factors that trigger a social response in members of the same species. There are alarm pheromones, food trail pheromones, sex pheromones, and other types that affect behavior or physiology of a plant, an insect, a reptile, or a mammal.

Some frogs are bluffing by inflating their body and standing on its hind legs, other 'scream' loudly or remain immobile. There are many forms of parental care in frogs.

It has been reported that many sea animals can sense the coming eruptions of submarine volcanoes, possibly through acoustic vibrations. These underwater fissures in the Earth surface located mostly near ocean ridges (tectonic plate movement areas) account for 75% of annual magma output. In the past, life on the seafloor was perishing because of the two huge volcanic eruptions; the age of dinosaurs might end about 65 million years ago for the same reason, especially due to the rise of the levels of carbon dioxide, carbon monoxide and other greenhouse gases followed by the climate warming and two mass extinctions.

THE SENSE OF NUMBERS: COMMUNICATION USING NUMERALS

In many studies and publications explorations about the sense of numbers take a form of a discussion about nonverbal counting and searching whether there is thought, especially math thought, without language. Human beings are capable of developing capacities of an exquisitely high order in many semi-autonomous intellectual realms. Howard Gardner thought of those processing capabilities in terms of environmental information processing devices, for example, he considered the perception of certain recurrent patterns, including numerical patterns, to be the core of logical mathematical intelligence (Gardner, 1983/2011; 2006). As stated by Rudolph Arnheim (1969, 1974), perceptual sensitivity is the ability to see a visual order of shapes as images of patterned forces that underlie our existence. The Arnheim's approach to perceptual sensitivity somehow corresponds to the way computer scientists talk about the codes in terms of for patterns. One may also ponder about programming languages such as HTML or Processing as the information and communication tools related to the numeral perception and serving for nonverbal communication between individuals and computers, as well as for HCI.

The Numerical and Verbal Cognition

There is an opposition between Symbolist theories (the thinking occurs in mental symbols) and Conceptualist theories (mental symbols are products of the thinking about conceptual and abstract entities). The numerical cognition involves mental processes of acquiring and processing knowledge and understanding of numbers. Adults, infants, and animals have already revealed a shared system of representing numbers as analog magnitudes (that contain information on a continuous scale not restricted to a specific set of values). Researchers look for correlations between verbal (language dependent) and nonverbal number knowledge. According to Bar-David, Compton, Drennan, Finder, Grogan, & Leonard (2009), the number concepts can be categorized into two major conceptual camps: language-independent model that claims that nonverbal number concepts are not shaped or created by verbal number understanding and abilities, and the language-dependent model that claims that the use and development of number in verbal activities enables children to become proficient (conceptually and procedurally) with numbers.

One hypothesis is that language is necessary to produce thought and children are unable to think about exact number quantities without first learning and mastering verbal counting. Gordon (2004) and then Frank, Everett, Fedorenko, & Gibson (2008) examined connections between language and numbers in a research with the Piraha people, an indigenous hunter-gatherer Amazonian tribe mainly located in Brazil. The Piraha have a counting system with words for "one," "two," and "many" but no exact verbal number words for numbers higher than two; the lack of language for exact large numbers affected the Piraha's numerical cognition and performance with numbers higher than three (according to Bar-David et al., 2009).

It is believed that, in contrast with the Arabic decimal system used today almost everywhere, the Mayans counted with fingers and toes and thus they used the Maya vigesimal system based on groups of twenty units (Maya Mathematical System, 2012). One may wonder whether the covering of toes with shoes might influence the developing of counting systems by limiting possibility to use this counting tool to ten. The Maya discovered a notion of zero, and thus the 20 units meant [0-19] with a significant zero, while in the decimal system we apply [0-9] as a placeholder. Each number, from zero to 19, had its name. They only used three symbols: an ovular shell for a zero, a dot, and a dash, alone or combined, and wrote them vertically or horizontally.

Ancient Egyptians performed division as multiplication in reverse: they repeatedly doubled the divisor to obtain the dividend. Actually, we often avoid using exact verbal number words when we say, some, less, a few, several, many, numerous, or a couple. A rosary designed as a string of beads for keeping count in practicing devotion, a calculating tool abacus, the secular strings of worry beads komboloi used in Greece and other countries, and Hindu prayer beads can be seen as examples of numerical cognition processed without verbalizing.

On the other hand, there is evidence for precise numerical representation in the absence of language, indicating that language is not necessary to represent small exact numbers or large approximate numbers, and one does not have to know language to understand number. Researchers explored the counting systems in native tribes of South America, and indigenous Australian Aborigines looking for languages that encode the linguistic as well as numerical information. The quipu, a system of knotted cords, is a numerical recording system that was used in the Inca Empire in the Andean region in the 15th and 16th centuries. Gary Urton, a specialist in Andean archaeology, posed that this combination of fiber types, dye colors, and intricate knotting contains a seven-bit binary code capable of conveying more than 1,500 separate units of information. According to Marcia and Robert Ascher (1980, 1997), most information on quipus is numeric, and these numbers can be read. Each cluster of knots is a digit, and there are three main types of knots: simple overhand knots; "long knots", consisting of an overhand knot with one or more additional turns; and figure-of-eight knots. Butterworth, Reeve, Reynolds, & Lloyd (2008) questioned claims that thoughts about numbers are impossible without the words to express them, that children cannot have the concept of exact numbers until they know the words for them, and adults in cultures whose languages lack a counting vocabulary similarly cannot possess these concepts. They have shown that children who are monolingual speakers of two Australian languages with very restricted number vocabularies possess the same numerical concepts as a comparable group of English-speaking indigenous Australian children.

Dunn, Greenhill, Levinson, & Gray (2011) pose that cultural evolution, rather than innate parameters or universal tendencies, is the primary factor that determines linguistic structure and diversity, with the current state shaping and constraining future states. Data collected by Gelman and Butterworth (2005) imply that numerical concepts have an ontogenetic origin (resulting from the development of an individual organism) and a neural basis that are independent of language. They question theories about the necessity of using language to develop numerical concepts, both in the case of the child developmental psychology and the counting systems in Amazonian cultures that have very restricted number vocabularies, with the lack of language for exact large numbers.

Preverbal humans and animals can represent number through analog magnitudes – nonverbal approximate representation of number values (Dehaine, 1999). Adult people, when told to tap

15 times with their finger while blocking out the counting system through verbal interference, can usually perform 15 taps, showing that humans can approximately map the number 15. As subjects try to tap higher numbers, responses become less precise (Carey, 2004). According to Varley, Klessinger, Romanowski, and Siegal (2005), the numerical reasoning and language are functionally and neuro-anatomically independent in adult humans. They demonstrated a dissociation between grammatical and mathematical syntax in patients with brain damage who suffered from severe aphasia, preventing them from understanding or producing grammatically correct language. Patients demonstrated proficiency in mathematical syntax, despite an inability to comprehend analogous syntax in spoken or written language. The authors describe the patients who were unable to differentiate between the statements "Mary hit John" and "John hit Mary," but these same patients successfully solved mathematical operations that were structurally dependent in this same general way, for instance, the difference between 52 - 11 and 11 - 52. Similarly, performance of patients who were unable to comprehend sentences with embedded clauses was unimpaired in computing expressions with embedding, for example, answers to a sequence such as $90 - (3 + 17) \times 3$. Language disorders and calculation abilities occur independently of each other (Gelman & Butterworth, 2005), which have been shown with the use of neuroimaging – techniques for picturing the structures of functions in the brain. Rhesus monkeys can represent the numerals one through nine on an ordinal scale (Brannon and Terrace, 1998). Models of number processing, which are language-independent, indicate that number concepts exist prior to verbal counting (Bar-David et al., 2009).

Several research studies investigated the senses and brain areas involved in numeral activities in children, first without naming, then with names spoken, read, or seen as written words. Young children have a limited understanding of a concept of number. Children start out as "subset-knowers:" 1-knowers (i.e., they can give one exact object when asked for one), then 2-knowers, 3-knowers, and 4-knowers (Wynn, 1992). Children that can give a requested number of objects (for example, yellow rubber ducks from a green bowl) within their productive count range are labeled as cardinal principle (CP)-knowers. They fall into two groups: mappers and non-mappers (LeCorre & Carey, 2006). Mappers are able to metaphorically map an accurate verbal number estimate to an observed array of objects while non-mappers can only do this for quantities under four. Mappers have an intuitive sense of quantity for each mapped number word and thus an understanding of the logic of the count list. The subset-knowers do not truly understand the concepts that the number words represent outside of a certain range, while CP-knowers do. Bar-David et al. (2009) explored the relationship between number language and the representation of exact large numbers. They analyzed the cognitive tools that enable children develop number concepts. They tested both subset- and CP-knowers on a nonverbal task involving numbers, both small (less than four) and large (greater than four) numbers. Among many other tests, they authors asked children to retrieve "just enough socks" for caterpillars with varying quantities of feet. The authors propose two interpretations of their results: (1) the accumulation of verbal knowledge by a subset-knower leads to the development of more precise nonverbal number structures which scaffolds and enables the induction to CP-status, or (2) the accumulation of verbal knowledge by a subset-knower enables the induction directly, and this induction facilitates the development of non-verbal number concepts.

According to Halberda, Mazzocco, & Feigenson (2008), adults, infants, and animals share a sense of a quantity, while competence in domains such as calculus emerges from a different system, as it relies on symbolic representations that are unique to humans who received instruction. An ancient evolutionary number system supports basic intuitions about the numerals; as a result, people and animals can represent the approximate number of items in visual or auditory arrays without verbally counting. Moreover, they can use this capacity to guide everyday behavior such as foraging, searching for and exploiting food provisions. We do not know whether some individuals have a more precise non-verbal 'number sense' than others, and how this system interfaces with the formal, symbolic mathematical abilities. Halberda et al. (2008) found large individual differences in the non-verbal approximation abilities of 14-year-old children, which correlate with children's past scores, starting from kindergarten, on standardized math achievement tests. The authors found such differences related to the number acuity – sharpness of an evolutionarily ancient, unlearned approximate number sense.

Elizabeth M. Brannon (2005) discussed the current understanding of the neural organization of numerical cognition in the light of the relationship between language and thought, and inquired whether thought occur in the absence of language. It was previously assumed (Whorf, 1956; Vygotsky, 1972) that the universal properties of language, specifically words and syntax, determine human thought processes; human cognition differs from animal cognition as a result of exposure to a natural language. According to Brannon, even in the light of the Varley's et al. finding that when language is lost complex thought is retained, the syntax of mathematics may be evolutionarily or developmentally derived from the syntax of language or possibly even vice versa. The author believes it is possible that the human brain may have been transformed by the evolution of the language faculty, thus paving the way for complex social reasoning and mathematical syntax.

EXPANDING THE SENSES THROUGH THE USE OF TECHNOLOGY

From the seventies, inventors searched for computational solutions for enhancing sensitivity and worked on creating wearable computers. A wearable computer is included into the personal space of a user and controlled by the user. Steve Mann is one of the pioneers of wearable, mobile, wireless computers. He built the WearComp with display and camera concealed in ordinary eyeglasses, so the wearer had increased awareness of the environment. WearCam was one of the first cameras transmitting images on the web. Mann's Cyborglogger continuously recorded processed, computationally interpreted, and shared personal day-to-day life. The WearComp could be used as an expressive medium for direct communication or the artistic production. Mann considers himself a cyborg, a person whose physiological functioning is aided by or dependent upon a mechanical or electronic device. His writings on wearable computer industry warn of dangers to our liberty, privacy, and democracy. Smart clothing described by Mann (2001) provides several functions including time/date/calendar/reminder (replaces wristwatch), voice communications (replaces cellular phone), messaging, email (replaces pager), Sound Blaster device, sound and video capture (replaces camcorder. Covert camera and viewfinder look like ordinary bifocal eyeglasses and are useful "for shooting documentary video, electronic newsgathering, as well as for mediated shared visual communications space" (Mann, 2001, p. 90). Motion video capture capability (e.g. AVI files, etc.) partially eliminates the need for a camcorder), pencigraphic image compositing (replaces camera and may be transmitted wirelessly, computation (replaces calculator), and measurement that provides logging of heart rate and ECG waveform, respiration, and the like, with automatic alert program. The developments in facial recognition systems, computer applications for identifying and automatically verifying a person on a video frame or other image source contributed to the security systems and other application areas.

In the 1990s, several authors had no more seen the wearable computers a harmless, perfect technology but a vector towards posthumanism (Mitchell, 2005). The concept of posthumanism was focusing on changes in the way humans perceived their place in the world. Katherine Hayles (1999, p. 288) wrote about posthumanism, "emergence replaces teleology; reflexive epistemology replaces objectivism; distributed cognition replaces autonomous will; embodiment replaces a body seen as a support system for the mind; and a dynamic partnership between humans and intelligent machines replaces the liberal humanist subject's manifest destiny to dominate and control nature." Then, an international intellectual and cultural movement, transhumanism declared that technology in the future would change the human condition, eliminate aging, limitations on human and artificial intellects, suffering, and our confinement to the planet earth (Boström, 2005).

Life in a city results in adaptation of our senses to the variety of disparate signals and diminishes their sensitivity. We become accustomed to the coexistence of closely compacted or scattered skyscrapers and small cottages. However, the combined effect of various sensory stimuli that are typical of a big city modifies our lifestyle preferences. Figure 6 presents a work "Visual Culture."

City patterns provide visual shortcuts
that define the ambience we are living in
and feed the marketing imagery.

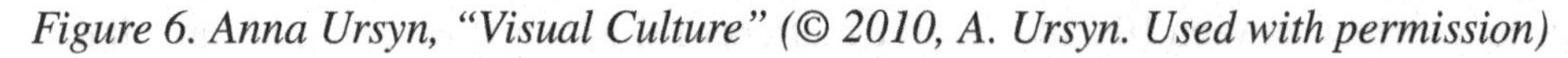

Figure 6. Anna Ursyn, "Visual Culture" (© 2010, A. Ursyn. Used with permission)

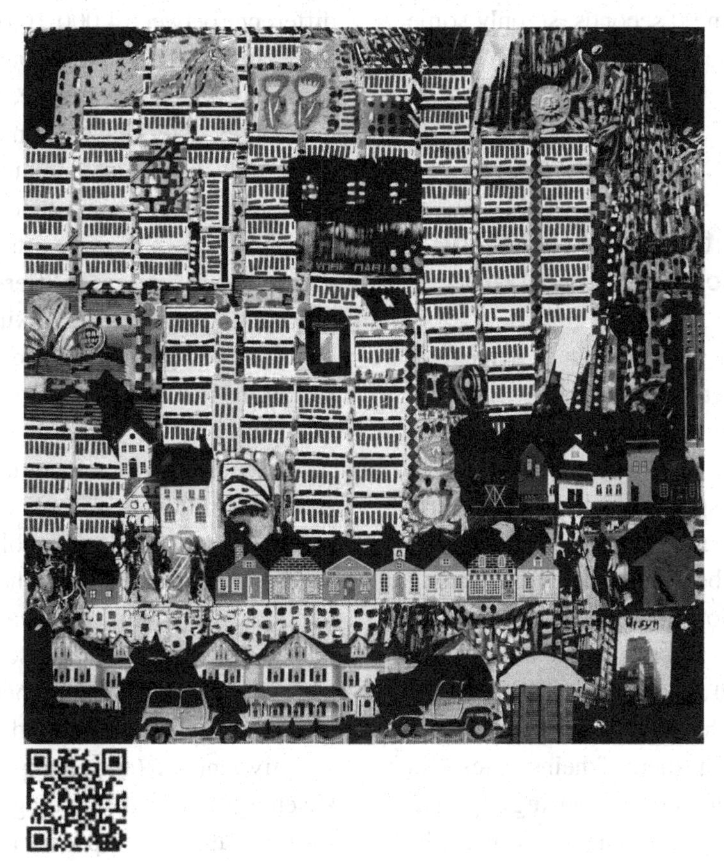

Applications designed to power human senses and improve emotional wellbeing of the user take now a form of small gadgets such as a bracelet (that may soon contain a smart phone) or a waterproof wristband synced to a smart phone, which tracks sleep patterns (deep versus light sleep, awake time, and overall sleep quality), has a silent vibrating alarm, a motion sensor (that counts your steps on a track), and a food journal (that encourages to snap a photo of the food and report how each meal makes you feel, and then uses responses over time to discover which meals make you feel best). Marshall, Chamberlain, and Benford (2011) designed a wearable audio artwork titled "I Seek the Nerves under Your Skin," which encourages people to run increasingly fast, pushing themselves physically and mentally, to mirror the intensifying performance of a poet heard on the headphones. If the listener goes faster, the poem runs for longer before fading. To hear the whole poem the listener must accelerate to a sprint (between 10 and 20 miles per hour) in 90 seconds, so only some listeners will hear the complete poem.

SENSITIVITY

Questions about the Threshold and the Source of Sensory Information

It seems we can look at the notion of perceptual sensitivity in two different ways. First, we can think of sensitivity from an angle of a threshold intensity of a stimulus that must be exceeded to evoke a reaction. In this respect, sensitivity to specific kinds of stimuli such as visuals, sounds, or smells varies greatly, both between species and among individual people. Not only most of animals have a lower threshold level to respond to a smell, but also they produce and release chemical substances called pheromones that cause physiological and behavioral reactions in others of their species or in different living beings, such as causing aggregation, alarming, attracting a mate, marking trails, or defining landmarks for territory boundaries. In case of pheromones the threshold level is usually so low that single molecules can evoke reaction. Also plants release pheromones to attract bees and other pollinators to their flowers. In case of human reactions, individual people differ very much in their sensitivity to smells (and to other kinds of stimuli). In most cases those living in big cities may desire to have a quality of smell displayed by people living in pristine conditions, such as members of tribes living in the Sahara Desert or the Aboriginal people living in Western Australia.

On the other hand, we can inspect the notion of perceptual sensitivity by looking for sources of the experience. From this angle, the quality of being sensitive may mean an ability to recognize and remember a stimulus and a signal it denotes without an external reference for making a comparison. This appears to be a case with people possessing perfect pitch and a perfect sense of color. As for the matters of smell and taste, people can tell the difference between 4,000-10,000 smells, and they have close to 10,000 taste buds in their mouths. One may say, some people, for example wine, beer, or tea testers have attained connoisseurship, as they can discriminate minute differences in the smell and taste of a substance. In a similar way, a connoisseur in the fine arts, for whom a small difference makes a great difference, is a person not only having a great deal of knowledge but also a quality of being sensitive to minute, incomparable variations in what they see or experience, giving them an insight.

There are also a great number of people having sensual abilities more acute than those of others – both in terms of the threshold and the source of sensory information. With the existing scientific methods, we cannot verify and appreciate their talents and successes because of lack of tools and methods of assessment. Whether one accepts it as true or not, some individuals claim they are sensitive enough (and equipped with an appropriate energy) to be able to diagnose others' illness, give one at least temporary relief of pain, find a lost person or an object, and detect ground water or underground pipes, aside from some harmful possible actions linked to black magic or voodoo. Other persons use sticks of wood, stone, ivory or metal to channel energy in the vicinity of water-courses, and detect possible water resources under ground. People prove their faculties in a pragmatic, matter-of-fact way, no matter what are the results of scientific assessment. There is also the U.S. Psychotronics Association founded in 1975 for people interested in the study of psychotronics – the science of mind-body-environment relationships.

Some hold that we use our senses poorly. For example, we are apt to develop electronic devices for gathering data about our external and internal environment, and patients usually receive medical examination obtained through the use of instruments, meters, and various kinds of laboratory equipment. In contrast with the teachings of ancient medicine that trained Chinese, Babylonian, Egyptian, or Greek physicians to feel and recognize symptoms through palpation, for example discern eighteen to thirty six types of pulse corresponding to particular diseases (Kaptchuk, 2000), present-day doctors not so often touch sick people to gain an insight coming from the careful examination of physiological condition of the patient's skin or from feeling their pulse. While we are happy we have so many options and do not have to rely on doctors' sensory information, many would agree that giving more attention to our faculties coming from the senses would enhance to a great extent our connection with the world and condition of communicating with others. Many artists are fully aware of this challenge and do their best to combine digital technologies with artistic means of expression.

Sensibility

In cognitive terms, sensibility to internal and external signals enhances our mental processes, focusing our attention, supporting remembering things, enhancing knowledge, and stimulating such cognitive processes as solving problems or making decisions. Sensibility enable us appreciate perceptions and sensory input, as well as contributions coming from cognitive thought material, so we can feel and understand things deeply. Sensibility helps us see and react to our own and others' aesthetic, emotional, or complex cognitive processes, and makes our responsiveness acute.

With sensibility, we can receive information about others' perception. Therefore we can take advantage of many ways to communicate not only by speaking or writing but also by using nonverbal means such as images (for instance signs, posters, or pictures), sounds (speech or music, for example), gestures (everyday expression through bodily actions with which we enhance our statements, mudra gestures in Hinduism and Buddhism), or artistic forms such as actor's gestures, body language and motions, among other possibilities. For more than seventy years Dorothy's powerful ruby slippers from the 1939 MGM musical fantasy film *The Wizard of Oz* based on the novel *The Wonderful Wizard of Oz* written in 1901 by L. Frank Baum (Baum et al., 2000) maintain their iconic status. The way Dorothy tapped her heels together became a time-based kind of a symbol. Remembering the significance of nonverbal means of communication is of utmost importance because we are constantly immersed in interactive contact with others through social networking that uses multimedia as a way to convey information through various combinations of content forms such as still images, animation, video clips, or interactive media.

One may say musicians, fine artists, performance artists, and physicians concerned with human brain and psychology, as well as many attentive, creative, intellectually gifted partners and productive workers are seen as sensible individuals. Ken Bernstein (2005), who performs automatic drawing on any occasion, created doodle-based artwork that was selected for a wall that insulates residential buildings from traffic noise and dust on a busy street in Boulder, Colorado. According to Bernstein, automatic drawing follows a sensibility rather than a thought process. As stated by Danny Coeyman (dannycoeyman.tumblr.com/) a portrait, as a vehicle for deep connection can witness the ways we relate and our shared humanity. To explore this further, from 2008 the artist draws portraits of people selling him products on the receipt tapes of these products, right after completing a purchase: he puts ink on a receipt tape made from sales and purchases in daily life.

We used to talk about a highly sensitive person who has a vision about some ideas not only in relation to the faculty of seeing but also in cognitive terms. Maybe when we describe someone's vision as an ability to think imaginatively about future actions or to experience a vivid mental image of future events, we may refer to someone's sensibility and acute responsiveness toward the environment or other people. Strong stimuli coming from previous disturbing or joyful events may lower the sensibility threshold; uncontrollable emotion or even allergic reaction may happen in response to a stimulus that had been traumatic in the past. Some hold that one may learn how to lower one's sensitivity threshold. In terms of the frames of mind discussed by Howard Gardner (1983/2011; 2006) in his study on multiple intelligences, Gardner described some particular abilities that require sensibility: the interpersonal intelligence oriented toward the understanding of other people, intrapersonal intelligence as an understanding of oneself, and naturalistic intelligence – an apprehension of the environment. An Italian semiotician, philosopher, and novelist Umberto Eco (2002) proposed that human sensibility to both beauty and ugliness has been different through the ages. The author argued that a correspondence between the public's tastes and artists' sensibilities must be assumed and studied accordingly: "Every age has its own poetic sensibility, and it would be wrong to use the modern sensibility as a basis for passing judgment upon the Medievals" (Eco, 2002, p. 61).

It could be worth one's while to check out how one's personal sensitivity depends on abstract thinking ability. A Swiss psychiatrist and analytical psychologist Carl Jung (1875-1961) wrote,

> *There is an abstract thinking, just as there is abstract feeling, sensation and intuition. Abstract thinking singles out the rational, logical qualities of a given content from its intellectually irrelevant components. Abstract feeling does the same with a content characterized by its feeling values; similarly with sensation and intuition. ... I put abstract feelings on the same level as abstract thoughts. Abstract sensation would be aesthetic as opposed to sensuous sensation and abstract intuition would be symbolic as opposed to fantastic intuition. (Jung, 1921/1971, p. 410, par. 678)*

Elaine Aron (2006) presented highly sensitive individuals (existing, according to the author, in about twenty percent of the population) in a positive light, in contrast to previous descriptions of this trait. However, being sensitive may make somebody easily hurt because even small signals are detectable and easily recognized by someone with heightened sensibility. Jung, who employed the concept of innate sensitiveness, had believed the process of art creation might alleviate emotional distress and feelings of trauma by revitalizing life-enhancing energies (Malchiodi, 2006). It doesn't mean sensitivity is a flaw, in spite of a fact that for some people sensitivity is considered a weakness, especially in boys. Some boys are often trained to be insensitive to be considered strong, and thus they may develop a wide spectrum of unfriendly attitudes toward highly sensitive people. In many instances boys are encouraged to shoot, hunt animals, and fish. They are trained to be insensitive to pain coming from sport games and competitions. Military training, especially intelligence services and espionage training comprise exercises focused on building insensitivity to own or others' pain. Many domains of art that require sensitivity (and sensibility) often evoke hostile reaction and a need to deny the meaning and value of these qualities.

Biology Inspired Computational Solutions for Enhancing Sensitivity

Animals are able to detect and locate many of environmental changes, so the scientists are working on finding biology-inspired computational solutions that improve our sensitivity and thus our ability to control our external and internal environment. For example, we apply echolocation, ultrasound imaging, and gather thermal information with which one can know, scanning from a helicopter or a satellite whether an object on the ground is alive, dead, or inanimate. Communication satellites include civilian and military Earth observation satellites, weather satellites, research satellites, GPS navigation satellites, space stations and other spacecraft orbiting a planet. Researchers and engineers use nano technologies for various purposes: to attract, to repel, to cure, or to enhance a feel of wellbeing. Research lines either aim to enhance our senses with the use of applications or to compensate for our lack of sensitivity to signals. Within this spectrum, sensors inform us about levels of substances that are crucial factors in our metabolism such as glucose level in our blood, or are vital components of our environment such as air pressure or information about time and position.

Bio-interfaces and brain-computer interfaces enable communication between humans and machines and/or humans-machines-humans (Zuanon, 2012). Bio-interfaces translate biological functions into numerical data that can be interpreted by computer systems. A brain-computer interface transforms the electrophysiological signals of a brain into the messages and commands, and produces movements associated with the hardware and software that translates them into actions. Bio-interfaces can be used as wearable devices that provide organic interaction between man and machine. This field of research involves different fields of knowledge, such as neurobiology, psychology, design, engineering, mathematics, and computer science. According to Zuanon (2012), electrophysiological and electroencephalographic activities, and other measures of brain functions, serve as means to non-muscular channels of communication for sending messages and commands to the external world governed by the biology of the users, without interrupting their activity. Biosensors are used as input channels for galvanic skin response sensor (measuring the electrical conductance from skin), blood volume pulse sensor (that uses photoplethysmography to detect existing blood pressure), breathing sensor (monitoring the individual's thorax or diaphragm activity), and electromyogram sensor (that captures the electrical activity produced by a muscle at the moment of contraction). The physiological information of the users acts as data to configure an interaction that responds to their emotional state in order to match the state of their body at that particular moment. This enhances potential of individuals with social and/or motor disabilities, severe muscle disorders such as lateral amyotrophic sclerosis, cerebral hemorrhage, and muscle damage. It also enables applications in the areas of design, art, and games. Within the scope of interaction with games, the wearable computers, "BioBodyGame" (Zuanon & Lima Jr., 2008) and "NeuroBodyGame" (Zuanon & Lima Jr., 2010) created by artists and designers Rachel Zuanon and Geraldo Lima, all allow the users to interact with digital games through their physiological and cerebral signals, respectively.

Synesthesia

Human imagination may be somehow related to synesthesia, a neurologically based condition that results from merging of two or more sensory or cognitive pathways. It happens when a sensory stimulus received by one sense triggers perception and evokes involuntary experience in another sense: one may taste some shapes or hear color. For example, visual perception may involuntary accompany sensations of taste, touch, pain, smell, or temperature. Synesthesia may be induced with drugs. The most common types of synesthesia is the word-to-color association, and colored hearing, when someone is seeing colors when hearing a musical tone, usually the lower a musical note, the darker the color. Some musicians interpret their color experience in terms of musical keys or associate their mood or feeling with musical keys. Therefore synesthesia can be used as a metaphor for transposing elements indigenous to one medium into another (Ox, 1999).

Synesthetic ways of expression are applied in electronic art, performance arts, data presentation, technical implementations, design, advertisement, visualizations and simulations for scientific and educational purposes; they address and interact with the varied audience. Several writers, such as Edgar Allan Poe or Arthur Rimbaud described synesthetic experience and even linked it with memory. Some poets state they hear sounds or musical tones when they see words, images, and colors, and musicians assert that they see colors assigned to particular tones or musical passages. Interactive multimodal data presentations are of interests to scientists. Many kinds of user interfaces are realized in visual, auditory, or tactile domains. The haptic/touch interfaces, for example pressure sensitive interfaces, exemplify another kind of multimodal data presentations.

Digital images, sounds, or animations can be analyzed as metaphors and seen as important sign systems working beyond the literary culture. Some artists create synesthetic metaphors that cross the sense modalities; they may seem incomprehensible, for example when someone (probably a musician) said that, in a mood of irritability, he bought a coat in the color of C sharp minor (Grey, 2000). The reduced division between the culture of the image and that of the text, for example, sending messages containing words, images, sounds and objects, makes synesthetic modes of cognition operate outside the language-based structures of thinking. Messages conveyed in the non-verbal mode of visual communication media are often being more abstract and more interactive with the viewer. By crossing senses, viewers construct their perception of the reality that is communicated by an artist in their own art-historical and cultural contexts. Strong aesthetic experience elicited by synesthetic art may result in a sensation of increased artistic potential and a need for doing one's own artwork. Examples of the approaches to visual presentation of music and sound include artwork sonification and environmental problems sonification. Examples of scientific sonifications include a pie chart sonification enabling the hearer to understand information, sonification of ocean buoy data (Sturm, 2005), or sonification of proteins (Dunn & Clark, 1999). Van Scoy and Gifu (2004) generated music by converting the relations in data sets in order to alert a viewer to the existence of hidden clusters of such kinds of data as mathematical functions and basketball data. The haptic/touch interfaces, for example pressure sensitive interfaces exemplify another kind of multimodal data presentations.

See Table 3 for Your Response.

SPATIAL PERCEPTION AND COGNITION

Spatial skills and abilities mean competence in spatial visualization or orientation. Spatial visualization is an ability to mentally rotate, twist, or invert pictorially presented objects. Spatial orientation is the understanding of the array of objects, utilizing body orientation of the observer. When we recognize something, we compare what we see with our memory representation. First, we build (encode) the structural schema of what we see in both visual and verbal memory, and make comparison to the canonical objects (those with already known top/bottom, front/back, and left/right differences). The split-brain research results (when cerebral hemispheres become disconnected) show that spatial information is common to both hemispheres.

Table 3.

Your Response: Synesthetic Abilities
You may want to check your own synesthetic abilities. In many cases synesthesia is genetically inherited and one does not perceive it as something experienced differently from others; many individuals are unaware of this condition. It may be worth a while to try your own ability to connect your visual skills with an awareness coming from other senses. For example, you may want to make simple line drawings singing at the same time notes or melodies and imaging tunes that harmonize well with each line you are drawing. Try to sketch your experiences or write about your connotations.

Spatial cognition is mental reflection and reconstruction of space in thought, with a distinction between the perception (what is seen, when we process sensory information) and cognition (what is assimilated by a person, based on person's cognitive structures). Spatial cognition and has been defined by Hart and Moore (1973) as an internalized reflection and reconstruction of space in thought. Spatial perception is a form of a processing of sensory information and immediate figurative knowing, whereas spatial cognition is considered operative knowing based on person's cognitive structures. Information processing in spatial cognition includes pattern recognition, that means comparing a stimulus pattern with memory representation. Pattern is encoded as the structural schema in both visual (iconic) and verbal (symbolic) memory codes. Concepts used in the recognition process consist of structural description of spatial concepts, the canonical objects (with assigned descriptions of easily identified differences in top/bottom, front/back, and left/right differences) and the non-canonical objects analysis. Writings on spatial cognition include several concepts that have been defined by the researchers.

Visualization has been sometimes characterized as explaining something visually to oneself, from the viewpoint of the mind's eye. Spatial skills are necessary in many activities and occupations as they allow the use of visualization. In such cases, mental images exist at mental spatial screens as concrete pictures, besides of the words. Spatial skills help to perform mental rotation, when objects must be rotated to be compared and check whether there is similarity or a difference. They also support transposition of three-dimensional objects to a two-dimensional paper.

The question has been around a long time whether mental images exist as the concrete pictures. Besides of the words and images, there is a third mode of representation that allows mental communication between words and images. Kosslyn (1978) analyzed mental spatial screens that were horizontal, rounded, fading around edges, with varied overflow and details. Mental rotation was considered necessary to compare objects; one must rotate objects to compare if there is angular difference (Shepard, 1978). Olson, Bialystok (1983) observed that canonical objects are easier to rotate. When a child rotates around an object, the egocentric system is used and no topological codes to evaluate connectedness and continuity are in use. In the egocentric stage, the frame of reference is fixed and mistakes are easily made, so another viewpoint is needed in rotations. Transformation of a 3-D object to the 2-D orthographic drawings need the mental rotation to a plane, changing scale, and bringing into existence an abstract, semi-iconic signs (small toys help to do it).

Cognitive mapping means mental transformations that are made in order to arrange information contained in everyday spatial environmental space. Environmental space may mean different things for different people. It can be pragmatic – related to where we live, perceptual – what we experience, existential –introducing social and cultural issues, cognitive – based on thinking, and logical – an abstract space. Cognitive mapping is not a map but a metaphor, a process rather than product, made by making routes, building a network, and defining and metrical descriptions of relative positions. We form an environmental cognitive map using schemata as ways to define a structure: paths, edges, districts, nodes, and landmarks. The characteristics of the environment are: barriers, number and type of landmarks, vertical/horizontal spaces, and their effect on the environment's representation. Piaget did not use the concept of the environment; he used the hand-sized items. However, gaining some motor experience, for example, by walking around the object, helps to grasp the concept of the environment. Using the aerial photos requires the rotation to another plane, reduction in scale, and abstraction of objects to semi-iconic signs (Downs & Stea, 1977).

Cognitive mapping is a process composed of a series of psychological transformations by which an individual acquires, codes, stores, recalls, and decodes information about the relative locations and attributes of phenomena in his everyday spatial environment (Downs and Stea, 1977). In cognitive mapping transformation of a 3-D object to the 2-D orthographic drawing requires the use of the mental rotation on a plane, changing scale, and bringing into existence an abstract representation. In a new environment, a person uses routes (which provides egocentric information), then landmarks (fixed objects), and finally builds a map (a framework with coordinated relationships). Methods of coding for a map includes, according to Kuipers (1982), (1) procedures for following routes, (2) topological network descriptions, and (3) metrical descriptions (relative positions vectors). Evans, Marrero, Butler (1981) defined it in another way, as (1) associated networks based on abstract representations, and (2) an analog view, where mental representations are rough isomorphic images.

Field dependence and independence has been investigated intensively since the seventies. Witkin included this concept into his theory on psychological differentiation that pertained to the ways people might differ in psychological and neurophysiologic behavior, in areas such as perceptual, intellectual, personality, and social domains. Field independent individuals are more likely to use internal referents as primary guides in information processing, and field dependent people use more external referents. For example, field independent people have an internalized frame of reference as a guide to self-definition, and use themselves as a reference. Field dependent individuals require externally defined goals and reinforcement. They tend to excel in interpersonal and social competencies as compared to the field independent people. Field dependence-independence has been examined as one of the constructs of personal epistemology, the extent and nature of personal knowledge. Wilkinson & Schwartz (1991) postulated a general belief system of external rigidity versus internal flexibility; hence the personal epistemology of relativism was developed that stressed the subjective truth and values, relative to a particular frame of reference. The field dependence-independence has not been associated with perceptual and cognitive systems. The Group Embedded Figures Test (Witkin, Dyk, Faterson, Goodenough, & Karp, 1974) was used to identify field independent students (Moore, 1985). Field independent persons scored higher on the visual location task that tested the subjects' ability to select a criterion picture from a group of three similar pictures after viewing three quadrants of the criterion pictures in random order. To avoid developing science misconceptions, one must consider different strategies of information processing, and the extent to which a person's perception of information is influenced by the context in which it appears.

The Development of Spatial Cognition

Spatial cognition is a developmental process. Piaget and Inhelder (1971) carried out a study of the development of spatial cognition and described mental manipulations: accommodation which means transformations induced in one's existing schemata, and assimilation which means transformation of the new object to correspond with existing schemata. Their nature determines what is actually perceived. When we recognize something, we compare what we see with our memory representation. First, we build (encode) the structural schemata, recognize and process visual patterns of what we see in both visual and verbal memory, and make comparisons to the recognized canonical objects (those with already known top/bottom, front/back, and left/right differences).

Spatial cognition depends on the child's cognitive level – sensorimotor, intuitive, concrete operational, and formal operational problem solving. At first, a child's view of the world (frame of reference) is egocentric, with no coordination of vision and grasp, and no relationships related to other figures (cannot imagine anything that is not visible because it is covered). Then, a child acquires a fixed system and knows about objects in space even when hidden. Children use iconic or picture representation at the preoperational level, and they develop the use of abstract symbols during the operational stage. Finally the coordinated system is developed. An adult can understand Euclidian geometry, be capable of using formal cognitive operations, and iconic modes of representation.

Thus, there are several areas of grow which have been taken into account in this description:

- **The Frame of Reference:** Which is first egocentric, then it shifts from egocentric to fixed, then the coordinated system is developed, depending on the child's cognitive level.

- **Spatial Relationship:** which is first topological, then projective, and finally Euclidian.
- **Operational Cognitive Levels:** Sensorimotor, intuitive, concrete operational, and logical (formal operational) problem solving. Thus, the adults should be capable of using a coordinated frame of reference, Euclidian spatial relationships, formal cognitive operations, and iconic modes of representation.

The stages of the spatial skills development have been usually described as follows.

Stage 0: the sensorimotor cognitive level (from birth to about two years). The child operates on a purely perceptual level, initially with no coordination of vision and grasp; it "recognizes various objects by sense of touch alone, or by what has been termed haptic perception (Piaget, Inhelder, 1956, p.xii). "The primitive, topological space is internal to the particular figure whose intrinsic properties it expresses, as opposed to spatial relationships ... related to other figures" (Piaget, Inhelder, 1956, p.153). Topological perception is governed by proximity, separation, order, succession, and enclosure. The child's view of the world (the frame of reference) is strictly egocentric.

Stage 1: the preoperational cognitive level (2-3 years). The child accomplished coordination of vision and grasping, and now is able to understand the consistency of a solid, its size, and shape. The spatial relationship is topological, based on properties of being connected or bounded, but not based on distance, angles, or straightness. The child's frame of reference is egocentric, accepted from its own viewpoint.

Stage 2: the transitional level, still at a preoperational cognitive level (4-6 years). Euclidian shapes are recognizable through tactile experiences, and are still topological. Some advance in frame of reference (other's viewpoint is accepted, but not imaginable).

Stage 3: concrete operational cognitive level (7-12 years). Children use perspective and link objects in a coordinated system. Free from egocentrism, they use a fixed frame of reference, and can perceive projective spatial relationship. Objects are viewed in relation to a number of points of view. Children can recognize similarities of parallel lines, angles, and reversed images.

Stage 4: both projective and Euclidian spatial relations (11-12 years). Earlier, there was the use of projective space, then, about 11, Euclidian relationships are understood and the distance and measurement are used, which require a coordinated system of reference with 3-D axes.

The developmental stages can be identified quantitatively. According to Coie, Costanzo & Farnill (1973), the sequence in the mastery of spatial perspective skills within non-egocentric stages begins with interposition (hidden areas), then moves to aspect (corner, view, back, etc.), and finally culminates in right and left; children start from the motor representation of information, then they use iconic or picture representation at a preoperational level, and finally they develop the use of abstract symbols during the operational stage. Besides this hierarchical, orderly arranged pattern of modes of representation, Paivio (1970) reported that this mode could vary with the situation; spatial information being best processed using the iconic mode. Spatial skills that are necessary for architectural drawing are: visualization, mental rotation, transposition of 3-D objects to 2-D paper, and cognitive mapping.

There are individual differences in spatial skills; causes for variation in such skills can be age-related, gender-related, genetic, hormonal, neurological, and induced by the environmental influences. Spatial abilities were extensively studied in the 1970s. Sociocultural influences act in favor of developing spatial abilities in males, e.g. by selection of toys and plays, such as Lego, games, and wooden or metal blocks used as a child's toy.

Variations in spatial ability can be caused by the age-related or sex-related factors, genetic, hormonal, neurological, and induced by the environmental influences; some evidence indicates a hereditary component transmitted via a sex-linked recessive gene and the estrogen-androgen balance necessary for development of superior spatial skills. Right hemisphere has been considered more specialized for spatial processing. Sociocultural influences have been acting in favor of males, because of selecting toys and plays such as Lego, games, and blocks. In spite of expectations of differences on spatial test results, there was no difference in performance.

Training can improve spatial skills, with mediated experience nearly as effective as active experience (Saloman, 1979; Olson & Bialystok 1983). People see something similar to a mental image that can be mentally rotated and changed in form. Such mental skills are necessary to draw orthographically. Olson and Bialystok (1983) posed that mental representations used for testing spatial skills require the clarifying of the unstated information (about the object's basic shape, size, and position), so a person must develop a schema that uses spatial information as a part of structured description of an object. For simple perception of the environment, the highest level of spatial cognition is not necessary: topological relations and egocentric or fixed frames are being used as the cues, but not necessary the Euclidian relationships or coordinated frames of reference. Karlans, Schuberhoff and Kaplan (1969) reported high correlation between spatial skills and creativity.

Learning with Computing, Computer Graphics, and Spatial Ability

A question has been around whether the use of computers may affect spatial abilities and is generally beneficial for the mind. The results of cognitive tests measuring the spatial visualization abilities are useful in estimating user performance in applying interfaces. Bio-interfaces allow participation and audience interaction with art works, design projects, and games. In contrast with previous forms of interaction – through the use of mice, keyboards, joysticks or touch screens, and participation by adding voice, changing position, face expression, or gesture – bio-interfaces connect participants' physiological and brain activities with the therapeutic or entertainment systems, so the participants feel and act as co-authors.

People with high spatial visualization ability perform faster at human-computer interaction tasks, information search and information retrieval. Visual navigation in a virtual reality strongly depends on spatial ability, and so the accuracy of sketches about a semantically organized spatial model (Chen, 2010). Therefore, in accordance with the Universal Design philosophy, computer system designers are expected to compensate for possible low spatial abilities of users by improving interfaces through creating site maps and site previews. Some hold that gaming and programming may improve spatial orientation by evoking motivation and providing exercise with a feedback. The improvement on spatial orientation, which was correlated with the improvement on the video games, evokes high motivation, provides exercise with a feedback and improves spatial orientation. The speed causes a non-verbal, non-analytic approach, and non-verbal sounds are processed in the right hemisphere. Video games can enhance such aspects of spatial ability as visual discrimination, coordination of horizontal and vertical axes, mental manipulation, and rotation. Computer and television imagery may be educating viewers to a new aesthetic appreciation of 3-D depth information, motion, and image transformation, so users benefit cognitively from their experiences and reach a new level of visual sophistication. Environmental experience grows with the development of the frame of reference and with an understanding of the Euclidian relationships that are not projective or topological. The use of an active movement, a film or TV projections, and an object manipulation are helpful for spatial development. Several kinds of computer graphics-based game software have been produced to improve spatial skills. Children as young as 3 can use both graphic editing programs and games, and children about 7 can create and transform representations of 3-D objects (Piestrup, 1982). Bolter (1987) stressed spatial character of writing and proposed that computer technology may foster change in the structure and symbolic character of writing.

Gaming may thus influence cognitive development and neural organization. The use of computers may influence cognitive development and neural organization but there are varied opinions about these effects. Some research data show that communication media (e.g. video games, digital television, two-dimensional computer graphics, and three-dimensional representations of geometric data) influence the viewer's perception, comprehension, and retention of computerized images. Such key factors as the shape, direction, and motion may act on viewer's comprehension and retention of computerized images, so the users may thus benefit cognitively from these experiences and reach a new level of visual sophistication. Another data indicate that the ability to recreate a complex figure from memory does not rise with the declared knowledge of software (Knipp, 2003). It may be caused by the fact that tedious tactile/kinesthetic hand skills and visual exercises executed by the designer's hand might be eliminated by the use of computers, but still they may be necessary for improving perceptual development. Traditional schooling had been mostly focused on developing memory skills. At present, a necessity of coping with large amount of information creates a need to expand abilities of higher order thinking, visualization, and understanding of abstract concepts.

Spatial Abilities and Imagination

Visual imagination helps meet challenges that require good spatial skills, that means, competency in the area of spatial visualization and spatial orientation. It can be improved with practice, but a need for developing spatial skills has long been overlooked by public education. According to the U.S. Employment Service, spatial ability is a strong predictor variable for the achievement in sciences and in many careers, so, several years ago, the U.S. Employment Service published a list of 84 occupations where spatial ability is important for the achievement in a career. It is possible to improve one's spatial abilities, especially if somebody is at the formal operational level. Practicing freehand drawing may improve spatial skills. However, there is unidirectional relationship between freehand drawing ability and spatial skills: somebody's high spatial visualization and orientation do not insure drawing ability (Arnheim, 1974) and does not lower the "drawing barrier" between a one's mental representation and what is actually drawn. Active learning and control over decisions in a learning environment improve spatial abilities. Improvement comes fast and happens just by doing mental exercises.

A Mental Exercise

You may now want to make a mental exercise with a cube, using your visual mind's eye. Try to solve this problem without sketching the cube, unless you really need to do so. Let's imagine a cube which is painted blue on the surface and it is red inside. It can be a big or a not-so-big cube, and the choice of the shade of blue is up to you. Now, cut this cube with two vertical planes parallel to the surface of the screen, then with two vertical planes perpendicular to it, finally with two horizontal planes. How many small cubes do you have now (nine of them can be seen on each side of the big cube)? Please tell:

How many cubes have three blue sides,
How many cubes have two blue sides,
How many have one blue side, and
How many cubes are all red?

Answer: there are 8 cubes with 3 blue sides, 12 cubes with 2 blue sides, 6 cubes with 1 blue side, 0 cubes with no blue side, and 1 small cube inside a big one would be all red.

Doing simple mental operations like this one, without sketching it first, is considered improving one's spatial abilities.

Measurement of Spatial Abilities

Piaget and Inhelder used models and an interview method. Paper and pencil tests are often used to determine spatial visualization and spatial orientation. Typically, spatial visualization abilities are measured with cognitive tests; mental rotation can be accomplished using spatial visualization. Cognitive tests used to measure spatial abilities are, among others: the Spatial Visualization Abilities Test (SVAT) that is suitable for 4th graders; the adult level Thurston's Paper Folding Test, VZ-2 (Paper Folding, with a sequence of folds and a set of holes punched), and VZ-3 (Surface Development, with a flat shape with numbered sides and a three-dimensional shape with lettered sides) – results were better at the formal operational level); a Form Board Test (with a shape and a set of smaller shapes); Baltista, Talsm, and Wheatly Test; Purdue Spatial Visualization Test on Rotation; Vandenburg/Kuse Mental Rotation Test (a test of 3-D spatial visualization based on stimuli constructed from figures used by Shepard and Metzler in 1971 (Thomas, 2007). The test comprised twenty problems, and each test item consisted of a criterion figure and four alternative figures, two of which were the same as the criterion and two incorrect. The two correct alternatives were rotated versions of the criterion figure. For example, the stimulus figure pairs used by Shepard and Metzler contained (Nigel J. T. Thomas, Stanford Encyclopedia of Philosophy):

- Identical objects differing by a rotation in the plane of the page
- Identical objects differing by a rotation in depth
- Mirror-image objects (also rotated in depth)

CONCLUSION

Both communication with others and exchange of knowledge, insight, and information can be done in non-verbal way. Our communication with the world and people goes through our senses in many ways, not only by sight, hearing, touch and haptic experience, smell, or taste. Several senses that gather information from the external and internal receptors take part in our defense systems. Explorations about the sense of numbers take a form of a discussion about nonverbal counting and searching whether there is thought, especially math thought, occurring without language. We are able to capture light and then recognize, name, and also categorize it with our senses. We are still not sure which senses we had already lost, and which ones we ignore or are not aware of, while they may be possessed by some other living creatures. We observe, research, and record nature in terms of making inventions, and also by creating art; we apply our senses for that. Issues discussed in this chapter are discussed in further chapters with diverse frames of reference, from physiological, physical, technological, to aesthetical.

REFERENCES

Arnheim, R. (1969). *Visual thinking*. Berkeley, CA: University of California Press.

Arnheim, R. (1974). *Art and visual perception*. Berkeley, CA: University of California Press.

Aron, E. N. (2006). The clinical implications of Jung's concept of sensitiveness. *Journal of Jungian Theory and Practice, 8*(2), 11-43. Retrieved January 6, 2012, from http://www.junginstitute.org/pdf_files/JungV8N2p11-44.pdf

Ascher, M., & Ascher, R. (1980). *Code of the Quipu: A study in media, mathematics, and culture*. Ann Arbor, MI: University of Michigan Press.

Ascher, M., & Ascher, R. (1997). *Mathematics of the Incas: Code of the Quipu*. Dover Publications.

Auson, K. S. (2012). 0h!m1gas: A biomimetic stridulation environment. In *Biologically-Inspired Computing for the Arts: Scientific Data through Graphics*. Hershey, PA: IGI Global Publishing. doi:10.4018/978-1-4666-0942-6.ch004.

Bales, K. L., & Kitzmann, C. D. (2011). Animal models for computing and communications: Past approaches and future challenges. In X. Yang (Ed.), *Bio-Inspired Computing and Networking*. CRC Press, Taylor & Francis Group. ISBN 1420080326

Bar-David, E., Compton, E., Drennan, L., Finder, B., Grogan, K., & Leonard, J. (2009). Nonverbal number knowledge in preschool-age children. *Mind Matters: The Wesleyan Journal of Psychology, 4*, 51–64.

Baum, L. F., Hearn, M. P., & Denslow, W. W. (2000). *The annotated wizard of Oz* (Centennial Ed.). New York: W. W. Norton & Company. ISBN 0393049922

Beltz, E. (2009). *Frogs: Inside their remarkable world*. Firefly Books.

Bernstein, K. (2005). Expression in the form of our own making. In *Proceedings of the Special Year in Art & Mathematics, A+M=X International Conference*. Univ. of Colorado.

Bolter, J. D. (1987). Text and technology: Reading and writing in the electronic age. *Library Resources & Technical Services, 31*(1), 12–23.

Bossomaier, T., & Snyder, A. (2004). Absolute pitch accessible to everyone by turning off part of the brain? *Organised Sound, 9*(2), 181-189. DOI 10.1017/S1255771804000263. Retrieved December 13, 2011, from www.centreforthemind.com/publications/absolutepitch.pdf

Boström, N. (2005). A history of transhumanist thought. *Journal of Evolution and Technology, 14*(1). Retrieved January 31, 2012, from http://www.nickbostrom.com/papers/history.pdf

Brannon, E., & Terrace, H. (1998). Ordering of the numerosities 1 to 9 by monkeys. *Science, 282*, 746–749. doi:10.1126/science.282.5389.746 PMID:9784133.

Brannon, E. M. (2005). The independence of language and mathematical reasoning. *Proceedings of the National Academy of Sciences of the United States of America, 102*(9), 3177–3178. doi:10.1073/pnas.0500328102 PMID:15728346.

Butterworth, B., Reeve, R., Reynolds, F., & Lloyd, D. (2008). Numerical thought with and without words: Evidence from indigenous Australian children. *Proceedings of the National Academy of Sciences of the United States of America, 105*(35), 13179–13184. doi:10.1073/pnas.0806045105 PMID:18757729.

Carey, S. (2004). Bootstrapping & the origin of concepts. *Daedalus*, 59–68. doi:10.1162/001152604772746701.

Chen, C. (2010). *Information visualization: Beyond the horizon*. Springer.

Chernoff, H. (1971). *The use of faces to represent points in n-dimensional space graphically*. Palo Alto, CA: Stanford University.

Coie, J., Costanzo, P., & Farnill, D. (1973). Specific transitions in the development of spatial perspective taking ability. *Developmental Psychology*, *9*(2), 166–177. doi:10.1037/h0035062.

Dehaine, S. (1999). *The number sense: How the mind creates mathematics*. Oxford, UK: Oxford University Press.

Downs, R. M., & Stea, D. (1977). *Maps in minds: Reflections on cognitive mapping*. New York: Harper & Row Publishers.

Dunn, J., & Clark, M. A. (1999). Life music: The sonification of proteins. *Leonardo*, *32*(1), 25–32. doi:10.1162/002409499552966.

Dunn, M., Greenhill, S. J., Levinson, S. C., & Gray, R. D. (2011). Evolved structure of language shows lineage-specific trends in word-order universals. *Nature*, *473*, 79–82. doi:10.1038/nature09923 PMID:21490599.

Eco, U. (2002). *Art and beauty in the middle ages* (H. Bredin, Trans.). New Haven, CT: Yale University Press.

Evans, G. W., Marrero, D. G., & Butler, P. A. (1981). Environmental learning and cognitive mapping. *Environment and Behavior*, *13*, 83–104. doi:10.1177/0013916581131005.

Frank, M. C., Everett, D. L., Fedorenko, E., & Gibson, E. (2008). Number as a cognitive technology: Evidence from Piraha language and cognition. *Cognition*, *108*(3), 819–824. doi:10.1016/j.cognition.2008.04.007 PMID:18547557.

Gardner, H. (1983/2011). *Frames of mind: The theory of multiple intelligences* (3rd ed.). Basic Books.

Gardner, H. (2006). *Multiple intelligences: New horizons in theory and practice*. Basic Books.

Gelman, R., & Butterworth, B. (2005). Number and language: How are they related? *Trends in Cognitive Sciences*, *9*(1), 6–10. doi:10.1016/j.tics.2004.11.004 PMID:15639434.

Goldsworthy, A. (2007). *The biological effect of weak electromagnetic fields*. Retrieved January 6, 2012, from www.electrosense.nl/nl/download/6

Gordon, P. (2004). Numerical cognition without words: Evidence from the Amazonia. *Science*, *306*, 496–499. doi:10.1126/science.1094492 PMID:15319490.

Grey, W. (2000). Metaphor and meaning. *Minerva, an Internet Journal of Philosophy, 4.*

Halberda, J., Mazzocco, M. M., & Feigenson, L. (2008). Individual differences in non-verbal number acuity correlate with maths achievement. *Nature, 455*, 665-668. doi 10.1038/nature07246. Retrieved January 6, 2012, from http://www.nature.com/nature/journal/v455/n7213/full/nature07246.html

Hardesty, L. (2012, January 24). 'Genetic programming': The mathematics of taste. *PhysOrg*. Retrieved March 4, 2013, from http://phys.org/news/2012-01-genetic-mathematics.html#jCp

Hart, R. A., & Moore, G. T. (1973). The development of spatial cognition: A review. In R. Downs, & D. Stea (Eds.), *Image and Environment*. Chicago: Aldine Press.

Herz, R. S. (2009). Aromatherapy facts and fictions: A scientific analysis of olfactory effects on mood, physiology and behavior. *The International Journal of Neuroscience*, *119*(2), 263–290. doi:10.1080/00207450802333953 PMID:19125379.

Herz, R. S., Schankler, C., & Beland, S. (2004). Olfaction, emotion and associative learning: Effects on motivated behavior. *Motivation and Emotion*, *28*, 363–383. doi:10.1007/s11031-004-2389-x.

Jarvis, E. D. (2004). Learned birdsong and the neurobiology of human language. *Annals of the New York Academy of Sciences, 1016*, 749–777. doi:10.1196/annals.1298.038 PMID:15313804.

Jenny, H. (2001). *Cymatics: A study of wave phenomena & vibration* (3rd ed.). Macromedia Press.

Jung, C. G. (1976). *Psychological types*. (Original work published 1921).

Kaptchuk, T. (2000). *The web that has no weaver: Understanding Chinese medicine*. McGraw-Hill.

Karlans, M., Schuerhoff, C., & Kaplan, M. (1969). Some factors related to architectural creativity in graduating architectural students. *The Journal of Genetic Psychology, 81*, 203–215. doi:10.1080/00221309.1969.9711286.

Kepes, G. (1995). *Language of vision.* Dover Publications. (Original work published 1944).

Knipp, T. (2003) Creative performance: Does the computer retard artistic development? In *Proceedings of the International Conference on Information Visualization*, (pp. 621-625). IEEE.

Kosslyn, S. M. (1978). Measuring the visual angle of the mind's eye. *Cognitive Psychology, 10*, 356–389. doi:10.1016/0010-0285(78)90004-X PMID:688748.

Kuipers, B. (1982). The map in the head metaphor. *Environment and Behavior, 14*(2), 202–220. doi:10.1177/0013916584142005.

LeCorre, M., & Carey, S. (2006). One, two, three, four, nothing more: An investigation of the conceptual sources of the verbal counting principles. *Cognition, 105*(2), 395–438. doi:10.1016/j.cognition.2006.10.005.

Lehrer, J. (2007). *Proust was a neuroscientist.* Houghton Mifflin.

Lohman, K. J., Lohmann, C. M., & Putman, N. F. (2007). Magnetic maps in animals: Nature's GPS. *The Journal of Experimental Biology, 210*, 3697–3705. doi:10.1242/jeb.001313 PMID:17951410.

Loizides, A. (2012). *Andreas Loizides research home page*. Retrieved February 12, 2012, from http://www.cs.ucl.ac.uk/staff/a.loizides/research.html

Loizides, A., & Slater, M. (2001). The empathic visualisation algorithm (EVA), Chernoff faces revisited. In *Conference Abstracts and Applications, Technical Sketch SIGGRAPH 2001*. ACM Press.

Loizides, A., & Slater, M. (2002). The empathic visualisation algorithm (EVA) - An automatic mapping from abstract data to naturalistic visual structure. In *Proceedings of iV02, 6th International Conference on Information Visualisation,* (pp. 705-712). Los Alamitos, CA: IEEE.

Malchiodi, C. A. (2006). *Expressive therapies.* The Guilford Press.

Mann, S. (2001). *Intelligent image processing.* John Wiley & Sons, Inc. doi:10.1002/0471221635.

Marshall, J., Chamberlain, A., & Benford, S. (2011). I seek the nerves under your skin: A fast interactive artwork. *Leonardo Journal, 44*(5), 401–404. doi:10.1162/LEON_a_00239.

Mäthger, L. M., Barbosa, A., Miner, S., & Hanlon, R. T. (2005). Color blindness and color perception in cuttlefish (Sepia officinalis) determined by a visual sensorimotor assay. *Vision Research, 46*, 1746-1753. Retrieved October 15, 2012, from http://hermes.mbl.edu/mrc/hanlon/pdfs/mathger_et_al_visres_2006.pdf

Mattison, C. (2007). *300 frogs: A visual reference to frogs and toads from around the world.* Firefly Books.

Maya Mathematical System. (2012). *Maya world studies center*. Retrieved November 1, 2011, from http://www.mayacalendar.com/f-mayamath.html

McLaughlin, J. (1998). Good vibrations. *American Scientist, 86*(4), 342. doi: doi:10.1511/1998.4.342.

Mitchell, L. (2005). *A posthuman methodology for nondual world*. (Unpublished Thesis). York University, Toronto, Canada.

Moore, D. M. (1985). *Field independence-dependence: Multiple and linear imagery in a visual location task*. Paper presented at the Annual Convention of the Association for Educational Communications and Technology. Anaheim, CA.

Olson, D. R., & Bialystok, E. (1983). *Spatial cognition*. Hillsdale, NJ: Lawrence Erlbaum.

Ox, J. (1999). Synesthesia. *Leonardo, 32*(5), 391. doi:10.1162/002409499553622.

Paivio, A. (1970). On the functional significance of imagery. *Psychological Bulletin, 73*, 385–392. doi:10.1037/h0029180.

Piaget, J., & Inhelder, B. (1956). *The child's conception of space*. London: Routledge & Kegan Paul.

Piaget, J., & Inhelder, B. (1971). *Mental imagery in the child: A study of development of imaginal representation*. New York: Basic Books Inc., Publishers.

Piestrup, A. M. (1982). *Young children use computer graphics*. Cambridge, MA: Harvard Univ.

Proust, M. (2011). *A fragment of the remembrance of things past: Swann way*. Retrieved January 17, 2012, from http://www.authorama.com/remembrance-of-things-past-3.html

Saffran, J. R. (2003). Absolute pitch in infancy and adulthood: The role of tonal structure. *Developmental Science, 6*, 35–47. doi:10.1111/1467-7687.00250.

Saloman, G. (1979). *Interaction of media, cognition, and learning*. San Francisco, CA: Jossey-Bass.

Shepard, R. N. (1978). The mental image. *The American Psychologist, 17*, 179–188.

Slobodchikoff, C. N. (2002). Cognition and communication in prairie dogs. In M. Beckoff, C. Allen, & G. M. Burghardt (Eds.), *The Cognitive Animal*, (pp. 257-264). Cambridge, MA: A Bradford Book.

Stapput, K., Güntürkün, O., Hoffmann, K.-P., Wiltschko, R., & Wiltschko, W. (2010). Magnetoreception of directional information in birds requires nondegraded vision. *Current Biology, 20*(14), 1259-1262. Retrieved October 17, 2012, from http://dx.doi.org/10.1016/j.cub.2010.05.070

Sturm, B. L. (2005). Pulse of an ocean: Sonification of ocean buoy data. *Leonardo, 38*(2), 143–149. doi:10.1162/0024094053722453.

Thomas, N. J. T. (2010). *Stanford encyclopedia of philosophy*. Retrieved November 18, 2012, from http://plato.stanford.edu/entries/mental-imagery/mental-rotation.html

UCLA. (2011). *Instructional research lab: Chladni plate*. Retrieved December 28, 2011, from http://www.physics.ucla.edu/demoweb/demomanual/acoustics/effects_of_sound/chladni_plate.html

Van Scoy, F. L., & Gifu, U. (2004). Sonification of remote sensing data: Initial experiment. In *Proceedings of the International Conference on Information Visualization*. IEEE.

Varley, R. A., Klessinger, N. J. C., Romanowski, C. A. J., & Siegal, M. (2005). Agrammatic but Numerate. *Proceedings of the National Academy of Sciences of the United States of America, 102*, 3519–3524. doi:10.1073/pnas.0407470102 PMID:15713804.

Vygotsky, L. S. (1972). *Thought and language*. Cambridge, MA: MIT Press.

Whorf, B. L. (1956). *Language, thought, and reality*. Cambridge, MA: MIT Press.

Wilkinson, W. K., & Schwartz, N. H. (1991, January). A factor-analytic study of epistomological orientation and related variables. *The Journal of Psychology*, 91–101. doi:10.1080/00223980.1991.10543274.

Witkin, H. A. (1954/1972). *Personality through perception*. London: Greenwood Press.

Witkin, H. A., Dyk, R. B., Faterson, G. E., Goodenough, D. R., & Karp, S. A. (1974). *Psychological differentiation: Studies of development*. Potomac, MD: Lawrence Erlbaum Associates.

Wueringer, B. E., Squire, L., Kajiura, S. M., Hart, N. S., & Collin, S. P. (2012). The function of the sawfish's saw. *Current Biology*, *22*(5), R150–R151. doi:10.1016/j.cub.2012.01.055 PMID:22401891.

Wynn, K. (1992). Children's acquisition of the number words and the counting system. *Cognitive Psychology*, *24*(2), 220–251. doi:10.1016/0010-0285(92)90008-P.

Yang, X. (2011). *Bio-inspired computing and networking*. CRC Press.

Zuanon, R. (2012). Bio-interfaces: Designing wearable devices to organic interactions. In *Biologically-Inspired Computing for the Arts: Scientific Data through Graphics*. Hershey, PA: IGI Global Publishing. doi:10.4018/978-1-4666-0942-6.ch001.

Zuanon, R., & Lima, G. C., Jr. (2008). *BioBodyGame*. Retrieved March 09, 2011, from http://www.rachelzuanon.com/biobodygame/

Zuanon, R., & Lima, G. C., Jr. (2010). *NeuroBodyGame*. Retrieved March 09, 2011, from http://www.rachelzuanon.com/neurobodygame/

Chapter 3
Essential Art Concepts

ABSTRACT

This chapter comprises a basic overview about visual literacy, discussion of art definitions, basic art concepts, elements and principles of design, differences between art, design, craft, technical issues, elements, and principles related to art, design, and many other disciplines, and the quality of display, among other topics. An example of translation from nature, with all resulting connotations, to the visual and the to the verbal and beyond the verbal takes the form of a project involving drawing an apple and then writing a short poem about the apple. The reader is invited to integrate meaning, visualization, and knowledge about an object.

INTRODUCTION: VISUAL LITERACY

Concepts and problems pertained to art and design emerge incessantly in the following chapters as the essential factors in nonverbal or not exclusively verbal productions such as creating visual solutions of abstract concepts, designing visualization, or a website. The central idea under this chapter's content is to provide those of the readers who did not find time or interest to study art-related problems with basic information that could be found conducive to creating meaningful projects suggested in the framed spaces, which are present in the following chapters. Short introduction of this kind may be useful for those focused on domains other than visual arts, and may be helpful for those readers who would like to be reactive to further themes. The advantage of visual display of information over speech or writing is in its nonlinear, flexible time of viewing, multiple dimensions, and possibility of restructuring of its content. The readers' projects resulting from reading this book will hopefully display information aesthetically, with visible traces of reasoning about concepts.

DOI: 10.4018/978-1-4666-4703-9.ch003

Visual literacy, which results from the ability to use a visual language to exchange information, is needed to design visual presentation of abstract concepts we are working on. When we apply strategies for visual problem solving, we communicate through visual images as icons, signs and symbols, as opposed to verbal symbols or words. One of the conditions for developing visual literacy is the visual intelligence, which makes provision for interchanging thoughts, opinions, or information without the use of spoken words. That means we apply visual processing in our thinking by creating images to communicate notions and ideas, gain insight into some difficult to analyze schemes and find patterns and order in complex structures. Scientists discovered time-related gains in cognitive performance of people from various countries (Flynn effect) along with growing visual intelligence. In his study of IQ tests scores for different populations over the past sixty years, James R. Flynn discovered that IQ scores increased from one generation to the next for all of the countries for which data existed (Flynn, 1987; 1994; 1999). Possible explanations of the IQ growth include increased years spent in formal education, the societal changes, better worldwide nutrition, and the increase in test taking skills. Visual intelligence of students is also increasing due to playing three-dimensional computer games that simulate real and virtual worlds. Since students are able to perceive, understand, and process progressively more complex visual messages, visual way of communication becomes more and more effective and important.

It may be sometimes difficult to draw a line dividing art, design, and craft because art works often serve as a basis for useful traditional and digital applications. In many cases we appreciate exceptional design or craft as equal to art. For this reason visual literacy is not a theoretical, abstract quality. For example, glass is usually considered a useful substance good for producing windows, optical instruments, glassware, or mirrors. A glass sculptor Dale Chihuly (www.chihuly.com) utilizes glass for creating transparent sculptures of thin glass, and large indoor and outdoor installations on land, water, or air of intricate colors, thus becoming not only a glass sculptor but also a successful entrepreneur. One may say Chihuly converted a traditional craft form into art. Moreover, an interactive application for the iPhone allows the user to make own forms by blowing into the speaker of the phone and reshaping Chihuly's ocean life- inspired creations (http://mashable.com/2012/05/12/chihuly.html). Users can download this free application, make their own glass-blown sculpture following the style of the artist, and use the touch technology in an unexpected way.

Figures 1a and b present works of students taking my 2012 Computer Graphics courses, where, in concert with the title of this book, students go beyond text with their science inspired solutions by applying their art works to design projects and practical applications. Below there are examples of converting art projects and exercises in pattern design into functional objects such as t-shirts.

Taylor Royal (Figure 1a) illustrated his verse (written as a limerick), and then used his work to design a t-shirt.

There once was a glamorous boat
That people said would always float
Till it hit some ice
Due to people vice.
To those dead this poem I devote.

While Cody Johnson (Figure 1b) applied for this purpose his science-based artwork and a verse.

Carbon is found in all things
From diamonds to airplanes to buildings
Without it nothing would exist
Earth would be a dark, dark abyss
Good thing we have carbon and all it brings.

Figure 1. (a) Taylor Royal, "A Titanic t-shirt" (© 2012, T. Royal. Used with permission) (b) Cody Johnson, "A Carbon t-shirt" (© 2012, C. Johnson. Used with permission)

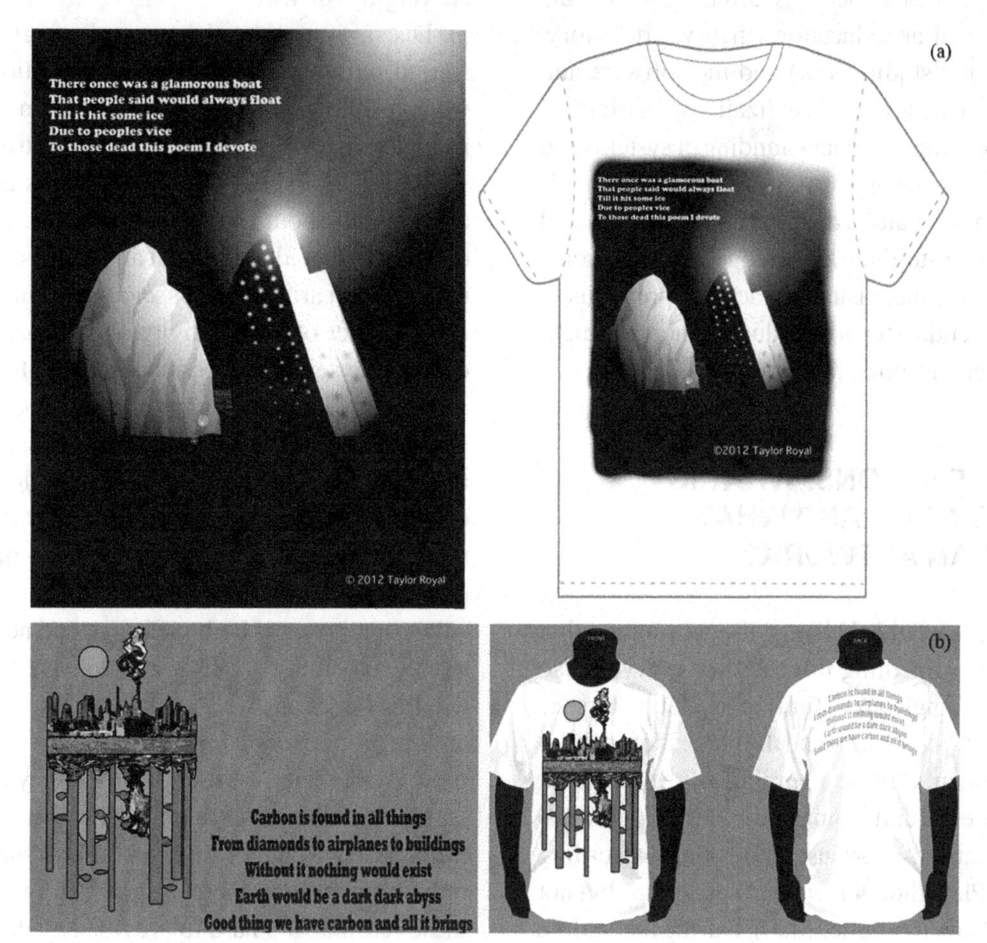

Visual competence is spatial rather than linear. Learning through visual and spatial thinking can produce meanings and connotations that cannot be achieved using language alone. Visual literacy allows expressing concepts with visual means and constructing meaning with the use of nonverbal communication, such as facial expressions, eye contact, gestures, body language, posture, and tone of the voice; it is an important component of personal business interactions. Clothing, hairstyles, and ambient factors such as architecture, interior design, and infographics are also important. Lengler (2006) lists the vision competencies as the abilities to:

- Speedily locate, identify and assess patterns
- Speedily assign complex shapes to visual categories
- Structure, store, and recall objects and paths in maps
- Reconfigure shapes into new objects
- Express concepts with visual means in a wide variety of ways
- Construct meaning by integrating different associated visual messages
- Imagine and rotate objects in 3D space
- Simulate the future behavior of objects, based on their pattern of change in a given time period
- Deduce the rules that govern patterns.

In education, the level of digital art literacy depends on the balance between the components of traditional art education (theory, art history, critique, and studio work) and the software and technological literacy. Visualizations, for example animations improve understanding of systems and processes that change over time. The growing visual literacy creates a demand for more advanced and very up-to-date knowledge visualizations. At the same time, visual literacy is a prerequisite and an essential step for producing more efficient knowledge visualizations (Lengler, 2006).

ART DEFINITIONS: WHAT IS AN ARTWORK AND WHAT IS NOT AN ARTWORK?

Few people would challenge the assumption that art means something different for everyone, so there is no single art definition available. First of all, relation between art and the process of creation has not been seen in a similar way in different times. Greeks and Romans did not talk or write about creation – they just made things according to rules. Plato thought artists could imitate but not create (Plato, Republic, 2000). During the Middle Ages creation was restricted to God and artists studied natural beauty to imitate it. Renaissance brought about freedom of creating un-natural forms; Leonardo da Vinci examined them by creating sketches and drawings. According to the 18th century beliefs, art creates beauty or follows beauty as a model.

Rigid art definitions applied by the art critics in the 19th century caused that prestigious art exhibitions refused to accept works of artists who chose a non-traditional subject matter. Art jurors refused an acceptance to the official 1863 Paris Salon sponsored by the French government of those artists who were fascinated by a relationship between light and color but not by the exact rendering of objects and people. Art galleries displayed works rejected by the Salon jurors, and then artists opened their own Salons des Refusés, starting in 1874.

Discussion about the essence of art can be started with a search for an art definition. Artwork resulting from writing a program is often as pleasing and impressive as nature itself. It is so, regardless of any possible remarks and disagreements about the features of art and about conditions that identify what art really is. Most of us agree that art means conscious human activity or a product of such activity. Art works act not only as the sensory experiences; they also evoke a stream of ideas about the art meaning that are determined in particular cultures and analyzed by art critics and historians. For many philosophers deciding what is art and what is not art seems to be impossible; many pose that everyone may construct one's own opinion. A question has been still around what is both sufficient and necessary to call something art. If we would want to create a definition of art, we should be able to tell what are the properties that all art works have in common. Alas, there is no such thing; every artwork is different. Also, we should tell if there are any properties we cannot find in non-art objects but only in art. Again, there are no such things. For these reasons, it's hard to provide any definition of art. As stated by Weitz (1956), here is no essence that would be common to all artworks and restricted only to artworks.

Discussion about what is art and what is not art continued, especially when Marcel Duchamp, created a readymade and called it "Fountain" (1917, en.wikipedia.org/wiki/Fountain_(Duchamp). In terms of readymade art, one can take any object and pronounce it art, for example, by taking it out of its place and setting it in a new, unfamiliar place. Duchamp brought a porcelain urinal to the gallery, signed with the pseudonym R. Mutt, and then entered it at the art exhibition. This way, Duchamp defined the concept of a 'ready-made' art and a 'found-object' art. Marcel Duchamp (1917) stated, "Whether Mr. Mutt with his own hands made the fountain or not has no importance.

He CHOSE it. He took an ordinary article of life, placed it so that its useful significance disappeared under the new title and point of view – created a new thought for that object." Therefore the idea and selection, not just the creation was important, and an artifact might become art. Artifact is usually described as an object produced by human craft (a tool, a weapon, or an ornament of archaeological or historical interest).

In his book titled "Definitions of Art" (1991) Stephen Davies discussed functional and procedural approaches to art definition – two strategies that might be adopted to define art. For the functionalist, an artwork performs functions distinctive to art, usually providing a rewarding aesthetic experience. For the proceduralist, an artwork is created in accordance with certain rules and procedures. Stephen Davies reviewed also two separate meanings of the word "artifact:" in its primary sense an artifact is something that is modified by work, while in an alternative meaning, an artifact is something which has significance in a given culture and invites interpretation. Morris Weitz said in his paper "The Role of Theory in Aesthetics" that even artifactuality (whether an object is man-made) is not decisive in defining art. Thus, Duchamp's "Fountain" may be both an artwork and an artifact. He presented an opinion that artists and aestheticians failed to define the nature of art because there is no essence that would be common to all artworks and restricted only to artworks. However, others counter with that opinion; for example, Raymond Lauzzana and Denise Penrose (1987; 1992), stated that natural things are not art because art must be made by humans. Also, many ask, can a monkey or an elephant create art?

A status of the work as a work of art depends on the opinion of the Artworld, the cooperation and recognition of others. Arthur Danto introduced the term "Artworld" in 1964 as informal groups of art-related people. Art facilitates communication. Artworks can be appreciated in relation to other artworks, whether they are providing enjoyment, aesthetic experience, and some social benefits. The interesting question is, whether artwork regarded now as art was considered art at the time it was made, and would it be art in the future.

At the present times, rules, formulas, and skills are valued, but also imagination and creativity. Also, novelty is important, even if existing only in part: "on an old tree, in an old way, new leaves are growing every spring," and "every car is new even if all its parts do not differ from other cars" (Tatarkiewicz, 1976, p. 302). Many authors and critics question whether or not art can be defined in terms of aesthetic and philosophic frameworks. Texts written by Matisse, Van Gogh, Klee, Tolstoy, Malraux, Albers, and so many others contain attempts to find some perceptible or imperceptible property that would be intrinsic to art works.

Digital art, which could not be created in previous epochs, differ from traditional art because of its interactivity, programmability of images, and capacity for modeling real objects due to computer functions. Sometimes it causes that the straight-from-the-monitor electronically generated art, which does not have physical material but only a file of numbers that controls the image, has poor recognition by the art world. There is no original in the digital world; there are equally valued copies. The edition of a digital artwork can be limited and numbered by the artist in order to raise the price of a printout.

See Table 1 for Your Visual Response.

BASIC ART CONCEPTS

Many concepts about art may stem from a generally accepted opinion that every artwork, as well as every project should fulfill general principles related to its design, editing, and analysis of presentation. For example, a good graphic project should enhance complexity, dimensionality, density, and beauty of communication (Tufte, 1983; 1992). We will talk about these qualities, especially beauty and aesthetics of every display.

Table 1.

Your Visual Response: What is Art for You?
Looking carefully at your everyday environment (home, workplace, neighborhood) select some objects that you would like to appreciate as art works, examine the process involved in their creation you believe was applied, the final product, and then justify your choice.

In order to better analyze and appreciate a work of art, it may be useful to focus on basic art concepts that can apply to any work of a visual art. These basic art concepts would include the type or form of art, subject matter, style, medium, and design.

If we'd like, for example create an artwork representing an apple, we'd need to make certain decisions (after acquiring an apple):

1. The type or form of art describes the kind of art or art products, the schemes used to classify art, and the functions art products serve; an artwork may be described as a drawing, a painting, a sculpture, etc. Would we paint, sculpt, draw an apple, or erect an architectural structure in the form of an apple?
2. The subject matter defines the meaning of the work of art; the theme, topic, or motif represented as a person or object; it may be a portrait, a landscape, a still life, an abstract work, etc. Would we represent an apple as a part of a still life, a portrait of a man or a woman eating an apple, a landscape such as an orchard with small, multiplied images of an apple, an abstract artwork about an apple, or an almost abstract image of a close-up of the apple skin or showing an apple's interior.
3. The style tells about the traits and resemblances within a group of works of art, the

visual similarities influenced by the time, place, or personal manner of the artist. We'd need to decide if an apple would be seen as a geometric form, as the cubist artists would see it; if we 'd work on the scenery issues, like in paintings from the Renaissance or Mannerism periods; if we'd focus on light and paint related work, as it was done by the Impressionist artists; or if we'd create a generative artwork using programming or software.

4. The medium informs about the materials, tools and procedures the artist uses to create the work of art. For example, medium can be described as: 'oil on canvas mounted on panel,' 'acrylic on canvas,' 'marble sculpture,' 'paper, pen and ink over chalk drawing,' or 'polychrome woodblock print on paper.' We need to secure proper materials, tools, and apply selected techniques for our representation of an apple. The medium can depict the differences resulting from a particular use of materials and tools. For example, an oil-on-canvas painting will look differently than an oil-on-paper painting because a canvas would repel paint, while paper would absorb it. That is while we refer to the artwork as 'oil-on-canvas painting' rather than just 'oil painting.'
5. The design specifies the visual elements and principles used in the artwork; the planned organization of the visual phenomena the artist manipulates. We can cut an apple in half and focus on pips (seeds) and how they are placed inside five carpels arranged in a five-point star. We can picture the repetition of lines and dots on the apple skin or exaggerate irregularities in its design.

We may also examine further properties of an artwork, for example:

- **Composition:** The arrangement of visual elements in a work of art, the construction and layout of a work that defines how everything is put together using thoughtful choices. In graphic design and desktop publishing, composition is usually called a page layout. Now we would decide on the arrangement and style treatment of our image of an apple.
- **Iconography:** Visual signs, symbols, and icons contained in the work. In case of an apple it is extremely rich set of meanings, connotations, and cultural traditions.

Figure 2 is an example of an analysis of a work of art: "Head of a Young Man" by Andrea del Sarto. On the page: http://commons.wikimedia.org/wiki/File:Andrea_del_Sarto_-_Head_of_a_Young_Man_-_WGA0384.jpg there is a frontal view of a "Head of a Young Man" by Andrea del Sarto (1486-1530) created around 1520. We can analyze this drawing in terms of the basic art forms.

Figure 2. Andrea del Sarto, Head of a Young Man (circa 1520, art in public domain)

- **Type or Form of Art:** A "Head of a Young Man" by Andrea del Sarto is a two-dimensional drawing.
- **Subject Matter:** Del Sarto drew a face of a young man but not a portrait of a particular individual.
- **Style:** Del Sarto, who worked during the High Renaissance and early Mannerism periods captured a realistic image of the young man and created the illusion of a three-dimensional space.
- **Medium:** Can be described in this case as chalk-on-paper; more specifically it was rendered in sanguine crayon on beige paper.
- **Design:** In a "Head of a Young Man" Andrea del Sarto used variety of lines: straight line, curved line, cross-hatching, and hatching, to create texture and a three-dimensional look in his character. Emphasis is given to the eye, the cheekbone, and the nose, where there is no cross-hatching.

PASTICHE

Pastiche is a dramatic, literary, or musical artistic work created in a way that refers to a work of other artist. One can find many pastiche examples on the Internet. The reason for creating many kinds of pastiche can be seen in applying various approaches to original works, including humor, symbolic or metaphorical way of interpretation. Satirical intent of the pastiche creator, or a need for visualization of one's needs and dreams can also be seen in works of this type.

We may find some pastiche features in the art works from different epochs. "Las Meninas" by Diego de Silva y Velazquez can be seen as an assortment of violations of the rules that were set at the times of its creation (Figure 3). Velazquez created work of art considered inappropriate and almost impossible to be accepted by the Royal family. He introduced several innovative approaches combined with a sense of humor while depicting the Royal family in this experimental work:

1. The viewer is placed side by side with the Royal Family (seen in the mirror) – it seems to be impossible in those times.
2. Portraits on the walls are showing works of the friends of the painter.
3. The painter's uncle (who was not a member of the royal family) is depicted, standing on a staircase.
4. The painter is shown as a self-portrait inside a main painting. He carries an honor cross – an award he was dreaming about; but, instead of being punished for violating so many rules, not only he was awarded with it after the completion of this painting, but the king took care of the artist's family even after his death.

Then, these traits have been exaggerated in a pastiche made by Pablo Picasso, one of the 58 interpretations of Las Meninas he created in order to study the original (Figure 4). Picasso created many copies of various masterpieces and played with the idea of a pastiche as a way of deciphering techniques used and concepts developed by other artists.

It is easy to find examples of pastiches; William Wegman (Simon & Wegman, 2006) created many pastiches using his dogs as models, in various costumes and poses. In a book "Sunday afternoon, looking for the car, the aberrant art of Barry Kite" (Bisbort & Kite, 1997) and also "Rude Awakening at Arles, the aberrant art of Barry Kite" (Burke & Kite, 2000/1996) we may find hundreds of familiar individuals and objects taken from famous works of art. In his collage-like works, Kite has integrated the media of photography and painting. The art of collage means creating artistic compositions of materials and objects set on a surface, often with unifying lines and color. Aberrant art is defined as 'that which strays from the right, normal, or usual course' and

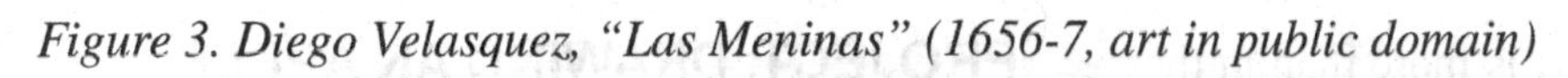
Figure 3. Diego Velasquez, "Las Meninas" (1656-7, art in public domain)

Figure 4. Pablo Picasso, recreation of "Las Meninas" by Diego Velasquez (1957, art in public domain: Creative Commos Attribution SharaAlike License)

so Barry Kite shows that art is more than in the eye of the beholder. A former software company named "Painter" packaged its software in paint cans decorated with pastiche works based on masterpieces in painting. Actually it was said the cans happened to be a less expensive packaging option than boxes.

An apple serves well explaining notions related to semiotics where an image of an apple often serves as an example of an object, the apple symbol is easy to draft and referred according to connotations used in an Ogden & Richards (1923) semiotic triangle.

See Table 2 for Your Visual Response.

PROJECT: DRAWING AN APPLE

An exercise involving drawing an apple and then writing a short poem about apple may serve as an example of translation from the visual to the verbal, and beyond the verbal. A choice of an apple as a theme may comes from the fact that an apple is undeniably an iconic object – an image that represents some object that has a symbolic meaning beyond the object itself. For example, a smiling apple logo can be found on some dentists' business cards. Apple and apple pie has been deemed an American icon (http://en.wikipedia.org/wiki/Icon_(secular)). Steve Jobs knew it well how many

Table 2.

Your Visual Response: Making a Pastiche
Maybe you'd want to select a painting of your choice and make your own pastiche by adding some personal, social, or political reality to the artwork.

connotations it builds when he created a logo for an Apple Company (Issacson, 2011). Apple has been painted by masters representing almost all styles in painting – such as Classical art, Impressionism, Mannerism, and Cubism, presented in many ways: as a still life with apples on a plate, in a basket, as a scene with a woman eating an apple, landscape with an orchard of apple trees, as an abstract, a cross section showing geometry of the seeds' placement, and in many other ways. An apple surely evokes connotations related to the old scripts (such as the biblical accounts), scientific research and experiments (Newton's apple), art works (Claes Oldenburg), legends (Wilhelm Tell), symbols (an apple for a teacher), fairy tales (Snow White), artistic exaggeration (Giuseppe Arcimboldo), companies (Apple computers), places (Big Apple), many literary and cinematic works, and colloquial expressions (an apple of my eye, an apple a day keeps the doctors away).

Figure 5 presents, as a computer graphics art, an image of an apple as a powerful icon. The coexistence of strong colors, dynamic composition, and the use of sharp shapes support the message of the artwork.

Apple is somehow related to Alan Turing, a British mathematical genius and a code breaker born 100 years ago. Turing died at the age of 41, allegedly because he committed suicide by biting an apple suffused with cyanide. However, at a conference held in Oxford in June 2012, Turing expert Prof. Jack Copeland questioned the evidence that was presented at the 1954 inquest; possibly, it was an accident related to chemical experiments, which Turing conducted in his house (Pease, 2012). Until recently, Turing and his legacy were virtually unknown to the public because of active persecution caused by his homosexuality. Turing's Pilot Ace – the Automatic Computing Engine was faster than other contemporary British computers by about a factor of five; it contained more technical detail than the American Edvac Report in dealing with software issues and in predicting future non-numeric applications of computers (Lavington, 2012). Turing broke the German wartime code Enigma that was used by the U-boats preying on the North Atlantic convoys. Germany's encrypted messages were intercepted within an hour, sometimes less than 15 minutes after the Germans had transmitted them (total of 84,000 Enigma messages each month - two messages every minute). Massive code breaking operation, especially breaking the U-boat Enigma code shortened the war in Europe by two to four

Figure 5. Matthew Rodriguez, "The Mighty Apple" (© 2012, M. Rodriguez. Used with permission)

years. At a conservative estimate, each year of the fighting in Europe brought on average about seven million deaths. The significance of Turing's contribution is in preventing the death of further 14 to 21million people that might have died if U-boat Enigma had not been broken and the war had continued for another two to three years (Copeland, 2012).

Drawing an Apple

The project 'Drawing an apple' is intended to cause that when we have a joyful, good feeling while looking at an object, which would really mean something for us, we can convey this feeling to our friends by sending a simple drawing with a short text or posting it online. The first task is to focus on our own ability to look and see, for example, to convey one's personal perception of the fruit and everything it could mean. There is a scene in a South Korean Chang-dong Lee's film "Poetry" (awarded the best screenplay prize at the 2010 Cannes International Film Festival) where an instructor of a poetry course in a city cultural center tells the participants they probably looked at an apple thousands of times but they did never see an apple yet, so they have to learn to see first, and then they will be able to write to write a short poem, by searching memories and emotions for inspiration. "Up till now, you haven't seen an apple for real" the teacher says. "To really know what an apple is, to be interested in it, to understand it, that is really seeing it."

When one aims at grasping the essence of the apple in one's drawing, apples painted by Cezanne may come to mind, for example http://images.metmuseum.org/CRDImages/ep/original/DT1940.jpg or http://www.thegardenerseden.com/wp-content/uploads/2009/11/cezanne-apples1878-79-.jpg.

Many years ago a French poet and screenwriter Jacques Prévert (1900-1977) wrote a verse, "Picasso's Promenade" – a collection of apple related connotations and addressed it to Pablo Picasso (1881-1973). The poem was set to music by Joseph Kosma and then sung by a prominent French vocalist Yves Montand in 1962. Below are short excerpts of this verse.

On a very round plate of real porcelain
an apple poses
face to face with it
a painter of reality
vainly tries to paint
the apple as it is
but the apple won't allow it
...
the apple disguises itself as a beautiful fruit in disguise
and it's then
that the painter of reality
begins to realize
that all the appearances of the apple are against him
...
then suddenly finds himself the sad prey
of a numberless crowd of associations of ideas
...
and the dazed painter loses sight of his model
and falls asleep
...
What an idea to paint an apple
says Picasso
and Picasso eats the apple
and the apple tells him Thanks

Here are short fragments of the French text (Paroles - Gallimard – 1949):

Le Peintre La Pomme & Picasso

Sur une assiette bien ronde en porcelaine réelle
une pomme pose
Face à face avec elle
un peintre de la réalité
essaie vainement de peindre
la pomme telle qu'elle est
mais

elle ne se laisse pas faire
...
la pomme se déguise en beau bruit déguisé
et c'est alors
que le peintre de la réalité
commence à réaliser
que toutes les apparences de la pomme sont contre lui
...
se trouve soudain alors être la triste proie
d'une innombrable foule d'associations d'idées
...
et le peintre étourdi perd de vue son modèle
et s'endort
...
Quelle idée de peindre une pomme
dit Picasso
et Picasso mange la pomme
et la pomme lui dit Merci

See Table 3 for Your Visual Response and Your Verbal Response.

Working on a Project

A work of a student from my Computer Art class is shown in a Figure 6. A poster announced an "Art + Math + X – a Special Year in Mathematics and Art Conference" Department of Mathematics, CU Boulder. The artist used an image of an apple because of its connotations related to the laws of physics, geometry, education, art, and many other domains. The two halfs of an apple visually comment the integrative quality of the Conference.

Figures 7a, b, c, and d show an apple quartet: four approaches of my students to present various iconic meanings ascribed to the image of an apple.

Table 3.

Your Visual Response: Drawing an Apple	Your Verbal Response: Playing with Words
From this point you may want to start building analogies and connotations and thus draw an apple, making this a drawing with a purpose. First focus on your own ability to look and see, and then convey your personal perception of the fruit and everything it could mean. It does not matter whether the drawing is done on an iPad, iPhone, on a computer screen with the use of any graphic software, or on paper. Whatever will be the end product, below is a space inviting you to start playing with a pencil. After that you may want to merge and combine text and image on a computer.	After collecting ideas, feelings, associations, and connotations carried by an apple and it's meaning, you may feel ready to write a short poem about an apple. This brief verse will become a translation from the visual mode of thinking to the verbal expression of your thought. Choose a form for your short verse: it may be rhymed or a blank verse with a few metrical lines and with no end rhyme; or, a free verse having no fixed meter and no rhymes. You may as well prefer to write a pun – a short, humorous word play that suggests two or more meanings, or create an apple out of letters and numbers.

Figure 6. Veronica Lucas, "Art + Math = X" poster (© 2005, V. Lucas. Used with permission)

Figure 7. (a) "Apple, A is for Apple" (© 2012, E. Sanchez. Used with permission) (b) Jason Johnson, "Forbidden fruits… What Entices You?" (© 2012, J. Johnson. Used with permission) (c) Preston Stone, "Apple" (© 2012, P. Stone. Used with permission) (d) Adam Smith, "Apple" (© 2012, A. Smith. Used with permission)

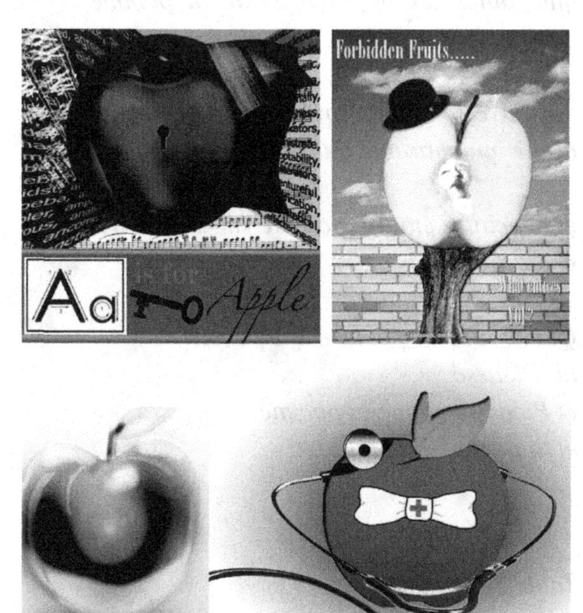

DESIGN, CRAFT, AND FOLK ART

Design

Dictionaries define the notion of 'Design' as a plan, layout, mockup, model, but also the product of such planning, a project, arrangement, and enterprise. Many hold that the design process is a problem-solving process. Making a good data graphics requires diverse skills – the visual-artistic, empirical-statistical, and mathematical. In his book entitled "Sciences of the Artificial" Herbert Simon (1996/1969) wrote,

The damage to professional competence caused by the loss of design from professional curricula gradually gained recognition in engineering and medicine and to a lesser extent in business (p. 112). One may consider art as a way to communicate and express ideas, concepts, and related emotions through their visual representation. The design instruction in the context of art uses skills and shares experiences to convey knowledge, produce aesthetic result, and engage learners' emotions. Thinking in terms of design is a creative process based around the building up of ideas, and a way to apply design methodologies to instruction (p. 55).

The use of different modes (pictorial, verbal, tactile) for presentation of essential information is an important part of instruction by making it simple, intuitive, and easy to understand, regardless of the students' experience, knowledge, language skills, and taking into account their economic, engineering, cultural, gender, and environmental concerns. We can attain it when we supply visuals along with verbal information, and use meaningful icons, as well as text. For example, design of a Maxwell House coffee large size container may attract customers and store managers at the same time with its slogan, "Good to the last drop," characteristic of this company dark-blue color, smart design of its unbreakable plastic box with a handle fitted into its cube shape, so the box can be easily picked up with one hand and at the same time more coffee containers would fit into a big cardboard box. The container would contain less coffee because of the design of the handle, yet it looks big.

Design skills became even more important in relation to the web culture. Fine art, functional (utilitarian) pictures, posters, commercials, and all kinds of web productions are evaluated in terms of the elements and principles of design, as all various forms of art use the same principles. Unlike speech, visual displays of information are controllable by the viewer, and encourage the viewer to use different styles of understanding and interpreting information. In accordance with Edward Tufte (1983; 1992), the good design means that information is effectively arranged and empty space is used properly due to contrast, comparison, and choice, to allow reasoning about information. In the words of Tufte,

We have capacities to select, edit, single out, structure, highlight, group, pair, merge, harmonize, synthesize, focus, organize, condense, reduce, boil down, choose, categorize, catalog, classify, list, abstract, scan, look into, idealize, isolate, discriminate, distinguish, screen, pigeonhole, pick over, sort, integrate, blend, inspect, filter, lump, skip, smooth, chunk, average, approximate, cluster, aggregate, outline, summarize, itemize, review, dip into, flip through, browse, glance into, leaf through, skim, refine, enumerate, glean, synopsize, winnow the wheat from the chaff, and separate the sheep from the goats.

See Table 4 for Visual Response.

Table 4.

Visual Response: Tapestry
Create a geometrical design for a tapestry. It could be done with the use of pattern (repetition of units), but the overall design should carry compositional elements. When buying a fabric in a store we ask for a specific length. If the fabric gets cut few inches shorter, it would not influence its composition. After creating an artwork on the computer screen and then leaving for a lunch, an artist would probably notice if someone would cut a fragment of the composition. It would change the artwork's balance and proportion. When creating art, we care for the placement of the elements in a way that would fit into the format of our composition. We can resize our frame by changing its image size, but we adjust the composition of the artwork afterwards. After finishing your abstract design, use your fabric for dressing a horse, for example making a carpet-like blanket. There are many occasions it can be done for, and it could mean covering the back of the horse, maybe also its head, legs, tail, etc. There are many descriptions of a long tradition of dressing horses, for example by Annisa Garrigues (2008).

Next are student solutions about the tapestry-changed-into-horse-dressing projects. Inspiration came from crystallography, chemistry, storytelling, and biology. Matthew Rodriguez (Figure 8a) utilized his work on geometry-based patterns (white crystalline structures, left) to design a tapestry and apply it for a horse cover, while Jon Furphy (Figure 8b) inspired himself with structural formulas (left) drawn from organic chemistry – aromatic compounds containing ring systems.

Figures 8d and e show two student projects, both including a sense of humor into their solutions. Travis Brandl projected his own visage onto his tapestry project to design a horse dress. Teddy Assuncion chose to leave the space of a dress for his horse 'Gigi' blank and open to interpretations, while the equipment for his horse such as snorkeling mask, snorkel, and fins, along with the horse's surroundings transfer everything into the oceanic environment.

Craft

Craft objects are primarily hand-made, as opposed to mass-produced objects. They are valued for their individual variations. Craft objects are usually both functional and decorative; they reflect traditions and social values of a given craft person. In a conscious or unconscious way, craftsmen apply the elements of art: line, shape, texture, color and form in designing objects. Craftsmen from many cultures have used the same subjects and themes in their works. Many times, the craft artist contributes a unique, individual creative element to their production. Craftsmen produce works, which are aesthetically pleasing and technically proficient in a similar way as are objects produced in "high technology" cultures. The history of crafts is basically cyclic with alternating periods of the development, great achievement, and decay of some solutions for craft objects. Clay,

Figure 8. (a) Matthew Rodriguez, "Horse Tapestry" (© 2012, M. Rodriguez. Used with permission) (b) Jon Furphy, "Horse Tapestry" (© 2012, J. Furphy. Used with permission) (c) Peter Arnegard, "Horse Tapestry" (© 2012, P. Arnegard. Used with permission) (d) Travis Brandl, "Horse dress" (© 2012, T. Brandl. Used with permission) (e) Teddy Asuncion, "Gigi, the sea horse" (© 2012, T. Asuncion. Used with permission)

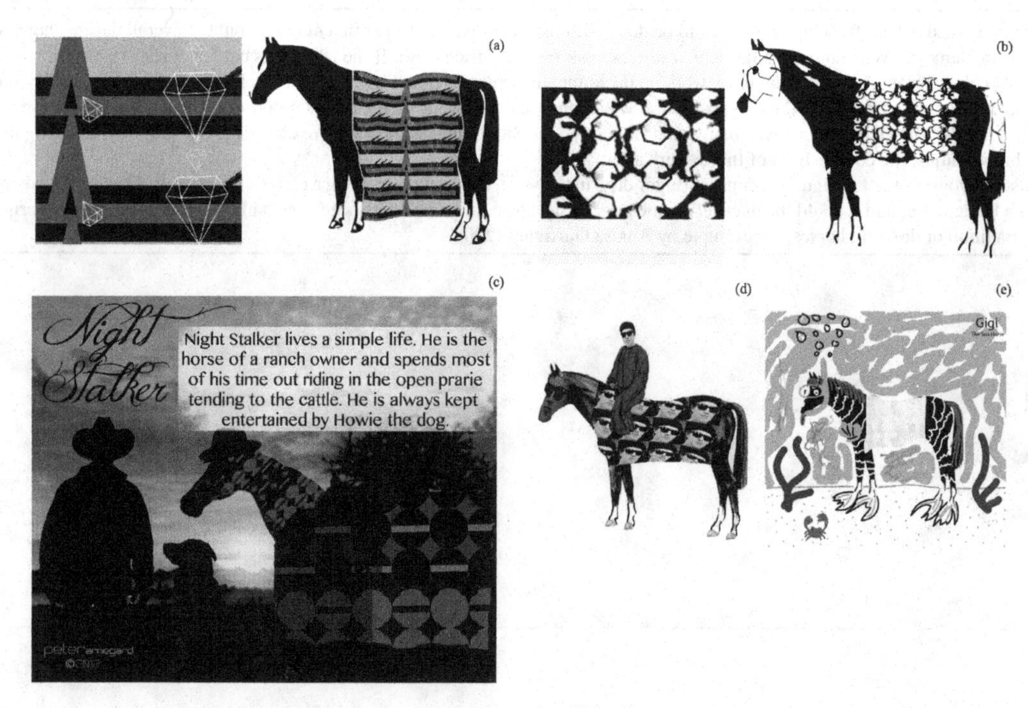

fiber, wood and metal are the primary media or materials used by craftsmen worldwide. Craftsmen develop a special sensitivity to the particular qualities of the materials with which they work. Crafts objects may be identified and classified by style: family resemblances based on culture, era, or artist. Crafts objects, especially in traditional societies are related to and reflect the pattern of the culture: its customs and values. Craftsmen have often worked within a cultural tradition, which limits possible forms and designs. Contemporary craftsmen, especially within cultural subgroups may draw upon and adapt the traditional crafts of their culture. Craftsmen in traditional cultures had to discover their own technical solutions while they were primarily relying on local materials and conditions. Craftsmen in contemporary society can draw upon worldwide resources and technologies for both media and design.

Folk Art

Folk artists create things that please them and fulfill their aesthetic needs, no matter what others think or say. Many times, folk art is intensely personal. Folk artists desire to make something that will last, they want to be remembered and loved for. Craft objects are often family oriented and passed down from generation to generation. They may be urban or rural in style. All kinds of materials are used. Many folk artists have no academic or professional training. Their art is often direct and unassuming, decorative and playful, uses interesting color combinations and inventive shapes. It is often created to pass time and relieve the feeling of isolation. Folk artists often make well-crafted items.

ELEMENTS OF DESIGN IN ART

The skillful use of elements and principles of design may enrich the work of art beyond just depiction of reality and invoke aesthetical and intellectual sensations. Elements of design refer to what is available for the artist/designer or any person willing to communicate visually, while principles of design describe how the elements could be used (Goldstein et al., 1986). Fine art, functional (utilitarian) pictures, posters, commercials, and all kinds of web productions are evaluated in terms of the elements and principles of design. These elements are known as the fundamentals for all works of art, because without them, art could not be created. For example, one may ask whether the sketched line (an element of design), is repeated, applied with the use of symmetry, spiral or radial, is it black, or of any other color. Also, they are applied for data presentation in all branches of knowledge. In the same way, basic kinds of picturing in art, such as still life, portrait, landscape, and abstract obey the design elements and principles. We usually refer color and value, shape and form, space, line, and texture to the design compositional elements. All of these elements exist in the world around us in nature and in the environments we create for ourselves.

Depending on the kind of an artwork and the message it conveys, particular elements, associated with the specific features to be exaggerated, take over the overall composition. Contrasting colors, distinctive types of line, and a bold use of space are explicitly set forth in politically involved art (such as election related posters), antiwar installations (such as messages displayed in public places), or the religious art (such as Russian icons). Furthermore, discourses, emotions, and artistic presentations related to the American flag resulted in a variety of the design solutions. Jasper Johns created several works about the American flag, such as his "Three Flags." Big companies' logos were displayed in the place of the flag stars during anti-war manifestations. Some of the manifestations from the nineties included dramatic effects of burning the flag.

Line

Line shows the shortest way between two points. Line can be a path of a point, which leads the eye through space. Thus, line is a record of movement, can create illusion of motion in a work of art. Lines define an enclosed space. In a drawing or a painting, line may be used both in a functional and imaginative way and may represent anything: an actual shape, a person, or a building.

Lines may be thick or thin, wavy, curved or angular, continuous or broken, dotted, dashed or a combination of any of these. There are many ways in which we can vary a line in art, by changing the line's width, length, the degree of curvature, direction or position, and/or by altering the texture of the line. The use of line in art involves selection and repetition, opposition, transition, and variety of length, width, curvature, direction, and texture. Movement shown by a line is considered a principle of art.

Norton Juster created "The Dot and the Line: a Romance in Lower Mathematics" (1963/2000) using line drawings for his amazing storytelling. The title is considered a reference to a book "Flatland: A Romance of Many Dimensions" by Edwin Abbott Abbott (1994/2008). This book is being constantly reprinted and easy to find online, in bookstores, or libraries. "The Dot and the Line" was adapted as a 10-minute animation by Chuck Jones who won for it the 1965 Academy Award for Animated Short Film. It can be seen online and attracts tens of thousands visitors. More information and the story of the book can be found at Wikipedia. One can also find surprisingly great amount of video responses to "The Dot and the Line: a Romance in Lower Mathematics" as online animations of the story.

For example,

- http://www.youtube.com/watch?v=OmSbdvzbOzY
- http://www.youtube.com/watch?v=OGh97__-uLA
- http://www.youtube.com/watch?v=FNXjUsJNiUM&feature=related

See Table 5 for Your Visual Response.

Table 5.

Your Visual Response: Playing with Lines
You may want to draw examples illustrating how the use of line involves selection and repetition, opposition, transition, and variety of length, width, curvature, direction, and texture. Draw some variations in lines. There are many ways for playing with lines. In various illustrated magazines, find lines that show movement. Also, examine short sketches, for example, The New Yorker cartoons, and explore by drawing how different lines can show emotion. Draw and title expressive lines: draw happy lines, excited lines, stressful lines, dramatic lines relaxed lines, pathetic lines, and boring lines.

Analysis of Art Works

The ability to perceive line separated from form can become a major aesthetical skill providing an increased awareness of the visual beauty and function of line. For example, Morris Louis, "Alpha Pi" (1961, http://www.metmuseum.org/toah/works-of-art/67.232) shows diagonal parallel lines (acrylic paint) like a Zen-like spirit of meditation. Louis resigned from shape and light in favor of pure color. El Lisitzky, "Composition" (1920, http://www.wikipaintings.org/en/el-lissitzky/composition) displays revolving geometric objects painted in colors that appear to be floating in the air, thus creating a sense of depth and an illusion of space. In "Unreal City" by Mario Merz (1968), http://www.guggenheim-bilbao.es/secciones/programacion_artistica/nombre_exposicion_claves.php?idioma=en&id_exposicion=69words are inscribed in neon light within a triangular framework. Is it an elusive and ambiguous metaphor of city life? "Composition" by Piet Mondrian (1929, http://www.artlex.com/ArtLex/d/destijl.html) is built from simple elements – straight lines and primary colors, in search of perfect balance and the order of the universe.

Color and Value

Color exists almost everywhere. In 1704, Sir Isaac Newton discovered that all the colors of the rainbow are contained in white light, such as sunlight. When the light passes through a prism, a band of colors is formed. This band is called a spectrum. Newton also invented a color wheel. He put the three primary colors: red, yellow and blue, and the three secondary colors: orange, green, and violet, in an outer circle. Black is the sum of all of these colors. Intermediate colors are the additional hues, which fall between the primary and secondary colors. The mixture of adjoining primary and secondary colors can produce intermediate colors. See Table 6 for an example color wheel.

Table 6. A color wheel: Color dimensions in pigment

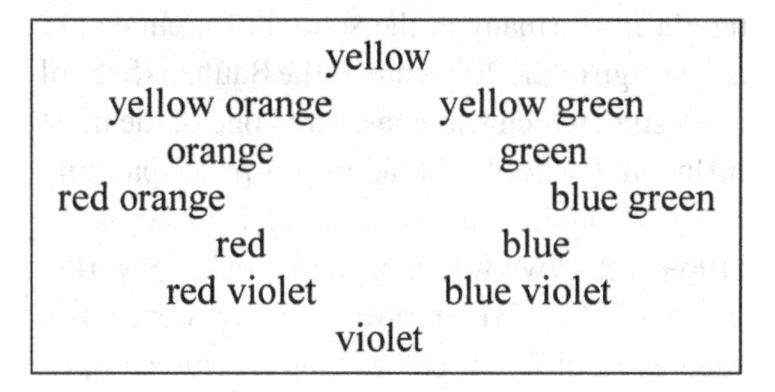

The source of color is light. We see color because light reflects from the object into our eyes. Light is visible radiant energy made up of various wavelengths. It is one of several electromagnetic waves listed in order of their frequency and length: long electric waves, radio, television and radar, infrared (felt as heat), visible light, ultraviolet (invisible), X rays, and cosmic and gamma rays.

Color is a property of the light waves reaching our eyes, not a property of the object seen. The white light of the sun contains all wavelengths of light. When light falls on a surface that reflects all white light, it appears white. When the surface absorbs all the white light, we see the object as black.

Color in Art Works

We can compare the use of color in paintings. For example, Michelangelo Buonarotti (1475-1564) applied in his work large areas of vibrant, contrasting colors and the play of light over human bodies (http://www.ibiblio.org/wm/paint/auth/michelangelo/). Henry Matisse (1869-1954) used in "The Dessert (Harmony in Red)" (1908, http://en.wikipedia.org/wiki/File:Matisse-The-Dessert-Harmony-in-Red-Henri-1908-fast.jpg) primary colors for creating a vibrant, unified pattern of pure color painted thickly, without brush marks. The Dutch/American artist Josef Albers (1868-1976) explored the perception of color both in his paintings and written works, such as *Interaction of Colors* (2010/1963). He had also studied, then taught in Germany at the school of architecture and design of the 20th century, the Bauhaus School of Design, which was considered one of the most influential schools. In his oil-on-panel painting titled "Homage to the Square: Soft Spoken" (1969, http://www.metmuseum.org/toah/works-of-art/1972.40.7) there are four almost concentric squares in different colors that provide an optical illusion about the painting's dimensionality (Albers, 1969). Art historians tend to include art works of Albers both into the Op Art and the Post-Painterly Abstraction movements. Op Art, short for Optical Art, was the abstract art movement from the 1960s focused on exploring the capabilities of the human eye and its tendency to be erroneous. With vibrant colors, artists evoked optical illusions of dimensionality, movement, and shimmering of forms. Abstract artists, working in the 20th century, abandoned the idea of art as imitation of nature, and painted forms that do not remind any specific objects. The paintings of the Russian/American Abstract Expressionist artist Mark Rothko (1903-1970, http://www.nga.gov/feature/rothko/classic1.shtm) present areas of glowing colors of considerable magnitude that provide the viewer with a color experience evoking intellectual and spiritual connotations.

The Primary Colors in Pigment and in Light

The pigment colors are derived from mixtures of pigment primaries. A pigment primary is caused by the reflection of two light primaries. The pigment primaries are red (magenta), yellow, and blue (a blue-green referred to as cyan). All light colors are derived from mixtures of light primaries. A light primary is caused by the reflection of two pigment primaries. The light primaries are green, red-orange, and blue-violet. This means that a pigment primary is a secondary color of light.

The value of color is its lightness and darkness. Colors can be made lighter or darker by adding either white or black. To lighten value, add white. Lightening (any color plus white) produces a tint. To produce a shade, add black. Any color plus black is a shade. Black plus white makes gray. A color plus gray is called a tone. Black, white and gray are called neutrals. Hue, a synonym for color, is a particular quality of a color (full intensity, tint, tone, or shade). In order to change the hue of a color, we add the neighboring color. Primary hues are: red, yellow, blue. Secondary hues are: orange, green, and violet. Intermediate hues are: yellow-green, blue-green, etc. Intensity means the purity or strength (also called chroma). To change intensity and produce a tone, add a complementary or gray color. Gray is a color without hue, made from black and white.

Psychological Aspects of Color

According to Tufte (1983; 1992), color is used in printing to (1) attract and gain attention, (2) to be legible and comprehensible, and (3) to make an impression. Tufte provides some helpful hints about using color:

1. **To Attract Attention:** Warm colors are higher in visibility than cool colors; contrast in values (light versus dark) is greater than contrast in hues (blue versus yellow); the darker the background, the lighter a color appears against it.
2. **To Produce Psychological Effects:** To convey coolness, warmth, action, purity, etc.
3. **To Develop Associations:** For example, do not use green when advertising fresh meet.
4. **To Build Retention:** Color has high memory value, especially in repeated messages.
5. **To Create Aesthetically Pleasing Atmosphere:** Too many colors may be worse than the use of no color at all. The use of elements and principles of design is effective for this purpose.

Table 7.

Your Reaction and Visual Answer: A Quick Drawing with Colors
You may now want to discuss the elements of design using an image of a city as a metaphor. You will transform your sketch in the following projects related to other design elements. In a city metaphor, you may want to sketch some features of an urban landscape to visualize elements of design. Imagine and picture the sky over the city. What can we tell about the color of the sky in the late summertime afternoon? How the color of the sky reflects in the windowpanes? Make a quick drawing of such a color study. Now, make another one, this time about colors of a changing weather over the city. Also, you may want o draw a nighttime in a big city. The rain just stopped and the cars leave lines of light reflected on the wet streets. Show the magic of this scene. What happens to the colors in your picture? Create another quick drawing, this time with different hues.

See Table 7 for Your Reaction and Visual Answer.

Shape and Form

All objects have shape or form. The shape of an object looks flat and two-dimensional. Shapes describe two-dimensional configurations; they have no volume. In an outline drawing we only show the shape.

Ambiguous Shapes

An ambiguous shape is doubtful, uncertain, or open to more than one interpretation. A favorite textbook example of ambiguous space is a vase/profile, and an old/young woman. Another classic example of an ambiguous image is a drawing of a cube with lines and squares. Our brain interprets the image as a three-dimensional cube. However, our brain is unsure where the closest part of the cube is. For many people, after looking for a while, the image seems to "jump" back and forth between two different cubes.

Form describes a three-dimensional object and gives a three-dimensional feel and look of an object. In drawing and painting, we may use shading and highlighting of an outline drawing to show this. Forms have volume – a word that describes the weight, density, and thickness of an object. The solidness or volume of the form could be obtained by using highlights on one side of each object and shading on the opposite side. Shapes are geometric (such as triangles, squares, circles, etc.) or organic (such as leaves). Forms also are geometric (such as pyramids, cones, cubes, spheres, etc.) or organic – natural (like trees). They can be irregular (like clouds).

Space is the void between solid objects (forms) and shapes. It is everywhere, all around us. Everything takes up space in one form or another, whether it's two-dimensional, like drawing and painting, or three-dimensional, like sculpture and architecture. Paintings, drawings, and prints take up two-dimensional space. In a painting, it is limited to the edges of the canvas. Sculpture and architecture take up three-dimensional space. Music and literature involve time. Some arts, such as film, opera, dance and theater take both space and time.

See Table 8 for Your Visual Response.

Table 8.

Your Visual Response: A Sketch of a City
Show space in your short sketch of a city. Draw shapes of windows and doors. Draw forms of skyscrapers and houses, lampposts, and benches. Draw geometric forms of roofs, then organic forms of trees and clouds. Show the volume of those by shading and highlighting. Are the streets broad or narrow? Is there a square? Are the houses spacious?

Positive and Negative Space

Space also describes the void between solid shapes and forms. The solid shape or form is called a positive space. The space within the drawn objects is a positive space; a doughnut has a positive shape or form. He space between the objects is a negative space; the doughnut hole is a negative shape or space.

See Table 9 for Your Visual Response.

Perspective

The appearance of depth or distance on a flat surface is called perspective. We live in a three-dimensional world of depth and distance, but we are picturing world on a two-dimensional surface of a piece of paper or a computer screen. The use of perspective enables us to show objects and scenes as they appear to the eye, with relation to implied depth on a flat surface of the picture. Designing a three-dimensional display may involve visual thinking in many ways. Perspective drawing is one

Table 9.

Your Visual Response: The Positive and Negative Space
To picture a coexistence of the positive and negative space on your drawing of houses you may now want to assign a selected color (for example, violet) to the negative space in the open windows and doors.

of the solutions, with orthographic projection or the bird's eye view often applied. Other methods of preparing a display in three dimensions may include presenting the data on cardboard models or on the surfaces of a polyhedron - a solid body bounded by polygons. Other presentation techniques include architectural miniatures, stereo illustrations and slides. Three-dimensional graphics are designed with the use of programming strategies or 3D graphics software packages; data can be presented as holograms, video, stereo illustrations, and slides. Maybe the most impressive way to show whatever you want is a computer-based immersive multi-wall virtual reality and interactive visualization environment. Thus we can show a 3-dimensional object as a 2-dimensional image when we draw a perspective drawing, a shaded drawing, or create a painting or a sculpture, but we do not achieve it by designing a map. We can consider a geometrical cardboard model a three-dimensional form of display, but we cannot say this about a musical score, a railroad schedule, a periodic system of elements in chemistry, or a tabular array of numbers.

Several methods help us to create an aerial perspective, the representation of objects and scenes according to their distance without drafting perspective: overlapping, vertical positioning, graying colors, varying details, varying size, and converging lines. By overlapping objects we may create an illusion that the partially hidden object is more distant. Vertical positioning lets us believe that objects placed higher are at a considerable distance and those positioned lower are closer in space. Gradation of the strength of light and colors of objects, and showing objects in grayed colors makes them look remote, if we take into account the quality of light falling on the objects, and the surrounding atmosphere through which they are seen. We may show less detail on an object to make it more distant than another, as well as drawing distant objects smaller and nearer objects larger. Converging lines is the most frequently used method that can be achieved by drawing lines closer and closer together in the distance.

The Dutch graphic artist M. C. Escher (1898-1972) explored a strange world of optical illusions, visual jokes, paradoxes, and visual puns. He drew ambiguous spaces to create illusions. In his work, often based on mathematical concepts, Escher combined humor, meticulous precision, logic, visual trickery, and beauty. Many of his works displayed metamorphosis, gradual transformation of one shape into another. Escher created impossible scenes by combining several kinds of distorted perspectives into a coherent whole. His work has been widely regarded representative of mathematical thinking in art.

See Table 10 for Your Visual Response.

Perspective in Painting

In his book entitled "Envisioning Information," Edward Tufte (1983; 1992) wrote that the world is complex, dynamic, multidimensional, and all the interesting worlds are inevitably and happily multivariate in nature, having many values and dimensions. The paper is static, flat. How do you represent the rich visual world of experience and measurement on mere flatland? How can an artist escape flatland?

We may examine how artists created perspective in different times. Florentine architects from the 15th-century Italian Renaissance perfected geometry and developed perspective drawing: for example, Fra Angelico "The Annunciation" (painted in 1441-3, http://www.artbible.info/art/large/255.html), and also Sassetta, "The Journey of the Magi" (Siena, painted in 1435, http://www.museumsyndicate.com/item.php?item=23630). Piero de la Francesca, "Portraits Federico da Montefeltro and Battista Sforza" (1465, http://www.wikipaintings.org/en/piero-della-francesca/portraits-federico-da-montefeltro-and-battista-sforza-1465). Frantisek Kupka, "Cathedral" (1913, http://en.wikipedia.org/wiki/František_

Table 10.

Your Visual Response: Showing Space in a 2D Drawing
You may want to create perspective in your sketch, using medium of your choice. Draw an interior of a video arcade. Try to create a false perspective and make some objects appear close and some farther away in your drawing, so the characters on a screen or on a paper interact visually with players: a) By overlapping objects b) By placing distant objects higher and closer objects lower c) By showing distant objects in grayed colors, shading distant objects d) By showing less detail on distant objects e) By sizing: showing distant objects smaller and closer objects larger f) By making lines come closer together in the distance g) By coloring back plane darker or lighter than the front part of the scene. Now rearrange the objects and spaces, so you may feel like being enclosed in an unreal space of a game.

Kupka). Vertical and diagonal geometric shapes are created in the pure abstraction style called Orphism, which was delivered from Cubism. Also, cubist artists, such as George Braque (1882-1963), Juan Gris (1887-1927), Pablo Picasso (1881-1973) showed simultaneous view of many viewpoints. Marcel Duchamp "Nude Descending a Staircase" (1912, http://en.wikipedia.org/wiki/Nude_Descending_a_Staircase,_No._2) was influenced by the geometric chronophotography of Marey. Etienne-Jules Marey (http://stage.itp.nyu.edu/history/timeline/marey.html) and the 'cinematography' of Eadweard Muybridge (from 1880s, http://commons.wikimedia.org/wiki/File:Eadweard_Muybridge_Gehender_Strauß_001.jpg) created more plastic cubism by representing a temporal fourth dimension on a two-dimensional canvas.

Cognitive Perception Addressed by Painters

Concepts related to photorealistic 3D rendering and to interactive visualization are often derived from an analysis of traditional media. Painters not always used regular perspective or orthographic projections, with lines at right angles to a plane of a painting. According to John Counsel (2003), many times the masters chose the kinds of perspective according to their cognitive perception. Sometimes, it was an isometric method of drawing, so that three dimensions were shown not in perspective, but in their actual size. For example, interiors painted by Vincent Van Gogh were not fully based on the familiar Euclidian perspective. Cezanne's paintings reflected the way the viewer perceives reality. Areas where the viewers were supposed to direct and focus their eyes were painted in a three dimensions, with greater detail and enhanced color. The rest of the painting addressed peripheral vision and so it was flat.

Figure 9 reflects the cognitive transfer resulting from the fact of traveling. Changes sceneries evoke new constructs in the driver's mind presented here as abstract, programmed quadrants. Irregular, diagonal geometric shapes correspond to the natural patterns in the fields.

We are told average American moves 14 times in ones life. This work leads the viewer from the cozy rural community to the big city in the way that may be felt by somebody who is just moving, depicting the rhythm and the rhythmical sound of a big city.

Figure 9. Anna Ursyn, "From the Village to the Big City" (© 1988, A. Ursyn. Used with permission)

Pattern

A pattern is an artistic or decorative design made of lines; thus pattern is a repetition of shapes. Patterns make a basis of ornaments, which are specific for different cultures. Owen Jones (1856/2010) made a huge collection of ornaments typical of different countries. He wrote a monographic book entitled "The Grammar of Ornament."

Texture

Texture is a general characteristic for a substance or a material. Texture exists all around us. It can be either actual (natural, invented, or manufactured) or simulated (made to look rough, smooth, hard or soft, or like a natural texture). Simulated textures are made to represent real textures such as a smooth arm or rough rock formation. But they are not actual textures, and if you touch the picture you feel only the paint or the pen, or pencil marks.

See Table 11 for Your Visual Response.

PRINCIPLES OF DESIGN IN ART

Principles of design that are most often used in visual arts are: balance, emphasis, movement, variety, proportion and unity. These principles may vary according to the person using them. For example, some textbooks discuss also contrast, rhythm, and repetition (Goldstein et al., 1986). The skillful use of elements and principles of design may enrich the work of art beyond just depiction of reality and invoke aesthetical and intellectual sensations of pleasure, appreciation, or repulsion.

Balance

Balance is an arrangement of lines, colors, values, textures, forms, and space, so that one section or side of the artwork does not look heavier or stronger than another. We can see in art three main types of balance: formal (or symmetrical), informal (or asymmetrical), and radial balance. Formal or symmetrical balance has equal weight

Table 11.

Your Visual Response: Some Textures and then a Booklet
Draw simulated textures that would enhance your sketch of a city: rough stonewalls, hard metal parts of a bridge, or a fluffy cloud. Now you can make a little booklet about the elements of design. First fold paper three times to make a booklet with 8 pages; number the pages. Using your drawings you have already made, design your own "Elements of Design" book.

on both sides (for example, two people of the same weight at the two ends of a see-saw). Informal or asymmetrical balance has a different weight placed differently on each side to maintain balance (for example, when a person on a see-saw who weights more sits closer to the center and the lighter person sits farther out on the end). Radial balance is a circular balance moving out from a center of an object to maintain balance (for example, when only one object is centered in a picture). Many times, we see a combination of formal and radial balance, or the informal balance with the unequal organization of elements; for example, several small shapes may juxtapose a large shape.

We may want to examine how artists created balance. For example, Anish Kapoor, Double Mirror 1998 http://art-glossary.com/definition/anish-kapoor/ achieved balance three dimensions, and Edward Hopper, People in the Sun, 1960 http://www.wikipaintings.org/en/edward-hopper/people-in-the-sun in a painting, with an informal balance that shows the stillness of the scene with strongly contrasting lights and shadows.

Emphasis

Emphasis is way of bringing dominance or subordination into a design or a painting. Major objects, shapes, or colors may dominate a picture by taking up more space (when they are larger and repeated more often), by being heavier in volume, or by being stronger in color and color contrast than the subordinate objects, shapes, or colors. There must be a balanced relationship between the dominant and subordinate elements; otherwise there is too much emphasis. Color and color contrast can be used to achieve emphasis in a work. Also texture contributes to emphasis.

Figure 10. Anna Ursyn, "Middleground" (© 2005, A. Ursyn. Used with permission)

Natural composition of shapes and forms we can see in a landscape acts as an incentive to seek a corresponding balance in a picture. One can see repetition of planes, find out how the position of a horizon alters depending on the terrain configuration, and how the density of objects and actions changes with the proximity of a city (Figure 10).

Analysis of Art Principles in Art Works

Artists sometimes exaggerate parts of their paintings to stress the importance of the depicted objects. For example, in the style of Symbolism, the mostly French artistic movement from the late nineteenth century, things were painted out of proportion at the cost of realism to convey otherworld ideas, unusual feelings, and states of mind. The giant's huge eye in "The Cyclops" created by the French painter Odilon Redon (1840-1916, http://en.wikipedia.org/wiki/File:Redon.cyclops.jpg) terrifies the viewer like in a nightmare. Artists painting in the Surrealist mode created figurative portrayals of dreams mixed with reality aimed to liberate the unconsciousness. An acrylic painting on panel "Death and Funeral of Cain" by David Alfaro Siqueiros (1896-1974, http://www.terminartors.com/artworkprofile/Siqueiros_David_Alfaro-Death_and_Funeral_of_Cain), influenced both by the Surrealists and the Mexican folk art, shows an enormous, out-of-proportion dead chicken.

Movement

The illusion of movement may be achieved by the use of lines, colors, values, textures, forms and space, to direct the eye of the viewer from one part of the picture to another. For example, the feeling of movement can be suggested by an arrangement of shapes. Some possibilities are circular, diagonal or vertical arrangements; we may also create illusion of movement that goes back into the distance, both by diminution of sizes and similarity of shapes.

See Table 12 for Your Visual Response.

Table 12.

Your Visual Response: A Scene with Balloons, Some of Them Escaping
Draw a scene with balloons, first a balanced one, and then you may want to emphasize your own balloon. You may choose to draw air balloons with people in consoles focusing on one your friends ride in. Sketch a clown invited to a birthday party to play tricks with balloons trying to hide one, or create some projects for a balloon logo to be used by dealers who sell homes or cars. In a scene with colorful balloons, show a balloon escaping out of control and disappearing in the sky. Show dynamics beyond someone's effort trying to catch them.

We may want to look over searchingly how artists employ some elements of art such as line or space, and apply emphasis, rhythm, or contrast to enhance the dynamics of their artwork and show motion in various styles in art such as: Symbolism, for example, in "Centaurs' Combat" (http://www.wikipaintings.org/en/Tag/Centaurs#supersized-search-260594) the Swiss painter Arnold Böcklin (1827-1901) depicted a mythological scene of combat to convey a savage and violent nature of ferocious fight. There are the contrasts in colors, contorted shapes, and unreal representation of the background space, all of them enhancing the sense of horror and drama. Cubism, for example, Fernand Léger (1881-1955) "The Builders" (http://www.amazon.com/Hand-Made-Oil-Reproduction-Builders/dp/B004LA7Q8E). Léger used horizontal and vertical lines defining the space of the ironwork skeleton. At the same time, he used colors and patterns to emphasize the dynamics of the scene. Artists belonging to the Futurism movement were intrigued with motion through time and space. Examples: Umberto Boccioni (1882-1916), "The Street Enters the House" 1911 (http://en.wikipedia.org/wiki/Umberto_Boccioni), Giacomo Balla (1871-1958), "Flight of the Swallows" (http://www.wikipaintings.org/en/giacomo-balla/flight-of-the-swallows-1913). In "Flight of the Swallows" the artist used repetition to enhance the feeling of movement. Works done in the Surrealism style, for example, Arshile Gorky (1904-1948), "The Waterfall" (http://www.wikipaintings.org/en/arshile-gorky/waterfall-1), which is almost abstract but conveys strong impression of water pouring through a rock. Semi-abstract art of Wassily Kandinsky (1866-1944), for example, "Cossacks" (http://www.wassilykandinsky.net/work-250.php), with colorful brush strokes, dynamic lines and shapes, and wonderfully balanced composition (Kandinsky, 2011/1911). Dynamic paintings of the American painter Thomas Eakins (1844-1916), for example, his scenes of wrestling (http://commons.wikimedia.org/wiki/File:Thomas_Eakins_-_The_Wrestlers_(1899).jpg).

Composition

In art, composition is orderly arrangement, a proper combination of distinct parts so they are presented as a unified whole. The whole arrangement of objects in a picture is made to the best advantage of its elements. To develop composition in a drawing or painting, we have to select the objects we want to show, then create a center of interest and find out balance among the objects. Good composition may involve movement, rhythm, and well-arranged positive and negative space.

It took many years to develop routine features of books, with page numbering, indexes, tables of contents, and title pages. Web documents undergo a similar evolution and standardization in order to define the way information is organized and made available in electronic form. Visual composition of a website and its graphics is an important part of the user's experience. In interactive documents, the interface design includes the metaphors, images, and concepts used to convey function and meaning of the website on a computer screen.

In accordance with Edward Tufte (1992), every graphic presentation, as well as every project should fulfill general principles related to its design, and should pass critical evaluation of its editing, analysis and critique of presentation. Edward Tufte taught that information should enhance complexity, dimensionality, density, and beauty of communication. Good information display should be: documentary, comparative, casual and explanatory, quantified, multivariate, exploratory, and skeptical; it should allow comparing and contrasting. When envisioning statistical information, such display should insistently enforce comparisons, express mechanisms of cause and effect quantitatively, recognize the multivariate nature of analytic problems, inspect and evaluate alternative explanations.

We may want to examine composition of some art works. For example, the sculpture by Constantin Brancusi "Bird in Space" (1911, http://en.wikipedia.org/wiki/File:Bird_in_Space.jpg) has the most basic, abstract primordial vitality. In Pierre Bonnard "The Dining Room in the Country" (1913, http://en.wikipedia.org/wiki/File:Bonnard-the_dining_room_in_the_country.jpg), composition of both the outdoor and indoor scene has been achieved by using broad areas of color.

See Table 13 for Your Visual Response.

Variety and Contrast

An artist uses elements of design to create diversity and differences in an artwork. Contrasting colors, textures, and patterns all add interest to the artwork. Highlights of color to the corners or edges of some shapes may be used to add contrast. To examine variety and contrast in some art works, we may look how the Dutch graphic artist M. C. Escher (1898-1972) used contrast to isolate some shapes from others and from a main body of his works (http://en.wikipedia.org/wiki/File:Escher_Circle_Limit_III.jpg), or Vassily Kandinsky, "On White II" (1923, http://

Table 13.

Your Visual Response: Working on Composition
Make a line drawing of your room. Think of what you selected for this sketch and what you decided not to include. What are some of the ways of making you a center of interest in this composition? Try to apply formal and then an informal balance. Introduce an impression of movement in your composition. What may add rhythm to your composition?

en.wikipedia.org/wiki/File:Kandinsky_white.jpg. Kandinsky's analytical book, "On the Spiritual in Art" (Kandinsky, 2011) was first published in December 1911. Kandinsky was one of the co-founders of The Blaue Reiter group. His work was shown in New York in 1913 at the Armory Show. He taught at the Bauhaus school of modern design. Born in Russia in 1866, Kandinsky became a German citizen in 1928. The Nazi government closed the Bauhaus in 1933 and later that year, Kandinsky settled in Neuilly-sur-Seine, near Paris; he acquired French citizenship in 1939. The Nazis in the 1937 purge of "degenerate art" confiscated fifty-seven of his works. Kandinsky died December 13, 1944, in Neuilly.

Proportion

The size of one part of an artwork in comparative relation its other parts is called proportion. Artists use proportion to show balance, emphasis, distance, and the use of space. Sometimes, proportion was used to add emphasis to an artwork. For example in medieval religious paintings, some rulers or saints were pictured out-of-proportion to emphasize their importance: important figures were painted bigger and humble donators smaller.

We may want to examine proportion in some art works. For example, in Cimabue, "The Santa Trinita Madonna" (c.1280) we may see small figures at the bottom (http://en.wikipedia.org/wiki/File:Cimabue_033.jpg). In Jasper Johns, "Three Flags" (1958, http://en.wikipedia.org/wiki/File:Three_Flags.jpg), flags of different sizes were superimposed on top of one another. We can see the use of proportion to evoke the feeling of dimensionality and elicit some optical effects characteristic for the American Pop Art style.

Unity

Unity is the result of how all elements and principles of design work together. All parts must have some relation to each other. They must fit together to create the overall message and effect.

See Table 14 for Your Visual Response.

Table 14.

Your Visual Response: Design a Booklet
A process of designing and printing a book involves planning. Typically pages are placed on a large sheet of paper. Your book may be black-and-white only or in color. One side of a sheet of paper may be black-and-white, while the other one might be in color. Printing in color is more complicated because often printer runs a sheet of paper four times to apply separate coatings of the cyan, magenta, yellow, and black color. This process requires an accurate paper registration, so each color is placed in the exactly same area of the page. Japanese woodblock prints Ukiyo-e were produced between the 17th and 20th centuries. Now you can make a little booklet about the elements and principles of design. First fold 8x10" paper horizontally. Then fold paper again three times to make a booklet with 8 pages per side. Number the pages. You will have 8 pages per side of your piece of paper. One page will serve as a cover page, and another as a back page of the book. Place your name, a graphic, and the title of the book on the cover page, while your bio, your photo, publisher name, ISBN number, QR code, and a book prize on the back cover page of the book. This will give you 14 pages for the content of your book. You will need to rotate the top row of your book while placing its content, so after folding the pages they will al show the same direction. Using your drawings you have already made, design your own "Elements of Design" book.

Analysis of Art Works

In the oil on canvas painting "Saint Francis of Assisi" created by the Spanish artist Francisco de Zurbaran (1598-1664), the expression of strong religious feelings is enhanced by both the ecstatic look on the saint's face and the dramatic effect of the *chiaroscuro* lightning, strong form the left side only. The artist chose a dark color of the background and an elongated form of the figure to place emphasis on an austere religiousness conveyed in the artwork. "The Treachery of Images" (1928-9, http://en.wikipedia.org/wiki/File:MagrittePipe.jpg) by Belgian artist Rene Magritte (1898-1967) brings a challenge to the ordered society by denying that a picture that is obviously an image of a pipe is a real pipe. A pipe has been pictured in a clear and sharp way expressing an awareness of an object as it really is. Thus, the artist declares, in accordance with the Surrealist movement in art, that all is not as it appears to be, and an image of an object cannot be confused with the tangible reality. Pablo Picasso (1881-1973) created a series of paintings entitled 'Weeping Women', along with his "Guernica," telling about the cruelty and violence imposed by the Spanish Civil War in 1936. The face of the "Weeping Woman" (1937, http://picassogallery.blogspot.com/2011_01_18_archive.html) is shattered in the style of the Cubism movement. Strong colors, firm paint strokes, and sharp lines create dramatic expression of the weeping woman. The same concern has been imparted quite differently by Robert Motherwell (1915-1991) in his acrylic on canvas painting entitled "Elegy to the Spanish Republic No. 134" (1961, http://www.metmuseum.org/Collections/search-the-collections/210009638). Strong, dynamic shapes are painted in black on a white background in the style of the Abstract Expressionism movement.

Elements and Principles of Design in Various Disciplines of Art and Science

Next is the record of students' comparative discussion of this theme. Add your own input.

Art
Music
Science
Mathematics

– are all various forms of art, and use a lot of the same principles of design.

Concepts like talent, creativity, thought, formulas, abstract symbols, and design – are common to all those areas. They all take a creative mind to be good at it, to see the spaces in between and all the good things that come out of them.

Both in art and music, we use metaphoric representation to tell a story or create emotional response. Both art and music follow a preset line, but can be distorted and abstracted earlier. We can find symbolizing numbers in math or science and symbolizing notes in music. Music has form and composition that can be symmetrical or asymmetrical (like art), and they usually have balance. Music involves science (physics of sound waves), and math (metered beats and measures, theories of music). One may see a difference in the use of unity and variety: in music, variety is offered by the use of altered rhythms, melody, and color. In visual arts, variety is offered by placement, theme, and color.

There is order in most of disciplines of art and science. Visual art and mathematics, especially geometry, are more spatial, while music is time based. Most of them hold a rhythm.

There are many common elements in art and architecture, for example: form, shape, color, repetition, composition, and pattern. Contrast is common – it evokes excitement and emotion. Different terms have similar outcomes. Visual representations of concepts in science and mathematics are better understood with contrast. Many of the techniques of visualization are common. In art, visual, societal and life-oriented messages are communicated with beauty. There is a common need of simplifying their meaning in visual display. There are also aesthetic dimensions of science. Elements and principles of design in art can be used to teach math and science, visually. Art works created in differing, sometimes conflicting styles bear visible design characteristics according to the messages they carry. We can analyze how the design-related issues were solved in previous times and in contemporary art, and how the style of an art work relates to the actions or behaviors of people, influences their thinking, keeps under control their emotions, and affects their reactions to the content of the image.

CONCLUSION

Concepts and problems pertaining to visual literacy, discussion of art definitions, basic art concepts, elements and principles of design, differences between art, design, craft, technical issues, elements, and principles related to art, design, and many other disciplines, and the quality of display, which made a content of this chapter, will be returning in the following chapters. Readers who want to find time or interest to examine art-related problems can find information conducive to creating meaningful projects suggested in the framed spaces existing in the following chapters. The advantage of visual display of information over speech or writing has been discussed, along with its nonlinear, flexible time of viewing, multiple dimensions, and possibility of restructuring of its content.

REFERENCES

Abbott Abbott, E. A. (1994/2008). *Flatland: A romance of many dimensions* (Oxford World's Classics). Oxford University Press.

Albers, J. (1969). Homage to the square: Soft spoken. In *Heilbrunn Timeline of Art History*. New York: The Metropolitan Museum of Art. Retrieved January 23, 2012, from http://www.metmuseum.org/toah/works-of-art/1972.40.7

Albers, J. (2010). *Interaction of color*. Yale University Press. (Original work published 1963).

Bisbort, A., & Kite, B. (1997). *Sunday afternoon, looking for the car: The aberrant art of Barry Kite*. Pomegranate Communications.

Burke, K., & Kite, B. (2000). *Rude awakening at arles: The aberrant art of Barry Kite: Postcard book*. Pomegranate Communications Inc. (Original work published 1996).

Copeland, J. (2012, June 19). Alan Turing: The codebreaker who saved millions of lives. *BBC News Technology*. Retrieved June 23, 2012, from http://www.bbc.co.uk/news/technology-18419691

Counsel, J. (2003). Pointing the finger: A role for hybrid representations in VR and video. In *Proceedings of the Seventh International Conference on Information Visualization*. IEEE. ISBN 0-7695-1988-1

Danto, A. C. (1964). The artworld. *The Journal of Philosophy, 61*, 571–584. doi:10.2307/2022937.

Davies, S. (1991). *Definitions of art*. London: Cornell University Press.

Duchamp, M. (1917). *The blind man*. Retrieved January 23, 2012, from http://sdrc.lib.uiowa.edu/dada/blindman/2/05.htm

Flynn, J. R. (1987). Massive IQ gains in 14 nations: What IQ tests really measure. *Psychological Bulletin, 101*, 171–191. doi:10.1037/0033-2909.101.2.171.

Flynn, J. R. (1994). IQ gains over time. In R. J. Sternberg (Ed.), *Encyclopedia of human intelligence* (pp. 617–623). New York: Macmillan.

Flynn, J. R. (1999). Searching for justice: The discovery of IQ gains over time. *The American Psychologist, 54,* 5–20. doi:10.1037/0003-066X.54.1.5.

Garrigues, A. (2008). *Dressing the horse and rider*. Retrieved December 10, 2012, from http://annisa.garrigues.net/classhandouts/Dressing%20the%20Horse%20and%20Rider.pdf

Goldstein, E., Saunders, R., Kowalchuk, J. D., & Katz, T. H. (1986). *Understanding and creating art* (Annotated teachers ed.). Dallas, TX: Guard Publishing Company.

Isaacson, W. (2011). *Steve Jobs*. Simon & Schuster.

Jones, O. (2010). *The grammar of ornament*. Deutsch Press. (Original work published 1856).

Juster, N. (2000). *The dot and the line: A romance in lower mathematics*. Chronicle Books. (Original work published 1963).

Kandinsky, W. (2011). *Concerning the spiritual in art*. Empire Books. (Original work published 1911).

Lauzzana, R., & Penrose, D. (1987, April 23). A 21st century manifesto. *FINEART Forum, 1.*

Lauzzana, R., & Penrose, D. (1992). A pre-21st century manifesto. *Languages of Design, 1*(1), 87.

Lavington, S. (2012, June 19). Alan Turing: Is he really the father of computing? *BBC News Technology*. Retrieved June 23, 2012, from http://www.bbc.co.uk/news/technology-18327261

Lengler, R. (2006). Identifying the competencies of 'visual literacy' – A prerequisite for knowledge visualization. In *Proceedings of 10th International Conference on Information Visualisation.* IEEE.

Ogden, C. K., & Richards, I. A. (1923). *The meaning of meaning: A study of the influence of language upon thought.* New York: Harcourt Brace and Co.

Owen, O. (2001/1856). *The grammar of ornament.* DK Adult. ISBN 10789476460

Pease, R. (2012, June 19). Alan Turing: Inquest's suicide verdict not supportable. *BBC News Technology.* Retrieved June 23, 2012, from http://www.bbc.co.uk/news/science-environment-18561092

Plato (2000). *The Republic.* Cambridge, UK: Cambridge University Press.

Simon, H. A. (1996). *The sciences of the artificial* (3rd ed.). Cambridge, MA: The MIT Press.

Simon, J., & Wegman, W. (2006). *Funney/strange.* New Haven, CT: Yale University Press.

Tatarkiewicz, W. (1976). *Dzieje szesciu pojec.* Warsaw, Poland: PWN.

Tufte, E. R. (1983/2001). *The visual display of quantitative information.* Cheshire, CT: Graphics Press.

Tufte, E. R. (1992/2005). *Envisioning information.* Cheshire, CT: Graphics Press.

Weitz, M. (1956). The role of theory in aesthetics. *The Journal of Aesthetics and Art Criticism, 15,* 27–35. doi:10.2307/427491.

Chapter 4
Creativity, Intuition, Insight, and Imagination

ABSTRACT

Discussion of the notion of creativity and the creative process seems necessary in a book about computational solutions going beyond text and numbers because notions such as art creation, creativity, and the creative process have been considerably broadened due to the input coming from computer science and computer technologies. Countless options of social networking provide fuel for many forms of online creative works. Comprehension of the role of creativity in new media art involving concepts beyond the 2D and 3D graphics such as interactive and time-based art, networking, the online, virtual, and Second Life presence, evoke initiatives taken in journals, books, college curricular programs, conferences, and the new options taken by artists and designers. This results in the quest of the new role of digital creativity and an emerging need for boosting digital creativity in schools. The further text looks at the role of creativity in a process of digital art image creation.

INTRODUCTION

Emphasis that is given in relation to digital creativity takes form of establishing journals, opening interdisciplinary academic degree programs, and developing software applications. Discussion of the notion of creativity and the creative process seems necessary in a book about computational solutions going beyond text and numbers because of the change in the meaning of this concept. The scope of this theme has been considerably broadened due to the input coming from computer science, computer applications, and their interfaces. The content of this chapter has been designed to invoke manifestations of digital creativity in the readers who interact with the text.

DOI: 10.4018/978-1-4666-4703-9.ch004

The role, function, and terminology related to creativity expanded onto the domains of technology and information technologies. This may result from the changing approach to art and art creation and current demand for art products as an inherent part of science- and technology-related solutions. These dynamic changes are going parallel to the developments in computer technology, with electronic computers being over 60 years old (the first-generation computer UNIVAC 1 was produced in 1951), personal computers existing for more than 30 years (microcomputers becoming popular in the late seventies), and the web accessibility to the public for more than 20 years: Tim Berners-Lee used the NeXT computer as the first web server and wrote the first web browser WorldWideWeb in 1990 (Berners-Lee & Cailliau, 1990). Because of all these events, the creativity demands are shifting from a need for proficiency in the traditional art forms to creating aesthetically challenging interactive digital content, which is essential for data mining and web visualization.

THE MEANING OF THE CREATIVE PROCESS

Creativity is often seen as the ability to create or design something useful or beautiful or novel: in science, to create even the simplest but own solution or invention; in art, the work that represents one's own flow of thought. As Robert Sternberg (1998/2011, p. 145) stated, "People are creative by virtue of a combination of intellectual, stylistic, and personality attributes. Sternberg & Lubart (1999, p. 3) wrote, it is "the ability to produce work that is both novel (i.e., original, unexpected) and appropriate (i.e., useful, adaptive concerning task constraints)." "First, creative ideas must represent something different, new, or innovative. Second, creative ideas are of high quality. Third, creative ideas must also be appropriate to the task at hand or some redefinition of that task. Thus, a creative response is novel, good, and relevant" (Kaufman & Sternberg, 2010, p. xiii).

Mihaly Csikszentmihalyi (2011) who in his own words, "devoted 30 years of research to how creative people live and work" stated that "creativity is a central source of meaning in our lives. Most of the things that are interesting, important, and human are the result of creativity." Csikszentmihalyi (1998) describes the autotelic activity as one we do for its own sake. Autotelic personality – "an individual who generally does things for their own sake rather than in order to achieve some external goal (Csikszentmihalyi, 1998, p. 117). Next are the notes excerpted from his article about the creative personality (Csikszentmihalyi, 2011).

Creative individuals are remarkable for their ability to adapt to almost any situation and to make do with whatever is at hand to reach their goals." Being creative provides "a profound sense of being part of an entity greater than ourselves... If I had to express in one word what makes their personalities different from others, it's complexity. They show tendencies of thought and action that in most people are segregated. They contain contradictory extremes...Creative people have physical energy, but they're also often quiet...they tend to be smart yet naive at the same time...they combine playfulness and discipline, or responsibility and irresponsibility...but this playfulness goes together with a quality of endurance and perseverance...most of them work late into the night and persist when less driven individuals would not...Creative people alternate between imagination, fantasy, and a rooted sense of reality...Creative people tend to be both extroverted and introverted...humble and proud at the same time...Creative people escape rigid gender role stereotyping: creative and talented girls are more dominant and tough than other girls, and creative boys are more sensitive and less aggressive than their male peers. Creative individuals are more likely to have not only the strengths of their own gender but those of the other one, too...Creative people are both rebellious and conservative... Most creative people are very passionate about their work, yet they can be extremely objective about it as well...Creative people's openness and sensitivity often exposes them to suffering and pain, yet also to a great deal of enjoyment... Being alone at the forefront of a discipline also leaves them exposed and vulnerable. Eminence invites criticism and often vicious attacks...Divergent thinking is often perceived as deviant by the majority, and so the creative person may feel isolated and misunderstood...Perhaps the most difficult thing for creative individuals to bear is the sense of loss and emptiness they experience when, for some reason, they cannot work...Yet when a person is working in the area of his of her expertise, worries and cares fall away, replaced by a sense of bliss. Perhaps the most important quality, the one that is most consistently present in all creative individuals, is the ability to enjoy the process of creation for its own sake.

There is no one generally accepted definition of creativity; one could possibly gather from the professional literature almost one hundred various definitions, especially when in addition to the established works in this field such as written by Robert J. Sternberg (Sternberg & Lubart, 1995, 1999; Sternberg, 2006, 2007), Joy Paul Guilford (1950, 1967), Howard Gardner (1984, 1994, 1997), and Michael Csikszentmihalyi (1996, 1997), creativity has been referred also to animals. Individual creativity in animals was manifested, for example in elephant paintings (Komar, Melamid, Fineman, Schmidt, & Eggers, 2000) or the mate selection in birds: the taste of the individual male bowerbird is visible when these birds gather collections of feathers, berries, and shells (Borgia, 1995; Dorin & Korb, 2012). Collective creativity, displayed for instance by swarms (Gloor, 2006) has been studied both in the animal kingdom (Miller, 2010) and in the domain of computing (al-Rifaie, Aber, & Bishop, 2012).

Semir Zeki (1993, 1999), professor of neuroaesthetics at London University College proposed a study of neuroesthetics, the neural basis of artistic creativity and achievement. Zeki (2001) started with the elementary perceptual process. He associated perception of the great works of art with the working principles of the brain. Artists, acting like instinctive neuroscientists, capture in their art works the essence of things in a similar way as the brain acts when it captures the essential information about the world from a stream of sensory input. Neuroesthetics explores the visual brain using anatomical, electro-physiological, psychological methods, and imaging techniques.

Semir Zeki (2009) believes that without understanding how the brain acquires knowledge it is difficult to understand its productive and creative actions. He states, beauty isn't in the eye of the beholder – it's in the brain (Lebwohl, 2011). Research on computational creativity, based on methods practiced in artificial intelligence, philosophy, and cognitive science, is aimed at constructing creative machines, programs, or systems; gaining a fuller, more formal description of human creativity; and design tools for making human endeavors more creative. International conferences on computational creativity (ICCC 2012) occur annually. A great part of computational creativity involves a combinatorial approach including evolutionary algorithms (that mimic processes existing in natural evolution such as inheritance, mutation, selection, and crossover) and artificial neural networks (that mimic the properties of biological neurons). Maybe for this reason many authors describing computational creativity tend to exaggerate the role of combinational creativity (which produces unfamiliar combinations of familiar ideas by making associations between them) and training in human creativity. As stated by Edmonds, Bilda, and Muller (2009), the evaluation of interactive art works in a public space is a part of the creative process. al-Rifaie, Aber, & Bishop (2012) discuss whether the hybrid swarm algorithms have the potential to exhibit 'computational creativity' in what they draw. The authors conclude that the combinatorial creativity of the hybrid swarm system, for example exhibited in the hybrid bird, ant, and blood vessel mechanism can be useful in generating interesting and intelligible drawing outputs.

Margaret Boden, who defines creativity as the ability to generate novel and valuable ideas (1998, 2007a, 2007b; 2006), describes three types of creativity – combinational (which produces unfamiliar combinations of familiar ideas by making associations between ideas that were previously only indirectly linked); exploratory (which rests on some culturally accepted style of thinking); and transformational (where the space or style are transformed by altering or dropping their defining dimensions, to generate ideas that could not have been generated before the change).

Boden (2010, p.16) also stresses that human minds are "capable of holding, integrating, following, and also abandoning general principles of behaviour. These may be moral, political, religious ... or aesthetic." Hence there is an ambiguous relationship between creativity and freedom: "A style is a (culturally favoured) space of structural possibilities: not a painting, but a way of painting, or a way of sculpting, or composing fugues... and so on. It's partly because of these thinking styles that creativity has an ambiguous relationship with freedom." Creativity and artistic personality have been often linked with the mood disorders, especially manic-depressive (bipolar) disorder. Traits of personality such as independence of judgment, self-confidence, attraction to complexity, aesthetic orientation, and risk-taking are used as measures of the creativity of individuals (Sternberg, 2008). Boden (2009) introduced a distinction between psychological and historical views of creativity. Psychological novelty (or P-creative idea) is new to the person who generated it, while a historical novelty (or H-creative idea) has never occurred in history before. The available tools may influence the working style; for example, an access to the Internet makes gathering information about other works quicker, and collaboration easier.

Several studies deal with the part played by creativity in the effectiveness of solving problems and the aesthetics of the solutions. Boden considers creativity a fundamental activity of human information processing, which can be modeled by artificial intelligence techniques; she also proposed that artificial intelligence ideas might help to understand creative thought. The combinational creativity strategies may include mixing familiar and strange settings, mixing genres, objects or styles, inserting an unrelated or unexpected fragment, or translating the meaning of iconic objects and images into new domains to evoke intended connotations. According to Boden (2010), this type of creativity seems to be difficult for artificial intelligence (AI) to model because of lack of the rich store of world and cultural knowledge.

APPROACHES FROM OTHER POINTS OF VIEW

A model of creativity is not universal because concepts of art and creativity differ in the world cultures (Sawyer, 2012). Moreover, creative capacities have been linked with many thinking activities. Complexity of the theme became even greater due to the studies about creativity conducted in other than art domains. Research in cognitive neuroscience, focused mostly on consciousness, memory, language, attention, decision-making, and learning alters the established notions about creativity. Advancements in the fields of computing, creativity of computers, artificial intelligence, computing technology in education, and many other approaches change these notions even more.

Several authors (for example, Wiggins, 2006; Ritchie, 2007) examined methods available for evaluation of computational creativity and discussed neuroaesthetics – the study of the neurological basis for aesthetic behavior, which results in creating art. In terms of the science of complex systems Galanter (2010, 2011) described the neuroaesthetic complexity model as characterized by a built-in balance of order and disorder, or expectation and surprise. According to Pablo Gervás (2009, p. 58) it is difficult to address creativity in the context of storytelling, especially inventing stories. Gervás reviewed the existing storytelling systems with respect to creative process and listed several points: who is the creator, what is the output, who is the audience (especially in terms of the evaluation of novelty of the output), expectations and whether the output is unexpected in some way, whether the output meets some goal, what are the inputs, whether feedback is being contemplated. Thomas S. Kuhn (1970) observed earlier that either very young individuals or the ones very new to the field made the great inventions, and they changed the paradigms. Maybe it is so because, according to Chi and Snyder (2011), thinking outside the box is difficult, especially for those with the most in-depth knowledge. Chi and Snyder, supported by the experimental results saw possible explanation in a structure of our thinking, which is driven by hypotheses, preconceptions, and expectations. The authors activated the side of brain associated with novel ideas (R, right anterior temporal lobe) and at the same time suppressed the side associated with maintaining existing hypotheses (L, left anterior temporal lobe). They posed they could free the brain from a fixed mental set and thus allow alternate solutions. In result of stimulation people were able to see the correct answer with a burst of insight from the right side of their brain that had been blocked by the left side.

Digital Portrait as Personification

With changing meaning and taxonomies of creative actions many may think about artistic productions in a new way, not only as the pictures but also the spatially-referenced and time-based events, or the interactive and interdisciplinary visual forms of storytelling. In the digital environment, both the notion of a portrait and the related aesthetic expectations may overcome the semiotic divide between material and fictional characters, or between realistic and virtual images (Heinrich, 2010). Falk Heinrich (2010) discussed digital portraits used for avatars in online worlds and communication networks. He noticed a shift toward understanding a digital portrait as performative acting, in place of the ontological understanding as an individual being to be contemplated in terms of its craftsmanship and aesthetic values. According to Heinrich, the avatar-portrait functions not only as representation but also as an embodiment, in a somewhat similar way as it operated in Eastern, Byzantine icons. It surpasses general notion of a technological medium by constructing a direct material and sensory relationship between people and transcending the humanistic concept of identity. As stated by Heinrich (2010, p. 9), "the digital iconic avatar seems to undermine the Western epistemic distinction between the human subject and pictorial representation, questioning the notion of the body as a mainly biologically defined entity."

In electronic art, we may discuss basic approaches to designing an artwork with all related freedom/precision compromises; for instance, by exploring "the line between analogue and digital" (Sullivan, 2000). In terms of the study of human speech, one may venture an analogy between the statements made with the use of speech or of visual art, and the statements in English or in Japanese language. The typical features of English language are seen as a digital mode of description, logical approach, linear structures, quantitative measure, and facile scientific description, while Japanese language is characterized by the analogous mode of description, emotional approach, qualitative evaluation, and facile poetic expression.

Project: Designing a Coin with a Visual and Verbal Side

The following project may involve using not only creativity but also ingenuity – applying your own solutions to problems or meeting challenges. When using one's own ingenuity one applies complex thought processes, brings together, often collectively, thinking and acting to overcome problems.

Social ingenuity means organizing ourselves differently, communicating and making decisions in new ways.

We may find a great variety of coins in various times and places. The oldest coin, more than 2,700 years old, was discovered in an ancient Hellenic city Ephesus on the coast of Asia Minor. The coin was made of electrum, a natural occurring alloy of gold and silver; it had a design on one side showing a head of a lion (Fleur de coin, 2012). Some regions in Africa used a large, round circular leg bracelets, denoting the wealth of a woman wearing it. The island of Aegina was the earliest state in ancient Greece to strike coins (Head, 2012/1886; Goldsborough, 2010/2004). Coins may have a form of a conventional disc or another form (for example, in some countries there are hexagonal or octagonal coins and also coins with a hole in a center; an Australian fifty-cent coin is twelve-sided, while a 15 cent coin of the Bahamas has been issued in a shape of a square).

The abstract way of thinking was observed a long time ago. Scientists recovered in a South African cave two pieces of red ochre (a form of iron ore) from the Middle Stone Age layers from at least 70,000 years ago (Limson, 2010; Wong, 2002). These archeological finds are considered an evidence of cognitive abilities allowing abstract thought. They bear crosshatched markings that appear to be symbolic engravings relied on syntactical language; they were made for transmission and sharing of the yet unknown meaning (Henshilwood et al., 2002). The earliest abstract representations found before this one were from the Eurasian Upper Paleolithic period mainly in France and dated to less than 35,000 ago.

You may now want to create a visual and verbal solution for designing a coin, and then design a coin in an abstract way. An image on your coin may show face of a person you want to commemorate, a place, an animal, or an artwork you create (Table 1).

SOME EARLIER STUDIES ON CREATIVITY

A great number of educators became aware of the need for developing creativity of students, but the notion itself seems to be different according to various sources. Jerome Bruner (1962) views the creative product as anything that produces "effective surprise" in observer, as well as a "shock of recognition" that the product or response, while novel, is entirely appropriate. For Joy Paul Guilford (1959), divergent thinking, characterized by the flexibility, originality, and fluency of thinking, is one of the most important factors of creativity. Eight types of creative contributions described by Robert Sternberg (2006) referred to his research works on psychometric studies of analytical, creative, and practical intelligence. Arthur Koestler (1964) considered imagination, humor, scientific inquiry, and art as sources of creativity; therefore he explored theories of play, imprinting, motivation, perception, and Gestalt psychology. According to Koestler humans are most creative when rational thought is abandoned during dreams and trances, while their creativity is suppressed by the everyday routines of thought and behavior. Teresa M. Amabile (1996) separated an advanced problem-finding developmental stage. It requires flexibility – changing the original approach and adopting a new approach in order to identify a new problem others do not see. She applied an operational definition of creativity: a product or idea is creative to the extent that expert observers agree it is creative. This definition has been used to channel the consensual assessment of creativity, which relies on the agreement of experts. According to the consensual assessment concept (Hennessey & Amabile, 1987), creativity depends both on temporary states and enduring traits. For John Young (1985) creativity is the skill of bringing about some new and valuable, actualizing of our potential. Contrary to the common notion of creativity as the ability to produce unique, remote associative responses to verbal and visual stimuli,

Table 1.

Visual Solution: A Coin	Verbal Solution: A Coin Description
Design a coin; An image on a coin may show a face of a person, a place, an animal, an invention, or an artwork you create. Think of your own, personal hero, for example somebody technology oriented or an inventor in computing. Choose a shape of your coin. Design both an obverse (the front face, a head) and a reverse (the back face, a tail) of your coin. Decide what orientation your coin would have: it would be a 'coin orientation' if you flip vertically to show the other side of the coin, or a 'medalic orientation' when one has to flip a coin horizontally to see the other side correctly oriented. Design your coin once again, this time in an abstract way. Find a metaphor, symbol, or a sign that would best characterize the theme of the coin you have already designed. If you have previously created a visage of somebody you admire, or whose work and achievements you respect, find a visual shortcut that signifies the object of your esteem. It may be a short formula, a segment of musical notation, a chunk of a computer program, an object that brought about a praise or fame (such as a book or a tool mastered by this person). If it is not a person, convey the essence of your message in a symbolic or abstract way.	Describe a design of a coin you want to create and the content of an image you would like to place on it. Explain your choice of this image and a way you would like to render it. If it is somebody's visage, tell about this person and the reason for minting a coin in this person's honor. If you would rather like to create another image, comment on the idea behind your choice and the issues you would like to disseminate or support. Later on, maybe you can summarize your statement in a concise form so it will fit for the back face of your coin. Also, remember to place a year of (virtual) minting on the head of your coin. Describe your coin once again, this time in an abstract way. Find a way to convey in a symbolic or abstract way the essence of a message you put on a coin. Put the shortest slogan on the coin. Write a short story about an experience that left a strong impression on you and influenced your choice of the theme for a coin. Describe how you have encountered a person, idea, object, or place you depicted on the coin and how your symbolic or abstract representation conveys the essence of this experience.

Herbert Walberg (1969) equated creativity with the winning of awards, prizes, or other recognition in competition. As it takes an interest in external effects or responses through the use of behavioral objectives, this attitude could be called a little bit Skinnerian. According to Victor Lowenfeld (1947) and then Lowenfeld & Brittain (1987) creativity is closely related to thinking abilities and to attitude development, and may have little to do with the processes of the intellect.

According to the Discipline Based Art Education (DBAE) program that was developed in the sixties at the J. Paul Getty Center for Education in the Arts (Dobbs, 1992) creativity has been conceived as "unconventional behavior that occurs as conventional understandings are attained" (Lipari, 1988, p.15). Many operational definitions of creativity focused attention on an internal mental process of creating knowledge and implied attributes of novelty and value recognized by experts and the general society (Bloom, Hastings, & Madaus, 1971). Elisa Giaccardi & Gerhard Fischer (2008) extend the traditional notion of design to include co-adaptive processes between users and systems that enable the users to act as designers in personally meaningful activities and be creative. But some recognized creators and artists merely recombine known bits of information. Manipulation of the paintings of great masters using computer-generated programs (Schwartz, 1985; Schwartz & Schwartz, 1992) could possibly serve as an example of this meaning of the notion of creativity.

Definition of creativity as communication of emotional genuineness can be found in the classification of art students that was made by the university art teachers (Hammer, 1984). They distinguished between facility and creativity, the facility being a sterile technique: the ability to paint or draw with ease but in unoriginal, stereotyped forms and colors, while creativity being a real feeling demonstrating (with or without facility) exquisitely sensitive powers of color, tone, form, and discrimination combined with personal imaginative expressiveness and an original approach. Thus creativity emphasizes visual and emotional authenticity, and is considered a process of calling forth inner emotions to put on canvas. In an early study on computer-based instruction Marianne Rash (1988) pointed at another feature of creative thinking that could be defined as functioning of metacognitive strategies in knowledge acquisition processes: not only creating ideas, but also defining the means by which these ideas could be judged. The creation of the criteria that could be used to determine the possible effectiveness of the solution appeared to be integral to the cognitive process of creativity (Tennyson, Thurlow, & Breuer, 1987).

Figure 1 presents a work created by the student from my "Introduction to Visual Communication Design" course. The artist paid homage to the inventor of a Polaroid camera (created in 1947) by collecting pictures of devices that he considered redeemable by one Polaroid camera: photographic cameras from 1947, projectors from 1960, 1977, and a copier from 2009; with this visual shortcut the artist signified the importance of the Edwin Land's invention.

Figure 1. Spencer Korey Duncan, Design for a coin commemorating Edwin Land, inventor of a Polaroid camera (© S. K. Duncan, 2012. Used with permission)

CONVERGENT VS. DIVERGENT THINKING AND PRODUCTION

Guilford (1959, 1967, 1968) discerned convergent production aimed at a single, correct solution and divergent production involving generation of multiple creative answers. Divergent thinking, mediated by the frontal lobe of the cortex, is not considered a synonym of creative ability; yet divergent thinking tests are used as estimates of creative potential. According to Runco (1986), divergent thinking seems to be stronger associated to creative performance in gifted children. As the convergent thinking is the capacity to direct one's attention on the expected target, intelligence of the student helps him to quickly see the teacher's point; as the divergent thinking dashes off to the unknown and comes up with the unexpected and the original, creativity helps the student to see beyond the teacher's point (Hammer, 1984), and generate some unique ideas. Teachers often regard divergent thinking in children distracting, and most teachers prefer the high IQ to the high creativity in children (Getzels & Jackson, 1962; Getzels & Csikszentmihalyi, 1970, 1976). Programs for the bright, good students are designed to accelerate those who manifest convergent thinking, and many times are producing people capable of learning anything easily, except how to think for themselves (Hammer, 1984). On the contrary, teachers-artists might assist students to break their creativity loose from their intelligence. Creativity is considered to be a process, not a product, a process of formulating and testing hypotheses and communicating results. According to Torrance (1962) training and stimulation are capable of increasing the quality of creative thinking of young people and adults: although no teaching would guarantee creativity, some conditions increase the probability that creative thinking will occur. Torrance provided empirical evidence for creativity stimulating effect of some teaching procedures by administering alternative forms of the Torrance Tests of Creative Thinking (1974, 1990) at the beginning and end of the workshop, and documenting statistically significant gains in ability to produce original ideas. This view provides a strong argument for encouraging talented individuals to accelerate in schools and thus increasing their chances for continued development.

CRITICAL THINKING AND CREATIVE THINKING

In accord with widely held positions critical thinking skills entail logical thinking and reasoning that include skills such as comparison, classification, sequencing, cause/effect, patterning, webbing, analogies, deductive and inductive reasoning, forecasting, planning, hypothesizing, and critiquing. Critical thinking, which traditionally has been ascribed to left-brain reasoning, is typified as analytic, convergent, verbal, linear, objective, judgmental, focused on a subject, and probability of its change. Scientific thinking is used in investigating processes and events, acquiring new information, and integrating previous scientific knowledge. However, research on scientific thinking (Dunbar 1997) revealed that much of the scientists' reasoning and over 50% of the findings resulted from interpreting unexpected findings that were very different from the hypotheses based on literature. It was also found that scientists use analogies from similar domains in proposing new hypotheses.

In some languages, one may notice a distinction between the act of creation from nothing (by the Creator, as described in the Bible) and the act of creation from something (by an artist) (Tatarkiewicz, 1999). If an artwork is a product of imagery, one may ask whether human imagery can create something non-existing in nature, or if an artist draws exclusively from natural sources. It is hard to indicate purely fantastic forms with no relation to natural forms. It is nature that arouses one's admiration and delight, thus inspiring art. Biological or physical forms, whether seen in a micro, a macro scale, or as blown-up pictures, are often used as a point of departure in traditional and computer art, while mathematical relations and formulas make a ground for algorithmic programs for graphics. Creativity is highly valued, but often misunderstood, because it is often mixed with originality: simply bringing together previously existing concepts in a new way. Creative thinking is based on the flexibility of mind, as well as on knowledge of previous works in one's field. It does not mean that erudition is the justification for teaching the art history and foundations, as it may not result in stimulating future research and creative work. It seems possible to develop ones ability to think intuitively and creatively. Thus creativity is sometimes described as the ability to see connections and relationships where others have not. The ability to think in intuitive, non-verbal, and visual terms has been shown to enhance creativity in all disciplines.

Visual way of thinking is strongly supported by our creative thinking. Traditionally ascribed to right-brain activity, it means the ability to think in non-verbal, visual terms, to see connections and relationships where others have not, and to imagine or invent something new. Creative thinking has been described as involving the skills of flexibility, originality, fluency, elaboration, brainstorming, modification, imagery, associative thinking, attribute listing, metaphorical thinking, and forced relationships. The aim of creative thinking is to stimulate curiosity and promote divergence. Creativity has been also described as an attitude: the ability to accept change and newness, a willingness to play with ideas and possibilities, a flexibility of outlook, the habit of enjoying the good, while looking for ways to improve it. With this approach, we may continually improve our ideas and solutions by making refinements to our work. Studies have shown that creative individuals are more spontaneous, expressive, and less controlled or inhibited. They also tend to trust their own judgment and ideas, they are not afraid of trying something new. Georg Wilhelm Hegel (Macdonald, 1970, p.372) wrote at the beginning of the 19th century, "The goal of art is to aid in the comprehension of the world's ideas."

Completely different outcomes follow the two types of thinking: precise or expressive. We may contrast the works of electronic design completed with first-rate tools with the expressive art works. For technology-oriented people teamwork with artists may bring a conflict between precision and accuracy of ready solutions and the artist's individual style. On the other hand, deformations in electronic art result in transforming the initial image aimed to depart from that what the eye can see. One can see many reasons for such transformations, either through programs or software. With purposeful deformations, we make mental shortcuts, react to synthetic signs, and imply connotations to symbols or icons. In advertising, messages are often successful due to impressive imaging that gives a shorthand summary of ideas.

CREATIVITY VS. INTELLIGENCE

Intelligence and creativity have been previously considered to be two distinctly different variables that often enough do not even tend to occur together (Wallach and Kogan, 1965, Torrance, 1962). Intelligence has been originally equated with the ability for convergent thinking, while creativity with the capacity for divergent thinking. Divergent thinking involves the ability to generate numerous and diverse ideas (fluency), the capacity to generate unusual and unique associations (originality), the capacity to shift perspectives or directions of thought and ideas (flexibility), and the ability to improve upon a simple idea through embellishment or detail (elaboration) (Runco, 1986, Rogers, 1985). Joy Paul Guilford (1959, 1967, 1968) described different types of creative abilities: sensitivity to problems, fluency factor, novel ideas, flexibility of mind, synthesizing and analyzing ability, reorganization or redefinition of organized wholes, complexity, evaluation, and also other abilities such as motivational factors and temperament. Donald Hoffman (2000, p. 202) declared, "Visual intelligence occupies almost half of your brain's cortex. Normally it is intimately connected to your emotional intelligence and your rational intelligence. It constructs the elaborate visual realities in which you live and move and interact."

Howard Gardner (1984) described distinct forms of intelligence – abilities to solve problems or design fashion products that are valued in at least one cultural setting or community. Gardner (1994, 1997) also investigated with cognitive approach the anatomy of creativity and discussed several kinds of creative mastery: the first is producing permanent works in a genre (like Mozart, Picasso, Virginia Woolf). The second way is in executing stylized performances (like Mozart, Martha Graham, Sarah Bernhardt). The third form of creation entails solving recognized problems (like James Watson and Francis Crick who solved the structure of DNA, mathematicians like Andrew Wiles who executed Fermat's last proof, and inventors like Wright Brothers who devised a flying machine). A fourth form of creation features formulating a general framework or theory (such as Freud, Darwin, or Albert Einstein). A final form of creation returns to the realm of performances, but this time performances of high stake, where an individual's ability to perform creatively under stress could spell the difference between life and death, escape and injury (Mahatma Gandhi, also, performances for career-making prizes in athletics or the arts). However, general expectation of conformity often results in cautious attitudes to the spirit of creativity and to the value of creative skills. Some pose that any individual who works diligently for a sufficient period of time should be able to become an expert and only practice separates the ordinary from the extraordinary. Yet, according to Garner, achieving expertise is not the same as achieving extraordinariness and most of individuals do not become extraordinary. Instruction is not designed that way, and most individuals lack the propensity to rebel, to become Makers of domains. Gardner found some regularity in the lives of Makers – those highly creative individuals who have invented or decisively altered domains:

- Born in a community close to intellectual life.
- Talented in a range of areas; they are youth at promise, contrary to youth at risk.
- Grown in a bourgeois ethic of regular, disciplined work.
- Given love and support from parents.
- Such individuals take off for a center of cultural life and make a selection of a domain. They need an alter ego to keep them on course when they explore ideas and concepts that make little sense to others.
- They arrive to new formulation that may transform the domain in which they work.
- They are able to make multiple representations of a problem think in a number of ways to illuminate it
- In their work, there may be further developments or dramatic shifts to new areas.

According to Abraham Maslow (1968/1998), who once offered an opinion, "It is as if Freud supplied us the sick half of psychology and we must now fill it out with the healthy half," self-actualizing people share several qualities including (Maslow, 1943, 1968/1998, p. 89): truth (honesty, completeness); goodness (benevolence, honesty); beauty (simplicity, completion); wholeness (integration, synergy); dichotomy-transcendence (acceptance, resolution); aliveness (spontaneity, self-regulation); uniqueness (individuality, novelty); perfection (nothing superfluous); necessity (inevitability); completion (fulfillment); justice (fairness, non partiality); order (lawfulness); simplicity (bluntness); richness (complexity, intricacy); effortlessness (ease; lack of strain); playfulness (fun, joy, amusement); and self-sufficiency (autonomy, independence, self-determining).

Creativity and artistic talent are strongly associated to specific traits of personality. This opinion has been advanced by Hammer (1984) following his research of traits, feelings, and attitudes that correlate creativity, where several projective techniques, like Rorschach Test, Thematic Apperception Test (TAT), House-Tree-Person Drawing Test (HTP), and Unpleasant Concept Test had been administrated to the high school students making art. Some students were considered "truly creative", another students "merely facile"; they were tested without researcher's knowledge of students' classification made by their art professors. The results were compared for personality variables. There was a high degree of similarity in personality pattern of the truly creative individuals. Reduction in personal spontaneity and retreat into an observer rather than a participant role was a typical response. Confidence and determination, ambition, striving for power and capacity to become aware of conflict, personal uniqueness and independence, tolerance for suffering and breadth of emotional horizons, were the traits of creative young artists personality.

The study of scientists (Taylor, 1959) has revealed some traits in common with artists, such as general quality of impulse control, repression of spontaneity, isolation and a low level of interpersonal involvement in human relations, together with a devotion to independence and autonomy. Scientists showed also a liking of toying with ideas for their own sake, and manipulation of ideas. According to Abraham Maslow's (1943, 1950, 1954/1987) exploratory study of productive people like Beethoven, Jefferson, Lincoln, and Einstein, these people uncovered such qualities as: a quality of detachment, autonomy in relation to culture and environment, freshness rather than stereotype in approach, and openness to mystic experiences, though not necessarily religious ones, most of them being consistent with the findings in the study of creative artists. A French abstract painter Georges Mathieu determined the creative process, as a sequence of actions aiming to pass on from what is conceptual to what is real, then from real to abstract, and from abstract to possible. Where is a place for artistic input? Art application in mathematics may relate to working

Figure 2. Sam Dailey, "Joy of Winning" (© 2005, S. Dailey. Used with permission). This work was created in 3D software to recreate an emotional response to winning and losing.

on mathematical formulas or programming. We still do not know what 'talent' means, especially 'talent' in a specific area. We do not know what cognitive processes make a mathematician or an artist a talented individual, for example, how some painters possess a sense of color. Maybe talent is needed to convey messages in essential, synthetic way.

Three factors that determine the development of creative potential are domain-relevant skills (something new, new combination of ideas), creativity-relevant skills (something extra, new cognitive pathways, working style), and intrinsic task motivation (trained in groups). Several factors may damage students' creativity; for example, having children focus on expected evaluation, using plenty of surveillance, and setting up restricted-choice situations. Dawson and Baller found a relationship between creative activity and the health of an elderly person (Hoffman, Greenberg, & Fitzner, Eds.,1980); a large amount of clinical and experimental results have confirmed this statement.

Figure 2 presents student's personal perception of the forces driving one at making a success. The choice of a chess game emphasizes the creativity and intellect components as crucial for going on to victory and the resulting elation.

The Allusionary Base

Is the imagined a clue to reality? Emancipation of the mind from the constraints of actuality may result in creating some things or ideas. For art and poetry, artistic creativity relates to our imagery. Many authors, for example Lakoff (1990) stressed the importance of images and accepted that the use of visual elements in teaching and learning yields positive results. A Polish/American specialist in aesthetic education Harry S. Broudy (1987, 1991) refers to imagination as the image-making function of the mind. Imagery plays a direct role in our perception because sounds, shapes, colors and motions convey meaning. Images play also an indirect role when they influence language, concepts, values, and ideals by making associations. Imagery constructed this way truly or falsely claims to be images of reality. In deconstructionist terms, viewers interpret those images. In his paper entitled "The Role of Imagery in Learning" Broudy (1987) introduced a notion of the allusionary base – those concepts, images, and memories that are available to provide meaning for the viewer. The allusionary base is a stock of meanings with which we think and feel. Images build the allusionary base: direct (concrete) perceptions of shapes, colors, and motions that convey meaning, and indirect (abstract) imagery that may influence the learning of languages, skills, concepts, and attitudes. The selection of images, words, and feelings from the associative store of the listener or the viewer is decisive in shaping the figurative language or visual representation. Selection of words in a non-educated language results from a lack of associative resources and a small power to make the cognitive uses of language and generate connotations in response to linguistic signals. Everybody is building an allusionary base during lifetime. While appreciating art, we are building it in our minds through associations and interpretations of images and works of art (Broudy, 1987, 1991).

INTUITION

Developing imagination seems to be an important task for teachers and instructional designers because problem-solving and intuitive decision-making abilities hinge to some degree on imaginative thinking. When the teachers and instructional designers work on creating new strategies for learning, they take into account not only the rules and formulas but they also envisage how the learners feel these rules out and can figure the best way to imagine, understand, and learn. Analyses of images, forms, and motions in interactive evolutionary design and art lead to new approaches in defining aesthetic criteria, paying attention to usability and efficacy. Intuition combined with logic enables us envisioning future moves of the opponent, which is a valuable ability in chess and not only in chess playing. For example, it is not enough to teach only the game rules when teaching to play chess. Imaginative approach to a task of this type involves developing intuition that allows learning from partial observations, solving problems without explicit reasoning and combinatorial complexity (Duch, 2007). Intuition helps us find effortlessly optimal thoughts and solutions, even if not justified because we often cannot explain why we know the answer. Intuition supports our strategies concerning choosing the best of possible options, for example an opening move in chess, but also envisioning strength distribution on the chessboard, anticipation of the other player's moves, and even endurance. Intuitive processes support our aesthetic preferences and may lead to scientific discoveries. Many times we apply predictions, when we need to plan our next steps but information about future is unavailable and we lack the data for inductive reasoning. A lot of attention is given to intuitive evaluation in medicine.

INSIGHT

Many agree that insight helps to achieve comprehension of complex concepts and solve difficult problems, but definitions and theories describing this capacity vary considerably. Some ascribe to insight the instants of sudden comprehension or illumination. Psychologists conducted several behavioral studies on this theme applying several kinds of tests such as analogies, series-completion problems, and the Remote Associates Test. For example, after conducting the Remote Associates Test, Ansburg & Hill (2002) argued that the creative and analytic thinkers differ in their use of resources requiring attention. While the creative thinkers use peripherally presented cues effectively and take advantage of incidentally presented cues, the analytic thinkers sustain directed attention and focus on the main problem elements. According to the authors, recording the seemingly irrelevant information may lead to insight.

Afterward, cognitive neuroscientists have been carrying out neuroimaging studies. They applied electroencephalography (EEG) and functional magnetic resonance imaging (fMRI) to examine neural correlates of insight. Sandkühler & Bhattacharya (2008) conducted an electroencephalographic study about the regions of the brain activity correlated with the insightful problem solving often resulting in discoveries. They described the components of the cognitive problem solving that correlate with the brain responses: an adjustment of selective attention resulting in a mental impasse or an immediate correct solution; the encoding and retrieval process of restructuring problem correlated with the right temporal region activity; then a deeper understanding of the problem ends with the sudden 'Aha!' feeling related to the problem solution correlated with a cortical excitation in the right prefrontal cortex. Kounios & Beeman (2009) assert that insight is the culmination of a series of brain states and processes going in the brain in sequence. However it is difficult to produce insight phenomena in the laboratory to conduct the brain-imaging studies and be certain that problems were solved by insight and not in a more stepwise manner; also, that the data obtained in laboratory research can be generalized (Luo, Knoblich, & Lin, 2009).

CURIOSITY

One may say curiosity is about emotion based on previous experiences with the resulting explorative behavior and anticipation. It has been considered a psychology-related term, but curiosity is also strongly connected with cognitive activities including investigation and learning. Animals, not only humans display behaviors caused by the emotion of curiosity. Such behaviors, which are more than only the fixed instinctive reflexes, have been widely perceived in apes, cats, rats (Berlyne, 1955), raccoons, insects, and other beings. Animal curiosity is focused not only on hunting and foraging; it seems there is an abstract mode and motivation, such as playfulness (that is important in the development of cubs, pups, and other offspring). Playfulness in some animals, for example cats or dogs, makes them pretend that lifeless objects are alive.

Intellectual curiosity drives people to research, study, and thus to knowing new things. Curiosity about ideas expressed in philosophy often leads to meta-curiosity – investigation into curiosity itself, conducted in terms of the abstract mode of thinking; people focus on broadening the limits of knowledge, applying imagination and creativity, exploring scientific and philosophical reasoning about world and consciousness. Some social and educational psychologists find curiosity an important factor that promotes intrinsic motivation to act because of one's own interest and enjoyment not depending on external rewards. Intrinsic motivation refers to curiosity about and interest in the task itself, which exists in a person and rewarded by individual enjoyment and satisfaction. However, there is also a need for attaining an outcome that is norm referenced, accepted by others, and rewarded, and such performance is driven by extrinsic motivation. Curricular and instructional design, testing, and grading do not come in mind when one seeks existential or empirical answers about reasons making sense of things or the human condition. We may say we owe a lot to curiosity in history, starting with indigenous tribes, then Greeks, other ancient schools, and all those who wanted to know much, know why, and avoid bad things happening for them because of that.

There is no doubt some infants (and also kittens, puppies, and other young animals including rats) are more curious than others, but the sources are scarce about what makes some infants curious, whether it is correlated with their intelligence and creativity, and how is curiosity in infancy correlated with achievement in later life periods. In the sixties and seventies researchers declared a need to develop a construct of science curiosity in children and to set up valid instruments for measuring curiosity (Berlyne, 1960; Harty & Beall, 1984); they developed several Likert-type instruments with items rated by children. Many self-report instruments followed (for example, Curiosity and Exploration Inventory by Kashdan et al., 2009), along with

Dictionaries define curiosity as a strong desire to know or learn something, inquisitiveness, and an interest leading to inquiry about unusual things. However, the notion of curiosity has been tied in the past with connotations such as nosiness – an inquisitive interest in others' concerns, and, even earlier as carefulness and fastidiousness, according to most of dictionaries such as Merriam-Webster or Dictionary.com. A proverb 'curiosity killed the cat' is about being inquisitive or even intrusive about other people's affairs that shouldn't concern others, which may get a nosy person into trouble. There is also a morbid curiosity that is built by desire to see hurtful events, accidents, death, blood, violence, and also monstrosities or abnormalities, for example a bearded woman in a circus or a gigantic skeleton in a natural science museum. Surprisingly, a search term 'curiosity' could not be found at the site 'Curiosity.com from the Discovery' website (2012) that claims it "explores life's most fascinating mysteries."

Curiosity is a name given several enterprises, bands, and online groups such as Curiosity Wonderment and Exploration group (http://curiosity-group.com/); for example, 'Curiosity' is a name of a NASA's car-sized rover that has been launched on Mars (NASA, 2012) to monitor space radiation. However, it is still difficult to find answers to many questions about current links and correlations between curiosity and knowledge. One may want to know, if curiosity causes a need for seeking knowledge (from the biblical times of Eden) and for explorations about events, discovering rules behind them, and devising new things, how is curiosity correlated with person's intelligence and creativity. Fostering curiosity might foster intelligence and vice versa. John Roder and scientists from the Samuel Lunenfeld Research Institute of Mount Sinai Hospital (no author, Medical Research News, 2009) have discovered a molecular link between intelligence and curiosity; some of the molecules acting on the brain regions that control learning and memory also control curiosity. This knowledge may lead to the development of drugs that improve learning. One may also want to know how does curiosity depend on one's knowledge and general education. One may reflect on another side of this question: why in history of mankind curiosity has been punished.

Curiosity may be focused on other people, things, or ideas. One may wonder in what ways curiosity about other people supports psychologists and psychiatrists, historians, writers, biographers, anthropologists, archeologists, geographers, and private investigators. Curiosity about things, processes, and events seems to be a great support, if not a condition, of successful research in biology, cosmology and astrophysics, to name a few. When curiosity singles out a need of searching for patterns and order in nature, one may ask how it actually supports mathematicians, philosophers, astronomers, theorists about science and religion. Curiosity makes one's mind active, so curious people exercise their intellect, ask questions, and search for answers; it makes one observant of new ideas; opens new worlds and brings excitement to one's life; as Albert Einstein wrote, "the important thing is not to stop questioning... Never lose a holy curiosity" (Lifehack, 2012). Curious people are quick to notice and recognize new ideas; they don't miss opportunities and discover new possibilities that attract their attention, cause their eagerness, and provide excitement.

Investigations into curiosity have usually been pertaining young children; not much thought can be found in literature about adult curiosity, as if a condition of being curious couldn't be considered serious enough to be pursued in a scientific way. However, someone who is curious about people, things, or ideas is on a quest of knowledge, on voyage of discovery about the reason and the best use of one's brain, while without intensive brainwork and mental effort many of us become unhappy. It seems true what a Spanish painter and printmaker Francisco Goya (1746-1828) wrote in one of his prints of the *Caprichos* (Goya, 2012), "The Sleep of Reason Produces Monsters." Due to the curious nature of most people, many explore further opportunities for their personal explorations and production. Some artists map numbers into sensory experience. For example, Dunn and Clark (1999) performed the sonification of proteins, Sturm (2005) recorded the ocean buoy data, while Evans (2011) created visualization of sound as an image, as well as sonification of an image into sound, thus creating the visual rendition of a music form as music for the eyes.

IMAGERY

Imagery plays an indirect role in making associations – mental connections between thoughts, feelings, ideas, or sensations related to different kinds of perception through the senses. We can remember (or even imagine) a feeling, emotion, or sensation linked to a person, object, shadow or even illusion of a shadow, sound, or idea. In an article "Seven Types of Visual Ambiguity: On the Merits and Risks of Multiple Interpretations of Collaborative Visualizations" Eppler, Mengis, & Bresciani (iV 2008, 391-396) examined visual ambiguity. Visuals catalyze collaboration (Sawyer, 2007). Ambiguity or the openness to multiple interpretations typifies visualizations such as sketches, diagrams, visual metaphors, etc., sometimes causing misunderstandings but often offering the

Figure 3. Michael Eaton, "Kingdom" (© 1999 M. Eaton. Used with permission)

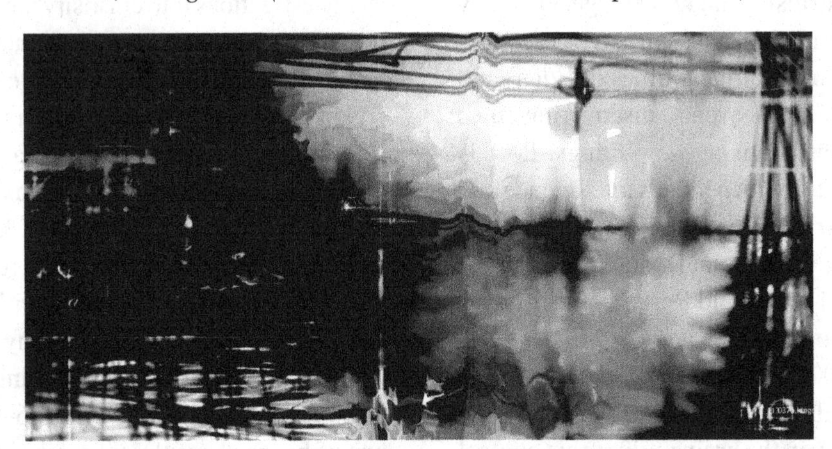

potential to reveal new insights, facilitate ad-hoc discoveries, reframe issues, increase identification, or stimulate group sense making. Thus, an arrow can be used to represent a vector (with position, orientation, and magnitude), a transition, a designator (i.e. pointing to an object), or a casual or temporal relationship. Bresciani, Blackwell, & Eppler (2008) and Bresciani & Eppler (2007) explored how conceptual visualizations (such as diagrams, visual metaphors, charts, sketches) can be constructed and used as cognitive artifacts that support collaborative knowledge work. A research study on this theme was described later (Bresciani &Eppler, 2008).

Visual imagery allows us to vividly imagine scenes in the absence of visual stimulation (Cichy, Heinzle, & Haynes, 2011). Many consider art a potent means of communication with the use of visual imagery; it could be the right moment to look at some images and art objects, for example, at the fantastic interpretation of sleep in the painting of Salvadore Dali (1904-1989) entitled "Sleep" (1904). Unreal imagery of the dream-like world seems depends on a fragile balance of the dreamer. The picture tends to evoke associations related to the unconscious activities of the viewer's mind. It can be seen online, for instance, at http://www.famousartists125.com/salvador-dali.htm. The amazing imagery of Giorgio de Chirico (1888-1978) fills his painting "The Uncertainty of the Poet" (1913) with unexpected scenery, which contains symbols, creates associations and a sense of mystery. This picture can also be seen online, for instance, at http://www.wikipaintings.org/en/giorgio-de-chirico/the-uncertainty-of-the-poet-1913. We may want to look at a color print by a British painter William Blake's (1757-1827) titled "Pity, second impression." Blake's poetic and symbolic paintings show his imaginative, mystical vision of the world (http://www.metmuseum.org/toah/works-of-art/58.603).

Figure 3 presents a computer graphic artwork "Kingdom." A word 'kingdom' may mean disparate, structured system with interdependent components. It may be a dynasty with a king that rules while being only a part of the system; a rank in biological taxonomy such as 'animalia' or 'plantae'; or systems in geometry and other math domains. This artwork shows the fuzzy boundaries existing between a system (the Kingdom) and its surroundings.

This is a picture of an imaginary land. Abstract works are considered by many the hardest ones to create. Having no base for visual cues, the creator of the abstract work still needs to follow the formal rules for composition, observe the elements of design, and apply the principles of design. The author of this artwork was inspired by the concept of combustion – the process of burning a fuel, related to acceleration of a car. He imagined distribution of forces caused by the intense chemical reaction between a fuel and an oxidant (oxygen in air), and presented them as lines of forces going away from the receding, already invisible car. The artist also showed the color-coded heat dispersion caused by the accelerating car. He decided to present his work as an abstract art. Some viewers may choose to see this work as a picture of an imaginary land.

The fantastic images in art works are often derived from artists' mental imagery – a nonverbal, cognitive representation of objects and concepts. The dual-coding process theory of memory proposed by Allan Paivio (1970, 1971, 1986, 1991) proposes that human cognition has separate but interconnected and interacting memory systems: the imaginal and the verbal. Images influence language, values, and ideals. Images and imagery are important in learning, reading comprehension, developing skills, concepts, as well as and in problem solving. Imagery often plays a role in explaining life. Imagery does not mirror perception: imagery and perception are different processes. We see an object, such as a pencil due our perception, and then we recognize it as a pencil because we compare it to a mental image (imagery) of a pencil previously experienced and known for its particular qualities. Thus knowledge is important in imagery.

When we ponder where our dreams come from, we may find partial explanation in the investigations made with the imaging techniques used to study the regions of the brain that are involved with cognition.

Imaging techniques provide pictures of sites where the working memory is in action. For example, it is possible to record how the short-term and the long-term memory are involved in recognizing an image of a familiar face. After a couple of days spent at a large gathering such as a conference one can easily recognize individuals one met in a big group of a thousand participants. Methods for researching imagery are, among others physiological recordings (e.g., cerebral blood flow measured with fMRI – functional magnetic resonance imaging, PET scans that measure cerebral blood flow using positron emission tomography, EEG – electroencephalography) and clinical neuropsychology (e.g., of the split-brain patients).

Research made by Stephen Kosslyn et al. (1993) showed that visual mental imagery activates topographically organized visual cortex. Using a combination of functional magnetic resonance imaging (fMRI) and multivariate pattern classification, Cichy, Heinzle, & Heines (2011) found that perception and visual imagery share cortical representations: fMRI response patterns for different categories of imagined objects can be used to predict the fMRI response patters for seen objects.

Mental images and perceived stimuli rely on the same type of representation and so are represented similarly (Borst & Kosslyn, 2008). According to Kosslyn (1991), mental imagery is a collection of functions occurring in short-term memory where we can redraw images from long-term memory. We can develop the 3D model in a long-term memory store. Visual imagery serves for generation, inspection, recoding, maintenance, and transformation of images. While generating images, one can "mentally draw" in one's imagery from a long-term memory, or produce images or patterns never actually seen. Interpreting the images allows 'zooming in' on isolated parts of them, or scanning across them. Preserving images goes by encoding the patterns of images into memory, remembering new combinations of patterns, or imaging new patterns. Maintenance of images requires effort to remember them, the more perceptual units that are included in an image, the more difficult it is to maintain. Transformation of images involves reshaping an image previously stored in long-term memory: rotating, enlarging, or shrinking them at will. This work goes in a visual buffer or a working memory – a mental space for manipulating information about

Table 2.

Visual Response
Maybe you may want to close your eyes and imagine a place where you liked to play when you were about ten year old (indoor or outdoor place). Then, scan to portions of this place that initially were off your mental screen. Draw what you have just visually recalled as a series of sketches, like a short storyline for animation.

objects, scanning and inspecting visual images. Visual buffer has limited resolution and it fades if not refreshed. The attention window selects a region within the visual buffer for detailed further processing. The size of the window in the visual buffer can be altered and it can be shifted. People can scan visual mental images even when their eyes are closed (Table 2).

IMAGINATION

We'll focus now on visual imagination; that means, our ability to create imaginary content and our capacity to use such components as mental shortcuts for understanding various concepts. Visual imagination refers to our ability to create imaginary content using signs, icons, symbols, metaphors, and analogies. We may say imagination is the ability of the mind to combine experiences, knowledge, ideas, and concepts into one's own, insightful representation and visual interpretation that supports creative activities. Imagery – the image-making function of the mind – is important for our perception: it makes that sensations (sounds, shapes, colors or motions) convey meaning. For example, the rustle of leafs means danger for a small animal. Thus, images and sounds become messages, and, in semiotic terms, a distinction is made between a signal and its referent. By thinking about our imagery, we can see relations of signals, symbols, and signs to their referents. Many times imagination relates to earlier sensory experiences. When confined to a common space and unable to satisfy their predilections, soldiers, sailors, and prisoners often invoke the taste of their favorite food. Imagination may enhance perception and communication through the senses. Bodily stimulation may also enhance imagination related to senses.

Imagination has an 'image' word inside of it; however, there are many ways we may use imagination without images, for example imagination of musicians does not need to be visual. Moreover, both the physical presence of objects and perception of material objects are not necessary to imagine them vividly. Imagination may relate to human capacities resulting from different kinds of perception through the senses: sight, hearing, taste, smell, touch, balance, kinesthesia that gives us a sense of motion, a sense of acceleration, proprioception that allows sensing the relative position of parts of the body, feeling direction, temperature, sensitivity to pain, and many other internal senses. We may ponder about self-awareness in animals and their capacity to imagine; it's a common knowledge that many animals can display fear, anger, affection, boredom, and playfulness, great apes learn sign language, dolphins understand iconic language and display sense of humor, some parrots can count, and some elephants paint pictures.

Imagination seems to be necessary to be able to feel empathy, which in turn, may make us more human and joyful. A former White House speechwriter Daniel Pink (2006, p. 159) described empathy as "the ability to imagine yourself in someone else's position and to intuit what that person is feeling." ...But Empathy isn't sympathy – that is, feeling bad *for* someone else. It is feeling *with* someone else, sensing what it would be like to be that person. Rats can show empathy, aptly unlocking a cage and liberating an incarcerated fellow rat that is held captive, even when lured to another place with a cache of chocolate cookies" (Sanders, 2011).

Sensory experiences stir and inspire our creative powers. However, it is sometimes not easy to see whether the imagined is a clue to reality; for example, if one can imagine flying, does it mean one could manufacture wings? Emancipation of the mind from the constraints of actuality may release freedom to create something: one can imagine what might be and then decide it ought to be, and thus conceive an idea. Basic instincts, sensations, emotions, and intuitions may combine and complement to support creativity. For art and poetry, artistic creativity relates to our imagery.

Imagination and creativity are needed in every professional or academic specialization and discipline. We live in more and more visual world because of the ways we learn (online interactive visuals, videos) and communicate (social network with exchange of videos and pictures, vimeo, Skype, Facebook, YouTube). Many agree that in order to become prepared for the changes in lifestyle and working habits, we need to expand our visual literacy and visual imagination, as they support our creativity, problem-solving, and problem-finding abilities.

It seems even more difficult to tell where is a place for talent in working on a formula or a code, or what makes one torus or knot looking dramatic and beautiful. How does a visualization of a formula become a way to convey an intense message? Why do some hand drawings with hardly any lines, make a gripping statement while other look arrogant or cynical? Mental shortcuts, synthetic signs, humor, caricature, or grotesque make a message even sharper. A realistically drawn dog has nothing to say to the viewer unless it becomes a character in a story where imaginary plot shows what is invisible in nature. Insightful metaphors address cognitive abilities to abstract the essence of the message.

The Role of Imagination in Developing Visualization and Computing

Imagination skills are important when one works on developing a computer-based data visualization, information visualization, or knowledge visualization. With imaginative thinking one can discover visual metaphors for abstract data, information, or concepts, and develop various kinds of visualization, for example, tag-cloud visualization of data. In computer science, imaginative approach to natural events and forces resulted in the development of many fields of research such as fractal geometry of nature, biology-inspired computing, and artificial life. A thorough study of nature results in creating metaphors for developing new computing methods, for example, artificial neuronal networks, evolutionary algorithms, swarm intelligence, and also genetic engineering techniques and bio-inspired hardware systems. In the arts, evolutionary computing resulted in creating generative art (Boden & Edmonds, 2009) and the developments in biology-inspired design, art, music, architecture, and other artistic fields. Analyses of the images, forms, and motions in interactive evolutionary design and art lead to new approaches in defining aesthetic criteria, not only in terms of the work beauty but also its effectiveness and usability.

Bio-interfaces allow participation and audience interaction with art works, design projects, and games. In contrast with previous forms of interaction – through the use of mice, keyboards, joysticks or touch screens, and participation by adding voice, changing position, face expression, or gesture –

bio-interfaces connect participants' physiological and brain activities with the therapeutic or entertainment systems, so the participants feel and act as co-authors.

Figure 4 presents an artistic response of a student from the Computer Art class to a lecture on astrophysics, specifically on a scientific theory of Big Bang, delivered by a physicist. The author of this project presented this abstract concept as an abstract artwork because nobody would know exactly how the Big Bang related events might look. He presented the lines of strong forces acting along the oblique lines, and the centers

Figure 4. Sam Dailey, "Big Bang" (© 2005, S. Dailey. Used with permission)

of the relatively denser matter that seed the new galaxies. He applied warm colors to convey the extreme temperature of the early universe. He has also signified the depth of the space with the exquisite composition of his artwork.

The Role of Imagination in Biology-Inspired Research

We strive to strengthen our imagination by understanding nature better, because biology-inspired research, computing, and engineering are the important directions of progress in science. The great part of new developments goes on the intersection of the biological sciences and the physical sciences and engineering. Imagination is essential for creating new materials and systems. Scientists are addressing problems lying at the intersection of biological sciences, physical sciences, and engineering. The mysteries of the biological world are being addressed using tools and techniques developed in the physical sciences. Collaborative research at the intersection of biology, chemistry, and physics centers on physical and chemical concepts such as dynamics of systems equilibrium, multi-stability, and stochastic behavior to tackle issues associated with living systems such as adaptation, feedback, and emergent behavior.

Committee on Forefront of Science and the interface of Physical and Life Sciences, one of the links in the National Research Council, published results of basic research and clinical practice aimed at advancing science (2010) and advised researching five areas of potentially transformative research:

1. Natural substances display remarkable architecture; identifying structures, capabilities, and processes that form the basis for living systems, and then to use that insight to construct systems with some of the characteristics of life that are capable, for example, of synthesizing materials or carrying out functions as yet unseen in natural biology?
2. The human brain may be nature's most complex system; building on the understanding how it works to predict brain function will require drawing on the resources of the physical sciences, from imaging techniques to modeling capabilities.
3. Genes and the environment interact to produce living organisms; from that understanding realize the promise of personalized medicine and access to better health care?
4. Earth interacts with its climate and the biosphere; how these mechanisms interplay and use that understanding to develop strategies that will preserve this heritage?
5. Living systems display remarkable diversity; how can we prosper while sustaining the diversity that allows life to flourish?

According to the authors of the Committee on Forefronts of Science report (2010), we often think of environmental challenges as being biological ("Save the whales!") or physical ("Limit greenhouse gases!") but this distinction between the disciplines is a distortion. The constant interplay between the biological sciences and the physical sciences is profound when Earth is viewed as an entire system (For this reason, it is important that we learn about the senses of animals, and translate this knowledge into ideas to feed the imagination in creating new, biology-inspired solutions. Some of animal senses that are different than human senses, for example echolocation inspired many technical solutions. Animal senses analogous to human ones but acting differently, such as a sense of smell, balance, vision, or hearing (e.g., ultrasounds) inspired scientists to develop devices that act on a similar basis. Yet many look at the same issues in the political, social, or cultural terms.

The more we investigate natural processes the more questions we have, many of them creating potential for biology-inspired research, computing, and engineering. In many instances we do not know for sure how birds, trout, earthworms, and dogs can navigate and use cues for finding a direction towards sometimes 200 miles distant home (humans rarely can do it without a GPS application or a cell phone). According to researchers from the National Park Service, animals can use mental maps, starlings orient themselves using the sun, Mallard ducks can find north using the stars, migratory birds, salamanders, salmon, or hamsters use the geomagnetic field for orientation, loggerhead turtles' hatchlings sense the direction and strength of Earth's magnetic field, while the subterranean Zambian mole rats have nerve cells that enable them to process magnetic information (No author, Migration basics, 2012). We are able to explain some other questions, for example, why ducks, seagulls, and other water birds can spend winter nights on ice in sub-zero temperature, and also we are able to design devices based on a similar principle of a counter-current heat exchange that distribute heat economically in machines cooled by water. According to a naturalist Tom Pelletier (2011), "It's all about heat exchange, and the smaller the temperature difference between two objects, the more slowly heat will be exchanged. Ducks, as well as many other birds, have a counter-current heat exchange system between the arteries and veins in their legs. Warm arterial blood flowing to the feet passes close to cold venous blood returning from the feet. The arterial blood warms up the venous blood, dropping in temperature as it does so. This means that the blood that flows through the feet is relatively cool. This keeps the feet supplied with just enough blood to provide tissues with food and oxygen, and just warm enough to avoid frostbite. But by limiting the temperature difference between the feet and the ice, heat loss is greatly reduced."

INITIATIVES TAKEN TO ADVANCE DIGITAL CREATIVITY

We can notice several actions responding to the emerging questions and needs, such as establishing new journals, opening new academic degree programs, and developing new software applications for art creating and art education. These initiatives include, among other initiatives, establishing an International Journal of Creative Interfaces and Computer Graphics IJCICG (an official publication of the Information Resources Management Association), IGI Global, Digital Creativity journal that has been published for 21 years, and Leonardo, Journal of the International Society for the Arts, Sciences, and Technology, MIT Press. A few books such as "Visual Complexity: Mapping Patterns of Information" by Manuel Lima (2011) and "Digital Creativity: A Reader" by Beardon & Malmborg (2010) analyzed digital media with this respect.

New, often interdisciplinary programs are being established at several universities, such as dual majors in Computer Science and Information Science, Creative Applications programs, or dual majors programs of CS with the arts that offer Game Design, Interactive Media, Music Technology, and Digital Art. Teaching methods apply the theory and practice of art to computing and problem solving and encourage students to express equations as pictures or stories. Aesthetic computing is a curricula-blending approach that applies the theory and practice of art to computing and problem solving (Shreve, 2010).

Responses to the current demands comprise also conferences, such as the 2005 Art+Math=X – a Special Year in Mathematics and Art Conference at the Department of Mathematics, CU Boulder, 2005, international annual conferences of Bridges: Mathematical Connections in Art, Music, and Science created in 1998 and ongoing, or Joint Mathematical Association annual meetings with Mathematical Art Shows and performances.

TECHNOLOGICAL OFFERINGS SERVING A CREATIVE PROCESS

Several factors cause that an artist needs now a substantial knowledge of computer technology, business, and other areas. Galleries, festivals, and shows are in many instances oriented to the media arts. Colleges are more often set out as the Art and Media rather than Art and Design Colleges. They prepare students to meet the job market expectations and enhance their career opportunities. To a certain extent, our education is market-driven. While there are job positions for visual artists and graphic designers, yet better opportunities are for interactive media developers who don't hesitate to explore current possibilities in computing.

Many believe that mutual inspiration between art and other disciplines can enhance creativity. Many kinds of software are designed to provide options to modify the code. Users can present a story graphically and create an action in an interactive and often interdisciplinary format. For example, web design software with the options: design, code, or split allows the user to design a website using iconic representations of each function; code it by writing a program in a computer language; or split, where the use of the icons makes a computer display commands. These actions may be supportive in learning programming. Some programmers often offer Java applets and free-to-download small packages, for example behaviors or actions built into the program.

Other programs for creating an interactive content allow the user to present a storytelling. A mode format makes possible to introduce diverse literary and graphic forms, such as a story, a poem, a manga, and an anime. Games represent another request from the market and the gaming industry, with a broad spectrum of the genre and media specifics. An Internet access and the presence of portable devices has additionally broadened the spectrum of available technologies, and also influenced the scope of the applied languages. Feasibility of interacting with a player from a distant, often unknown location, and the existence of the Internet-based forums, supporting groups, and social networks create possibilities for the new methods of sharing interests and information exchange. For example, with Livemocha (2012), in the online community that is learning a foreign language may find someone speaking a desired language who is eager to learn your language, so you can share interests and discuss ideas in both languages; both people would benefit from one another and yet enjoy their conversations.

Figure 5 presents student response to opportunities coming with the advances in technology. Several planes showing biological structures with different magnification and the overall dynamic

Figure 5. Jerod Wilson, "Just Do Something" (© 2001, J. Wilson. Used with permission)

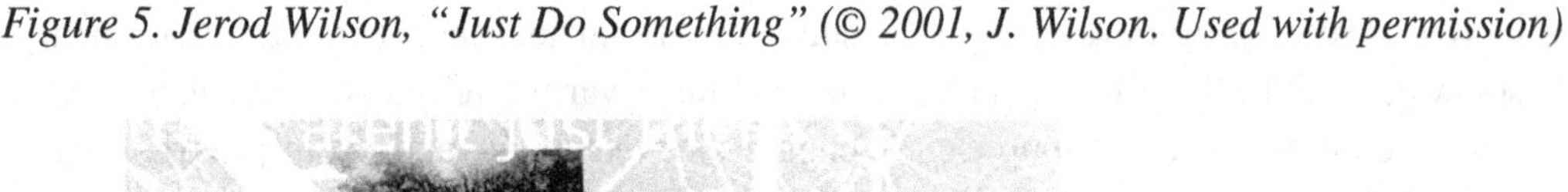

composition of the artwork reflect possible actions in interactive and interdisciplinary format. A biology-related pattern shows up and returns, like a theme or a refrain in a musical composition.

Several institutions such as Colleges of Art or Art Institutes offer courses that teach one kind of software, providing competence about the tool. However, needs for creativity require more: that one can search for what is available, understand constraits and options, jump between applications, combine them not being afraid of introducing legally acceptable changes in usage, acquire a holistic insight into the entire project, all the time having in mind its aesthetics, functionality, and usability. Job opportunities often require that the applicant is creative not only artistically but also technologically. This means this person is open to use pre-made tools in a creative way, not necessarily having a strong knowledge of math, computer science, and sciences. With that perspective, the Education Committee at the ACM/SIGGRAPH International Conferences on Computer Graphics and Interactive Techniques used to invite the highschool students and ask the faculty members to arrange informative tours at the conference and exhibition venues. They developed this mentoring program stressing the need to involve girls and minority students.

Another concept that deserves our attention is an Open Source approach to designing, developing and distributing software, which allows for a peer production of a source code for an open-source software that is available for public collaboration. A user has accessibility to a software source code, and may introduce legally, with relaxed or non-existing copyright restrictions, the small changes in an already existing code, to adapt the program to one's own needs, and thus use, change, or improve the software. Through the Internet, it provides access to various production models, communication paths, and interactive communities. A number of posting services offer opportunities for building a website without knowing software or having expertise in the field, while other applications allow even more sophisticated content management, which otherwise would require technical thinking.

THE ROLE OF IMAGINATION IN LEARNING AND TEACHING

Using one's imagination can enhance the process of learning or creating a project. One can achieve it with two approaches in mind: through applying design strategies into organizing and understanding concepts, or through artistic creation inspired by the material one needs to learn or to apply in a project. One may create an artistic project in order to present intended material in a pictorial way and then use it for a creative product.

Imagination skills are important when one works on developing a computer-based data visualization, information visualization, or knowledge visualization. With imaginative thinking one can discover visual metaphors for abstract data, information, or concepts, and develop various kinds of visualization, for example, tag-cloud visualization of data. Imaginative approach to natural events and forces resulted in the development of many fields of research such as fractal geometry of nature, biology inspired computing, and artificial life. A thorough study of nature results in creating metaphors for developing computing methods such as artificial neuronal networks, evolutionary algorithms, swarm intelligence, and also genetic engineering techniques and bio-inspired hardware systems. In the arts, evolutionary computing resulted in creating generative art and the developments in biology-inspired design, art, music, architecture, and other artistic fields. Interactive evolutionary design and art developed through analyzing natural images, forms, and motions lead to new approaches in defining aesthetic criteria, not only in terms of the beauty but also the effectiveness and usability of an artwork.

Developing imagination seems to be important for teachers and instructional designers because problem-solving and intuitive decision-making abilities hinge to some degree on imaginative thinking. When the teachers and instructional designers work on creating new strategies for learning, they take into account not only the rules and formulas but they also have to envisage how the learners feel these rules out and can figure the best way to imagine, understand, and learn. For example, it is not enough to teach only the game rules when teaching to play chess. Imaginative approach to a task of this type involves developing intuition that allows finding effortlessly optimal thoughts and solutions (even if not justified), for example an opening move in chess, but also envisioning strength distribution on the chess-board, anticipation of the other player's moves, and even endurance. Imagination combined with logic enables us envisioning future moves of the opponent, which is a valuable ability not only in chess playing. Imagination supports the strategies we choose, the options possible, what and how we anticipate or predict, and even our endurance. Imaginative processes often lead to scientific discoveries. Many times we apply predictions, when we need to plan our next steps but information about future is unavailable and we lack the data for inductive reasoning.

Implications for Learning and Teaching

These circumstances create urgent implications for reformatting educational programs at both the K-12 and the college level, such as including programming into the curriculum and introducing supportive programs. Free applications aimed at teaching children programming serve as a learning support, student skills enhancer, and student interest builder. Web-based applications enable students to start comfortably at varying degrees of technological competence. Instruction Technology and Art Education programs in colleges are also changing, stressing the importance of the usability and efficiency in visualizing a user-friendly instructional design process, as well as including aesthetic and artistic elements into interactive, technology-based instruction.

Carol Dweck (2006) pose that chances to reach one's creative potential depend on one's mindset, that means the way one think about one's own intelligence and talent. As claimed by Dweck, individuals can be placed on a continuum according to their implicit views of where ability comes from. Some believe their success is based on innate ability; these are said to have a "fixed" theory of intelligence. Others, who believe their success is based on hard work and learning, are said to have a "growth" or an "incremental" theory of intelligence. Individuals may not necessarily be aware of their own mindset, but their mindset can still be discerned based on their behavior. It is especially evident in their reaction to failure. Fixed-mindset individuals dread failure because it is a negative statement on their basic abilities, while growth mindset individuals don't mind failure as much because they realize their performance can be improved. Dweck argues that the growth mindset will allow a person to live a less stressful and more successful life. Adopting either a fixed or growth attitude toward talent can profoundly affect all aspects of a person's life. Dweck researched the role of two kinds of reward in learning, when she praised half of her students for their intelligence, and a second half for their effort. The 'smart' compliment was devastating in comparison with those praised for hard effort. Research results shown that praising children for their effort (e.g., "good job, you worked very hard") provides better motivation than praising them for their intelligence (for instance, "good job, you're very smart") (Mangels, Butterfield, Lamb, Good, & Dweck, 2006).

Using one's imagination can enhance the process of learning. Moreover, drawings may reveal what are the students' preconceptions. We can support learning with two approaches in mind:

Figure 6. Ben Hobgood, "Playfish" (© 2006, B. Hobgood. Used with permission)

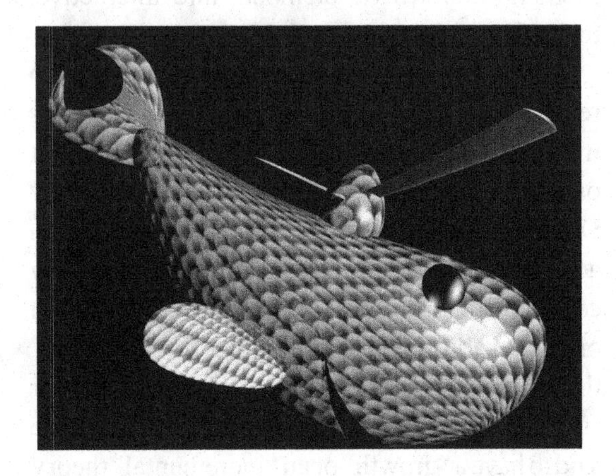

1. Through applying design strategies into organizing and understanding concepts.
2. Through artistic creation inspired by the material student needs to learn. As an instructional designer, you may create artistic projects or design instructional content in order to present various concepts in a pictorial way and then use them for a creative product.

The work "Playfish" created by my student was an imaginative answer to a subject matter defining the possible living or artificial life forms in Cosmos (Figure 6). The artist combined biological and technological features of his creation, which was executed in a 3D animation software.

CONCLUSION

The changing meaning of the creative process is evolving along with the developments in computer science, computer applications, and their interfaces. Nevertheless, creativity is widely considered the ability to generate novel and valuable ideas giving the sense of meaning in our lives. Neuropsychological and psychological background for examining the creative faces of the brain may involve studies on convergent versus divergent thinking and production, critical thinking versus creative thinking, and correspondence between creativity and intelligence. Creativity and imagination is deemed important in developing visualization, computing, and conducting biology-inspired research. Countless options of social networking provide a fuel for any form of an online creative work. All these factors bring about implications for learning and teaching.

REFERENCES

al-Rifaie, M. M., Aber, A., & Bishop, J. M. (2012). Cooperation of nature and physiologically inspired mechanisms in visualisation. In *Biologically-Inspired Computing for the Arts: Scientific Data through Graphics*. Hershey, PA: IGI Global. doi:10.4018/978-1-4666-0942-6.ch003.

Amabile, T. M. (1996). *Creativity in context: Update to the social psychology of creativity*. Boulder, CO: Westview Press.

Ansburg, P. I., & Hill, K. (2002). Creative and analytic thinkers differ in their use of attentional resources. *Personality and Individual Differences*, *34*(7), 1141–1152. doi:10.1016/S0191-8869(02)00104-6.

Beardon, C., & Malmborg, L. (Eds.). (2010). *Digital creativity: A reader (innovations in art and design)*. Routledge.

Berlyne, D. E. (1955). The arousal and satiation of perceptual curiosity in the rat. *Journal of Comparative and Physiological Psychology*, *48*(4), 238–246. doi:10.1037/h0042968 PMID:13252149.

Berlyne, D. E. (1960). *Conflict, arousal, and curiosity*. New York: McGraw-Hill Book Company. doi:10.1037/11164-000.

Berners-Lee, T., & Cailliau, R. (1990). *World-WideWeb: Proposal for a hypertexts project*. Retrieved May 1, 2012, from http://www.w3.org/Proposal.html

Bloom, B. S., Hastings, J. T., & Madaus, G. F. (1971). *Handbook on formative and summative evaluation of student learning*. New York: McGraw-Hill Book Company.

Boden, M. (2007a). Creativity in a nutshell. *Think*, *5*(15), 83–96. doi:10.1017/S147717560000230X.

Boden, M. (2007b). *How creativity works*. Retrieved April 23, 2012 from cii.dmu.ac.uk/resources/maggie/Boden.pdf

Boden, M. A. (1998). Creativity and artificial intelligence. *Artificial Intelligence*, *103*, 347–356. doi:10.1016/S0004-3702(98)00055-1.

Boden, M. A. (2006). *Mind as machine: A history of cognitive science*. Oxford University Press.

Boden, M. A. (2009). Computer models of creativity. *AI Magazine*, *30*(3).

Boden, M. A. (2010). *Creativity and art: Three roads to surprise*. Oxford University Press.

Boden, M. A., & Edmonds, E. A. (2009). What is generative art? *Digital Creativity*, *20*(1-2), 21–46. doi:10.1080/14626260902867915.

Borgia, G. (1995). Complex male display and female choice in the spotted bowerbird: Specialized functions for different bower decorations. *Animal Behaviour*, *49*, 1291–1301. doi:10.1006/anbe.1995.0161.

Borst, G., & Kosslyn, S. M. (2008). Visual mental imagery and visual perception: Structural equivalence revealed by scanning processes. *Memory & Cognition*, *36*(4), 849–862. doi:10.3758/MC.36.4.849 PMID:18604966.

Bresciani, S., Blackwell, A. F., & Eppler, M. J. (2008). A collaborative dimensions framework: Understanding the mediating role of conceptual visualizations in collaborative knowledge work. In *Proceedings of the Hawaii International Conference on System Sciences*. IEEE.

Bresciani, S., & Eppler, M. J. (2007). Usability of diagrams for group knowledge work: Toward an analytic description. In *Proceeding I-KNOW 07*. Graz, Austria: I-KNOW.

Bresciani, S., & Eppler, M. J. (2008). Do visualizations foster experience sharing and retention in groups? Towards an experimental validation. In *Proceedings I-KNOW 08*. Graz, Austria: I-KNOW.

Bridges. (2010). *Mathematical connections in art, music, and science*. Retrieved October 1, 2012 from http://www.bridgesmathart.org/

Broudy, H. S. (1987). *The role of imagery in learning*. Malibu, CA: The Getty Center for Education in the Arts.

Broudy, H. S. (1991). Reflections on a decision. *Journal of Aesthetic Education*, *25*(4), 31–34. doi:10.2307/3332900.

Bruner, J. S. (1962). The connotation of creativity. In *Contemporary Approaches to Creative Thinking*. New York: Atherton Press. doi:10.1037/13117-001.

Chi, R. P., & Snyder, A. W. (2011). Facilitate insight by non-invasive brain stimulation. *PLoS ONE*, *6*(2), e16655. doi:10.1371/journal.pone.0016655 PMID:21311746.

Cichy, R. M., Heinzle, J., & Haynes, J.-D. (2011). Imagery and perception share cortical representations of content and location. *Cerebral Cortex*, *22*(2), 372–380. doi:10.1093/cercor/bhr106 PMID:21666128.

Committee on Forefronts of Science at the Interface of Physical and Life Sciences. National Research. (2010). *Research at the intersection of the physical and life sciences*. Retrieved December 28, 2011, from http://www.nap.edu/catalog.php?record_id=12809

Csikszentmihalyi, M. (1996). Society, culture, and person: A systems view of creativity. In R. J. Sternberg (Ed.), *The nature of creativity: Contemporary psychological perspectives* (pp. 325–339). New York: Cambridge University Press.

Csikszentmihalyi, M. (1996). *Creativity: The work and lives of 91 eminent people*. HarperCollins Publishers.

Csikszentmihalyi, M. (1997). *Creativity: Flow and the psychology of discovery and invention*. Harper Perennial.

Csikszentmihalyi, M. (1998). *Finding flow: The psychology of engagement with everyday life*. Basic Books.

Csikszentmihalyi, M. (2011). *The creative personality*. Retrieved October 1, 2012, from http://www.psychologytoday.com/articles/199607/the-creative-personality

Curiosity.com. (2012). *Discovery*. Retrieved January 12, 2012, from http://curiosity.discovery.com/search?query=curiosity

Dobbs, S. M. (1992). *The DBAE handbook*. Los Angeles, CA: Getty Center for Education in the Arts.

Dorin, A., & Korb, K. (2012). Creativity refined. In *Computers and Creativity*. Berlin: Springer. doi:10.1007/978-3-642-31727-9_13.

Duch, W. (2007). Intuition, insight, imagination and creativity. *IEEE Computational Intelligence Magazine*, *2*(3), 40–52. doi:10.1109/MCI.2007.385365.

Dunbar, K. (1997). How scientists think: Online creativity and conceptual change in science. In T. B. Ward, S. M. Smith, & S. Vaid (Eds.), *Conceptual structures and processes: Emergence, discovery and change*. APA Press.

Dunn, J., & Clark, M. (1999). Life music: The sonification of proteins. *Leonardo*, *32*(1), 25–32. doi:10.1162/002409499552966.

Dweck, C. S. (2006). *Mindset: The new psychology of success*. New York: Random House.

Eppler, M. J., Mengis, J., & Bresciani, S. (2008). Seven types of visual ambiguity: On the merits and risks of multiple interpretations of collaborative visualizations. In *Proceedings of iV, 12th International Conference on Information Visualisation*. IEEE.

Evans, B. (2011). Materials of the data map. *International Journal of Creative Interfaces and Computer Graphics*, *2*(1), 14–26. doi:10.4018/jcicg.2011010102.

Ferri, F. (2007). *Visual languages for interactive computing: Definitions and formalizations*. Hershey, PA: IGI Global. doi:10.4018/978-1-59904-534-4.

Fleur de coin, your online guide to coin collecting. (2012). Retrieved April 27, 2012, from http://www.fleur-de-coin.com/articles/oldestcoin.asp

Galanter, P. (2010). Complexity, neuroaesthetics, and computational aesthetic evaluation. In *Proceedings of 13th Generative Art Conference GA2010*, (pp. 399-409). GA. Retrieved April 27, 2012, from http://www.generativeart.com/on/cic/GA2010/2010_31.pdf

Galanter, P. (2011). Computational aesthetic evaluation: Past and future. In *Computers and Creativity*. Berlin: Springer.

Gardner, H. (1984). *Art, mind, and brain: A cognitive approach to creativity*. Basic Books.

Gardner, H. (1994). *Creating minds: An anatomy of creativity seen through the lives of Freud, Einstein, Picasso, Stravinsky, Eliot, Graham, and Gandhi*. New York: Basic Books.

Gardner, H. (1997). *Extraordinary minds: Portraits of exceptional individuals and an examination of our extraordinariness*. Basic Books, Harper Collins Publishers.

Gervás, P. (2009). Computational approaches to storytelling and creativity. *AI Magazine*, *30*(3), 49–62.

Getzels, J. W., & Csikszentmihalyi, M. (1970). Concern for discovery: An attitudinal component of creative production. *Journal of Personality*, *38*, 91–105. doi:10.1111/j.1467-6494.1970.tb00639.x PMID:5435830.

Getzels, J. W., & Csikszentmihalyi, M. (1976). *The creative vision: A longitudinal study of problem finding in art*. New York: Wiley.

Getzels, J. W., & Jackson, P. W. (1962). *Creativity and intelligence, explorations with gifted students*. London: Wiley. doi:10.2307/40223437.

Giaccardi, E., & Fischer, G. (2008). Creativity and evolution: A metadesign perspective. *Digital Creativity*, *19*(1), 19–32. doi:10.1080/14626260701847456.

Gloor, P. A. (2006). *Swarm creativity: Competitive advantage through collaborative innovation networks*. New York, NY: Oxford University Press.

Goldsborough, R. (2010/2004). Article. *The Numismatist*. Retrieved May 2, 2012, from http://rg.ancients.info/lion/article.html

Goya, F. (2012). *Grabados de goya caprichos*. Retrieved from http://arte.laguia2000.com/pintura/neoclasicismo-2/las-series-de-grabados-de-francisco-de-goya

Guilford, J. P. (1950). Creativity. *The American Psychologist*, *5*, 444–454. doi:10.1037/h0063487 PMID:14771441.

Guilford, J. P. (1959). Traits of creativity. In *Creativity and its cultivation*. New York: Harper and Row.

Guilford, J. P. (1967). *The nature of human intelligence*. McGraw-Hill.

Guilford, J. P. (1968). *Intelligence, creativity and their educational implications*. San Diego, CA: Robert Knapp, Publ..

Hammer, E. F. (1984). *Creativity, talent, and personality: An exploratory investigation of the personalities of gifted adolescent artists*. Malabar, FL: R. E. Krieger Publishing Company.

Harty, H., & Beall, D. (1984). Toward development of a children's science curiosity measure. *Journal of Research in Science Teaching*, *21*(4), 425–436. doi:10.1002/tea.3660210410.

Head, B. V. (2012/1886). *Historia numorum: A manual of Greek numismatics*. Retrieved April 27, 2012, from http://www.snible.org/coins/hn/aegina.html

Heinrich, F. (2010). On the belief in avatars: What on earth have the aesthetics of the Byzantine icons to do with the avatar in social technologies? *Digital Creativity, 21*(1), 4-10. Retrieved May 3, 2012, from http://www.tandfonline.com/doi/abs/10.1080/14626261003654236

Hennessey, B. A., & Amabile, T. M. (1987). *Creativity and learning: What research says to the teacher*. Washington, DC: National Educational Association.

Henshilwood, C. S., d'Errico, F., Yates, R., Jacobs, Z., Tribolo, C., & Duller, G. A. et al. (2002). Emergence of modern human behavior: Middle stone age engravings from South Africa. *Science*, *295*(5558), 1278–1280. doi:10.1126/science.1067575 PMID:11786608.

Hoffman, D. D. (2000). Visual intelligence: How we create what we see. New York: W. W. Norton & Company. ISBN 0-393-04669-9 0393319679

Hoffman, D. H., Greenberg, P., & Fitzner, D. (Eds.). (1980). *Lifelong learning and the visual arts*. Reston, VA: National Art Education Association.

ICCC. (2012), *International conference on computational creativity*. Retrieved October 1, 2012, from http://computationalcreativity.net/iccc2012/

Kashdan, T. B., Gallagher, M. W., Silvia, P. J., Winterstein, B. P., Breen, W. E., Terhar, D., & Steger, M. F. (2009). The curiosity and exploration inventory-II: Development, factor structure, and psychometrics. *Journal of Research in Personality*, *43*, 987–998. doi:10.1016/j.jrp.2009.04.011 PMID:20160913.

Kaufman, J. C., & Sternberg, R. J. (Eds.). (2010). *The Cambridge handbook of creativity*. Cambridge University Press. doi:10.1017/CBO9780511763205.

Koestler, A. (1964). *The act of creation*. New York, NY: Penguin Books.

Komar and Melamid, Fineman, M., Schmidt, J., & Eggers, D. (2000). *When elephants paint: The quest of two Russian artists to save the elephants of Thailand*. Harper. ISBN-10: 0060953527

Kosslyn, S. M. (1991). A cognitive neuroscience of visual cognition: Further developments. In R. H. Logie, & M. Denis (Eds.), *Mental Images in Human Cognition*. Amsterdam: North Holland, Elsevier Science Publishers B.V. doi:10.1016/S0166-4115(08)60523-3.

Kosslyn, S. M., Alpert, N. M., Thompson, W. L., Maljkovic, V., Weise, S. B., & Chabris, C. F. et al. (1993). Visual mental imagery activates topographically organized visual cortex: PET investigations. *Journal of Cognitive Neuroscience*, *5*(3), 263–287. doi:10.1162/jocn.1993.5.3.263.

Kounios, J., & Beeman, M. (2009). The aha! moment: The cognitive neuroscience of insight. *Current Directions in Psychological Science*, *18*(4), 210–216. doi:10.1111/j.1467-8721.2009.01638.x.

Kuhn, T. (1970). *The structure of scientific revolutions*. Chicago: University of Chicago Press.

Lakoff, G. (1990). The invariance hypothesis: Is abstract reason based on image-schemas? *Cognitive Linguistics*, *1*(1), 39–74. doi:10.1515/cogl.1990.1.1.39.

Lebwohl, B. (2011, July 25). Semir Zeki: Beauty is in the brain of the beholder. *EarthSky*. Retrieved October 19, 2012, from http://earthsky.org/human-world/semir-zeki-beauty-is-in-the-brain-of-the-beholder

Lifehack. (2012). *4 reasons why curiosity is important and how to develop it*. Retrieved January 12, 2012, from http://www.lifehack.org/articles/productivity/4-reasons-why-curiosity-is-important-and-how-to-develop-it.html

Lima, M. (2011). Visual complexity: Mapping patterns of information. New York: Princeton Architectural Press. ISBN 978 1 56898 936 5

Limson, J. (2010). Abstract engravings show modern behavior emerged earlier than previously thought. *Science in Africa*. Retrieved April 21, 2012, from http://www.scienceinafrica.co.za/2002/january/ochre.htm

Lipari, L. (1988). Masterpiece theater: What is discipline based art education and why have so many people learned to distrust it? *Artpaper*, *7*, 14–16.

Livemocha. (2012) Retrieved October 1, 2012, from http://www.livemocha.com/

Lowenfeld, V. (1947). *Creative and mental growth*. London: Collier Macmillan.

Lowenfeld, V., & Brittain, W. L. (1987). *Creative and mental growth*. New York: Macmillan Publishing Company.

Luo, J., Knoblich, G, & Lin, C. (2009). Neural correlates of insight phenomena. *On Thinking, 1*(3), 253-267. doi: 10.1007/978-3-540-68044-4_15

Macdonald, S. (1970). *The history and philosophy of art education*. New York: American Elsevier Publishing Company, Inc..

Mangels, J. A., Butterfield, B., Lamb, J., Good, C. D., & Dweck, C. S. (2006). Why do beliefs about intelligence influence learning success? A social-cognitive-neuroscience model. *Social Cognitive and Affective Neuroscience, 1*, 75–86. doi:10.1093/scan/nsl013 PMID:17392928.

Maslow, A. (1943). A theory of human motivation. *Psychological Review, 50*, 370–396. doi:10.1037/h0054346.

Maslow, A. H. (1950). Self-actualizing people: a study of psychological health. *Personality Symposia, 1.*

Maslow, A. H. (1954/1987). *Motivation and personality*. New York: Harper.

Maslow, A. H. (1968/1998). *Towards a psychology of being* (3rd ed.). New York: Wiley.

Medical Research News. (2009). Molecular link between intelligence and curiosity discovered. *Medical Research News*. Retrieved January 11, 2012, from http://www.news-medical.net/news/20090917/Molecular-link-between-intelligence-and-curiosity-discovered.aspx

Miller, P. (2010). *The smart swarm: How understanding flocks, schools, and colonies can make us better at communicating, decision making, and getting things done*. Avery.

NASA. (2012). *Mars science laboratory*. Retrieved January 11, 2012, from http://www.nasa.gov/mission_pages/msl/index.html

Paivio, A. (1970). On the functional significance of imagery. *Psychological Bulletin, 73*, 385–392. doi:10.1037/h0029180.

Paivio, A. (1971). *Imagery and verbal processes*. New York: Holt, Rinehart, and Winston.

Paivio, A. (1986). *Mental representations: A dual coding approach*. Oxford, UK: Oxford University Press.

Paivio, A. (1991). Dual coding theory: Retrospect and current status. *Canadian Journal of Psychology, 45*(3), 255–287. doi:10.1037/h0084295.

ParkWise. (2012). Migration basics. *ParkWise*. Retrieved January 11, 2012, from http://www.nps.gov/akso/parkwise/students/referencelibrary/general/migrationbasics.htm

Pelletier, T. (2011). *Ask a naturalist*. Retrieved January 8, 2012, from http://askanaturalist.com/why-don't-ducks'-feet-freeze/

Pink, D. (2006). *A whole new mind: Why right-brainers will rule the future*. Riverhead Trade.

Rasch, M. (1988). Computer-based instructional strategies to improve creativity. *Computers in Human Behavior, 4*, 23–28. doi:10.1016/0747-5632(88)90029-5.

Ritchie, G. (2007). Some empirical criteria for attributing creativity to a computer program. *Minds and Machines, 17*(1), 67–99. doi:10.1007/s11023-007-9066-2.

Rogers, K. G. (1985). The museum and gifted child. *Roeper Review, 7*(4), 239–241. doi:10.1080/02783198509552906.

Runco, M. A. (1986). Divergent thinking and creative performance in gifted and nongifted children. *Educational and Psychological Measurement, 46*, 375–384. doi:10.1177/001316448604600211.

Sanders, L. (2011, December). He's not a rat, he's my brother: Rodents exhibit empathy by setting trapped friends free. *Science News*, 16. doi:10.1002/scin.5591801414.

Sandkühler, S., & Bhattacharya, J. (2008). Deconstructing insight: EEG correlates of insightful problem solving. *PLoS ONE*, *3*(1), e1459. doi:10.1371/journal.pone.0001459 PMID:18213368.

Sawyer, R. K. (2007). *Group genius: The creative power of collaboration*. Basic Books.

Sawyer, R. K. (2012). *Explaining creativity: The science of human innovation*. Oxford, UK: Oxford University Press.

Schwartz, L. F. (1985). The computer and creativity. *Transactions of the American Philosophical Society*, *75*, 30–49. doi:10.2307/20486639.

Schwartz, L. F., & Schwartz, L. R. (1992). *The computer artist's handbook: Concepts, techniques, and applications*. New York: W.W. Norton & Company.

Shreve, J. (2010). Drawing art into the equation: Aesthetic computing gives math a clarifying visual dimension. *Edutopia*. Retrieved April 21, 2012, from http://www.edutopia.org/drawing art-equation

Sternberg, R. J. (Ed.). (1998/2011). *The nature of creativity: Contemporary psychological perspectives*. Cambridge University Press.

Sternberg, R. J. (2006). The nature of creativity. *Creativity Research Journal*, *18*(1), 87–98. doi:10.1207/s15326934crj1801_10.

Sternberg, R. J. (2007). *Wisdom, intelligence, and creativity synthesized*. New York: Cambridge University Press.

Sternberg, R. J. (2008). *Cognitive psychology*. Wadsworth Publishing.

Sternberg, R. J., & Lubart, T. I. (1995). An investment perspective on creative insight. In R. J. Sternberg, & J. E. Davidson (Eds.), *The nature of insight* (pp. 386–426). Cambridge, MA: MIT Press.

Sternberg, R. J., & Lubart, T. I. (1999). The concept of creativity. In R. J. Sternberg (Ed.), *Handbook of creativity* (pp. 3–15). New York: Cambridge University Press.

Sternberg, R. J., & Lubart, T. I. (1999). The concept of creativity: Prospects and paradigms. In R. J. Sternberg (Ed.), *Handbook of creativity* (pp. 137–152). Cambridge, UK: Cambridge University Press.

Sturm, B. L. (2005). Pulse of an ocean: Sonification of ocean buoy data. *Leonardo*, *38*(2), 143–149. doi:10.1162/0024094053722453.

Sullivan, K. (2000). Between analogue and digital. *Computer Graphics*, *34*(3), 5.

Tatarkiewicz, W. (1999). *History of aesthetics*. Thoemmes Press.

Taylor, C. W. (1959). The identification of creative scientific talent. *The American Psychologist*, *14*, 100 102. doi:10.1037/h0046057.

Tennyson, R. D., Thurlow, R., & Breuer, K. (1987). Problem-oriented simulations to improve higher-level thinking strategies. *Computers in Human Behavior*, *3*, 151–165. doi:10.1016/0747-5632(87)90020-3.

Torrance, E. P. (1962). *Guiding creative talent*. Englewood Cliffs, NJ: Prentice-Hall, Inc. doi:10.1037/13134-000.

Torrance, E. P. (1974). *Torrance tests of creative thinking: Norms and technical manual*. Lexington, MA: Personnel Press.

Torrance, E. P. (1990). *Torrance test of creative thinking*. Benseville, IL: Scholastic Testing Service, Inc..

Walberg, H. J. (1969). A portrait of the artist and scientist as young man. *Exceptional Children*, *36*(1), 5–11.

Wallach, M. A., & Kogan, N. (1965). *Modes of thinking in young children*. New York: Holt, Rinehart, & Winston.

Wands, B. (2001). *Digital creativity: Techniques for digital media and the internet*. New York: John Wiley & Sons, Inc.

Wiggins, G. (2006). Searching for computational creativity: New generation computing, computational paradigms and computational intelligence. *Computational Creativity*, *24*(3), 209–222.

Wong, K. (2002, January 11). Ancient engravings push back origin of abstract thought. *Scientific American*. Retrieved April 21, 2012, from http://www.scientificamerican.com/article.cfm?id=ancient-engravings-push-b

Young, J. G. (1985). What is creativity? *The Journal of Creative Behavior*, *19*(2), 77–87. doi:10.1002/j.2162-6057.1985.tb00640.x.

Zeki, S. (1993). Vision of the brain. *Wiley-Blackwell., ISBN-10*, 0632030542.

Zeki, S. (1999). Art and the brain. *Journal of Conscious Studies: Controversies in Science and the Humanities*, *6*(6/7), 76–96.

Zeki, S. (2001, July 6). Artistic creativity and the brain. *Science*, *293*(5527), 51–52. doi:10.1126/science.1062331 PMID:11441167.

Zeki, S. (2009). *Splendors and miseries of the brain: Love, creativity, and the quest for human happiness*. Wiley-Blackwell. ISBN 1405185570

Section 2
Visual Cognition

Chapter 5
Cognitive Processes Involved in Visual Thought

ABSTRACT

Cognitive thinking is discussed here in terms of processes involved in visual thought and visual problem solving. This chapter recapitulates basic information about human cognition, cognitive structures, and perceptual learning in relation to visual thought. It tells about some ideas in cognitive science, cognitive functions in specific parts of the brain, reviews ideas about thinking visually and verbally, critical versus creative thinking, components of creative performance, mental imagery, visual reasoning, and mental images. Imagery and memory, visual intelligence, visual intelligence tests, and multiple intelligences theory make further parts of the chapter. This is followed by some comments on cognitive development, higher order thinking skills, visual development of a child, the meaning of student art in the course of visual development, and the role of computer graphics in visual development.

INTRODUCTION

In a book of an Italian writer, semiotician, and philosopher Umberto Eco (1990) a human person is having a conversation with a robot named Charles Sanders Personal. This person says to a robot,

To think means to have internal interpretations corresponding to the expressions you receive or produce. You have told me a lot about your memory. Well, your memory is inside you. You process the sentences you receive according to your internal encyclopedias. The format of these encyclopedias is inside you (Eco, 1990, p. 281).

And the robot responds,

I do not know whether my memory is the same as that of my masters. According to my information, they are very uncertain about what they have inside

DOI: 10.4018/978-1-4666-4703-9.ch005

them (as a matter of fact, they are not even sure that they have an Inside). That is the reason why they set me up. They know what I have inside me, and, when I speak in a way they understand, they presume that they have the same software inside them. (Eco, 1990, p. 281)

This chapter links cognitive processes with actions involved in visual thought and visual problem solving. They will be applied in interactive projects offered in this book. The content of this chapter provides basic information about human cognition and cognitive science, which relates to visual reasoning, aesthetic emotions, and art. Translation of scientific concepts to the realm of visual interpretations necessitates activation of processes involving cognitive structures, and perceptual thought. Themes discussed here will return in the following chapters tying reasoning about science and computing with mental stimulation to perceptual thinking. Scientists are developing cognitive computing theories and working on constructing cognitive computers that perceive, conclude, and learn. For this reason they need to study the cognitive potential of the brain. Gaining knowledge about the cognitive ability of the human brain and intelligence fosters investigations on information-processing mechanisms in computing and supports cognitive informatics.

BASIC INFORMATION ABOUT HUMAN COGNITION AND COGNITIVE SCIENCE

It is a common knowledge that investigations into visual intelligence, visual thinking, and actions like visual reasoning, problem solving, and decision-making all belong to a domain of cognitive science. Scientists explore how intelligence is implemented in animals and humans, along with applying essential features of intelligence to computing in order to develop artificial intelligence. Cognitive science studies intelligence as the ability to perform intellectual tasks by humans, intelligent organisms, or intelligent programs. It examines how people perceive, represent, and communicate information, both visually and verbally. Cognitive science evolved from the study of intelligence that was first based of a study of animal and human behavior developed mostly by the physiologist Ivan Petrovich Pavlow (1849-1936; awarded the Nobel prize in 1904) and the psychologist Burrhus F. Skinner (1904-1990). The launching of the cognitive movement is credited to the linguist Noam Chomsky (1967). In contrast with the Skinnerian behaviorist principles of associations, Chomsky built his theories on the concept of language and its complex internal representations encoded in the genome, which cannot be broken down into a set of associations. According to Chomsky, language faculty is a part of the organism's genetic endowment in the same way as other physiological systems. Then, the computational neuroscience analyzed complex biological systems such as brain or visual system. David Marr (1982/2010) considered them as an information processing systems. He described the modular organization of the visual processing system at three levels: the computational level that defines what the system is doing considering sensory information as an input, the algorithmic level describing processes that convert that input into the output, and the implementation level explaining how the information processing is physically realized by the system.

Figure 1 shows a work resulting from readings on human cognition that push us to make telling observations about our own mental and emotional processes and make predictions about future. The city scene repeats the pattern of encounters with individuals and groups, and the recurring division of time between the duty and leisure, social and private.

Lost and grounded in the City,
Sole and single, yet doggedly responding
To rhythmic patterns of urban life,
To threatening and disturbing City matters,
And finding fun in it.

For a cognitive scientist Michael Eisenberg, a notion that the mind can be modeled as a computational system is a leading idea in cognitive science. Indeed, several information-processing models of human cognition visualize relations

Figure 1. Anna Ursyn, "City Matters" (© 2002, A. Ursyn. Used with permission)

between perception, memory, control processes, and human response. This domain derives concepts from various scientific domains, such as neurophysiology, experimental and evolutionary psychology, behavioral neuroscience, psycholinguistics, computer science, artificial and human intelligence, and philosophy. Also, developments in cognitive science support these fields of science. Another question is, how is the Internet changing the way people think. When asked this question by the World Question Center (2010; Brockman, 2011), a cognitive psychologist and neuroscientist Stephen M. Kosslyn argued that "relationships can become so close that other people essentially act as extensions of oneself, much like a wooden leg can serve as an extension of oneself. When another person helps us in such ways, he or she is participating in what I've called a 'Social Prosthetic System'."

Kosslyn stated (Brockman, 2011),

The Internet has extended my memory, perception, and judgment. Regarding memory: Once I look up something on the Internet, I don't need to retain all the details for future use ... the Internet functions as if it is my memory. It's become completely natural to check facts as I write. Regarding perception: Sometimes I feel as if the Internet has granted me clairvoyance: I can see things at a distance ... the world really does feel smaller. Regarding judgment: The Internet has made me smarter ... it helps me to distill the essence of its meaning ... and I then compare and contrast what I think with what others have thought. ... Moreover, I use the Internet for "sanity checks," trying to gauge whether my emotional reactions to an event are reasonable, quickly comparing them to those of others.

Margaret Boden (2010) — a researcher in artificial intelligence, psychology, philosophy, cognitive, and computer science — describes cognitive science as "the interdisciplinary study of mind, informed by theoretical concepts drawn from computer science and control theory" (Boden, 2006, p. 12); a study that "deals with all mental processes. Cognition (language, memory, perception, problem solving, ...) is included, of course; but also motivation, emotion, and social interaction – and the control of motor action, which is largely what cognition has evolved *for*" (Boden, 2006, p. 10).

Fields of enquiry explored in computer science include perception, the use of language, thinking, thought processing and learning, knowledge representation, and modeling of those processes on a computer. Cognitive psychologists draw their results from studies of the activities focused on gaining knowledge, such as problem solving, attention, creativity, memory, and perception. Several centers conduct research on cognitive neuroscience, biopsychology, and cognitive informatics, e.g. the SAGE Center for the Study of the Mind, with results published in a number of scientific journals such as Journal of Cognitive Neuroscience, Cognitive Neuropsychology, and International Journal of Cognitive Informatics and Natural Intelligence. A wide spectrum of research methods that have been used for cognitive studies include naturalistic observation and experimental studies on animal behavior; introspection, clinical interviews and clinical observation; research on neural processes derived from studies of brain injuries, brain lesions, autopsies, brain stimulation, and other experimental studies on patients and animals; psychological and psychiatric studies; imaging brain structures and functioning; creating computational models with several information processing approaches: using simulations, visualizations, data mining, neural networks, cloud computing, biologically and evolutionary inspired computation, among other approaches. Margaret Boden (2006) examines the relation between neurophysiology and computational neuroscience:

If neuroscience looks at the brain (and the rest of the nervous system) and asks, 'What does this bit do?', computational neuroscience asks, 'How does it manage to do it?' And that 'How?' is computational.(Boden, 2006, p. 1111)

COGNITION AND SOME OF THE STRUCTURES OF THE BRAIN

Cerebral hemispheres function alone but with different abilities. The anterior parts of the left and right cerebral hemispheres control specific features of thought, action, and memory. They process information coming from visual fields situated on the other side and control movement of hands and fingers situated on the other side. Human cognition is assumed to consist of separate but interconnected verbal and imaginal systems and the left-right perceptual asymmetries. Verbal, visual, and spatial information is processed separately: the left hemisphere shows a tendency toward being dominant for language and speech, controlling and categorizing information, while the right hemisphere is better at visual motor tasks; it is mostly responsive for motor functions in vision, responses to novel events, and expression of emotional reactions. The separate processing of verbal and visual information confirms that verbal and imaginal symbols are coded independently, with mental imagery as an important form of nonverbal processing. It can be seen when cerebral hemispheres are disconnected in the split-brain operations. Split-brain research has advanced our understanding of functional lateralization in the brain (Gazzaniga, 1997, 2005). Specialization of hemispheres is present in other than human vertebrates such as fish, frogs, reptiles, birds, and mammals; for example, a left hemisphere controls feeding behavior, while attack and escape depend on the right hemisphere (Vallortigara & Rogers, 2005). The authors pose that lateralization of functions enhances cognitive capacity and efficiency of the brain. Moreover, in the parietal region of the brain that generally integrates sensory information, three circuits control organization of the number-related knowledge, quantity processing and calculation (Dehaene, Piazza, Pinel, & Cohen, 2003). Data obtained with the fMRI (functional magnetic resonance imaging that measures brain activity by detecting related changes in blood flow) show that the first structure (horizontal segment of the intraparietal sulcus) is involved in number manipulation independent of number notation; the second one (left angular gyrus area) supports verbal form of number manipulation, while the third circuit (bilateral posterior superior parietal system) supports attention aimed at the mental and spatial number line.

METHODS AND TOOLS FOR STUDYING NEURAL STRUCTURES AND FUNCTIONS

Researchers study brain functions and their links with structures; they also examine links between brain functions in specified regions and cognitive activities. Experimental methods include both animal and human brains, with the use of several approaches including the *postmortem* (after death) examinations detecting brain anomalies or lesions linked to diseases such as Alzheimer's disease (that causes dementia and memory loss), Parkinson's degenerative disease (causing movement disorder), or the memory and speech problems, and *in vivo* (on living organisms) studies focusing on functioning of a living brain. Studies of electrical activity of the brain include recordings of the responses evoked by the electrical or other stimuli, often involving single cells and their event-related potentials. Several angiography (arteriograms and venograms) techniques are aimed at visualizing the blood vessels' interior, often with the use of catheterization and radio-opaque substances. Results are often correlated with the static and functional techniques for imaging the neural structures. With computing and analyzing electromagnetic changes in energy of the subatomic particles contained in the brain molecules researches get metabolic imaging, functional patterns, and information about abnormalities due to a disease or brain damage. Also, computers produce three-dimensional pictures of the brain structures.

Functional neuroimaging techniques show what's inside the brain. They provide pictures of sites where the working memory is in action. For example, it is possible to record how the short-term and the long-term memory are involved in recognizing an image of a familiar face. Brain imaging techniques, including physiological recordings (e.g., cerebral blood flow, electroencephalography) and clinical neuropsychology (e.g., split-brain patients), serve also for researching cognitive imagery. Several techniques make possible mapping mental activity of the brain. Neuroimaging techniques allow structural imaging for diagnosing diseases and injuries of the brain, and functional imaging used to diagnose metabolic diseases and lesions, visualize fine structure of brain tissue, conduct research on neurological, cognitive, and psychological processes, and also develop interfaces between a brain and a computer. As listed in textbooks (e.g., Sternberg, 2011, Sternberg & Kaufman, 2011) and Wikipedia, types of functional neuroimaging include several techniques.

- **Computed Tomography (CT) and Computed Axial Tomography (CAT) Scans or Computed Axial Tomography (CAT) Scanning:** Use x-ray beams according to a computer program; CT scan technique provides three-dimensional representations telling how tissues absorb radioisotope energy.
- **Diffuse Optical Imaging (DOI) or Diffuse Optical Tomography (DOT) and High-Density Diffuse Optical Tomography (HD-DOT):** Are tomography based techniques that use near infrared light and rely on the absorption spectrum of hemoglobin depending on its oxygenation status (Eggebrecht, White, Chen, Zhan, Snyder, Dehlgani, & Culver, 2012).
- **Magnetic Resonance Imaging (MRI):** Uses magnetic fields and radio waves to quickly construct a two- or three-dimensional image of the brain structure on a computer, and its changes over time, without use of x-rays or radioactive tracers.
- **Functional Magnetic Resonance Imaging (fMRI) Technique:** (Logothetis, Pauls, Augath, Trinath, & Oeltermann, 2001) Show images of changing blood flow in the brain. Cognitive activity evoked by various stimuli results in changes in the amount of blood flow in different regions of the brain. Changes in blood flow are associated with perception, thought, and action related to different tasks: it reflects reasoning, the processing of emotions, conflict resolution, making moral judgments, or feeling reward and pleasure after making a proper conclusion.
- **Positron Emission Tomography (PET):** Is the functional brain imaging method that tracks radioactively labeled chemicals (such as glucose with radioactive atoms) in the blood flow. Glucose is metabolized in proportion to the brain activity, so radioactivity is concentrated in the most active areas and provides multicolored 3D images of brain areas in action. It serves for detection of brain tumors and diseases that cause damage to neurons and following dementia.
- **Single Photon Emission Computed Tomography (SPECT):** Uses gamma ray radioisotopes and a gamma camera to construct two- or three-dimensional images of active brain regions. The brain rapidly takes up injected radioactive tracer, reflecting cerebral blood flow at the time of injection. SPECT is used for epilepsy imaging and to differentiate diseases causing dementia.

- **Electroencephalography (EEG):** Is an early functional technique that detects a summary of the electrical activity of all cortex dendrites located under electrodes placed on a scalp; a recorder encephalogram can be compared with a normal activity of cortex.
- **Magnetoencephalography (MEG):** Directly measures the magnetic fields produced by electrical activity in the brain using extremely sensitive superconducting quantum interference devices (SQUIDs). Neural activity measured by MEG is less distorted by surrounding tissue (particularly the skull and scalp) compared to the electric fields measured by EEG. Uses for MEG include assisting surgeons in localizing pathology, researchers of the function of the brain, neuro-feedback, and others.

SOME IDEAS ABOUT COGNITIVE THINKING AND MEMORY

Human cognition and memory have separate but interconnected verbal and imaginal systems. Images, as mental models for thinking, influence language, concepts, and values, and support reading comprehension. Images and mental imagery are important in developing skills, concepts, problem solving, explaining life, and may assist in mentally combining forms and patterns in a visual image and making discoveries. The physical world that exists in our environment builds our mental inner reality and causes the generation and manipulation of images. Our survival in the real world depends to a great extend on the accuracy and completeness with which the mental models used by our mind represent actuality. In psychology and everyday conversation imagery refers to mental images – cognitive processes involving experiences in the mind that may be auditory, visual, tactile, olfactory, gustatory, or kinesthetic. People use imagery when they think about previous or upcoming events; for example, to tell about the feel of a pasture in the spring, one retrieves from memory images and other sensory experiences. Visual reasoning helps to overcome difficulties in comprehension of logical structures and assist in intelligent learning. Harry S. Broudy (1991, p. 33) stated that, "sensory images can often convey meaning directly with a clarity that formal analysis and reasoning cannot rival." Depending on one's right or left brain dominance, one may be inclined to create visually appealing imaginary art, or may prefer to construct a concept map or a visualization of a scientific concept. This approach may help the artist to reprocess the data in another way than it was presented, to look at the world from a different vantage point, and to look in a way that one is more comfortable with.

A Canadian psychologist Allan Paivio proposed that the human mind can transform information into two cognitive codes, to be transferred to long-term memory, and then compared to something previously learned, stored, and organized as a cognitive structure. The two cognitive codes include (1) A visual or iconic code, a visuo-spatial scratchpad for images and (2) An auditory or semantic code with verbal language rules – semantic, syntactic, and orthographic.

Semantics, usual described as a study of meaning, includes several sub-divisions. Lexical semantics concerns with language: the relation of words, phrases, or signs and their denotations, with the sets of symbols known as alphabets, sets of formal rules known as grammar, and transformation rules of inference. Formal semantics concerns with logical systems and structures of meaning in natural and formal languages, logical forms, implications, and references. Conceptual semantics examines formal systems of cognitive mental structures of meaning, explanatory language representations, and conceptual structures of understanding (Jackendoff, 2007/2009).

Paivio (1971, 1986/1990, 1991) put forward a dual-coding theory of cognition. It postulates that performance in memory and other cognitive tasks is mediated not only by linguistic processes but also by a distinct nonverbal imagery model of thought. Human sensory systems receive both verbal and imaginal stimuli and form representational associative structures. Verbal associations and visual imagery can both expand on learned material; visual and verbal information, while processed in different ways, create separate representations of information. Thus, researchers hold that mental imagery supports recollection of verbal material, when words evoke corresponding images.

Cognitive psychologists used to describe human memory as several stores (Baddeley, Thompson, & Buchanan, 1975; Baddeley &Hitch, 1974; Baddeley, Eysenck, & Anderson, 2009): the sensory register that is engaging perceptual mechanisms, the tasks of reasoning, decision-making, and coordinating incoming data; the short-term memory store with an phonological, also called articulatory (related to speech sounds) loop for temporary retention of speech-based material; the visuo-spatial scratch pad for temporary retention of visual and/or spatial material; the episodic buffer (Baddeley, 2000); and the long-term store. The short-term working memory is limited in the amount of information that it can store and the length of time it could store information. This store can simultaneously hold for a limited time (around twenty seconds) only a small number (seven, plus or minus two) of information chunks. Much of what we perceive is never recorded as perceptual experiences in the sensory storage; only a part of experiences absorbed through our senses are transferred to the short-term working memory. Some information, such as a stop sign, is instantly perceived and unconsciously remembered. It stores over few minutes as a temporary scratch pad our mental imagery, perceptions, potent emotions, conscious thoughts, feelings, and perceptions.

The long-term memory system allows for making comparisons. A long-term memory stores over long periods information that is partially transferred from the short-term memory. New links are created between neurons and old links are strengthened in this process. In order to comprehend words, old memory must be used. In the process of the language organization, information flows from visual (reading) and auditory (listening speech) reception to areas in the left temporal lobe for comprehension and then to frontal areas for speech production. The questions have been around, whether or not the long-term storage of information may involve the separate modes of representation, and if there is some common integrative mechanism for concrete and abstract materials.

In effect of studies using computerized techniques, memory has been divided into declarative and procedural forms. Declarative memory includes semantic memory – one' store of primary verbal knowledge, and episodic memory that refers to one's specific meanings and experiences. Procedural form of memory contains one's knowledge of actions, motor functions and abilities. Cognitive scientists explore, simulate, and model the structure of semantic memory, pattern recognition in processing data, attention, problem solving, decision-making, and abstract thinking that is coherent and logical. Traditional schooling had been focused on developing memory skills. At present, a necessity of dealing effectively with large amount of information creates a need to expand abilities of higher order thinking, visualization, and understanding of abstract concepts.

Thought, Language, and Visuals

Imagery and verbal codes present two distinct classes of mental representation, and thus verbal memory and image memory are located in two independent but interacting stores. As a form of nonverbal processing, mental imagery seems to be contained within the posterior lobe of the right cerebral hemisphere. Both these systems are involved in processing verbal material when it is done with the use of imagery, and are presumed to produce two memory codes. In a process of comprehension of abstract concepts in sentences, concrete words are coded in both verbal and visual memory, while it is presumed that the abstract ones are coded in verbal memory. Many authors agree that memory is better for learning concrete materials than for abstract materials and is better with the use of visual imagery. When people compare sizes of objects, pictures work better than verbal comparisons. Mental representations of visual objects may involve unconscious computations of moving objects and conscious knowledge of meaning of the object.

We can visualize our thinking as well as verbalize it. We can think in pictures, not in a linear, sequential fashion that is typical of talking. With images, linear information can be translated into spatial metaphors. Graphic imaging is considered a means to reason about an arrangement of data, to communicate, document, and preserve knowledge. An approach to visual thinking as the interaction of seeing, imaging, and idea sketching was strongly influenced by Rudolf Arnheim, especially by his book "Visual Thinking" (1969/2004). Interrelation of thought and speech can be seen as two interlocking circles, because not all of our thoughts are spoken and sometimes we speak without thinking, just repeating the thoughts of others thoughts (Vygotsky, 1986). Also, we can present visual thinking and graphic display as two interlocking circles; not every graphic (visual) language involves visual thinking and not always visual thinking is in the form of the graphic display; for example, we use symbols without thinking. Humor, caricature, or grotesque make a message even sharper. A realistically drawn dog may be seen as having nothing to say to the viewer unless it becomes a familiar dog or a character in a story line where imaginary and fantastic plot shows what is invisible in nature. Also metaphors address some cognitive abilities to abstract the essence of the message.

Visual thinking, defined sometimes as generation and manipulation of images that come both from imagery and from the abstract systems, involves visual processing beyond the definitions of language. It often cannot be translated into verbal, linear manner. Imagery is the essence of thinking because we generate and manipulate images when we think. We produce graphic images by applying graphic language to visual thinking, attaining synthesis, using perceptual and mental imagery, relying on intuition, and working at various levels of consciousness (such as dreaming) outside the realm of language thinking. As stated by Arnheim (1969/2004), perceptual sensitivity, the ability to see a visual order of shapes as patterned forces that underlie our existence, helps the most gifted minds with intuitive wisdom to avoid troubles with the formalistic thought operations due to their brilliant cross-circuits. According to Denis (1989), there is an opposition between Symbolist theories (thinking occur in mental symbols and representations, as media) and Conceptualist theories (mental symbols are products of cognitive thinking about conceptual and abstract entities). Robert Sternberg (2007) discussed the interrelationships among intelligence, creativity, and wisdom in adapting to, shaping, and selecting environments. Interaction with environment involves aspects relative to creativity such as knowledge, styles of thinking, personality, and motivation, with balance of these personal attributes necessary to do creative work. According to Sternberg,

Wisdom results from the application of successful intelligence and creativity toward the common good through a balancing of intrapersonal, interpersonal, and extrapersonal interests over the short and long terms. Wisdom is not just a way of thinking about things; it is a way of doing things. If people wish to be wise, they have to act wisely, not just think wisely. We all can do this. Whether we do is our choice. (Sternberg, 2007, p. 188)

Andrew Targowski (2011) argues, "wise civilization cannot function without wise people and vice versa, that wise people cannot function without positive conditions for the development of wise civilization." In terms of cognitive informatics, the author offers the following premises (Targowski, 2011, preface):

1. Every mentally healthy individual has some level of wisdom in thinking and making decisions.
2. Wisdom is not knowledge; it is a virtue. However, there is knowledge about wisdom, which is just in *status nascendi*.
3. Wisdom, briefly defined, is *prudent judgment and choice*. Hence, one can perceive a person to be knowledgeable but not necessarily wise, and vice versa.
4. Wisdom is not a synonym or an extension of intelligence. Intelligence is the ability to solve problems while wisdom is the final touch in prudent judging and choosing a good solution among available options.
5. Wisdom can be practical, theoretical, global, and universal.
6. Wisdom can be taught. Left to practice only, it is usually applied too late to impact the right course of action. Wisdom is like a plant, which must be nurtured to grow.
7. Wisdom should be monitored in civilization like strategic resources because it is the most important human resource on Earth.

Figure 2 presents a visual response to some questions about comprehension of abstract concepts. Independent but interacting events in nature may affect our actions; just the same, visual and verbal load coded in memory feeds our curiosity and forces our actions.

Figure 2. Anna Ursyn, "Rondo" (© 2006, A. Ursyn. Used with permission)

Table 1.

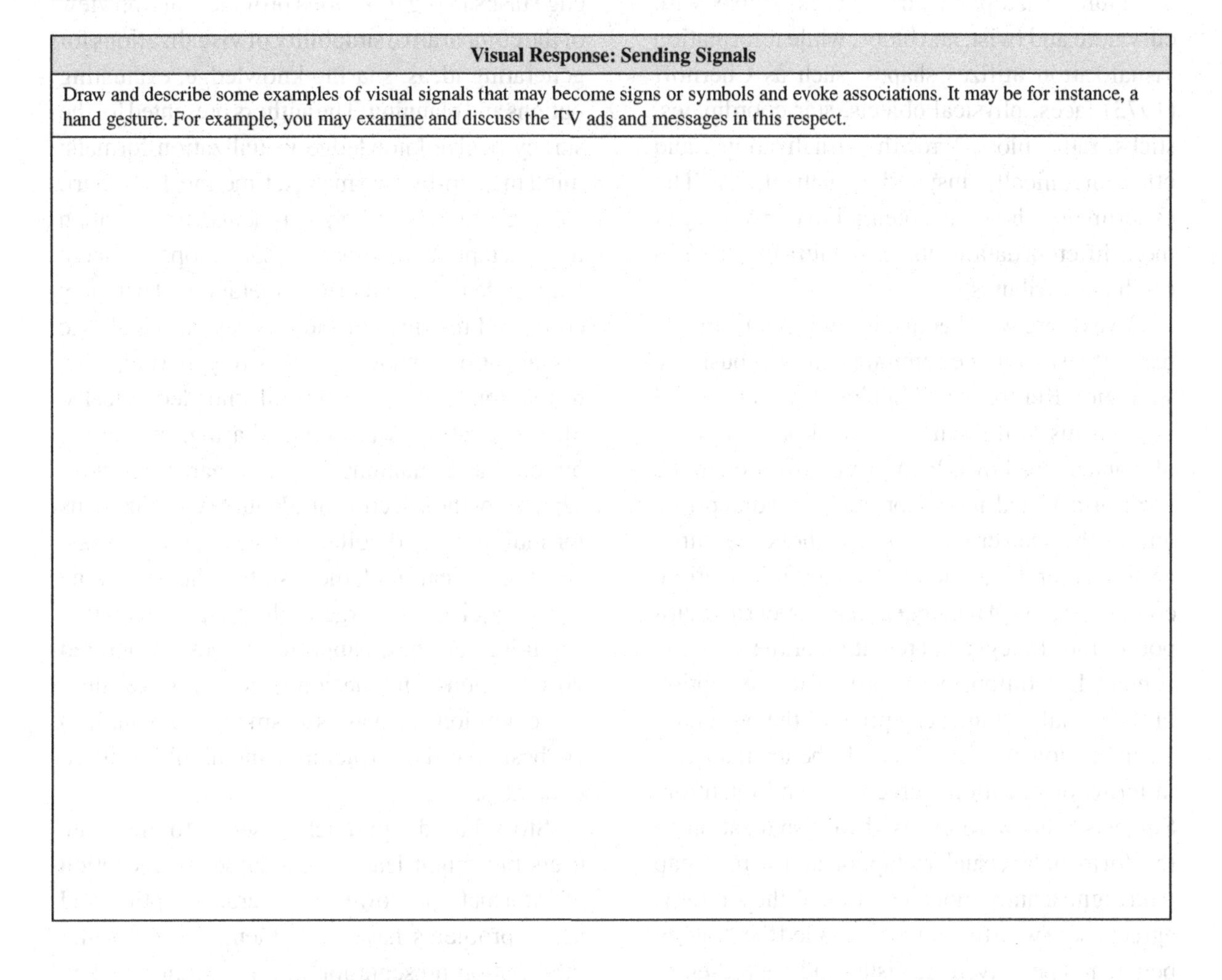

Visual Response: Sending Signals
Draw and describe some examples of visual signals that may become signs or symbols and evoke associations. It may be for instance, a hand gesture. For example, you may examine and discuss the TV ads and messages in this respect.

Shades of blue, coolness of clouds, and direct sunlight
Repeat their recurring themes on the ground
With the power to influence things that just happen there.

See Table 1 for Visual Response.

Examples of Research Studies on the Visual vs. Verbal Presentation

We may utilize various shapes to represent multiple data values and thus create learning environment with the use of information visualization. The shape characteristics include closure, curvature, corner angle and type, edge and end-type, notch, whiskers, holes, intersection, and local warp (Brath, 2009). They can be used separately or together to convey data, as opposed to icons, numbers, common symbols, or compound glyphs. The use of multiple shape attributes increases the expressive range and the information density of visualizations (Brath, 2010). According to Brath, shapes used for data presentation through scientific visualization differ from those used in information visualization. Scientific visualization represents physical phenomena and is therefore restrained to a spatial context, while information visualization often uses shape to represent only a single data attribute. Scientific visualization often uses curvature-based parametric shapes, glyphs with curvature and twist, and blobs, while information visualization utilizes shapes, such as Chernoff (1973) faces, physical objects, star coordinates, sticks, radar plots, 'growth' visualizations, and other organically inspired visualizations. The experiments show the potential to convey ten or more different data attributes within a glyph based on shape attributes.

To explore whether the use of visualization is better than text in the communication of business strategies, Kernbach and Eppler (2010) presented three forms of the same presentation to a group of managers: a PowerPoint, a visualization in the form of a visual metaphor, and a roadmap that guides the viewer's eye. The authors measured awareness and attention to strategy information, comprehension of the strategy, agreement and support of the strategy, and retention of the strategic content. In addition, they measured the perception of the visual and the perception of the presenter. Visualization was significantly better than text in terms of attention, agreement, and retention. Subjects who were exposed to visualization in the form of a visual metaphor and a roadmap paid significantly more attention to the strategy, agreed more with the strategy, recalled the strategy better, and perceived the visual and the presenter significantly better than did subjects who saw a text in the form of PowerPoint.

Bresciani and Eppler (2010) explored the positive impact of employing conceptual visual representation for individual reasoning, communicating and facilitating meetings in organizations. The study was aimed to advance the understanding of the use of visual representations to support managerial cognition and the relationship between the structure of the visual forms and the type of convergent (logical, analytical, deductive, aimed at a single correct answer) or divergent (exploring possible original solutions) knowledge task type. Questionnaire responses about the usefulness of 12 common business visualizations for typical knowledge tasks in organizations provided an overview of the comparative suitability of visualizations for generating ideas, sharing knowledge, evaluating options and planning. The authors presented for the survey twelve knowledge visualization formats: mind map, to-by-two matrix, timeline, flowchart, iceberg visual metaphor, argument map, mountain trail metaphor, roadmap, casual loops, concept map, slide ruler, and bridge metaphor. Then they compared rankings of these types of knowledge visualization, made by the study participants, in relation to the four typical knowledge tasks: idea generation, knowledge sharing, evaluating options, and planning. This research provides support for the selection of adequate visualizations for individual and collaborative cognitive tasks. At a theoretical field, the results indicate that the tasks requiring divergent thinking (generative cognition) are best supported by less structured visualizations, and the convergent tasks (evaluative cognition, such as assessment and planning) are best served by structured and highly codified templates.

Story-based approaches seem to motivate users but might lead to less intensive reception of information. However, several reception and usage problems have been identified regarding information presentation and interaction. When overloaded with data, interactive information graphics tend to overwhelm users, so they may disregard well-known principles and rules of the old media and web design (Burmester, Mast, Tille, & Weber, 2010).

Eppler and Pfister (2010) analyzed sketchmarks – hand-drawn, simple, ad-hoc annotations, modifications, or additions to diagrams – not only as creativity catalysts but also as analytic, collaborative thinking tools. The collaborative use of sketchmarks can provide numerous benefits to managerial discussions. Chart annotations can support management teams in their decision making based on quantitative charts by visually eliciting and capturing interpretation processes, clarifying basic assumptions, stimulating different perspectives, and extrapolating trends into the future. Sketch-based annotations combine the simplicity and immediacy of drawing with the clarity and richness of charting. The authors provide the rationale for this under-researched visual management and communication practice, illustrate it through examples, and – as their main contribution – provide a first overview classification of the different sketchmarks that management groups can use in their discussion of quantitative charts (for such contexts as strategy or project reviews).

It seems some people do not appreciate a visual way of presentation or even a metaphorical way of presentation of an idea. For example, when professor at the Columbia University James Hone visualized the strength of the graphene (a one-atom-thick sheet of carbon atoms packed in a honeycomb lattice) telling, "It would take an elephant, balanced on a pencil, to break through a sheet of graphene the thickness of Saran Wrap [cling film]" (Hudson, 2011), a columnist ridiculed this statement in a concrete operational thinking mode, telling how difficult it would be to get an elephant onto a pencil, how could the elephant balance on the tip of a pencil, and how a pencil would puncture the elephant (Mirsky, 2011).

A Long Way from Generic to Meaningful: Concrete and Abstract Way of Thinking and Imaging

Recognition (just by looking) not always means comprehension that is bound to thinking: we may remember we have seen something before, but we may not know what does it mean. In the process of comprehension, we perceive the relationships of the object to other categories, which have irreducible properties (for example, knife and fork belong to 'silverware' – it does not matter if it is silver, green, or wooden). Images, symbols, and words lie along a continuum from the concrete to the highly abstract. Abstracting is considered a thought process occurring on non-verbal level. Due to abstracting, on a long way from generic to meaningful, we take in those features that are good for creating categories, and suppress those features that are not generic (basic to comprehending it). This way we show the scissorness of the scissors. In order to give structure and meaning to our experience, we get rid of similarities (called defining features) and unimportant features (visible or semantic) that are not crucial and suppress non-basic features. Alfred Korzybski (1879-1950) a Polish/American linguist who initiated the movement called General Semantics drew attention to a difference between a thing and a word. According to Korzybski (1933/1995), language comes between someone and the objective world, sometimes causing the confusion between the signifier and the signified. Because of that, we allow language to take us up the 'ladder of abstraction'. A Canadian/Japanese linguist and semanticist Samuel Ichiye Hayakawa (Hayakawa & Hayakawa, 1941/1991) followed the ideas of Alfred Korzybski and built the abstraction ladder (S. I. Hayakawa's term) of categories, with four up to eight levels, which could be applied to various areas of our experience. In 1938/39 he wrote a book "Thought and Action" and then "Language in Action." The following example (Chung, 2012) shows abstraction ladders having four levels, based on Hayakawa's ladder of abstraction.

Level 1: Tells about specific, identifiable nouns, such as my blue Levi 501 jeans, Tina's newborn sister, a three-bedroom house on Hollis Street, African violets, Mina;

Level 2: Identifies noun categories as more definite groups; for example, teenagers, middle class, clothing industry, parents, a college campus, a newborn child, houseplants;

Level 3: Defines noun classes as broad group names with little specification, e.g., people, men, women, young people, everybody, nobody, industry, we, goals, things, television;

Level 4: Describes abstractions, for instance, life, beauty, love, time, success, power, happiness, faith, hope, charity, evil, good.

Abstraction ladders on four levels can be applied to several areas of life, like the three examples shown below, which are applied to the:

- **Society:** Level 1 – my sister, Tracy; level 2 – spoiled child; level 3 – most people; level 4 – society,
- **Human Endeavors:** Level 1 – Max Factor, Inc., level 2 – cosmetic company; level 3 – industries; level 4 – human endeavors,
- **Economy:** Level 1 – Bessie the cow; level 2 – cattle; level 3 – farm assets; level 4 – economy (Chung, 2012).

The power of abstract thought was further examined in the eighties, to reveal the relevance of reasoning performance to chronological maturation. To improve the reasoning processes required understanding the abstract relationships Causey introduced as early as in 1987 interactive computer programs that simulated abstract relationships and required students to use inductive reasoning at the courses on philosophy of science delivered at college level. Student development in art is more related to transaction processes than to chronological age. A positive impact of the modified methods of teaching chemistry with the use of diagrams and explanation sheets was reported, for the high school students functioning at the concrete operations stage of cognitive development; there were 46 percent of concrete level students, as measured with the Arlin Test of Formal Reasoning (Boyd, 1989). In philosophy concepts can be formed by their abstraction from a set of common features in individuals. In ontological terms one can say about abstract relations or properties when they do not exist in particular space or time but there may be instances they exist in many different times and spaces. However, teaching abstract concepts, for example in science, increases cognitive load related to the working memory we possess and may cause communication problems.

Representational vs. Abstract Images in Graphics and Art Works

In terms of cognitive thinking we can examine an image as a representation of what is seen, a record of an object or scene produced in the mind. The image can be also understood as the mental picture evoked by words or experiences, many times a vivid or concrete picture. It may be also a pictorial phrase used to express some abstract notion, thus the concept of imagery may refer to figurative language, particularly metaphor and simile. In a metaphor (which means in Greek "to carry over"), a feature of an object or a word belonging to a particular object is applied to another thing, concept, or idea. Metaphor is a means to increase a range of the concept. It is often easier to grasp concepts when they are expressed by analogy with a concrete thing. Many words, such as a leg of a table or a hood of a car are metaphors. Metaphors are also used as the whole sentences, for example, "It rains cats and dogs" that was first recorded as used in 1738 by the Anglo-Irish satirist and writer Jonathan Swift (1667-1745).

Drawings and graphics can become instruments of thinking. Images, symbols, and words lie along a continuum from concrete and representational to highly abstract ones. Understanding of the structured thought requires the ability to perform abstract thinking and to apply problem solving. This may involve performing the problem finding, logically defining one's goal, and making conclusions. Drawing in abstract way may refer to any simplification or ideation of form, for example, when achieved by converting it into regular, geometric form. One of the ways to distill essential elements to be a revealed in an artwork is in transforming an exact realistic representation into an abstract work that has no model in nature at all, for example, into the art of pure geometry. Thus, we can organize our knowledge, perceive relationships, select only irreducible properties, and use only what we need. Artwork that does not imitate natural forms in recognizable way is called non-objective or non-representational art.

Going from concrete to abstract can be also seen in the realm of art. Highly abstracted drawings, which are not representational and thus have their representational power reduced, become symbols (they become word-like). Due to abstraction, we enhance those features that best identify it, and suppress those features that are not crucial to comprehending it. Symbolic drawings may convey messages about universal forces of which we are not always conscious. The intention of many modern paintings is not to render the naturalistic image of natural world but to bridge the distance between object and spectator; the value of such art is in its impact. Symbolic drawings may convey messages about universal forces of which we are not always conscious. Abstract art and non-representational art often uses symbols and symbolic drawings and thus convey messages about universal forces.

Graphic drawings provide many ways to view an idea. We can mentally combine geometric forms as a visual means to present abstract concepts and make discoveries. Graphic images may be concrete, when they show physical reality, or they may be abstract. Concrete graphic languages may include several kinds of perspective projection. An orthographic projection shows a three-dimensional object in two dimensions using parallel lines to project the shape of this object on a plane at the right angles. Oblique projection is a type of a parallel graphical projection; this time parallel lines intersect the projection plane at the oblique angles. An isometric projection shows a three dimensional object in two-dimensional technical and engineering drawings applying the same foreshortenings of the three coordinate axes at 120^0 angles. Concrete graphic languages may also utilize working models for engineering simulations (that may involve software products), and also digital mockups that present products in 3D.

Examples of abstract graphic languages may include charts (bar chart, pie chart, organization chart, flow chart, etc.), graphs, diagrams, link-node diagrams, schematics, or pattern languages that group symbols into logical relationships. Thus, the photography of a desk lamp or a car is a concrete representation, while an electric circuit or a blueprint are the abstract ones. Developing abstract thinking abilities is an essential issue in science education. Teaching abstract concepts, for example in science, increases cognitive load and causes many communication problems. According to Howard Gardner (1993/2011, p. 204), Napoleon held that individuals who think only in relation to concrete mental pictures are unfit to command. Chess masters think in relation to gestalt-like, abstract pattern-like sensory image, with a formless vision of the positions as lines of forces (such as Queen going straight and diagonal, and a Tower – straight only). The chess masters seem to 'see' good moves due to their visuospatial concept based perception. In chess players' thinking, visuospatial working memory is believed to be the central processing system and the place of advanced calculations. A fluency of thought (time needed to generate one move) differs among skilled and moderately skilled players.

Visual Reasoning, Aesthetic Emotions, and Art

Visual reasoning, that helps to overcome difficulties in comprehension of logical structures and assist in analytical learning, is somehow akin to the methods of simulation and visualization because it makes easier to perceive complex systems. We can witness how the progress in many disciplines, such as graphic presentation of quantitative and qualitative information, structural analysis, semiotics, computing, and also art inspired with generative algorithms, cellular automata, emergent systems, or the A-life systems, originates from the cognitive

Figure 3. Anna Ursyn, "Two Moons" (© 1992, A. Ursyn. Used with permission)

approach to visual thinking. Education, computer science, business, and marketing, all offering both visual and verbal types of communication, are becoming increasingly visual, especially in the web space context. Product design and advertising depend on visual thinking. The changing fields of interest in product design used to approach images as models for thinking and discuss imagery as a medium for thought. It is probably one of the most integrative fields, combining literature, poetry, sociology, psychology, social anthropology, art, and music with any discipline related to the media techniques.

As Arnheim put it, man, in perceiving the complex shapes of nature, creates for himself simple shapes, easy on the senses and comprehensible to the mind. The artist creates non-mimetic art of pure shapes through a grasp of perceptual structure, and through the magic and challenge of transformation of simple forms, with their connotations of order, size, and pattern, and quantitative relations to nature. In such transformations, numerical qualitative, or spatial relations between shapes are inseparable from their function in the whole artwork of which they are a part (Arnheim, 1969/2004).

Figure 3 is about finding an order behind the constructs and models.

A cycle of the city life (with day and night time, cars parked and cars driven, red lights and green lights, windows open and windows closed) is presented in the form of its rhythmical structure and organization. The overall character of the particular city gets organized around it's own elements, blended with bigger projection window, and depending on one's own focus of attention.

An artist may know when this set of objects is complete without knowing their number or without counting them. Jean Piaget showed that young children, when asked to copy a figure with counters, do justice to the shape of the figure without using the correct number of counters. In a similar way, the artist creating a computer artwork, when thinking with pure shapes, may cause numbers to become filled with life and reflect natural shapes. When a sound structure needs a trinity to represent an intertwinement rather than a contrast, two objects would induce duality, while four would be redundant.

One may discern at least two approaches to creating art, each of them often evoking a dislike toward another one as undeserving to be named art. Several artists who work in the fields both art and mathematics (e.g., such electronic artists as Manfred Mohr or Helaman Ferguson) found a way to develop their individual artistic style through applying mathematical rules that are seemingly reverse to artistic expression. Some computing savvy individuals look for artistic solutions using applications to respond to the market demands; they search the Internet and use plenty of tools, applications, and solutions found on websites.

On the other hand, many times artistically talented people are compelled to create art, determined to do it with whatever is at hand: a computer or a broom. Their powerful visual reasoning evokes an outburst of strong emotions. Moreover, while looking at the artwork consisting from a few simple strokes everybody feels they could do the same easily.

Completely different outcomes may result from these two approaches. We may compare perfect works of electronic design, completed with first-rate tools, with intentional deformations characteristic of traditional mainstream art, for example created by Francisco Goya, Alberto Giacometti, or Amedeo Modigliani. Geometry has been often neglected in works by Paul Cézanne, Pablo Picasso, or Vincent Van Gogh, proportion was often changed by Henri Matisse, Jonathan Borofski, or Fernando Botero, and so on. For the technology-oriented people, involving in teamwork with an artist may bring a conflict between precision and accuracy of ready solutions and the artist's individual style. On the other hand, we may find purposeful deformations in electronic art resulting from translating, scaling, slanting, and otherwise distorting the initial image, aimed to depart from what the eye can see. One can imagine many reasons for such transformations and distortions. For example, some advertising messages are successful due to impressive imaging that gives a shorthand summary of patterns of ideas. With purposeful deformations, we may address some cognitive processes or make mental shortcuts, reacting to synthetic signs, and implying connotations to symbols or icons.

Visual reasoning has been considered a cognitive activity because it leads up to concept formation, helps to comprehend logical structures, and assists in analytical learning. According to Roger Pouivet (2000), also an aesthetic experience is a function of cognitive activity, and knowledge is an emotional process: aesthetic pleasure and the cognitive dimensions of aesthetic experience have a direct connection.

Certain emotions are cognitive and may be experienced in the field of science as well as in the field of aesthetic experience. In his "Critique of Judgment," German philosopher Immanuel Kant (1724–1804) (2007) wrote about judgment of knowledge (conceptual judgment) and aesthetic judgment (non-conceptual judgment). Therefore, emotion accompanies scientific research, verification of knowledge and the feeling of surprise, when discoveries contradict our beliefs. A Greek scientist and philosopher Aristotle (384 BC–322 BC) wrote about Tragedy, "Thus comes the element of Thought, i.e. the power of saying whatever can be said, or what is appropriate to the occasion ... the older poets make their personages discourse like statesmen, and the modern like rhetoricians" (2013, Poetics 1450b). Aesthetic pleasure is entailed by cognitive activity, depends on it, though it cannot be reduced to it, as it is "...an aesthetic kind of cognitive activity consisting of mastering the particular functioning of symbolic systems and enacting certain relations between symbols and what they stand for" (Pouivet, 2000, p. 53).

Computer art images have been considered helpful in knowledge comprehension. To comprehend a structure, our brain compares information contained in an image to something previously learned and stored in memory. Information from the past is grouped and organized there in a cognitive structure. This allows for our perception, which means that we may recognize images and decipher them on a basis of our memory. When drawing computer-generated graphics, one can notice topological relations – qualitative characteristics of arrangement. At the same time, by grasping the quantitative character of the subject one can make productive transformations. According to classical works led by the educational psychologist Benjamin Bloom (1913–1999) (1956) intellectual behavior in learning occurs in the cognitive, affective and psychomotor domains, with the levels of cognitive activity including knowledge, comprehension, application, analysis, synthesis, evaluation, induction and deduction. This, in turn is conducive to creating an image telling about personalities of individual people: who they are, what they want to be, and what they are expected to be. Resulting images may embody an aesthetic factor (an image one has of him or her or thinks other people have), a knowledge component, a skill part, and a value element that one would like to create.

See Table 2 for Your Visual Response.

Table 2.

Your Visual Response: Organizing Actions
Select and present some activities you like and consider important, and then and your skills of bringing them about. Choose themes related to your work or interests. Put them into meaningful relationship. For this project, you may prefer to write a short computer program or make a concept map in the form of a sketch. Your program or the concept map will show a hierarchical set of your preferences and their justification. Now make this short project as a fast, inspiring interaction between your concepts and the visual references. Transform your preferred activities into pictures that would not only record what you have selected but also show the meaning resulting from this presentation, which is important for you.

SOME PHILOSOPHICAL APPROACHES TO PERCEPTUAL THINKING

Before the onset of the constructivist approach, there was a general belief in an objective reality: all things exist objectively around the observing person, reality is independent of an observer or culture, all structures, objects, codes and laws are ready to be described.

In constructivist terms, not an objective reality but the understanding of one's own experiences is important. Reality is not residing outside or independent of an observer. Reality is starting within circular process of perception, understanding and making things, and through experience constructing individual understanding of the world. Knowledge results from a construction of reality by a person. Making constructions of the cognitive schemes, categories, concepts, structures, rules, mental models, and assigning meaning to individual experiences are important in learning – the process of adjusting our mental models to accommodate new experiences. For example, the prints on the sand on a beach become meaningful for the thought through perception. A notion of a gift becomes a social structure that arises out of social practices. And thus, constructivism is based on the principles telling that learning is a search for meaning, and meaning requires understanding both wholes and parts in the context of wholes. Learners construct mental models, the assumptions that support those models and their own meaning (someone else's meaning).

In the process of learning, the impact of visuals on cognitive activities and learning gains more attention every day because of the facilitating effect of visual learning on understanding configurations and relationships described by formulas. In terms of pedagogical constructivism, learning is an interpretive process leading to the construction of the individual's subjective reality, not identical with the knowledge of the teacher. The value of the works of art, as well as of visual educational materials, might be in their emotional impact, which would bridge the distance between the object and the viewer. Thus cognitive and expressive meanings conveyed by visuals, for example, the use of line for the visual exchange of the shared symbols, may improve the capacity for learning.

With the phenomenological approach, which is focused on the study of essence connecting our consciousness with our thought, science tends to be considered as only one of possible ways of knowing, not necessarily securing access to truth. Perceptional thinking may help to connect what we perceive as real with our awareness of the constructed notions about universal properties of phenomena experienced and essences of things thought. Phenomena experienced may differ from what we know or believe.

Cognition is often described as the act of knowing. Theorists often refer cognitive thinking to problem solving, hypothesis testing, and concept acquisition. Human mind is structured to select and analyze information, organize it into memory store, and then retrieve from memory information picked up by the senses to use it in various ways, not necessarily consciously, for example, for use in decision-making. Cognitive thinking is used in experimental research for gathering data and information, testing hypotheses, interpreting results, and providing scientific evidence for new theories. Cognitive scientists connect cognitive reasoning with rationality (Stanovich, West, & Toplak, 2011); they recognize two types of rationality: instrumental rationality, which denotes "behaving in the world so that you get exactly what you most want, given the resources (physical and mental) available to you" and epistemic rationality, which "concerns how well beliefs map onto the actual structure of the world" (Stanovich, West, & Toplak, 2011, p. 795).

We build our knowledge on conceptual structures that result from our own experience of the real world, prove them experimentally, and then test them against experience shared with others and by consistency with our beliefs about what we hold as true. Visual input induces the construction of new information that can enhance our knowledge base. Learning this way depends largely on one's perceptual skills that are involved in producing more effective responses to stimuli. Conceptual structures include also our beliefs that we accept as true, and imaginative activities leading to the construction of new conceptual structures that may not match actuality but instead shape the real world the way we want. Imagination adds to what makes life meaningful; it is also a necessary part of scientific thinking. Intelligence is creative and self-creative within our inner realities, in a process of self-actualization.

Computer programs can be used for creating shapes and space as visual equivalents of quantitative descriptions, giving perceptual form to thoughts, and providing visual knowledge. They are used to control the relation of shapes to images at an abstract level. Visual thinking applied to practice serves as an information tool. Even bank reports are often shown as pictures, sometimes as an artwork. For example, Yasuhiko Saito (Ursyn & Banissi, 2003) applied visualization technique for analyzing financial data as a tool for generating artistic images. He produced art works by defining portfolio textures to gracefully visualize dry financial data. The time-based data about Japanese automobile companies contained warm (red, orange and yellow dots) or cold (green and blue dots) textures depending on the tide in the Japanese stock market. In effect, portfolio textures generated from stock price data of about 2,500 Japanese companies looked like a tapestry or an abstract painting.

COGNITIVE INFORMATICS

Cognitive informatics is an interdisciplinary field encompassing natural systems such as living organisms, and natural sciences such as cognitive science, neural science, psychology, computer science, information sciences, and intelligence science to examine human internal information processing systems along with their engineering implementations in computing and distributed collaborative work. It has been defined as the science of cognitive information that investigates into the internal information processing mechanisms and processes of the brain, natural intelligence, and their engineering applications in computing via an interdisciplinary approach (Wang, 2011). Cognitive informatics specialists search for the potential applications of information processing and natural intelligence to cognitive computing. Another goal is to extend, through the use of technology, information management capacity and reduce limitations in attention, memory, learning, comprehension, visualization abilities, and decision-making (Pacific Northwest National Laboratory, 2008). As a part of cognitive informatics, affective computing studies interactions between animals, humans, and computational agents from the psychological, cognitive, and neuroscientific perspective (Gökçay & Yildrim, 2011). Goal-directed behavior, survival, and adaptation involve conscious and unconscious emotional processes. Agents in affective control mechanisms influence performance, both in basic tasks and social encounters. Researchers examine computational nature of affective states. At a psychological level, emotions combine into three components: physical sensation, emotional expression, and subjective experience; however their coordinated cortical and sub-cortical neural mechanisms originate from disparate anatomical structures and thus they cannot be ascribed to specific anatomical structures (Erdem & Karaismailoglu, 2011). We can perceive nonverbal communication by facial expressions, vocalizations, gestures, and postures as a channel for inner life of others. Nonverbal communication can be used as a viable interface between computers and social emotions or attitudes (Vinciarelli, 2011).

A need to study the cognitive potential of the brain results from the understanding of the great capacity of human memory as compared to the amount of information stored by a computer (Bancroft, & Wang, 2011). A storage space on a computer and the memory is usually measured in bytes (binary digits as the smallest increments of data on a computer). Some hold that with about 10^{11} neurons, and several thousands synaptic connections on each neuron, the human brain's capacity can be estimated in petabytes (10^{14} bytes). Some hold that a number of potential (not necessary real) neural connections may be bigger than the number of stars in the universe. With about 1.5 GB of data stored in a human genome, and ten to one hundred cells existing in human body, about a zettabyte (10^{21} bytes) of genetic data is stored in the cells of an average human body (Grigoryev, 2012). As a consequence, information-processing mechanisms in computing are investigated, along with the cognitive ability of the human brain and intelligence. Generally, the number of neurons does not change over the life span of an adult; information in the brain may grow through the creation of synaptic connections. Wang and his colleagues developed a mathematical model to find that the maximum capacity of human memory, i.e., the possible number of synaptic connections among neurons in the brain is up to $10^{8,432}$ bits (Bancroft, & Wang, 2011). According to the authors, the amount of information the human brain can hold, the accessibility of that information, and speed it can be sorted to recall specific knowledge reveal how humans are still much better at understanding patterns than a computer. The authors conclude that the brain be studied so that current computers may be improved and the future generation of cognitive computers may be developed. Indeed, scientists work on developing cognitive computers that perceive, deduce, and learn, as well as on establishing cognitive computing theories and methodologies. A wide range of applications of CI are identified such as in the development of cognitive computers, cognitive robots, cognitive agent systems, cognitive search engines, cognitive learning systems, and artificial brains. It is recognized in CI that information is the third essence of the natural world supplementing to matter and energy. Informatics is the science of information that studies the nature of information, its processing, and ways of transformation between information, matter and energy (Wang, 2011). Scientists create cognitive models of different kinds (ranging from diagrams software programs) to study specific cognitive phenomena, processes, and approximate their possible interactions and behavioral predictions for selected tasks. By developing models of cognitive architecture scientists examine functional properties of the structures under study.

EARLIER INVESTIGATION ABOUT COGNITIVE DEVELOPMENT

Models of cognitive processes developed in the seventies and eighties derived out of the rationalistic philosophical tradition; the development is seen as a process toward greater complexity and differentiation. Thinking skills development is usually analyzed in terms of the development of higher cognitive levels described by Piaget; intellectual processes, such as problem solving, decision making, synthesizing, as described by Bloom; and intellectual activities such as cognition and memory (Guilford, 1967). In constructivist terms, knowledge is a process of becoming "resulting from a construction of reality through the activities of the subject" (Inhelder, 1977, p. 339), so it is construction of the cognitive schemes, categories, concepts, and structures that is important in learning (Bettencourt, 1989).

Swiss developmental psychologists Jean Piaget and Bärbel Inhelder (Piaget and Inhelder, 1971; Inhelder & Piaget, 1958) wrote about structural schema developed in the mind, to recognize and to process visual patterns. The structural schema are developed in the mind, to recognize and to process visual patterns. They described mental manipulations as: accommodation, which means changing existing schemata, and assimilation, which means changing a new object to correspond with existing schemata. Piaget discerned developmental stages in children mental activities: children (age 2-8) perform symbolic or semiotic functions; children 7-12 years old make concrete mental operations, they define classes and relations; the age of 11- adolescence a stage marks the beginning of conceptual thought and formal operations, when hypotheses are formed, not only representable realities.

Jean Piaget (1896-1980) constructed a model of child development and learning that showed how children develop cognitively by building cognitive structures, mental maps, schemes of their environment. When the experience is new or different, children alter their cognitive structures to maintain their mental equilibrium, accommodate the new conditions, and to set up more adequate cognitive structures.

Piaget assumed there are several cognitive developmental stages:

- **A Sensorimotor Stage (Birth–2 Years Old):** When children interact with their environment and build concepts about reality
- **A Preoperational Stage of Concrete Mental Operations (Age 7–12):** When children form symbolic and semiotic classes and relations, conceptualize upon their experiences, and create logical structures, and
- **A Stage of Formal Operational Thinking and Conceptual Thought (11–Adolescence):** When children cognitive structures include hypotheses, and conceptual reasoning, not only the representable realities. A stage of formal operational thinking was considered to be the final equilibrium status (Inhelder & Piaget, 1958); it was meant as the ability to engage in abstract thought, generate hypotheses, subject them to investigation, make decisions, synthesize information, understand abstract concepts (such as electromagnetism or ß-radiation). Thus, the formal stage implied the ability to deal with propositions and to employ proportionality and combinatorial systems in problem solving (Arlin, 1984). The formal operational stage characterizes a considerable part of population; however, about 50% of the adult population under study (college seniors) never attained the Piagetian stage of formal operational thinking called the problem solving level (Arlin, 1984).

Inquiry about Higher Order Thinking Skills

Results of investigations from the seventies suggested that the adult thought structures might extend beyond the equilibrium level, so the adults might develop progressive changes in their thought structures (Gruber, 1973). Some adults extend beyond this level and enter a problem-finding stage that would include the creative formulation of problems, raising questions and developing scientific thought. This most complex stage in the development of cognitive thinking is characterized by adaptability, flexibility, the use of concepts and generalizations, drawing logical conclusions from observations, making hypotheses and testing them. The Piagetian notion of formal operations includes the convergent or problem solving phenomena in response to specific tasks. Arlin (1984) hypothesized a fifth, problem-finding stage going beyond formal operations and including the creative thought on problems described by other authors as the 'discovered problems' (Getzels & Csikszentimihalyi, 1970, 1976), the formulation of generic problems (Taylor, 1972), the raising of general questions from ill-defined problems (Mackworth, 1965), and the cognitive growth represented in the development of significant scientific thought (Gruber, 1973).

According to the accepted criteria for a stage model, there is a hierarchical nature of changes; therefore fulfilling the requirements of the earlier stage is necessary but not sufficient for acquiring the new stage. Thus, all subjects who are successful in problem finding should also be characterized as formal operational thinkers, but not all subjects being in the problem-solving stage should also be in the problem-finding stage (Arlin, 1984). Arlin and Getzels tested separately problem solving ability (formal operations), the three Piagetian stages, and problem finding ability. The expectation was that high problem solvers would ask questions that structurally required combinatorial and systematic operations. Johnson (1987) suggested a cognitive-structural framework for developmental studies of adult inventiveness due to some limitations imposed by the genetic epistemological paradigm. He examined hypothesized post-formal thought structures and their relevance to the understanding of adult creativity, and in accordance with the Arlin's (1974, 1984) hypothesis about the fifth stage of the cognitive development postulated that the problem-finding stage of cognitive development was a critical component in the creative problem.

VISUAL DEVELOPMENT

It is generally accepted that developing the visual thinking starts with learning the symbolic thinking, which is not intuitive. We have to gain an ability to understand pictures as symbolic objects; that means, to become aware that one object can stand for another. To grasp the meaning of a symbol, we must achieve dual representation. All symbolic objects have dual meaning. We think symbolically when we can see an object both as itself and as depicting something else. Thus, we must mentally represent the object and the relation between the object and what it stands for. Then we may create and manipulate various symbolic representations to making knowledge and culture, even using abstract concepts without direct experience.

Before they develop symbolic thinking, children learn to recognize objects, pictures of objects, and their meaning. They learn to understand what pictures really mean and how they differ from the things depicted – the referents. Joy and excitement experienced by an eight-month old baby when spotting a picture of the familiar face is a well known in a field of developmental psychology example of such recognition. At first, a baby perceives without any emotion two circles that later on will stand for eyes. Then, the circles gain the meaning of the eyes, when the baby recognizes a semicircle put below these circles as a smiling mouth. According to Judy S. DeLoache (2005), nine-month old babies see the pictures as real objects; they try to reach and pick a depicted apple, and rub or scratch the paper. The confusion seems to be conceptual, not perceptual, because when the infants have a choice between a picture and a real object, they choose the real thing. By 18 months, infants know that a picture stands for a real thing. They point pictures and name an object or ask for its name. But two and a half year old children still mistake photographed objects for the real thing. For example, they try to put their foot into a photographed shoe. Three-year-olds can find a toy in a real room when they see a miniature toy in a little model of this room, while two-and-a half-year-olds cannot do this because they do not grasp a relation between the model and the room. When they are told (and believe) that a magical machine has shrunk the room, they have no problems in finding toys in same places as they have seen them before shrinking. But until they are four they do not fully understand pictures, for example, think that turning a picture of a bowl of popcorn upside down will result in the depicted popcorn falling out the bowl. They also make the scale errors, the Gulliver's errors; they treat small toys as if they were much larger and try to get into a miniature car (DeLoache, 2005). According to DeLoache, very young children might not be able to relate their own body to a doll; for example, they cannot place a sticker on a doll in the same place it was placed on their own body. Therefore, anatomically detailed dolls should not be used in forensic situations in cases of suspected abuse. In educational practice, using wooden blocks designed to represent numerical quantity, as "manipulatives", could be counterproductive in teaching subtraction. Similarly, old-fashion alphabet books may serve better than manipulative books designed to interact with the book (flaps, levers to animate images, etc.).

The concept of the visual developmental stages has been applied to describe the process of evolving skills and capacities in the visual domain (Gardner, 1978, 1983). Analogies have been drawn between mental development and the visual development stages of representation. The levels of student visual and cognitive development may be important in their problem solving and types of expression and perception. Several authors provided their descriptions of the visual development stages. Sir Ciril Burt (1922/2009) defined the stages as: Scribble (2-5 years), Line (4 years), Descriptive Symbolism (5-6 years), Descriptive Realism (7-8 years), Visual Realism (9-10 years), Repression (11-14 years), Artistic Revival (early adolescence). Helga Eng (1931) defined three stages up to eight years – scribbling, transition, and formalized drawing. Historically, the most prevailing and widely used description of the visual development stages was developed by Victor Lowenfeld (1947), who defined the stages as: scribbling (2-4 years), pre-schematic (4-7 years), schematic (7-9 years), dawning realism (9-11 years), pseudo-naturalistic (11-13 years), and the period of decision (adolescence). Lowenfeld and Brittain (1987) further reworked this model. Using the concept of schema, the fundamental basis of the Piagetian psychology, the authors described the stages of visual development as: scribbling – manipulation stage (up to 3 - 4 years); generalization (early symbolic, preschematic stage, 3-4 up to 6 years); characterization (symbolic, schematic stage, 6-7 up to 9 years); and visualization stage (9-12 years).

Most studies on visual development were performed in the context of the child's cognitive development. However, it must be realized that students functioning at varied levels of visual and formal operation stages can be also found at the college level (Arlin, 1974). Using Arlin's test of formal reasoning, Boyd (1989) found that only 54 percent of secondary students functioned at the formal operation stage. For this reason Boyd stressed the need for introducing modified methods in teaching chemistry for students functioning at the concrete stage of development. In respect to visual development, Wohlwill (1988) described the problems and frustrations faced by students approaching the visualization level. The meaning of student art in the course of visual development of young and adult students can be discussed from at least four different points of view.

- The psychoanalytic approach has some background in clinical psychology. The production of art objects is used as a projective technique (through observation of the drawings, especially those of human figures). The main concern is to develop a mentally healthy person. Art is a means of discovering the internal conflicts and disturbing experiences. Art activities for mentally ill persons are sometimes considered therapeutic.
- In behavioral psychology, student's environment is responsible for the development of the student. Art activities are considered reinforcing and shaping behavior. Art both changes and reflects thinking processes. For behaviorists, art becomes an indication of a subject's understanding of the task at hand.
- With the developmental approach, art is examined to see how the children measure up to what is expected of them at any particular age. The children cannot change until they are ready; the developmental level predetermines the change and

the student cannot be pushed into the next stage. According to this approach, the best thing is to stay out of the way. One should remember that the creative work must be understood individually. One should also remember that it is not the lack of formal education that makes the child a child.

- Within a frame of reference of cognitive psychology, a student needs to develop a vocabulary of verbal and pictorial symbols. These symbols are executed by the student in mastering artistic skills before the development of expression. Therefore, an art product is the record of a student's preparation. This approach became a starting point for developing the Discipline Based Art Education (DBAE) program, which taught and explored the art experience through art production, art history, art criticism, and aesthetics.

Figure 4. Anna Ursyn, "Report from Colorado" (© 1986, A. Ursyn. Used with permission). Printout of the 3D computer program: VAX mainframe, FORTRAN 77, Interactive Graphic Library (IGL) and PPC.

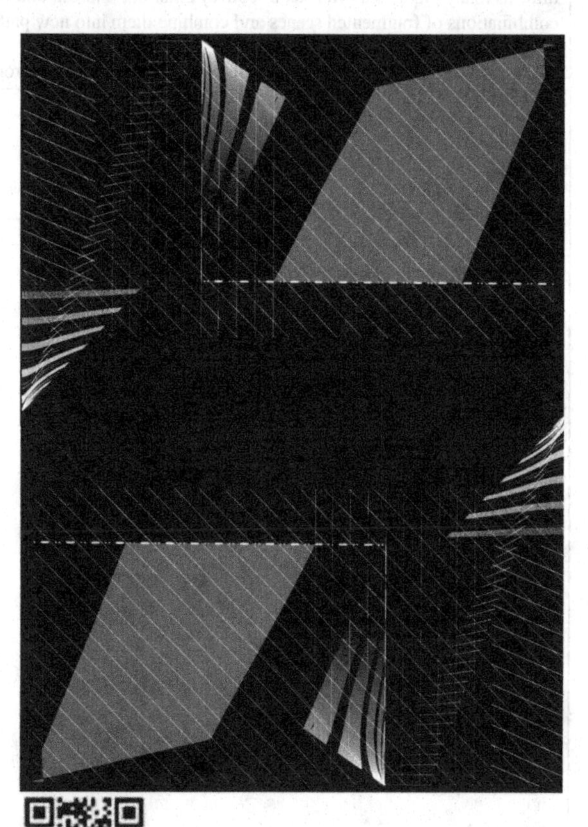

MENTAL IMAGERY

Mental imagery is usually described in cognitive psychology as one's nonverbal, cognitive representation of objects, concepts, and perceptual experiences (visual, auditory, tactile, olfactory, gustatory, or kinesthetic) from the past, or images anticipated in the future that are desired or feared. Input from the eyes induces a pattern of activation during perception. We can comprehend an image because our brain compares information contained in an image to previously learned and stored in memory information that is organized there in a cognitive structure. Memory of visual material allows for building our visual imagery and perception, which means that with the use of memory we may recognize and decipher images and their meaning. Mental images help to compare objects. Our perception of a face is possible because we compare it to a mental imagery of a face previously viewed. Thus, a mental image is a representation of what is produced in our mind – a vivid, concrete mental picture evoked by words or experiences. Imagery preserves relations among external objects, not necessarily in a concrete way. Thus mental images help to compare objects. When people compare sizes of objects, pictures work better than verbal comparisons. Mental representations of visual objects may comprise unconscious computations of moving objects or conscious knowledge of meaning of the object. Cognitive scientists stress the importance of the unconscious (Stanovich, West, & Toplak, 2011).

Table 3.

Your Visual Response: Old Skills and Interests
Make a short manual or tutorial about how to perform something you knew well how to do it when you had been very young and proud of knowing it. After that, inspect mentally images preserved in your old memory by zooming and scanning them in your mind, and then draw as many pictures as you can to convey emotion, interest, and engagement you felt while doing it in the past or right now. Create new combinations of fragmented scenes and combine them into new patterns.
Along with sketches, you may choose to write a short computer program, a graph or a table.

According to Lewicki, Hill, & Czyzewska (1992), "Data indicate that as compared with consciously controlled cognition, the unconscious information-acquisition processes are not only much faster but are also structurally more sophisticated, in that they are capable of efficient processing of multidimensional and interactive relations between variables" (p. 796).

We may think about several spaces for creating and navigating images: mental imagery occurs when we call forth in our mind an imaginary object or a scene; digital images, which can be animated or interactive, and can be rendered on a computer screen; another pictorial representations of mathematical and process-driven operations can be seen in a virtual space; also in an Internet-based world of Second Life where avatars act for the remote users as graphic representations. An avatar in Hindu and Sikh religions means a deliberate descent of a deity (mostly Vishnu) on Earth, sometimes as an animal reincarnation (Vivekjivandas, & Dave, 2011). We can contemplate (as in the inner, mystical dimension attained by the Sufi practitioners) an imaginal space existing independently between the physical and spiritual

Table 4.

Your Visual Response: Object Changes its Meaning
Part 1 Draw or sketch a picture showing your room. It may be your favorite room in your home, your bedroom, office in your workplace or home, or another space you used to work. Then, create a drawing of objects you see in this room. It does not have to be a photographical representation; we have a camera for that. Focus on some significant features of that object that are crucial for the recognition of its function and mechanics. If you are drawing on a computer, or a tablet computer such as iPad, name your file: 'room' and save the file.
Part 2 Change the meaning of your picture and its message: convert your room into countryside scenery. Use a copy of your original file. Open your first assignment and rename it so you have two same looking files. Modify the content of your first file ('room') so the room becomes a landscape. You may want to use transforming tools, and maybe filters. Select a fragment of the composition and use all kinds of transformations: scaling, slanting, perspective, distort, etc., so it begins looking like a landscape. Change the meaning of objects in the room, so they become another things with different function. You may convert each item into some element of the landscape or modify it differently, for example by using filters. For example your desk becomes a giant rock or a mountain. You may want to convert a computer to a rock, if you are creating a nature-based landscape; or a building, if you are interested in a manmade cityscape. The curtain may become a river, a cloud, etc. Work with portions of your composition till the viewer would feel being completely immersed in an outdoor environment, but you have used the elements from you indoor scene. The idea is that you'll recycle the components of your picture to use them as the transformed building blocks of the landscape. If it happens to be wintertime just now, and you look out of the window, you'll probably need to change the colors to white:-). You may prefer to copy and paste the components from the first file to a blank document, and then copy them one by one onto the new file (saved as the 'landscape'), while thinking about each and every object how to modify it. You do not have to use all the parts. Some people like to draw it as a "bird's eye view" or a plan for a room. The overall character of the image will become personal, maybe showing beautifully done shading techniques. It might be interesting to enrich the new meaning of your work by assigning new connotations to the art works, maps, or photographs hanging on the walls of the room.

planes (Lambert, 2011). Before they fade, mental images may occur without external stimuli.

Imagery plays some role in making associations – mental connections between thoughts, feelings, ideas, or sensations. We can remember (or even better, vividly imagine) a feeling, emotion, or sensation that is linked to a person, object, or idea. For this reason mental imagery plays a pivotal role in building memory and motivation; one feels more interested and more emotionally involved in the work by creating mental images of objects and concepts one has been working on. With the use of visual imagery we may mentally draw images from short-term memory representations or even generate images that we have never seen. We can then mentally inspect images by zooming and scanning them, encode the patterns of images into memory, remember new combinations of patterns or imagine new patterns.

"Report from Colorado" is an appreciative record of the celebration of community life with its colorful events (Figure 4). In an abstract way, it relates to our symbolic representations that build our knowledge and culture.

In a similar way as we may create and manipulate various programs, we may use abstract concepts without direct experience.

See Table 3 for Your Visual Response.

As stated by Stephen Kosslyn (Borst & Kosslyn, 2008), mental imagery serves for generation, inspection, recoding, maintenance, and transformation of images:

- **Generation:** Activation of information stored in long-term memory and construction of a representation in short-term memory; we do not have images all of the time. Images come and go, through short-term memory representations. One can "mentally draw" in imagery, producing images of patterns never actually seen.
- **Inspecting the Object in the Image:** We must have a way of interpreting the patterns of images, 'zoom in' on isolated parts of them, or scan across them.
- **Recoding:** We can encode the patterns of images into memory, remember new combinations of patterns or imaging new patterns.
- **Maintaining the Image Over Time:** However, mental images require effort to remember them. The more perceptual units that are included in an image, the more difficult it is to maintain.
- **Transforming the Image:** It lies at the heart of the use of imagery in reasoning. For example, we can rotate patterns in images, also in the third dimension, so that we 'see' new portions as they come into view. We also can imagine objects growing or shrinking, add or delete their parts, or change the color.

Investigations from the seventies and eighties provided data about the correspondence between imagery and perception. Researchers are examining whether perception and imagery are different processes. Several areas in the brain depict phenomenal experience during perception by encoding image representations. Physiological recordings such as using EEG show alpha waves (8-12 Hz) across the occipital regions after imaging visual stimuli, across the parietal region after imaging tactile stimuli, and the occipital region when imaging visual references of words. Cerebral blood flow increases in the occipital region during daydreaming.

Imagery seemed to be unrelated to the right hemisphere only (Kosslyn, 1980). Left hemisphere can also generate images; they are further transformed in right hemisphere. Split-brain patients generate mental images mainly in a right hemisphere, in waking life and also in dreams; they imagine letters mostly with right hemisphere, and the configuration and spatial pattern of imagined objects mostly with left hemisphere. Electrical

Figure 5. Anna Ursyn, "Yellow Pages" (© 1986, A. Ursyn. Used with permission)

stimuli applied to temporal lobes or to a limbic system evoke memory-like hallucinations. Brain areas involved during perception could be also involved in mental imagery (Borst & Kosslyn, 2008). Research conducted with the use of fMRI (Slotnick, Thompson, & Kosslyn, 2012) has shown that visual long-term memory and visual mental imagery rely on highly similar – but not identical – cognitive processes: they recruit common control and sensory regions of the brain.

Imagery is important for our perception because it makes that sensations (sounds, shapes, colors, tastes, and motions) convey meaning (Table 4). In the sixties, Rudolf Arnheim described artistic activity as a form of reasoning, with perceiving and thinking being indivisibly intertwined (Arnheim, 1969/2004). Thus, productive thinking is perceptual thinking that takes place in the realm of imagery, as visual perception lays the groundwork of concept formation. Word and picture cannot be split up into parts that have any meaning separately (Arnheim, 1990). No Museum of Cognitive Art has been opened yet, even though mental imagery plays the significant role in cognitive thinking and communication through art. However, a Museum of Mathematics has been opened on December 15, 2012, at 11 East Street in Manhattan (http://momath.org).

GESTALT PSYCHOLOGY

An Austrian philosopher, poet, and dramaturg Christian von Ehrenfels (1859–1932) developed in 1890 the gestalt psychology focused on a whole shape. This theory of mind tells about a brain as a parallel, analog, self-organizing holistic structure. According to the gestalt psychology, perception requires the grasping of the essential structural features. For example, a melody can be recognized even when played on various instruments or in different keys. Von Ehrenfels argued that the whole is not simply the sum of its parts but a total structure. The perceiving eye and the mind are looking for pattern and simple whole shapes. When we look at more complex visual images such as paintings we can see that they convey visual information. When our early ancestors, and also contemporary aboriginal artists have created paintings on sand, a rock, their own body, or a bark, they created a variety of meanings, sometimes hard to decipher, often providing the evidence of abstract thought as early as in the Middle Stone Age, at least 70,000 years ago (Limson, 2010; Wong, 2002). Gestalt systems, which have been elaborated with the therapeutic applications in mind, discussed key principles of the emergence (formation of complex patterns from simple rules); reification (which happens due to the generative, constructive ability of mind, which allows more perceptual information than is provided in the sensory data); multistability (ambiguous perceptual experiences with more than one interpretation possible, visual illusions such as a Necker cube, which can be seen from above or from below, or the Rubin vase that can be also recognized as two human profiles); and invariance (experienced when we recognize simple geometrical objects in spite of their rotation, translation, scale, elastic deformations, or lighting). Modules of perception have been modeled with the use of computing and examined in terms of the computational theory of vision, to explore the ways in which perception is involved in generation of computer programs and applications (Sternberg, 2011). Computer imagery, which can be easier described by its structural characteristics than by visual ones (Lambert, 2011), has been described by Timothy Binkley (1996) as both abstract and concrete because various different originals can be derived from the same collection of numbers included to the program. Moreover, according to Binkley (1990), the computer is not a medium, as it does not have physical material for image production but only a file of numbers that controls the image. Lambert (2011) examined virtual space as a Platonic construct. As he pointed out, "in a Platonic sense, all these forms already exist (in potential or in actuality), and the computer-using artist is exploring this space of potential forms to bring them into being" (2011, p. 442).

Knowledge is important in imagery because our tacit knowledge about physical relations in the world supports information about objects that is depicted in a visual buffer or working memory – a mental space for manipulating, scanning and inspecting visual images. Both visual mental imagery and visual working memory are processing visual information in comparable ways; both rely on depictive representations of the same format (Borst, Ganis, Thompson, & Kosslyn, 2011). The visual buffer activates mental images induced by stored information; we may redraw maps from memory. Visual buffer has limited resolution, can be rotated and scaled at will. It fades if not refreshed. It may take information from long-term memory or develop a 3D model representation in a long-term memory store. Kosslyn provided a computer model of the cognitive processes involved in visual imagery. The attention window selects a region within the visual buffer for further detailed processing. The size of the window in the visual buffer can be altered and its location can be shifted. We can scan to portions of visual mental images that initially were off the screen, even when our eyes are closed, and the farther we scan across the imaged object, the more time we need. There may be a three-way link between visual perception, visual memory and control of action. Visual attention control is important in designing mobile robots; for example, when a four legged robot tries to shoot a ball into the goal (Mitsunaga & Asada, 2004).

People can work well with mental images only, but for difficult tasks external representation is pivotal. Thinking is the manipulation of mental representations: both the long term (lasting knowledge) and temporary representations (new information). In order to use imagery in reasoning, we can transform, rotate patterns in three dimensions, or imagine objects in motion. Thus, imagery is used to recall information about previously perceived objects and events, to represent concrete objects, reason about spatial information in the form of mental maps, compare metaphorical non-spatial relations, or learn abstract information. Visual imagery may facilitate problem solving, allow avoid the mechanical use of algorithms elicited by verbal formulation of the problem, and promote parallel processing of information (without examining single elements sequentially). Using brief visual presentations of letters and numbers, George Sperling (1960) demonstrated the existence of a special visual information store (called iconic memory) with almost unlimited capacity but very short duration. It is now supposed that recoding from iconic memory to more lasting forms of storage takes place by both verbal and nonverbal means. Further research with the use of the fMRI technique revealed that the parietofrontal network of selective attention is reportedly relevant to readout from iconic memory (Saneyoshi, Niimi, Suetsugu, Kaminaga, & Yokosawa, 2011).

Artwork presented in Figure 5 reflects possible cognitive maps linking the actuality with forgotten facilities we use only in rare moments. Yellow Pages, before the advent of the networked world, had been an indispensable source of information of vital importance.

This is the space of transition from rural to urban,
From production to consumption,
From essentials to ornamental detail:
Odd houses of obscure use, old viaducts,
Sources for our flourishing.

INTELLIGENCE AND VISUAL LITERACY

We talk about general intelligence when we study the behavior of intelligent organisms or intelligent programs and their ability to perform intellectual tasks. People are considered behaving intelligently when they choose courses of action that are relevant to their goals, reply coherently and appropriately to questions that are put to them, and solve problems.

Intelligence has often been defined as the ability to perform intellectual tasks, and can be studied from the behavior of intelligent organisms or intelligent programs. A theory of intelligent processes or the computational principles can be constructed on the basis of contributing disciplines concerned with intelligence such as psychology (brain research, experimental psychology with behaviorism, gestalt psychology, psychometrics, neuropsychology), artificial intelligence (within computer science), linguistics and psycholinguistics, philosophy and neuroscience.

Visual intelligence is involved in thinking through visual processing, which is spatial rather than linear and allows gaining insight into difficult to analyze schemes and finding patterns and order in complex structures. The use of visual language by creating images to communicate concepts makes possible the visual exchange of information. Learning through visual and spatial thinking can produce meanings and connotations that cannot be achieved using language alone.

Ralf Lengler (2006) brought into the researchers' attention the so-called "Flynn Effect: "Since the eighties it became apparent that average IQ scores, especially those tested with non-verbal IQ tests, have been increasing steadily, particularly in industrialized countries (Flynn, 1980; Flynn & Dickens, 2001). The collected data suggest that the "Flynn Effect" – the successive increase in scores with each new birth cohort – is continuing (Flynn, 1994). Visual processing capabilities of students are increasing even more with playing up-to-date computer games that simulate 3D worlds" (Lengler, 2006). The growing visual literacy creates demand for more advanced and very up-to-date knowledge visualizations and the understanding of visual intelligence; therefore, visual literacy is an essential step for producing more efficient knowledge visualizations (Lengler, 2006). This author has also stressed that the growth in visual processing skills resulting from the exposure to visual media where video, TV, and movie productions provide rich information with the decreasing length of the average shot. Since students are able to perceive, understand, and process increasingly complex visual messages, visual way of communication becomes more and more effective and important. According to Lengler, visual literacy is a characteristic that depends upon visual intelligence. He lists the vision competencies as the abilities to:

- Speedily locate, identify and assess patterns.
- Speedily assign complex shapes to visual categories.
- Structure, store, and recall objects and paths in maps.
- Reconfigure shapes into new objects.
- Express concepts with visual means in a wide variety of ways.
- Construct meaning by integrating different associated visual messages.
- Imagine and rotate objects in 3D space.
- Simulate the future behavior of objects, based on their pattern of change in a given time period.
- Deduce the rules that govern patterns.

It is usually assumed that the vision competencies and skills advance thinking, problem solving, decision-making, discovering solutions, creating and communicating messages, and learning. Some hold that visual literacy involves visual thinking where critical review of a visual content includes asking the descriptive, analytical, interpretative, and evaluative questions; visual communication that enables effective presentation of knowledge or a creative product; and visual learning that facilitates construction of knowledge. In education, the level of digital art literacy is considered dependent on the balance between the components of traditional art education (theory, art history, critique and studio work) and the software and technological literacy. By expanding visual imagery students improve their skills in finding out and interpreting patterns, transforming and transmitting visual information. Visualizations such as animations improve understanding of systems and processes that change over time. Since students are able to perceive, understand, and process progressively more complex visual messages, visual way of communication and learning becomes more and more effective and important. The growing visual literacy creates a demand for more advanced and very up-to-date knowledge visualizations. At the same time, visual literacy is a prerequisite and an essential step for producing more efficient knowledge visualizations (Lengler, 2006).

Computer Graphics and the Development of Visual Literacy

A rich supply of imagery and a strong exposure to visual media create the environmental feedback resulting both in richer information delivery by the media and spectacular growth in visual processing skills (Flynn and Dickens, 2001). Video, TV, and movie productions, with their decreasing length of the average shot, contribute to this effect. Visual processing capabilities of students are increasing even more with playing up-to-date computer games that simulate 3D worlds (Lengler, 2006). In education, the level of digital art literacy depends on the balance between the components of traditional art education (theory, art history, critique and studio work) and the software and technological literacy. Visualizations such as animations improve understanding of systems and processes that change over time.

The visual and cognitive development of the students may be an important factor in the effectiveness of the computer art treatment. The Piaget's model of cognitive development delineated the formation of logical, mathematical, and scientific thought but not the growth of creative activity as related to cognitive development. Computer graphics have usually been applied to further students' comprehension of spatial relations, their symbolic development, and their response to computer-generated feedback (Wohlwill, 1988). They also allow for facilitating the creative process in which the students choose and define the problem. McWhinnie (1989) discussed the use of computers in developing visual literacy and stressed the importance of the right brain in various areas of creative behavior and in developing drawing skills. Wohlwill found that the computer permits one to capture the creative process of constructing one's own imagination, so "this medium may shed light on the basis of the alleged decline in children's imaginative and artistic productivity during the school years" (Wohlwill, 1988, p.130), a decline which may be caused by their concern for realism and objective literalism at the expense of expression and free imagination. The lack of control that students can exert in their artistic freehand drawings may be another reason for discomfort. The medium of computer graphics may offer a strategy to counteract this decline. With its precision and clear definition, it may provide a balance between the figurative and operative, or between the analytical and the intuitive. The process of creating computer art graphics is more determinate in its form than the freehand drawing that is analogically structured. Drawing on the computer provides the students with cognitive control over their creations.

Visual Intelligence Tests

The understanding of visual intelligence is essential to advance visual science. Non-verbal intelligence tests that tell about visual abilities have been applied for more than 80 years with high reliability and internal consistency (Rezaei and Katz, 2004). Raven's Progressive Matrices (Raven, 1981) is a widely used non-verbal, multiple-choice test of reasoning. In each test item, one is asked to find the missing part required to complete a pattern. Each set of items, designed as progressively more complicated matrices, require greater cognitive capacity to encode and analyze. According to the Berlin model (BIS-4, Bucik and Neubauer, 1996), a structure of the measurable general intelligence factor G measures working memory capacity. The model includes four Operations components (processing speed, memory, creativity, processing capacity), and three Contents components (verbal, numerical, spatial-figurative ability). The model was constructed with the use of cluster and factor analysis out of 2,000 tests. Visual competencies can be evaluated with relevant tests, such as processing speed, memorization, figural analogy, continuation, image rotation, pictorial reasoning, and other tests.

Multiple Intelligences

In the early eighties Howard Gardner (1993, 1993/2006, 1993/2011) defined intelligence as the capacity to solve problems or make things that are valued in a culture (at least one cultural setting or community). For him, it is more important to discover areas of strength and to build upon them than it is to fret too much about areas of weakness. In the eighties, Gardner developed a theory of multiple intelligences – a system of classifying human abilities, and posed suggestions about how to encourage learning in ways that respect the individual interests and strengths of children. He indicated in his research that human cognitive competence could be described in terms of abilities, talents or mental skills, which we call intelligences. Human beings are capable of developing capacities of an exquisitely high order in many semi-autonomous intellectual realms. He thought of those processing capabilities in terms of environmental information processing devices, for example, he considered the perception of certain recurrent patterns, including numerical patterns, to be the core of logical mathematical intelligence (Gardner, 1983). His list of intelligences include:

1. **Linguistic Intelligence:** The ability to use language to express meaning, tell a story, react to stories, learn new vocabulary or languages. Poets exhibit this kind of ability in its fullest form.
2. **Logical/Mathematical Intelligence:** Individuals that have these two kinds of intelligence strong are prized in schools and especially in school examinations.
3. **Spatial Intelligence:** Appreciation of large spaces and/or local spatial layouts, the ability to form a mental model of a spatial world and to be able to carry out that model. For example, it is the ability to form an image of large (a block building) and local (a home) spatial layouts and to find one's way around a new building.
4. **Musical Intelligence:** Capacity to create and perceive musical patterns.
5. **Bodily-Kinesthetic Intelligence:** The ability to use the body or parts of the body (hands, feet, etc.) to solve problems or to fabricate products, as in playing a ballgame, dancing, or making objects with hands. Dancers, athletes, surgeons and crafts people all exhibit highly developed bodily-kinesthetic intelligence.

Two forms of personal intelligence one oriented toward the understanding of other persons, the other toward an understanding of oneself.

6. **Interpersonal Intelligence:** One of the forms of personal intelligence is oriented toward the understanding of other persons, and means the ability to understand other people: what motivates them, how they work, how to work cooperatively and effectively with them. Successful salespeople, politicians, teachers, clinicians, and religious leaders are all likely to be individuals with high degree of interpersonal intelligence.
7. **Intrapersonal Intelligence:** An understanding of oneself, is the ability turned inward to form an accurate model of oneself, to understand things about oneself, how one is similar to, different from others; remind oneself to remember to do something; know how to soothe oneself when sad and to be able to use that model to operate effectively in life.

This list has been then enlarged by including the eighth form of intelligence (Gardner, 1997, pp. 35 6):

8. **Naturalistic Intelligence:** Apprehension of the natural world, as epitomized by skilled hunters or botanists; the ability to recognize species of plants or animals in one's environment; for example, to learn the characteristics of different birds, and other capacities.

Gardner regarded mastering of symbolic systems as the principal mission of modern educational systems (Gardner, 1988). He stressed the need for studying the relation between artistic and scientific forms of knowledge, because art is seen in a cognitive view as a matter of mind, and artistic production involves the use of symbols. His view of intelligences suggested that the arts might be helpful in facilitating intellectual potential. Gardner investigated also the extraordinariness. He described four forms of extraordinary individuals: a master who gains complete mastery over one or more domains of accomplishment; a maker who devotes energies to the creation of a new domain; one who introspects and explores his/her inner life; and an influencer who has an impact on other individuals.

The Structure-of-Intellect Model

Joy Paul Guilford (1967, 1968) developed the structure-of-intellect (SI) model consisting of cells depicting the intellectual abilities classified in three intersecting ways. The three dimensions of the model represent: operation (evaluation, convergent production, divergent production, memory, cognition); product (units, classes, relations, systems, transformations, implications); and content (figural, symbolic, semantic, behavioral).

In his triarchic theory of intelligence, Robert Sternberg (1985, 2011, p. 584) depicted intelligence as comprising three aspects dealing with the relation of intelligence:

1. How intelligence relates to the internal, inner world of the person, through three types of processes: metacomponents or executive planning, monitoring, and evaluation processes; performance processes; and knowledge-acquisition processes allowing to learn, to comprehend, to remember information (a componential subtheory).
2. How intelligence relates to experience (the experiential subtheory).
3. How intelligence relates to the external world of the individual (the contextual subtheory).

Sternberg (1985, 1999) explored higher-order reasoning in post-formal operational thought and defined orders of relations in terms of what is related: first-order relations are between primitive terms, second-order relations are relations between relations, and so on. He described the stages of cognitive development in terms of his theory: the concrete operational stage applies first-order relations; transition to formal means second-order relations; post formal may mean the third-order analogies in a 3-D semantic space, analogies between two analogies: (A_1: B_1:: C_1: D_1):: (A_2: B_2:: C_2: D_2). Examples of the third-order analogy are: (Bench: Judge:: Pulpit: Minister):: (Head: Hair:: Lawn: Grass) or (Machine Language: Hardware:: Programming Language: Software):: (Language of the Brain: Neural Connections:: Ordinary Language: Cognitive Structures).

With the cognitive approach, Robert J. Sternberg (2011) defined intelligence as a mental activity directed toward purposive adaptation to, selection, and shaping of, real-world environments relevant to one's life. He discerned meta-components (used in problem-solving and decision-making), performance components (perceiving problems in long-term memory, perceiving relations between objects, and applying them in new sets of terms) and knowledge-acquisition components in the work of mind (used in obtaining information). However, different contexts and tasks require different types of giftedness: componential/analytical, experiential/creative, and practical/contextual. In order to understand adult intelligence, main aspects of intelligence may be explored in terms of the relation to the internal or mental world of the learner, its relation to experience, and its relation to the surrounding world. Sternberg characterized the differences between everyday problems and academic or test-taking problems. The academic problems are pre-recognized, predefined, and well structured; most school problems have one right answer; academic problems provide relevant information; in school settings, there is clear feedback; and schools emphasize individual problem solving. Sternberg's theory of intellectual styles, based on a notion of mental self-government, concerns the ways in which people use their intelligence.

Artificial Intelligence

Computer scientists construct intelligent systems able to process information in an intelligent way. This part of computer science, which began about 1956 when the computer scientist John McCarthy coined this term, focuses on the study and design of intelligent agents, which perceive their environment and take the most possible successful actions. Tools for accomplishing these goals include mathematically based search and optimization, logic, probability methods based on computing for economics, and others.

Advancing the artificial intelligence domain involves creating agents that think humanly and rationally as well as act humanly and rationally (Russel & Norvig, 2009). Systems of agents and their societies may include software, robots, and humans. Multi agent systems (MAS) may interact with each other and/or with an environment to exchange data and support human social activities such as communication, cooperation, coordination, negotiation, sharing, or competition. Intelligent agents may observe through sensors to act rationally (learn and use knowledge) by taking autonomous actions aimed at performing their design objectives (Trajkovski, 2009, 2010).

As stated by Boden (2006),

Cognitive science uses abstract (logical/mathematical) concepts drawn from artificial intelligence (AI) and control theory, alias cybernetics.

- *AI tries to make computers do the sorts of things that minds can do. These things range from interpreting language or camera input, through making medical diagnoses, and constructing imaginary (virtual) worlds, to controlling the movements of a robot.*
- *Control theory studies the functioning of self-regulating systems. These systems include both automated chemical factories and living cells and organisms. (Boden, 2006, p. 4)*

Creating or simulating intelligence require several computer's capabilities (Luger, 2004) including reasoning, perception (with object and facial recognition), deduction, problem solving, creativity and imagination, general and social intelligence and skills, knowledge representation (including ontologies representing concepts and relationships), planning of actions (including multi-agent planning, emergent behavior as displayed in evolutionary algorithms and in swarm intelligence) machine learning, natural language processing (including text mining for information retrieval and machine translation), robotic motion and object manipulation, among other problems.

Scientists working on biologically inspired artificial intelligence (Floreano & Mattiussi, 2008) are studying theories and observations of biological systems that result from evolutionary processes: populations' diversity, heredity, and selection processes leading to genetic evolution and mutations. They develop methods to combine engineering with technologies aimed at constructing evolutionary algorithms to create collective systems presenting self-organization, particle swarm organization, ant colony organization, and swarm robotics, along with developing models of evolutionary dynamics, competition, and cooperation. Artificial systems characterized by the bio-inspired artificial intelligence include evolutionary systems and artificial evolutionary developmental systems, cellular systems and cellular automata, artificial and hybrid neural networks, artificial immune systems, and behavioral systems, among other approaches.

Singularity

The writer Vernor Vinge popularized the term singularity; he believed that artificial intelligence, human biological enhancement, and computer cognitive abilities would surpass these of any human being. Some researchers hypothesize the future emergence of greater-than-human super intelligence, partly because people would undergo intelligence explosion (a term coined in 1965 by I. J. Good) and partly because "computer engineers would possibly build an entity with intelligence at a level able to compete with human intelligence" (Joshua Fox, a blog at the Singularity Institute website http://singularity.org/blog/2012/08/). In 2009, Ray Kurzweil and Peter Diamandis established a Singularity University in Silicon Valley, CA. Since then, the theme is being vividly discussed by the computer scientists, artificial intelligence researchers, and robotic engineers, for example at the Association for the Advancement of Artificial Intelligence conferences and at the Singularity Institute for Artificial Intelligence and the Future of Humanity Institute forum.

Collective Intelligence

Some animals who form organized groups, such as schools of fish, flocks of birds, ant colonies, or swarms of insects, display swarm intelligence where interactions between individual members lead to emergence of behavior that is often optimal for the whole group. Members of a school, flock, colony, or swarm exhibit collective behavior that is decentralized and self-organized. This shared or group intelligence involves cooperation, coordination and collective actions coming from the consensus decision making. Basic model of flocking behavior simulated by Craig Reynolds (2011), who in 1986 developed boids – an artificial life program that simulates the flocking behavior of the birds, describes individual creatures as controlled by three simple rules:

- **Separation:** Steer to avoid crowding neighbors (short range repulsion).
- **Alignment:** Steer towards the average heading of local flock mates.
- **Cohesion:** Steer to move toward the average position of neighbors (long range attraction).

With these three simple rules, the modeled flock moves in an extremely realistic way, creating complex motion and interaction that would be hard to create otherwise. Collective intelligence appears in computer networks, business, political science, sociobiology, mass communication, and many other domains.

Several games such as the Sims series, and also Second Life depend on collective intelligence. Data mining models designed for particular games and the behavior learning models based on evolutionary optimization of mixed-games can serve to predict outcomes in the real-world time-series data analysis such as the stock market (Yu Du, Dong, Quin, &Wan, 2011). One may imagine future sculptures placed in public spaces to allow the passer-by people interact with these sculptures by changing their appearance and behavior, maybe even competing for attaining the final result according to their will.

CONCLUSION

Basic information about human cognition and cognitive science, which relates to visual reasoning, aesthetic emotions, and art, is considered essential to developing cognitive computing theories, investigations on information-processing mechanisms in computing, and working on constructing cognitive computers that perceive, conclude, and learn. Studying neural structures and functions of the brain supports theories about cognitive thinking, memory, and philosophical approaches to perceptual thinking. Cognitive informatics encompasses cognitive science, neurophysiology, psychology, and information sciences such as computer science, information sciences, and intelligence science. It enables designing engineering applications. Cognitive informatics specialists search for the potential applications of information processing and natural intelligence to cognitive computing. This chapter links cognitive processes with processes involved in visual thought and visual problem solving, which will be applied in interactive projects offered in this book.

REFERENCES

Aristotle (2013). *Poetics*. Oxford University Press.

Arlin, P. K. (1974). *Problem finding: The relation between selective cognitive process variables and problem-finding performance*. (Unpublished doctoral dissertation). University of Chicago, Chicago, IL.

Arlin, P. K. (1984). Adolescent and adult thought: A structural interpretation. In M. L. Commons, F. A. Richards, & C. Armon (Eds.), *Beyond formal operations: Late adolescent and adult cognitive development*. New York: Praeger.

Arnheim, R. (1969/2004). *Visual thinking: Thirty-fifth anniversary printing*. University of California Press.

Arnheim, R. (1988). *The power of the center - A study of composition in the visual arts*. Berkeley, CA: University of California Press.

Arnheim, R. (1990). Language and the early cinema. *Leonardo*, 3–4.

Baddeley, A. (2000). The episodic buffer: A new component of working memory? *Trends in Cognitive Sciences*, *4*(11), 417–423. doi:10.1016/S1364-6613(00)01538-2 PMID:11058819.

Baddeley, A., Eysenck, M. W., & Anderson, M. C. (2009). *Memory*. New York: Psychology Press.

Baddeley, A. D., & Hitch, G. J. L. (1974). *Working memory*. Academic Press.

Baddeley, A. D., Thompson, N., & Buchanan, M. (1975). Word length and the structure of memory. *Journal of Verbal Learning and Verbal Behavior*, *1*, 575–589. doi:10.1016/S0022-5371(75)80045-4.

Bancroft, J., & Wang, Y. (2011). A computational simulation of the cognitive process of children knowledge acquisition and memory development. *International Journal of Cognitive Informatics and Natural Intelligence*, *5*(2), 17–36. doi:10.4018/jcini.2011040102.

Bettencourt, A. (1989). *What is constructivism and why are they all talking about it?*. Journal Announcement: RIEMAR91. (ERIC Document Reproduction Service No. ED325402).

Binkley, T. (1990). Digital dilemmas. *Leonardo*, 13–19.

Binkley, T. (1996). Personalities at the salon of digits. *Leonardo*, *29*(5), 337–338. doi:10.2307/1576396.

Bloom, B. S. (1956). *Taxonomy of educational objectives*. Boston, MA: Allyn and Bacon.

Boden, M. A. (2006). *Mind as machine. A history of cognitive science*. Oxford, UK: Clarendon Press.

Boden, M. A. (2010). *Creativity and art: Three roads to surprise*. Oxford University Press.

Borst, G., Ganis, G., Thompson, W. L., & Kosslyn, S. M. (2011). *Representations in mental imagery and working memory: Evidence from different types of visual masks*. DOI 10.3758/s13421-011-0143-7. Retrieved May 6, 2012, from http://isites.harvard.edu/fs/docs/icb.topic561942.files/Borst_Ganis_Thompson_Kosslyn_2011.pdf

Borst, G., & Kosslyn, S. M. (2008). Visual mental imagery and visual perception: Structural equivalence revealed by scanning processes. *Memory & Cognition*, *36*(4), 849–862. doi:10.3758/MC.36.4.849 PMID:18604966.

Boyd, R. B. (1989). *Identifying and meeting the needs of students functioning at the concrete operations stage of cognitive development in the general chemistry classroom*. RIEJAN90 (ERIC Document Reproduction Service No. ED309984).

Brath, R. (2009). The many dimensions of shape. In *Proceedings of the International Conference on Information Visualization*. IEEE.

Brath, R. (2010). Multiple shape attributes in information visualization: Guidance from prior art and experiments. In *Proceedings of the 2010 14th International Conference Information Visualisation*, (pp. 433-438). IEEE.

Bresciani, S., & Eppler, M. J. (2010). Choosing knowledge visualizations to augment cognition: The managers' view. In *Proceedings of the 14th International Conference on Information Visualisation*, (pp. 355-360). IEEE.

J. Brockman (Ed.). (2011). *Is the internet changing the way you think? The net's impact on our minds and future*. New York: Harper Perennial.

Broudy, H. S. (1991). Reflections on a decision. *Journal of Aesthetic Education*, *25*(4), 31–34. doi:10.2307/3332900.

Bucik, V., & Neubauer, A. C. (1996). Bimodality in the Berlin model of intelligence structure (BIS), a replication study. *Personality and Individual Differences*, *21*(6), 987–1005. doi:10.1016/S0191-8869(96)00129-8.

Burmester, M., Mast, M., Tille, R., & Weber, W. (2010). How users perceive and use interactive information graphics: An exploratory study. In *Proceedings of the 14th International Conference on Information Visualization,* (pp. 361-368). IEEE.

Burt, C. L. (2009). *Mental and scholastic tests*. Cornell University Library. (Original work published 1922).

Chernoff, H. (1973). The use of faces to represent points in k-dimensional space graphically. *Journal of the American Statistical Association*, *68*, 361–368. doi:10.1080/01621459.1973.10482434.

Chomsky, N. (1959). A review of B. F. Skinner's verbal behavior. *Language*, *35*(1), 26–58. doi:10.2307/411334.

Chung, D. N. (2012). *Language arts*. Western Washington University. Retrieved May 6, 2012, from http://faculty.wwu.edu/auer/Resources/Hayakawa-Abstraction-Ladder.pdf

Dehaene, S., Piazza, M., Pinel, P., & Cohen, L. (2003). Three parietal circuits for number processing. *Cognitive Neuropsychology, 20*(3/4/5/6), 487-506.

DeLoache, J. S. (2005, August). Mindful of symbols. *Scientific American*, 72–77. doi:10.1038/scientificamerican0805-72 PMID:16053140.

Denis, M. (1989). *Image and cognition*. Paris: Presses Universitaires de France.

Eco, U. (1990). *The limits of interpretation*. Bloomington, IN: Indiana University Press.

Eggebrecht, A. T., White, B. R., Chen, C., Zhan, Y., Snyder, A. Z., Dehlgani, H., & Culver, J. P. (2012, February 10). A quantitative spatial comparison of high-density diffuse optical tomography and fMRI cortical mapping. *NeuroImage*. doi:10.1016/j.neuroimage.2012.01.124 PMID:22330315.

Eng, H. (1931). *The psychology of children's drawings - Form the first stroke to the coloured drawing*. London: Kegan Paul, Trench, Trübner, & Co.

Eppler, M. J., & Pfister, R. A. (2010). Drawing conclusions: Supporting decision making through collaborative graphic annotations. In *Proceedings of the 14th International Conference on Information Visualization*, (pp. 369-374). IEEE.

Erdem, A., & Karaismailoglu, S. (2011). Neurophysiology of emotions. In Affective Computing and Interaction: Psychological, Cognitive, and Neuroscientific Perspectives (pp. 1-24). IGI Global. ISBN 1616928921

Floreano, D., & Mattiussi, C. (2008). *Bio-inspired artificial intelligence: Theories, methods, and technologies*. The MIT Press.

Flynn, J. R. (1980). *Race, IQ, and Jensen*. London: Routledge and Kegan Paul.

Flynn, J. R. (1994). IQ gains over time. In R. J. Sternberg (Ed.), *The encyclopedia of human intelligence* (pp. 617–623). New York: Macmillan.

Flynn, J. R., & Dickens, W. T. (2001). Heritability estimates versus large environmental effects: The IQ paradox resolved. *Psychological Review, 108*(2), 346–369. doi:10.1037/0033-295X.108.2.346 PMID:11381833.

Gardner, H. (1978). *Developmental psychology*. Boston: Little, Brown and Co..

Gardner, H. (1983). Artistic intelligences. *Art Education, 36*(2), 47–49. doi:10.2307/3192663.

Gardner, H. (1988). Toward more effective arts education. *Journal of Aesthetic Education, 22*(1), 157–167. doi:10.2307/3332972.

Gardner, H. (1993c). *Art, mind, and brain: A cognitive approach to creativity*. New York: Basic Books, A Division of Harper Collins Publishers.

Gardner, H. (1993b/2006). *Multiple intelligences: The theory in practice*. New York: Basic Books. ISBN 978-0465047680

Gardner, H. (1993a/2011). *Frames of mind: The theory of multiple intelligences*. New York: Basic Books. ISBN 0465024335

Gazzaniga, M. S. (1997). The split brain revisited. *Scientific American, 279*(1), 50–55. doi:10.1038/scientificamerican0798-50 PMID:9648298.

Gazzaniga, M. S. (2005). Forty-five years of split-brain research and still going strong. *Nature Reviews. Neuroscience, 6*(8), 653–U651. doi:10.1038/nrn1723 PMID:16062172.

Getzels, J. W., & Csikszentmihalyi, M. (1970). Concern for discovery: An attitudinal component of creative production. *Journal of Personality, 38*, 91–105. doi:10.1111/j.1467-6494.1970.tb00639.x PMID:5435830.

Getzels, J. W., & Csikszentmihalyi, M. (1976). *The creative vision: A longitudinal study of problem finding in art*. New York: Wiley, a Wiley-Interscience Publication.

D. Gökçay, & G. Yildrim (Eds.). (2011). *Affective computing and interaction: Psychological, cognitive, and neuroscientific perspectives*. IGI Global.

Grigoryev, Y. (2012, March 16). How much information is stored in the human genome? *BitesizeBio*. Retrieved May 9, 2012, from http://bitesizebio.com/articles/how-much-information-is-stored-in-the-human-genome/

Gruber, H. E. (1973). Courage and cognitive growth in children and scientists. In M. Schwebel, & J. Ralph (Eds.), *Piaget in the classroom*. New York: Basic Books.

Guilford, J. P. (1967). *The nature of human intelligence*. New York: McGraw-Hill.

Guilford, J. P. (1968). *Intelligence, creativity and their educational implications*. San Diego, CA: Robert Knapp, Publ..

Hayakawa, S. I., & Hayakawa, A. R. (1941/1991). *Language in thought and action* (5th ed.). Harvest Original.

Hudson, A. (2011, May 21). Is graphene a miracle material? *BBC News*. Retrieved October 22, 2011, from http://news.bbc.co.uk/2/hi/programmes/click_online/9491789.stm

Inhelder, B. (1977). Genetic epistemology and developmental psychology. *Annals of the New York Academy of Sciences, 291*, 332–341. doi:10.1111/j.1749-6632.1977.tb53084.x.

Inhelder, B., & Piaget, J. (1958). *Growth of logical thinking from childhood to adolescence*. New York: Basic Books. doi:10.1037/10034-000.

Jackendoff, R. S. (2007). *Language, consciousness, culture: Essays on mental structure*. Cambridge, MA: MIT Press.

Johnson, S. H. (1987). *A cognitive-structural approach to adult creativity*. Paper presented at the Annual Symposium of the Jean Piaget Society. Philadelphia, PA.

Kant, I. (2007). *Critique of judgement* (Oxford World's Classics). Oxford University Press.

Kernbach, S., & Eppler, M. J. (2010). The use of visualization in the context of business strategies: An experimental evaluation. In *Proceedings of the Information Visualisation (IV) 14th International Conference*. ISBN 978-1-4244-7846-0

Korzybski, A. (1995). *Science and sanity: An introduction to non-Aristotelian systems and general semantics* (5th ed.). Brooklyn, NY: Institute of General Semantics. (Original work published 1933).

Kosslyn, S. M. (1980). *Image and mind*. Cambridge, MA: Harvard.

Lambert, N. (2011). From imaginal to digital: mental imagery and the computer image space. *Leonardo*, *44*(5), 439–443. doi:10.1162/LEON_a_00245.

Lengler, R. (2006). Identifying the competencies of 'visual literacy' - A prerequisite for knowledge visualization. In *Proceedings of the Tenth International Conference on Information Visualisation (iV'06)*. iV.

Lewicki, P., Hill, T., & Czyzewska, M. (1992). Nonconscious acquisition of information. *The American Psychologist*, *47*, 796–801. doi:10.1037/0003-066X.47.6.796 PMID:1616179.

Limson, J. (2010). Abstract engravings show modern behavior emerged earlier than previously thought. *Science in Africa*. Retrieved May 9, 2012, from http://www.scienceinafrica.co.za/2002/january/ochre.htm

Logothetis, N. K., Pauls, J., Augath, M., Trinath, T., & Oeltermann, A. (2001). Neurophysiological investigation of the basis of the fMRI signal. *Nature*, *412*, 150–157. doi:10.1038/35084005 PMID:11449264.

Lowenfeld, V. (1947). *Creative and mental growth*. New York: Macmillan Publishing Company.

Lowenfeld, V., & Brittain, L. (1987). *Creative and mental growth*. Prentice Hall.

Luger, G. F. (2004). *Artificial intelligence: Structures and strategies for complex problem solving* (5th ed.). Addison Wesley.

Mackworth, N. H. (1965). Originality. *The American Psychologist*, *20*, 51–66. doi:10.1037/h0021900 PMID:14251990.

Marr, D. (1982/2010). *Vision: A computational investigation into the human representation and processing of visual information*. The MIT Press.

McWhinnie, H. J. (1989). *The computer & the right side of the brain*. Journal Announcement: RIESEP90. (ERIC Document Reproduction Service No. ED 318461).

Mirsky, S. (2011, November). Article. *Scientific American*, 96.

Mitsunaga, N., & Asada, M. (2004). Visual attention control by sensor space segmentation for a small quadruped robot based on information criterion. *Lecture Notes in Computer Science*, 2377.

Pacific Northwest National Laboratory. (2008). *Cognitive informatics*. Retrieved April 2, 2013, from http://www.pnl.gov/coginformatics/

Paivio, A. (1971). *Imagery and verbal processes*. New York: Holt, Rinehart, and Winston.

Paivio, A. (1986/1990). *Mental representations: A dual coding approach*. Oxford University Press.

Paivio, A. (1991). Dual coding theory: Retrospect and current status. *Canadian Journal of Psychology, 45*(3), 255–287. doi:10.1037/h0084295.

Piaget, J., & Inhelder, B. (1971). *Mental imagery in the child: A study of development of imaginal representation*. New York: Basic Books Inc., Publishers.

Pouivet, R. (2000). On the cognitive functioning of aesthetic emotions. *Leonardo, 33*(1), 49–53. doi:10.1162/002409400552234.

Raven, J. (1981). *Manual for raven's progressive matrices and vocabulary scales*. San Antonio, TX: Harcourt Assessment.

Reynolds, C. (2011). Interactive evolution of camouflage. *Artificial Life, 17*(2), 123-136. doi:10.1162/artl_a_00023. Retrieved May 13, 2012, from http://www.mitpressjournals.org/doi/abs/10.1162/artl_a_00023?journalCode=artl

Rezaei, A. R., & Katz, L. (2004). Evaluation of the reliability and validity of the cognitive style analysis. *Personality and Individual Differences, 36*(6), 1317–1327. doi:10.1016/S0191-8869(03)00219-8.

Russell, S., & Norvig, P. (2009). *Artificial intelligence: A modern approach* (3rd ed.). Prentice Hall.

Ryder, A. P. (n.d.). *Artcyclopedia*. Retrieved May 12, 2012, from http://www.artcyclopedia.com/artists/ryder_albert_pinkham.html

Saneyoshi, A., Niimi, R., Suetsugu, T., Kaminaga, T., & Yokosawa, K. (2011). Iconic memory and parietofrontal network: fMRI study using temporal integration. *Neuroreport, 22*(11), 515–519. doi:10.1097/WNR.0b013e328348aa0c PMID:21673607.

Slotnick, S. D., Thompson, W. L., & Kosslyn, S. M. (2012). Visual memory and visual mental imagery recruit common control and sensory regions of the brain. *Cognitive Neuroscience, 3*(1), 14-20. Retrieved May 6, 2012, from http://www.tandfonline.com/doi/abs/10.1080/17588928.2011.578210

Sperling, G. (1960). The information available in brief visual presentations. *Psychological Monographs, 74*, 1–29. doi:10.1037/h0093759.

Stanovich, K. E., West, R. F., & Toplak, M. E. (2011). Intelligence and rationality. In *The Cambridge Handbook of Intelligence*. Cambridge, UK: Cambridge University Press. doi:10.1017/CBO9780511977244.040.

Sternberg, R. (2011). *Cognitive psychology*. Wadsworth Publishing.

Sternberg, R. J. (1985). *Beyond IQ: A triarchic theory of human intelligence*. New York: Cambridge University Press.

Sternberg, R. J. (1999). The theory of successful intelligence. *Review of General Psychology, 3*, 292–316. doi:10.1037/1089-2680.3.4.292.

Sternberg, R. J. (2007). *Wisdom, intelligence, and creativity synthesized*. Cambridge University Press.

Sternberg, R. J., & Kaufman, S. B. (Eds.). (2011). *The Cambridge handbook of intelligence*. Cambridge University Press. doi:10.1017/CBO9780511977244.

Targowski, A. (2011). *Cognitive informatics and wisdom development: Interdisciplinary approaches*. Hershey, PA: IGI Global.

Trajkovski, G. (2010). *Developments in intelligent agent technologies and multi-agent systems: Concepts and applications*. Hershey, PA: IGI Global. doi:10.4018/978-1-60960-171-3.

Trajkovski, G., & Collins, S. G. (2009). *Handbook of research on agent-based societies: Social and cultural interactions*. Hershey, PA: IGI Global. doi:10.4018/978-1-60566-236-7.

Ursyn, A., & Banissi, E. (Eds.). (2003). Visualizing data sets. *The YLEM Journal: Artists Using Science & Technology*, *23*(10).

Vallortigara, G., & Rogers, L. J. (2005). Survival with an asymmetrical brain: advantages and disadvantages of cerebral lateralization. *The Behavioral and Brain Sciences*, *28*(4), 575–633. doi:10.1017/S0140525X05000105 PMID:16209828.

Vinciarelli, A. (2011). Towards a technology of nonverbal communication: Vocal behavior in social and affective phenomena. In *Affective Computing and Interaction: Psychological, Cognitive, and Neuroscientific Perspectives* (pp. 133–156). Hershey, PA: IGI Global.

Vivekjivandas, S., & Dave, J. (2011). *Hinduism: An introduction*. Ahmedabad, India: Swaminarayan Aksharpith.

Vygotsky, L. S. (1986). *Thought and language* (A. Kozulin, Ed.). The MIT Press.

Wang, Y. (2011). Towards the synergy of cognitive informatics, neural informatics, brain informatics, and cognitive computing. *International Journal of Cognitive Informatics and Natural Intelligence*, *5*(1), 75–93. doi:10.4018/jcini.2011010105.

Wohlwill, J. F. (1988). Artistic imagination during the latency period revealed through computer graphics. In G. Forman, & P. B. Pufall (Eds.), *Constructivism in the Computer Age* (pp. 129–150). Hillsdale, NJ: Lawrence Erlbaum Associates, Publishers.

Wong, K. (2002, January 11). Ancient engravings push back origin of abstract thought. *Scientific American*. Retrieved April 21, 2012, from http://www.scientificamerican.com/article.cfm?id=ancient-engravings-push-b

World Question Center. (2010). Retrieved October 27, 2012, from www.edge.org/question-center.html, http://www.edge.org/q2010/q10_3.html#kosslyn

Yu, D., Dong, Y., Qin, Z., & Wan, T. (2011). Exploring market behaviors with evolutionary mixed-games learning model. In *Proceedings of the Computational Collective Intelligence: Technologies and Applications*. ICCCI. doi:10.1007/978-3-642-23935-9_24

Chapter 6
Semiotic Content of Visuals and Communication

ABSTRACT

The semiotic content of visual design makes a foundation for non-verbal communication applied to practice, especially for visualizing knowledge. The ways signs convey meaning define the notion of semiotics. After inspection of the notions of sign systems, codes, icons, and symbols further text examines how to tie a sign or symbol to that for which it stands, combine images, and think figuratively or metaphorically. Further text introduces basic information about communication through metaphors, analogies, and about the scientific study of biosemiotics, which examines communication in living organisms aimed at conveying meaning, communicating knowledge about natural processes, and designing the biological data visualization tools.

INTRODUCTION

Concepts about semiotics pertain to a great deal of facts and events described in this book, such as art, computing solutions, social interactions, interaction with technology, machines, and practically everything else, so they will return as themes for discussion in the chapters that follow. Themes examined in this chapter such as natural language, communication with signs, symbols, icons, codes, spoken and written communication tools and conventions allow describing communication through computer graphics and art, literature, and computational solutions, and so they will be useful in working on knowledge inspired projects.

DOI: 10.4018/978-1-4666-4703-9.ch006

We can see rich semiotic content in descriptions of processes and events occurring in natural or technology-induced settings. Maybe it's good to remember that "meaning" is always the result of social conventions, even when we think that something is natural or characteristic, and we use signs for those meanings. Therefore, culture and art can be seen as a series of sign systems. Semioticians analyze such sign systems in various cultures; linguists study language as a system of signs, and some semioticians examine film as a system of signs. A sign tells about a fact, an idea, or information. It takes a form of a conventional shape or form. Albert Camus (1913-1960), a French-Algerian philosopher and author wrote (Camus, 1951/1992, *The Rebel*, part 4), "Just as all thought signifies something, so there is no art that has no signification."

Applying semiotics to practice, for example of the industrial design, entails making the choices between objectivism – a knowledge system that represents the world objectively and constructivism, which postulates that knowledge is actively built up. The studies on semiotic were based on objectivism, with the concepts of physically existing sign, its meaning, a symbol showing what it stands for, and an artifact telling what it expresses. Objectivism is widely seen as a belief in an observer-independent or culture-independent reality, with structures, codes, and laws ready to be described. Objects reside outside the observing person, so the theory may describe an objective reality. However, objectivist semiotics of the design products, which refer to something that exists independently, may easily disrespect the cognitive autonomy of other individuals and of other cultures. The objectivist and constructivist approaches may lead to different social practices.

Radical constructivism was developed by Chilean biologist Humberto Maturana (1970), Heinz von Foerster (2002), Ernst von Glassersfeld (1989, 1991), Klaus Krippendorff (1990), and several other authors. Maturana – the founder of constructivist epistemology (dealing with meanings and structures underlying constructivism) and biological constructivism, together with Francisco Varela introduced the concept of autopoiesis, which means self-creation and provides a basic dialectic between structure and function (Maturana & Varela, 1987/1992). Heinz von Foerster, one of originators of cybernetics and initiator the second-order cybernetics, contributed to the constructivist theory.

Ernst von Glasersfeld proposed two principles of radical constructivism: (1) Knowledge is not passively received but actively built up by the cognizing subject; (2) The function of cognition is adaptive and serves the organization of the experiential world, not the discovery of ontological reality (Glasersfeld 1989, p. 162). Glasersfeld calls his version radical constructivism because he claims that constructivism has to be applied to all levels of description. "Those who ... do not explicitly give up the notion that our conceptual constructions can or should in some way represent an independent, 'objective' reality, are still caught up in the traditional theory of knowledge" (Glasersfeld, 1991, 2007).

Klaus Krippendorff (2011a,b) proposed a radically social constructivism, which disagrees with the cognitivism put forward by von Glasersfeld, Heinz von Foerster, and Humberto Maturana. Klippendorff studies the meaning of designed objects, making critical choices between objectivism (a belief in an observer- or culture-independent reality) and constructivism (arising out of social practices based on understanding of one's own experiences and the understanding of participating individuals). The author considers participation in conversation and cooperative constructions of reality more important than observation and a representational theory of language because he finds conversation being the starting point of his conceptualizations of being human. Constructivism takes reality as residing neither outside of an observer, nor inside human mind, but emerging in practice or in social practice as a circular process of perception and action of conceiving and making things. The sand on a beach becomes meaningful for the thought: through sensory participation it a reality becomes manifest in practice. Key to the constructivist approach is not an objective reality but the understanding of our own experiences. The semantic discourse on meaning should be embedded in human understanding and embrace others' understanding of the social practices. For example, a notion of a gift becomes a social construction that arises out of social practices. A Dutch historian and cultural theorist Johan Huizinga (1872-1945) analyzed the notion of gift as a social construct in his book from 1938 "Homo Ludens: A Study of the Play Element in Culture" (Huizinga, 1950). For Krippendorff (1990) a notion of a gift becomes a social construction that arises out of social practices. Key to the constructivist approach is not an objective reality but understanding, own understanding of one's own experiences. Social practice becomes a dynamic unfolding of the constructions in the understanding of participating individuals (Figure 1).

Those with similar research interests may connect without leaving their physical space.

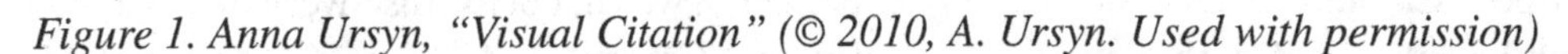

Figure 1. Anna Ursyn, "Visual Citation" (© 2010, A. Ursyn. Used with permission)

SIGN SYSTEMS, CODES, AND SEMIOTICS

The core of the domain of semiotics stems from philosophy and psychology, starting from Plato, Aristotle, Augustine, and then the medieval scholastic thought. In his paper "Eight Historical Paradigms of Cultural Studies and Their Semiotic Explication" Roland Posner (2009) described how the reinvention of humanities was motivated in the past and at present "by trying to account for new types of signs and sign processes that had become necessary for a successful life in its time," from the Greek cities of the 5th century BC to now, when "in the context of globalization at the turn of the third millennium, many European universities are now re-organizing the humanities into faculties of cultural studies grouped around *media studies.*"

A Swiss linguist Ferdinand de Saussure (1857-1915) devised his model of semiology and laid a scientific foundation for structuralist methodology of semiotics by introducing his core concepts of the sign, signifier, signified, and referent. De Saussure (1915) assumed a position that reality does not exist before language defines it. Hence, all signs are arbitrary; they have meaning because a community has agreed upon what they signify, not because they have some intrinsic meaning. According to de Saussure's first dyadic model, a two-part model of the sign is composed of a 'signifier' – the form that the sign takes and a 'signified' – the concept it represents. Thus the sign results from the association of the signifier with the signified. At the same time an American mathematician, scientist, pragmatist philosopher and logician Charles Sanders Peirce (1839-1914) described three types of signs: icons, indexes and symbols, which allow communication. As discussed by Marcello Barbieri (2010), the iconic, indexical and symbolic processes can be also described in animals. A sign becomes an icon when there is similarity between it and an object.

Figure 2 presents my student's work from a Computer Art Class. A tree became a strong icon existing in our nature- and computing-related iconography – the use of images as symbols. Our biology-inspired imagination compelled us to shape a widely used data structure as a hierarchical tree with linked nodes, starting from a root node, that resembles a living tree.

We can recognize an object as a tree because its features are common with other trees. That means we perform pattern recognition to assign an object to a particular mental category. When there is a physical link with an object, a sign becomes an index. Indexes allow humans and animals to infer the existence of something, for example learn about correlation existing between dark clouds and rain, find out that a pheromone is an index of a mating partner, and the smell of smoke is an index of fire. We can possibly consider links on the web pages as indexes, when a specially marked word refers to a related piece of information somewhere else on the Internet. A sign can also be a symbol when there is a conventional link is established between a sign and an object, often without any

Figure 2. Steffanie Sperry, "The Meaning of Tree" (© 2012, S. Sperry. Used with permission)

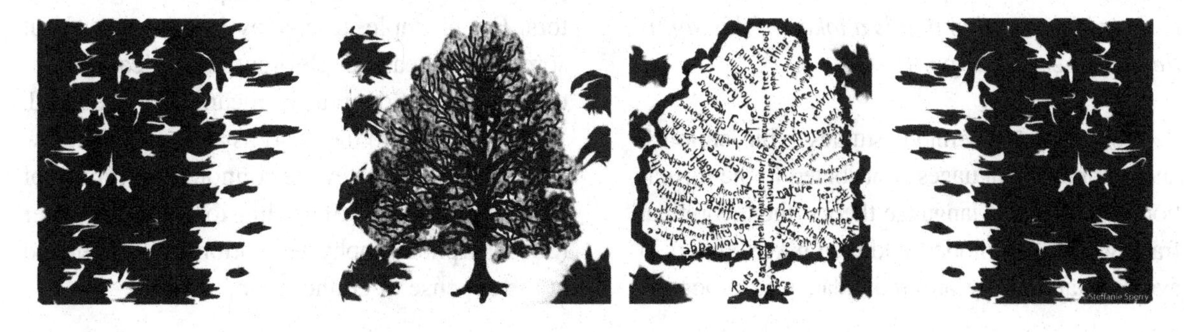

similarity or a physical link between them. For example, relationships between a flag and a country or a name and an object result from a convention. Peirce developed a semiotic set of principles with a triadic model of a sign consisting of the representamen – the form (material or abstract) that the sign takes; the interpretant – the sense made of the sign; and the referent – the object to which the sign stands for. Peirce (1931-1935/1958) considered logic as formal semiotics, a study of signs that is relevant to the universe, not only to linguistic studies. The semiotic triangle of Ogden and Richards (1923/1989) became a well-known model of the semiotic triangle that uses the concepts of: a sign with physically existing sign vehicle; meaning as a referent, the actual thing or event to which thinking or symbolization refers; a symbol and what it stands for, or an artifact and what it expresses.

Later on, semioticians examined natural language, communication with signs, symbols, icons, codes, spoken and written communication tools, and conventions that allow communication, mainly in art and literature. They also studied semantic relations between a sign and its denotation – its primary meaning. According to a semiotician Roland Posner, semiotics deals with sign processes (semiosis) and involves the following factors:

There is a sender, who intends to convey a message to an addressee and makes sure that he or she is connected with him or her through a shared channel. In preparing the intended message, the sender chooses an appropriate code and selects from it a signified (a meaning) that includes the message. Since the signified is correlated through the code with a corresponding signifier, the sender then produces a sign that is a token of this signifier. (Posner, 2009, pp. 10-11)

Application of semiotic studies to biosemiotics and computer languages contributed to the formation of the formal language theory, the theoretical framework of semiotics, and then played a part in avoiding ambiguity and redundancy in programming or markup languages. Further implementation to computer languages As stated by Tanaka-Ishii (2010), studies on programming languages and computer programs in terms of the semiotic discussion of signs help to situate certain technological phenomena within human networks and explain the process of creating meaning by both computers and humans. Research on symbolic systems of interconnected symbolic meanings (for example, at the Stanford University, 2012) focuses on dynamic relationships between symbols, human and machine intelligence, and includes natural language, programming languages, mathematical logic, and non-verbal communication, such as through sign languages.

"The Meaning" (Figure 3) symbolizes the language of gestures, which helps to impart, communicate, and convey what we want to say with sign languages, pantomime, acting ballet, and other means of non-verbal performance. A hand can tell a lot about a person who can be aged, tired, weary, exhausted, or fatigued. Hands may be well kept, and may be skilled, and thus you can recognize a good designer by the way they handle paper.

Signs

The semiotic content of visual design is important for non-verbal communication applied to practice, especially for visualizing knowledge or communicating online. People use signs to communicate or express something. For this purpose they assign denotations to signs and words – fixed, relevant, and logical meanings, and connotations – more subjective and immaterial, context-dependent meanings, all of them dependent on the anthropological, social, cultural, and psychological factors. For example, to convey information about somebody's heart people draw anatomical illustration and depict denotative meaning of the word. When referring to the heart as an icon, they draw a symbol of heart to evoke connotations typical of their cultural group. According to Daniel Chandler (2001), in photography denotation is foregrounded at the expense of connotation.

Figure 3. Anna Ursyn, "The Meaning" (© 2008, A. Ursyn. Used with permission)

Pablo Picasso's (1942) sculpture of a bull's head, which is composed of a bicycle handlebars and a saddle, has a form that functions as a sign. This sign is composed of three parts:

- The material form of the sculpture carried by the 'sign vehicle' (the saddle–handbars combination),
- Its meaning (a bull head), and
- Someone who interprets it.

One may see some tension between standardization of signs and the creative approach to design. Standardization of signs and their meanings makes possible their global, culture-independent understanding, facilitates communication and social orderliness, which is important in designing computer interfaces, icons used for manipulating computer screens, or public traffic sign systems. On the other hand, interacting with technology in an engaging, playful way helps to create meaning and construct individual worlds in the design process.

Icons

Art or design form can be approached in iconic and symbolic categories because icons and symbols help compress information in a visual way. A term 'icon' usually relates to an object or its graphic representation that has a symbolic meaning or refers to something by resembling or imitating it; thus a picture, a photograph, a mathematical expression, or an old-style telephone may be regarded as an iconic object. Thus, an iconic object has some qualities common with things it represents, by looking, sounding, feeling, tasting, or smelling alike. Visual artists use signs as vehicles of meaning and often use icons to bring to mind thoughts, represent, and organize perception of the world. Icons displayed on a computer screen are not physical objects but represent some objects such as file folders, documents, and applications present within a graphical user interface. Literary semioticians analyze texts in this respect, while computational and algebraic semiotic is a tool for the theory and practice of programming languages for computing.

A cultural icon is an easy to identify image of an object or concept with great cultural meaning and significance to a wide cultural group; it can be a symbol, person, logo, artwork, building, or other cultural artifact. In the media, many describe as "iconic" some well-known manifestations of popular culture. For example, Andy Warhol revealed and demonstrated general fascination with the celebrity image (for example, http://en.wikipedia.org/wiki/Marilyn_Diptych). Some art works and also objects have become the accepted American icons, such as a painting entitled "American Gothic" created by Grant Wood (1930), or an apple pie. One may say such everyday objects as recycling bins, disposable cups, credit cards, cell phones and even smartphones, which combine mobile phones with the mobile computing platform and the media players, can be considered iconic objects telling about our current civilization; intense globalization processes make them meaningful all over the world.

A group of people may accept a cultural icon as a symbol representing their shared beliefs, a period in history, social status, or a life style. "Symbols in Nature" (Figure 4) tells how many objects and creatures became for us the cultural icons, which we use in everyday conversations. At the same time, they are fundamental in biology-inspired computing, which knits together many fields of study (e.g., genetic algorithms, cellular automata, communication networks, or artificial life) with their biological counterparts.

To describe love, do not use forbidden words:
Heart, rainbow, moonlight.
After absorbing concepts of necessity, radiance, and spectrum
You will no longer need them.

Figure 4. Anna Ursyn, "Symbols in Nature" (© 2003, A. Ursyn. Used with permission)

Symbols

A notion of a symbol represents an abstract concept, not a thing, and is comparable to an abstract word (for example, a symbol of the ying-yang – an interplay of contrasting and complementary forces which engender and sustain the universe). As stated by Barbieri (2010, p. 212), "Symbols allow us to make arbitrary associations and build mental images of future events (projects), of abstracts things (numbers), and even of non-existing things (unicorns)."

A symbol does not relate to a specific object or picture but only contains meaning by itself. Symbols, for example letters, numbers, words, codes, traffic lights, and national flags do not resemble things they represent but refer to something by convention, like the word "red" represents red. We must learn the relationship between symbols and what they represent, especially when we are immersing in another culture. A symbol represents an abstract concept and is comparable to an abstract word. Highly abstracted drawings that show no realistic graphic representation become symbols. Symbols are omnipresent in our life, for example:

- An electric diagram that uses abstract symbols for a light bulb, wire, connector, resistor, and switch.
- An apple for a teacher or a bitten apple for a Macintosh computer.
- A map – typical abstract graphic device.
- A 'slippery when wet' sign.
- A plan for a house.

We encounter icons, symbols, and indexes on the web. Most web pages contain symbolic images – conventional signs, usually at the top of a page, that help users navigate by assigning their moves through the website and across the network. Many times these symbols on a website, as well as in electronic mail communication take form of the iconic, conventional signs; they develop into a specific language, often conveying personal messages and feelings. In a similar way communication through the use of avatars in a virtual social space applies ready expressions, gestures, and actions conventionally and comfortably collected at the action panels to animate avatars in an expressive way. Not only websites and e-mails but also game spaces are designed in an iconic way, so the players can identify optimal moves and links to additional tools, helpful hints, or other spaces.

Another semiotically rich area of activity relates to tattoos. A report by the Food and Drug Administration estimated that as many as 45 million Americans have tattoos, and they spend $ 1.65 billion annually on tattoos (Tattoos, 2012); However, applying tattoos may be sometimes a controversial issue – people with tattoos are excluded from some jobs. There may also be a removable tattoo picked from a book for body decoration, or placed in concealed places. Tattoos convey messages that are conventional (such as hearts pierced with an arrow), artistic (modern tattoo art displayed in flash art flash magazines) or aesthetically pleasing (such as flowers and angels popular in the 1960s), religions (Christian tattooing in Bosnia and Hercegovina was widespread, intended to prevent forced conversion to Islam), representing social or cultural groups (e.g., showing group membership), or otherwise (such as in case of the Japanese crime syndicate yakuza) restricted to specific circles. In other cases the tattoo semantics becomes helpful, as a non-verbal communication (in connecting to others). Individual reasons to display signs, symbols, and icons may include a need to commemorate a cherished person (such as Elvis Presley), apply a camouflage (to blend with the environment and become invisible for the enemy), to offer storytelling (like some Aboriginal hunters who honor animal spirits), document a rite of passage (e.g., among Freemasons), present oneself as a member of a prestigious formation (sometimes through choosing images of mermaids by marines and people in military service). The message can become dreadful (like tattoos showing numbers ascribed to the prisoners of the Nazi concentration camps during the WWII). In some cases tattoos convey offensive and blasphemous messages (like the Last Supper made out of frogs) or make a form of protest. Tattoo culture and its semiotic contents have been documented in many books and websites.

Symbolic drawings may convey messages about universal forces of which we are not always conscious. Colors are often used as symbols, for example, they signal the cold-warm water on faucets. Abstract art often uses abstract symbols, as well as idea sketches are abstract. We use abstract symbols when we order parts for electrical installation. When we have a map of an island, we can understand symbols and write a fiction about events on that island without being there. Writing programs requires using abstract thinking about described concepts. Visual communication occurs through visual symbols, as opposed to verbal symbols or words. This way of thinking is related to the methods of simulation and visualization because it brings forth an ability to perceive complex systems. Current approaches to product design apply images for visual thinking and use imagery as a medium for thought. We need to recognize a difference between concrete and abstract images or concepts because this way we can organize our knowledge and use only what we need. Semiotics is useful in knowing how it is all connected and make a distinction between a signal and its referent. We use symbols and link-node diagrams for abstract concepts in graphic languages. Symbolic shortcuts in human perception make a meaning of what we see or hear. And signals of that kind are grounded deeply in neural system. Signs, symbols and their meanings are the focus of art education. While cognitive scientists are usually focused on looking for the sources of human communication that are grounded in precise thinking, artists often use purposefully transformed simple signs to direct the thoughts of the viewers.

See Table 1 for Visual Response.

Figure 5 shows a work of my student who presented her daily activities as a part of a "Monopoly" board. The author of this work combined old, traditional symbols characteristic of the Monopoly game with pictorial symbols indicating current actions and circumstances.

To examine an example of art that conveys a rich symbolic content, we may look at an old calendar that combines naturalism of detail with overall decorative effect, thus conveying facts and meanings that have not been lost for contemporary viewers. The calendar is a part of the manuscript "Très Riches Heures" (c. 1413); it is the art treasure of France. Images from the calendar can be seen at the site of Les Trés Riches Heures du Duc de Berry: http://www.ibiblio.org/wm/rh/ or: http://sunsite.icm.edu.pl/wm/rh/. This medieval book of hours contains texts for each liturgical hour of the day - hence the name. The original is stored in the Chantilly museum, but is so degraded that it is no longer available to the public, except for the Web museum visitors. The "Très Riches Heures" was painted by the Limbourg brothers, Paul, Hermann, and Jean, born in the 1370s or 1380s. By 1408 the Limbourg brothers began to serve as artists for Jean Duc de Berry, one of the art lovers in France. Most of their other works have been lost. In 1416 all three Limbourg brothers died before the age of thirty, apparently killed by an epidemic. Their work is a landmark of the art of book illumination and an example of the International Gothic style.

See Table 2 for Your Response.

Images, sounds, or animations can be analyzed as metaphors and seen as important sign systems. A philosopher and educator Harry S. Broudy (1987) stressed the role of images in the development of the educated mind. He considered the relation of signals, symbols, and signs to their referents (the objects or ideas to which they relate) as a subject of thinking metaphorically by combining images at will. He perceived the imagined as a clue to reality: if one can imagine flying, one could manufacture wings. The mind that is set

Table 1.

Visual Response: Symbols of Daily Life
Draw symbols you use in your daily life, which represent you best. Then, draw your own sign to be placed on your door, the sign that sends a message to the visitors. Try to make a sign informing a passerby without any text that this is your room. This could even form your own logo. Caveat (a warning or caution): • Some symbols become a cliché and evoke an enormous amount of connotations, hints, and associations, for example, the "heart." symbol. Symbols are not the same in different parts of the world. For example, a sign "WC" ('water closet') is used in some parts of Europe for a restroom. In the US, an answer to a question, "Where is WC?" could often be, "WC? I do not know him." • Symbols may change their meaning. For example, an image of smoke going out of factory chimneys was considered a sign of progress in the times of industrial revolution and was even pictured on some currency bills. Nowadays, it means smog and pollution. • Some usages of symbols result in misunderstandings. For example, children can often see a bulb drawn over Einstein's head is depicted as a symbol of his genius. Thus for some children, Einstein + the light bulb = a bright idea, so Einstein was the one who invented a bulb.

Figure 5. Cayden Osley, "A Day" (©2012, C. Osley. Used with permission)

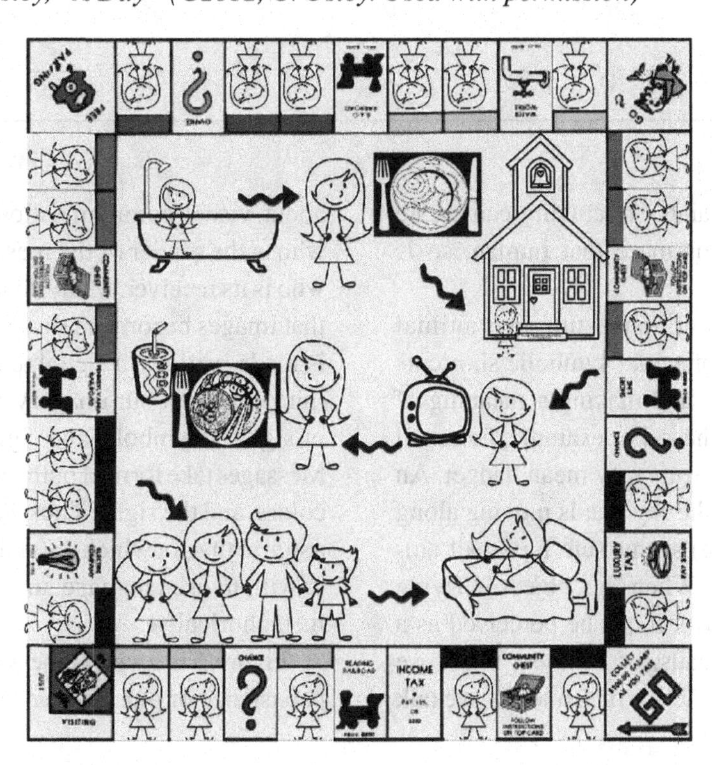

Table 2.

Your Response: Writing about Images
Go to the site of Les Trés Riches Heures du Duc de Berry: http://sunsite.icm.edu.pl/wm/rh/ or http://www.ibiblio.org/wm/rh/ and write an analysis of Elements and Principles of Design you can recognize while looking at these works of art.

free from the constraints of actuality can create something that seems more that human, so is feared and revered.

Ethologists: Naturalists studying animal behavior are aware of actual symbolic shortcuts in animal perception and making a meaning of what animals see or hear. For example, for small animals the rustle of leafs may mean danger. An elongated object on the sky that is moving along its axis (like a goose) seems safe for small animals on the ground. When this object is flying sideways (like a hawk), it can be perceived as a predator by frogs, snails, or bunnies. When we think about the arts in terms of semiotics, we talk about visual communication between the artist who is the sender of the message, and the viewer who is its receiver. Many concur with the opinion that images become messages when a distinction is made between a signal and its referent. When thinking about our imagery, we can see relations of signals, symbols, and signs to their referents. Messages take form of both the visual stimuli, e.g., colors, and the signs. Thus we can tie a symbol or a sign to that for which it stands to combine images at will, to use language and think figuratively or metaphorically.

Figure 6 is a visual message telling about the persistence of our approaches toward nature and environment. We visit pristine swamps and lakes to contemplate the static and dynamic life and build our memories. We list endangered species, not necessarily recognizing them. We do extensive research by cataloging, sketching, making X-rays, and learn about their habits, fears, and food chains. We protest against mining, changing water systems, cutting trees, and installing pipelines. However, we eat animals that are not on the endangered species list.

Frog Issues
A whole landscape can change when the frog leaps. We visit pristine swamps and lakes to contemplate the static and dynamic landscape and build our memories according to the frog issues.

A number of art works may come to the mind when thinking on translating semiotic content into visual means of communication, many of them reproduced in textbooks and albums.

Figure 6. Anna Ursyn, "Frog Issues" (© 2008, A. Ursyn. Used with permission)

- In his oil on canvas depiction of the artist-performer Senecio, http://www.wikipaintings.org/en/paul-klee/senecio-1922, Paul Klee (1879-1940) applied geometric shapes within a circle and subtle color planes to convey connections between his style in art, a psychological portrait, and the dramatic performance of the subject. Paul Klee taught at the Bauhaus school of design and published books (e.g., Klee, 1948/1979; 1969) about his theory of art and principles of applying graphic elements. In 1933 the Nazis exiled Klee from Germany and removed his art works from German galleries.
- Alphonse Mucha (1860-1939) created art and design objects, such as posters, jewelry, stained glass, and costumes applying decorative, subtle, linear drawings; he became one of proponents of the Art Nouveau style. For example a poster "F. Champenois, Imprimeur-Éditeur," http://en.wikipedia.org/wiki/File:Alfons_Mucha_-_F._Champenois_Imprimeur-Éditeur.jpg conveys notions that suggest elegance and grace.
- A painting by Rene Magritte (1898-1967) "The Treachery of Images" (1928-9), http://en.wikipedia.org/wiki/File:MagrittePipe.jpg is an artwork with a rich semiotic content. The artist shows an image of a real thing, and then denies it's meaning in writing, thus confusing a thing, its depiction, and it's reference.
- "The persistence of memory" (1931) by Salvador Dali (1904-1989), http://en.wikipedia.org/wiki/File:The_Persistence_of_Memory.jpg, is an iconic work of Dali significant because of its surrealistic style that conveys beyond the words its semiotic content. "Night fishing at Antibes" (1939) by Pablo Picasso, http://www.pablo-ruiz-picasso.net/work-190.php painted in the times when the artist supported those who fought against the nationalist dictator Franko.
- "Madonna" (1984-5) by Edvard Munch (1863-1944), (http://en.wikipedia.org/wiki/File:Edvard_Munch_-_Madonna_(1894-1895).jpg, painted in expressionistic style, has been often interpreted as a depiction of sensuality and the duality of love and pain. The painting, along with "The Scream," was stolen in 2004 and recovered in 2006.
- "Reservoir" (1961) by Robert Rauschenberg, http://americanart.si.edu/collections/search/artwork/?id=20593 is a work combined in oil, pencil, fabric, wood, and metal. Rauschenberg (1925-2008) was bringing together images of objects "to act in the gap between art and life," and so his works contained rich semantic content alluding to the viewer's everyday experience. The artist's work influenced a shift from the Abstract Expressionism to the Pop Art style.

"Discretion Advised" (Figure 7) resulted from examining facts about the biology inspired research conducted in a growing number of fields. Biology inspired art has been gaining importance, as much as biology-inspired computing. We explore what happens among living creatures and draw conclusions as we see it fit.

Flocks of birds, schools of fish, and man-made constructions make us think of shortcuts, metaphors and analogies to communicate faster. I wrote the fish image in Fortran with three-dimensional wireframe.

Let's learn to be silent from schools of fish,
Let our actions be sound without noise.

Figure 7. Anna Ursyn, "Discretion Advised" (© 1999, A. Ursyn. Used with permission)

THE SEMIOTIC CONTENT OF PRODUCT DESIGN

Product design and product advertising in various media, as a part of material culture, can be constructed with distinct approaches and then evaluated in the semiotic framework. One may see design as a tool for marketing that conveys messages about the valuable features of products. Another approach, derived from the 19th and 20th century movements, draws attention of the users to the artistic context of material culture, and thus design is intended as Art. Semiotic content of product design, its artistic quality, and marketing effectiveness depend on its technical solutions, socio-economical messages, and aesthetic-communicative features. The design of computer graphics interfaces for industry, business, and media make a significant part of material culture. Creation and discussion of the new programs for software employs conceptual structures of semiotics, while its essential notions such as images, signs, icons, or metaphors are effective for explaining conceptual models of abstract functions.

Klaus Krippendorff (1990) examined epistemological difficulties in applying traditional semiotics to the industrial design and product semantics. According to Krippendorff, semiotic signs, symbols, and linguistic expressions are the products of human consciousness, so they cannot be considered the same for everyone. In constructivist way, he would approach a new product with curiosity, as a variation from what is already known and what he wants it to be. Krippendorff wrote that designers are facing challenges resulting from the progression of artifacts that add new design criteria. This trajectory of artifacts include products (with their utility, functionality, and aesthetics); goods, services, and identities (marketability, symbolic diversity, and folk or local aesthetics); interfaces (interactivity, understandability, adaptability), multiuser networks (informativeness, connectivity, accessibility); projects (social viability, directionality, commitment); and discourses (generativity, re-articulability, solidarity).

By comparing the designer's intentions and user's perceptions of some technologically innovative products, one can optimize the quality of new products by reducing its design mismatches. Product semantics means an inquiry about the meaning of objects, their symbolic qualities and their psychological, social and cultural context. For example, a search was made about street benches, for a user oriented, optimal combination of formal elements for street furniture, this means, benches placed in a town environment (Vihma, 1992). Semantic analysis concerned the similarity of the benches' style, weight, price and solidity and the users' preferences in terms of comfort, ergonomics, personal space, and material. This search provided information about how people experience new furniture and served as a starting point for the new design and the production of benches.

BIOSEMIOTICS

The study of biosemiosis examines communication and information in living organisms (Emmeche, Kull, Stjernfelt 2002). Biosemiotics has been studying biology not exclusively in terms of physical and chemical processes; researchers interpret living systems as sign systems, which generate meaning (Emmeche, Kull, Stjernfelt 2002: 26). Biosemiotics is built on a system of signs, symbols, and icons to convey meaning and communicate knowledge about natural processes or events with the metaphors, analogies, and a variety of the biology data visualization tools. In the arts, pictorial signs may be arbitrary, for example showing resemblance of a stimulus to its symbol, or conventional, which are often unrelated. Biosemiotics thus examines communication and signification in living systems in terms of the production, action, and interpretation of signs.

In terms of semiotics signs and signification transfer information, meaning, and interpretation between people only. Biosemiotics is the study of signification about generating content, meaning, and interpretation of signs not only in humans but also in other living organisms that produce and receive signs. Liz Stillwaggon with Swan and Louis Goldberg (2010) explored why and how certain features of environment become meaningful to the organism. They address neurobiological intermediary function between the environment and the consequent behavior of the organism and discuss what constitutes a biosemiotic system in terms of their concept of meaning-making brain-objects.

According to Jesper Hoffmeyer (2008), brains are supplementary organs that grew relatively late in evolution in a few species to support psychological life. The bodily existence depends on transfer of intracellular signals, for example of the skin that makes a huge landscape of membranes with ever reconstituting semiotically controlled dynamics, and the semiotic mechanisms control homeostasis and the psycho-neuro-endocrine integration in the body.

Jesper Hoffmeyer wrote in 1993,

Biosemiotics can be defined as the science of signs in living systems. A principal and distinctive characteristic of semiotic biology lays in the understanding that in living, entities do not interact like mechanical bodies, but rather as messages, the pieces of text. This means that the whole determinism is of another type. ... The phenomena of recognition, memory, categorization, mimicry, learning, communication are thus among those of interest for biosemiotic research, together with the analysis of the application of the tools and notions of semiotics (text, translation, interpretation, semiosis, types of sign, meaning) in the biological realm. However, what makes biosemiotics important and interesting for science in general, is its attempt to research the origins of semiotic phenomena, and together with natural sciences, culture with nature, through the proper understanding of the relationships between 'external and internal nature. (Hoffmeyer, 1993, p. 155; in Kull, 1999, p. 386)

Processes going in living systems do not interact like mechanical bodies but rather as messages. Sign processes in living organisms are intrinsic to their highly organized physical and chemical processes that go at all levels, from molecules, cell organelles, tissues, organisms, to ecosystems. As Donald Favareau states, "Biosemiotics is the study of communication and signification observable both within and between living systems. It is thus the study of representation, meaning, sense, and the biological significance of sign processes – from intercellular signaling processes to animal display behavior to human semiotic artifacts such as language and abstract symbolic thought" (Favareau, 2008, p. 10).

The data (such as physical dimensions or electromagnetic properties) is meaningless *per se* but is potential information for a living being, which receives and records information. With this approach, the ability to record, store, and process information from the data contained in matter, and then to convey it further may distinguish life from inanimate objects. Complex, self-organizing living systems display functional meanings that can be studied with regard to biosemiosis – sign action in living systems called by Thomas A. Sebeok (1991, p. 22) the process of message exchanges. "Semiosis is the specific feature which distinguishes the living from the inanimate" (Battail, 2009), as the living world is the unique place where semiosis takes place. Both information theorists and neuroscientists agree that any operation that involves information needs to be physically implemented; human thought is physically inscribed in neurons. Information is the means by which ideas can reside in the physical world, interact with it, and provide the link between physical-chemical processes and those involving meaning (Battail, 2009). Living organisms contain symbolic information inscribed into DNA molecules; information links abstract and the concrete (Battail, 2011). Marcello Barbieri (2010) asserts that biosemiotics put forward by Thomas Sebeok as a non-human communication system, and biolinguistics as developed by Noam Chomsky, are in the case of language studying the same phenomenon. The convergence of these two domains may lead to a unified framework for research with a new approach to the origin of language.

Studies conducted in terms of biosemiotics involve a growing number of varied domains including, for example, architecture and cell biology. Maria Isabel Aldinhas Ferreira (2011) discusses architectural forms as context-dependent semiotic objects with functional and/or aesthetic values in their specific physical, social, and cultural frameworks, which are all engaged in a semiotic interactive relationship with the natural and man-made environment perceived as locus, place, site, or a part of a mental map. Mehmet Ozansoy and Yagmur Denizhan (2009) examine the semiotic analysis of the eukaryotic secretory protein synthesis, which is accordant with the Peirce's chain of signs. In the evolutionary development from the prokaryotes to the eukaryotes, the eukaryotic cells developed a system of compartments involved in the synthesis process, the so-called endomembrane system, which consists of the nuclear envelope around the nucleus; the rough and smooth endoplasmic reticulum serving for transport and synthesis of biochemical compounds; the Golgi apparatus for transforming proteins and processing secretions; vesicles, lysosomes, vacuoles, peroxisomes, all providing storage space, transport, and protecting barriers; and the cell membrane. According to the authors "the modification in the secretory protein synthesis during the evolutionary transition from prokaryotes to eukaryotes constitutes an interesting case from a semiotic perspective because it inspires an intuitive description employing the concept of representation (Ozansoy & Denizhan, 2008, p. 265): the eukaryotic endoplasmic reticulum membrane is a representation of the prokaryotic plasma membrane.

CONCLUSION

Semiotic content of visual design has been described as important for non-verbal communication, especially for visualizing knowledge or social communication with the use of media. Sign systems, codes, icons, and symbols discussed in this chapter pertain to concepts related to semiotics such as art, computing solutions, social interactions, and reciprocal relations between languages, images, and technology. The semiotic content existing in descriptions of processes and events occurring in natural or technology-induced settings has been described in terms of objectivism and contrasted with radical constructivism. Biosemiotics examines communication and information in living systems not exclusively in terms of physical and chemical processes but also as sign systems, which generate meaning. This approach supports inquiries in further chapters.

REFERENCES

Barbieri, M. (2010). On the origin of language: A bridge between biolinguistics and biosemiotics. *Biosemiotics, 3*, 201-223. doi 10 1007/s12304-010-9088-7

Battail, G. (2009). Living versus inanimate: The information border. *Biosemiotics, 2*, 321–341. doi 10.1007/s12304-009-9059-z. Retrieved December 17, 2011, from http://www.springerlink.com/content/r376x87u5mk68732/fulltext.pdf

Battail, G. (2011). An answer to Schrödinger's what is life. *Biosemiotics, 4*, 55–67. doi:10.1007/s12304-010-9102-0.

Broudy, H. S. (1987). *The role of imagery in learning*. Malibu, CA: The Getty Center for Education in the Arts.

Camus, A. (1951/1992). *The rebel: An essay on man in revolt*. Vintage.

Chandler, D. (2001). *Semiotic for beginners*. Retrieved June 1, 2012, from http://www.aber.ac.uk/media/Documents/S4B/sem08a.html

De Saussure, F. (1915). *Ferdinand de Saussure, 1857-1913*. Genève, Switzerland: Kundig.

Emmeche, C., Kuli, K., & Stjernfelt, F. (2002). *Reading Hoffmeyer, rethinking biology*. Tartu University Press.

Favareau, D. (2008). Iconic, indexical and symbolic understanding: A commentary on Anna Aragno's the language of empathy. *Journal of the American Psychoanalytic Association*, *56*(3), 783–801. doi:10.1177/0003065108322687 PMID:18802128.

Ferreira, M. I. A. (2011). Interactive bodies: The semiosis of architectural forms a case study. *Biosemiotics Online First*. doi 10.1007/s12304-011-9126-0. Retrieved December 20, 2011, from http://www.springerlink.com/content/?Author=Maria+Isabel+Aldinhas+Ferreira

Hoffmeyer, J. (1993). Biosemiotics and ethics. In *Culture and Environment: Interdisciplinary Approaches*. Oslo, Norway: University of Oslo.

Hoffmeyer, J. (2008). The semiotic body. *Biosemiotics, 1*(2), 169-190. doi 10.1007/s12304-008-9015-3. Retrieved December 20, 2011, from http://www.springerlink.com/content/03752h27v677377l/

Huizinga, J. (1950). *Homo ludens: A study of the play element in culture*. New York: Roy Publishers.

Klee, P. (1948/1979). *On modern art*. New York: Faber & Faber.

Klee, P. (1969). *The thinking eye: Paul Klee notebooks*. G. Wittenborn.

Krippendorff, K. (1990). Product semantics: A triangulation and four design theories. In *Product Semantics '89*. Helsinki: University of Industrial Arts.

Krippendorff, K. (2011a). Conversation and its erosion into discourse and computation. In T. Thellefsen, B. Sørensen, & P. Cobley (Eds.), *From First to Third via Cybersemiotics* (pp. 129–176). Frederiksberg, Denmark: SL Forlagene.

Krippendorff, K. (2011b). Discourse and the materiality of its artifacts. In *Matters of Communication: Political, Cultural, and Technological Challenges to Communication Theorizing*. New York: Hampton Press.

Kull, K. (1999). Biosemiotics in the twentieth century: A view from biology. *Semiotica*, *127*(1/4), 385–414.

Maturana, H. (1970). *Biology of cognition*. Urbana, IL: University of Illinois.

Maturana, H., & Varela, F. (1987/1992). *The tree of knowledge: The biological roots of human understanding*. Boston: D. Reidel.

Ogden, C. K., & Richards, I. A. (1989). *Meaning of meaning*. Mariner Books. (Original work published 1923).

Ozansoy, M., & Denizhan, Y. (2009). The endomembrane system: A representation of the extracellular medium? *Biosemiotics*, *2*, 255–267. doi:10.1007/s12304-009-9063-3.

Peirce, C. S. (1958). *Collected papers of Charles Sanders Peirce*. Cambridge, MA: Harvard University Press.

Posner, R. (2009). *Eight historical paradigms of cultural studies and their semiotic explication*. Retrieved June 3, 2012, from http://www.ut.ee/CECT/docs/CECT_II_PPT/I_session/Posner_CECT09.pdf

Sebeok, T. A. (1991). *A sign is just a sign*. Bloomington, IN: Indiana University Press.

Stanford University. (2012). *Symbolic systems*. Retrieved January 20, 2012, from http://www.stanford.edu/dept/registrar/bulletin/6141.htm

Stillwaggon, L., & Goldberg, S. L. (2010). How is meaning grounded in the organism? *Biosemiotics*, *3*(2), 131–146. doi:10.1007/s12304-010-9072-2.

Tanaka-Ishii, K. (2010). *Semiotics of programming*. Cambridge University Press.

The New York Times. (2012, June 5). Tattoos. *The New York Times*. Retrieved June 5, 2012, from http://topics.nytimes.com/top/reference/timestopics/subjects/t/tattoos/index.html

Vihma, S. (Ed.). (1992). *Objects and images: Studies in design and advertising*. Helsinki: University of Industrial Arts.

von Foerster, H. (2002). *Understanding understanding: Essays on cybernetics and cognition*. New York: Springer-Verlag.

von Glasersfeld, E. (1989). Constructivism in education. In T. Husen, & T. N. Postlethwaite (Eds.), *International encyclopedia of education*. Oxford, UK: Pergamon Press.

von Glasersfeld, E. (1991). Knowing without metaphysics: Aspects of the radical constructivist position. In F. Steier (Ed.), *Research and reflexivity*. London: Sage.

von Glasersfeld, E. (2007). The constructivist view of communication. In A. Müller, & K. Müller (Eds.), *An unfinished revolution?* (pp. 351–360). Vienna: Echoraum.

Wood, G. (1930). *American gothic*. Retrieved May 30, 2012, from http://upload.wikimedia.org/wikipedia/commons/7/71/Grant_DeVolson_Wood_-_American_Gothic.jpg

Chapter 7
Pretenders and Misleaders in Product Design

ABSTRACT

Meaningful messages may be conveyed in product design with the use of pretenders as the carriers of hidden messages; they refer to visual practices in design, architecture, and visualization. For this reason, they may be useful for working projects in further chapters. The notions of iconic objects, or iconcity of an object, make a basis of product semantics. Proper design versus pretenders, misleaders, informers, double-duty gadgets, and multitasking tools are discussed and then contrasted with the notion of camouflage.

INTRODUCTION

Themes contained in this chapter describe the concerns that may be common for perceptual representation of scientific concepts and the product design. For this reason concepts related to product semantics may be useful for working projects up in further chapters. When we communicate online, send a voicemail, or use visualization to convey our message or information, we often interact with shortcuts. We use signs, symbols, icons and metaphors, connotations and associations to create open or closed messages depending on our intention: whether we want to be understood exactly as intended, or we'd allow for imaginative interpretation.

Object design has to fulfill functional, ergonomically oriented, aesthetic, material, and space related demands. Various trends in product design aim for comfort, simplicity, elaboration,

DOI: 10.4018/978-1-4666-4703-9.ch007

or a foolproof and easy use, along with aesthetical considerations. The best design is often self-explanatory. However, matching the area of joy of the user is equally important. Some objects are designed to inform and entertain at the same time by mimicking other objects' characteristics. Some of them are made to mislead us. The thinking behind this design is aimed to entertain, make the day brighter, or make a product more attractive, while fulfilling its purpose.

CANONICAL OBJECTS

Some items should always look in an obvious way and be easy to recognize. For example, a fire extinguisher should be easy to find and then used fast without any instructions. Scissors must fit a hand and a hammer should be easy to use. Such objects that have an easy recognizable shape are called canonical objects. In spite of the new line in a design of cellular phones, we still draw an old-style telephone with a round dial to signal where we can find much more modern touch-tone ones in the phone booths (however, phone booths began to disappear).

See Table 1 for Your Visual Response.

Table 1.

Your Visual Response: Canonical Objects
Draw as many examples as you can that show the use of images of canonical objects on various signs: at the mall, an airport, or on a road. How the sign designers use visual symbols and an iconic way of communicating information?

Iconcity

Designing commercial or public display, such as billboards, posters, or placards involves the use of signs, symbols, and icons, is collectively called signage. A designer looks for visual shortcuts that are powerful and effective. In semiotic terms, iconcity is meant as powerful icons; hence a notion of scissorness. For example, a designer may look for an icon showing the essence of the meaning related to scissors. Iconic references may refer to the product's own form, color, material, the metaphorical likeness to another form, its stylistic belonging to similar forms (such as period styles, fashion styles, and role models), and the product's belonging to a special environment (kitchen, office). Iconicity in product design may relate to a whole object or its details, with possible features such as naive simplicity, high-tech qualities, decorativeness, or designer's subjectivity (Vihma, 1992). Design with powerful icons can help memory and the learning of using complicated products.

See Figure 1 for examples of iconic objects.

It we try to redesign the shape of scissors and still produce an ergonomic tool intended to provide comfort without stress or injury, we may end up going back to its original shape in order to avoid blisters. When we look at some items of ceramics, jewelry, or craft objects, we can see some common features characteristic for these products' design – we look for iconcity of these products. Iconcity of a complicated product or its detail may help memorize and learn how to use this product. We may discuss the iconicity of the design products, such as furniture or appliances. We may analyze:

- What are the product's own form, color, and material?
- Does it have a metaphorical likeness to other form: does it 'pretend' to be something else?
- What style does it have (period style, company's or designer's style)?
- Can we see that this product belongs to a special environment (such as kitchen or office)?

These ideas have been adopted in product design. For example, in the Apple company ensures that the design of the connectors does not allow to plug them in the wrong way. Ecology and preservation ideas are also included in the design concepts. For example, keyboards for Mac computers are often made from the material that was cutout from the monitor to make an opening for a screen.

Figure 1. Examples of iconic objects (© 2012, photographed by A. Ursyn. Used with permission)

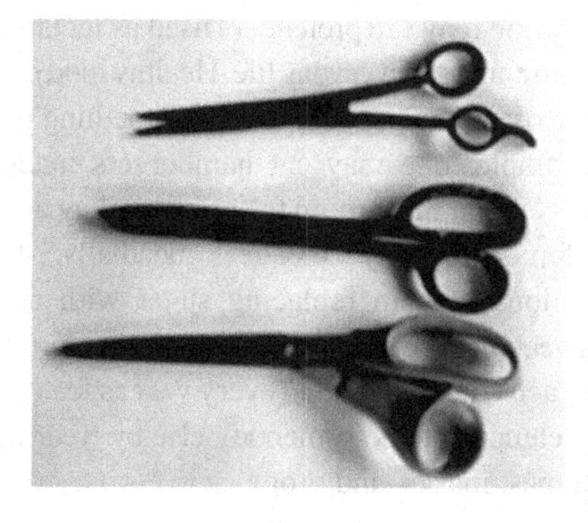

A product design that are provided with signs, symbols, and linguistic expressions may acquire additional meaning by:

- Imitating something else (a telephone in a shape of a duck decoy);
- Adding symbols of value from another semantic domain (non-functional additions);
- Replacing materials and still maintaining valued appearances (Navaho jewelry out of plastic); or
- Associating products with prestigious individuals or designers.

The design or use of signs and symbols is often called signage. Signage may include billboards, posters, placards, etc. It may refer to a number of signs thought of as a group. Symbols, iconic and canonical objects are used in commercials. They often cause that we take a suggestive commercial message for granted, without thinking. Also, body language may encompass some iconic features. A big problem may arise in another country where the same signs may mean something different, and one does not even know that one is conveying strong messages. To assist the tourists, Dieter Graf and several other authors designed the traveler's language kits in the form of wordless booklets containing pictures of items one may need when travelling in another country.

In product design, it often pays to obey traditional solutions. The choice of material may be an important factor; material that has been traditionally serving for a product is often simulated, for example by designing imitated knots on a plastic surface of a table. In case of a remote control, our habituation requests consistency in application of the usual solution; operating an unusually designed remote control would be tedious and tiresome. Therefore, many of the well-known producers offer their customers small innovations added to the familiar designs. For example, a bitten apple logo of Apple computers, originally rendered in rainbow colors, became white without any other changes in design; a girl with an umbrella drawn on cylindrical boxes of the Morton Salt has been replaced with another but similar girl with an umbrella; on the other hand, boxes of the Quaker oats and grits remain almost unchanged since tens of years.

PRODUCT SEMANTICS IN PRODUCT DESIGN AND MARKETING

Designers and advertisers consider visual imagery as non-verbal processing. By applying icons in design designers can support our learning of using a complicated product, so it's easy to answer a question, for example, "How to open this thing?" Specialists in marketing and advertising are ever more aware of the importance of aesthetics in marketing. Color is considered important in product semantics. Many times color sends a message, using generally accepted associations, such as green = calm and soothing, red = hot and violent, etc.

See Figure 2 for examples of product design.

To convey a message visually with an icon, we take in those features that best identify an object, and suppress those features that are not basic to comprehending it. For example, in a "slippery when wet" road sign even a driver has been removed from cars, to make the road sign easy to understand fast (Figure 3).

One may see pretenders used as metaphors in many areas of everyday life. Healthy food products are often designed to pretend something else, for example, one may eat hamburgers made with portabella mushroom to 'pretend' they are made of meat, or buy "meatless meatballs." Similar actions help in replacing sugar with artificial sweeteners. One may also see human pretenders in activities celebrated every year as festivities of pretending to be somebody else by wearing costumes, masks, and props.

Figure 2. (a) This ice-cream container is made of glass but pretends to be edible (© 2013, photographed by A. Ursyn. Used with permission). (b) This brush looks like a mushroom because it is designed for cleaning mushrooms (© 2012, photographed by A. Ursyn. Used with permission)

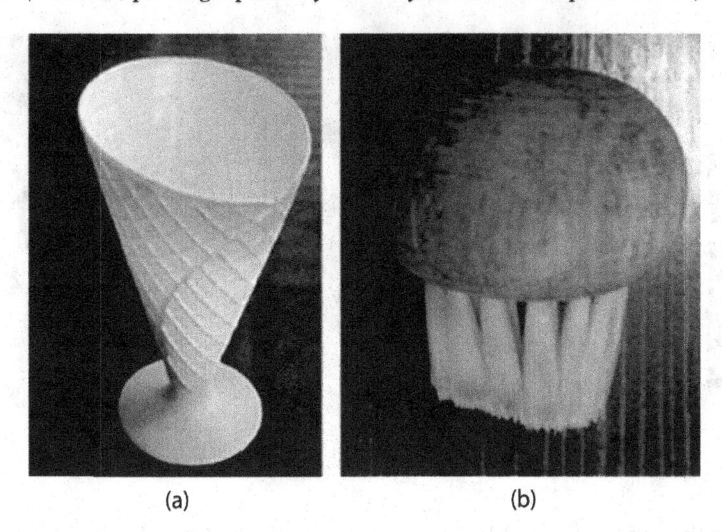

Figure 3. "Slippery when wet" signs used in Australia, the UK, and the USA (© 2013, photographed by A. Ursyn. Used with permission)

PRETENDERS, MISLEADERS, INFORMERS, DOUBLE-DUTY GADGETS, AND MULTIFUNCTIONAL TOOLS

Pretenders

In contrast to generally accepted philosophy of design, everyday we encounter objects that challenge our sagacity, entertain us by mimicking other things, mislead us about their function, or about the material they are made of. It happens in spite of the fact that the best design is self-explanatory. Some of these objects pretend they are something else, while some deceive us (Ursyn & Lohr, 2010). Pretenders represent a special kind of design. We may call "pretenders" those products that show metaphorical likeness to another forms and 'pretend' to be something else. Their function is hidden and reveals quite unexpectedly, so you say, oh, it does do something. A candle may be shaped like a cactus, a cat, or anything else. Some pretenders are used as props that decorate or create scenery for a holiday event. Sometimes they are toys, for example, a miniature kitchen appliance to be put as magnets on a refrigerator, or a kids' cup in a shape of an animal. Many times, ample semiotic content is needed in product design to make it easy

Figure 4. Pretenders: (a) A candle pretending to be a cactus in a pot; (b) a dog and a cat, salt-and-pepper shakers; (c) a candle pretending to be a garlic; (d) a scaled down glass flower pretender (© 2012, photographed by A. Ursyn. Used with permission)

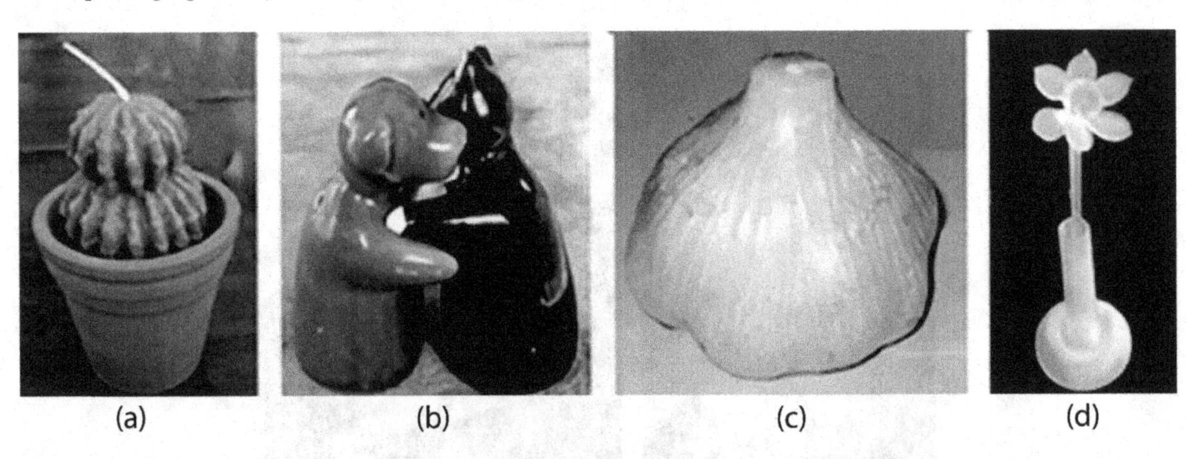

to understand, rather than metaphorical likeness of a product to something else. In other cases designers accentuate metaphorical side of the product, for example, a colander like a sea sponge designed by Anna Rabinowicz. In fact, there are a lot of gift products with the appearance of another well-known object and it is difficult to guess what it is for or how to use it. The deceptive design of such objects seems to lessen their iconcity. Pretending can be also seen in typography, when by applying fonts as metaphors it is possible to make one believe that the publication comes from another time; readers make connotations related to past times and events. One may say about Las Vegas, Nevada that a whole Las Vegas Strip is a big pretender, with Paris, Venice, and ancient Egypt recreated inside of it.

See Figure 4 for examples of pretenders.

Pretenders as a Cognitive Tool

We can see pretenders as a kind of metaphors, which may support visual communication through product design. Understanding of products and processes can be easier when they are presented through visual metaphors – a basic structure for communicating a message. One may ponder whether pretenders as visual metaphors may help us convey meaning efficiently. To design an effective product, we have to send a message with clarity, and precision, without embellishment, ambiguity, unnecessary details, and other non-functional characteristics. Visual metaphors may help the designers to convey the essence of a product. However, a series of questions comes next:

Can pretenders support information visualization aesthetics, usability, and efficiency?

Can pretenders ease our learning and reduce the heavy cognitive load on working memory by helping to break information into individual blocks? Visual presentation of data with the use of pretenders may possibly help to shift a part of explaining from the abstract to the meaningful parts that may be easier to learn and remember. Thus, we may make a mental comparison of something that is not easily understood to something visible. Understanding of abstract ideas can be easier when difficult concepts are presented through visual metaphors.

How can we use strong points of pretenders when we create avatars and design their body language and face expressions? Also, how can we assess the aesthetics of such avatars and visual metaphors they carry in terms of communication efficiency and usability.

Misleaders

With respect to all these deceptions, apart from pretenders we may think about some products as misleaders – objects designed to look like one thing while they are serving quite another purpose, and giving the appearance different from the true one. Misleaders are things that look like something else, for example, you may think it's a book but it is a CD cover. Some souvenir products lack the product semantics, as they have an appearance of another well-known object as a disguise; it prevents the viewer from understanding what it is for or how to use it. For example, candles that are made in a shape of fruits or toys, and therefore do not look like candles, are misleaders.

The most dramatic example of a misleader comes from the Bronze Age, when a huge figure of a wooden horse, with a force of men hidden inside, allowed the Greeks to enter the city of Troy (Figure 5). After the Greeks pretended to sail away, the Trojans pulled the wooden horse into their city as a gift and a victory trophy. Then the hidden troop opened the gates of the city for the Greek army.

Many times one material pretends to be another, for example, traditionally designed laces threaded in a web-like pattern are actually made of a synthetic material, not even knitted with thread but polymerized as a whole. A 'crystal' vase is often made of plexiglas. A countertop in a kitchen may be made of another plastic material that pretends to look like marble. It's often a diversion of the case when the same object has been made of different materials to improve its functionality, such as a bucket made of wood, metal, and then plastic.

See Figures 6 through 8 for more examples of misleaders.

Figure 5. A misleader – a wooden toy designed according to old depictions of a Trojan horse (© 2012, photographed by A. Ursyn. Used with permission)

Figure 6. Misleaders: (a) this character pretends to be someone else and hides that he is just a tool: a bottle opener; (b) this box misleads us, as it pretends to be a Wii controller. In fact, it contains a chewing gum. (© 2012, photographed by A. Ursyn. Used with permission)

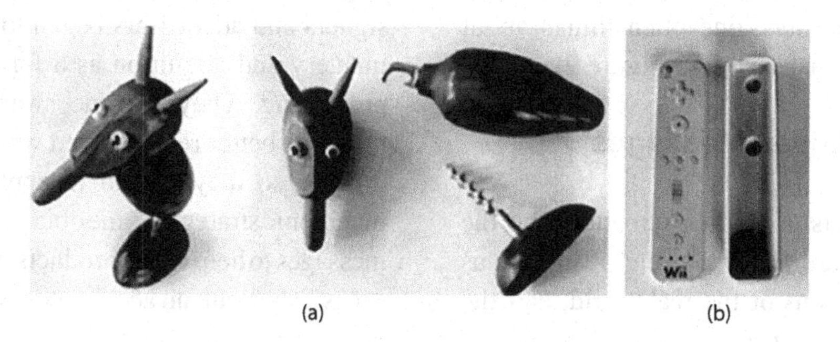

Figure 7. USB memory sticks as a misleader: Lego block and eraser (© 2012, photographed by A. Ursyn. Used with permission)

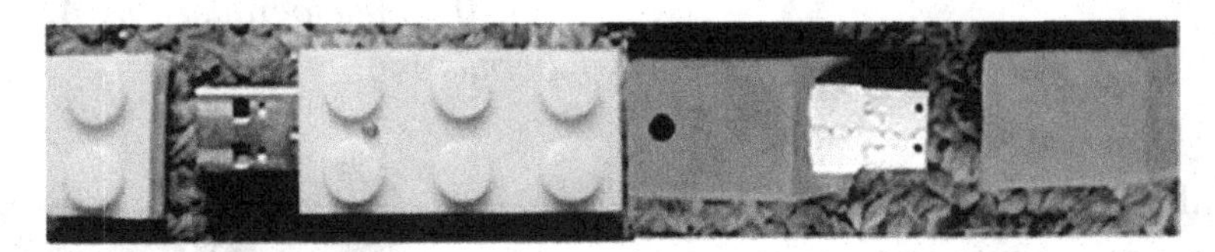

Figure 8. Misleaders: (a) this silverware is made of plastic, while pretending to be made of steel or silver; (b) Another one "pretends" to be made of bamboo twigs (© 2012, photographed by A. Ursyn. Used with permission)

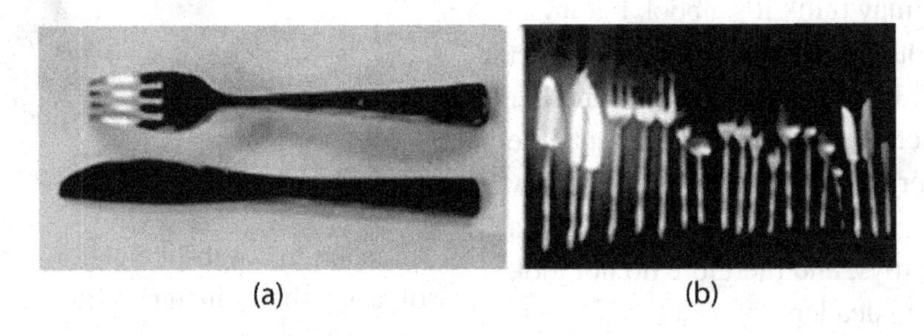

(a) (b)

Figure 9. Misleaders: (a) false metal pretender; (b) rocks made of a sponge (© 2012, photographed by A. Ursyn. Used with permission)

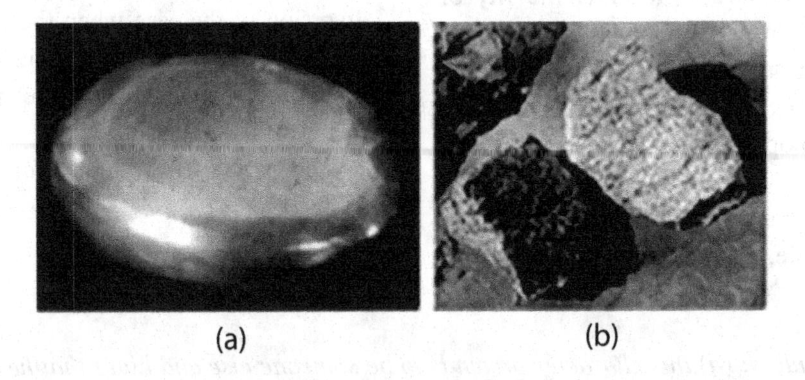

(a) (b)

With pretenders, one may think about function, while with misleaders, one often thinks about material used to make a thing (Figure 9).

Informers: Hidden Messages

The developments in product semantics put the designers on guard about the role of images as cognitive reflections of the real world, and the importance of aesthetics in marketing. Thus, designers and advertisers began to consider visual imagery and cognition as a form of non-verbal processing. They became aware that products might be better remembered when easy to evoke images, so they started to apply imagery as a mnemonic strategy. Some objects carry the hidden messages to help to use products; many pretenders act as a shortcut, an accelerated way of conveying

Figure 10. Informers: (a) a garlic container: a message embedded in an object, such as in a garlic-shaped garlic container, helps to operate in a new place, e.g., when one is a guest in a house for the first time; (b) this is a low-message garlic container because only a small garlic on the lid suggest the usage; (c and d) glass and porcelain dishes for fish; (e) potato peeler (© 2012, photographed by A. Ursyn. Used with permission)

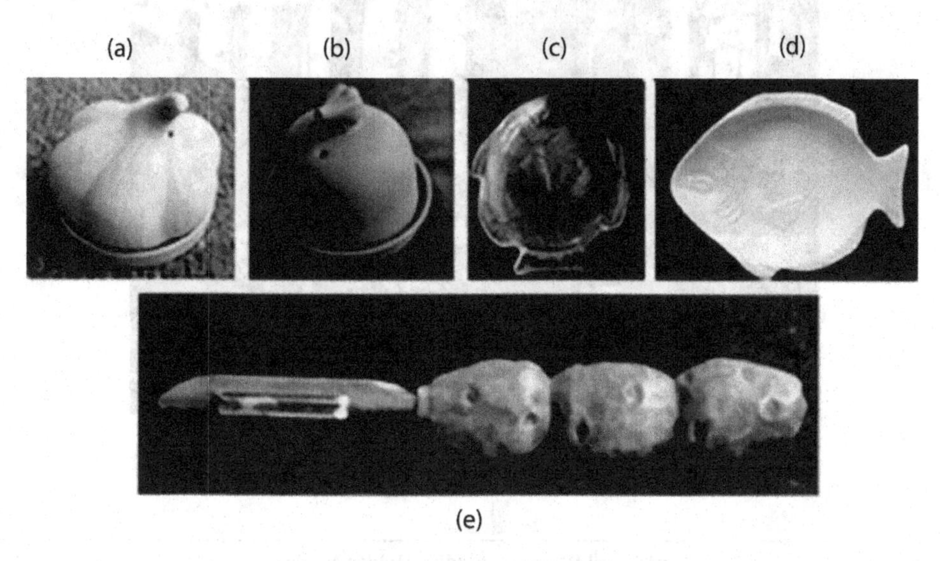

something, because they contain a hidden message. Designing informers and pretenders may be somehow related to entertainment; objects one can play with and laugh at may also be useful in marketing.

See Figure 10 for examples of informers.

One may also find informers in contemporary architecture. For example, The Longaberger Basket Company headquarters in the form of a seven-story gift basket informs about the company's profile, while the Big Duck building, Flanders, New York, Long Island informs that it was built (in 1931) to sell ducks and duck eggs.

Figure 11 shows another example of an informer in contemporary architecture.

Quite often, objects carry hidden messages that are loosely connected to the product. For example, many times car advertisers show a lion or a tiger to associate a fast, fearsome animal with a car. Driving it makes you feel powerful, and make you transfer this feeling on a car. Thus, there is a link from iconic, well known content to a hidden content of a commercial, but designers must know the audience they are talking to evoke expected reaction. Also, CD covers communicate hidden messages through images, sometimes political, sometimes sexual in nature, and sometimes they address visually a selected group of listeners. Many times color sends a message, when designers use generally accepted associations, such as green = calm and soothing, red = hot and violent, etc. In this style of the product advertisement, commercials are comprised of a presentation of a product along with a hidden message that draws the viewer into aesthetic sensations, emotional associations, and the producer's promises that supplement the product itself.

See Table 2 for Your Visual Response.

Many pretenders act as a shortcut, an accelerated way of conveying something, when they contain a hidden message. For example, we may think about a specific message design when the product characteristic tells the user what is the purpose of it. Sometimes a unified, standardized design, such as in case of electric switches and sockets, helps the users to successfully operate in

Figure 11. Façade of the Kansas City Library, built as a row of books, informs about the content of this building (© 2012, A. Ursyn. Used with permission)

Table 2.

Your Visual Response: Finding Hidden Messages
Find examples of hidden messages conveyed through product design. Examine objects you are using everyday. Find details that may convey some directives in an unnoticeable way or contain a meaning, so it is easier to use this gadget or you like it more than other ones.

Figure 12. Informers: salt-and-pepper shakers pretend they are parts of hardboiled eggs (© 2013, photographed by A. Ursyn. Used with permission)

Figure 13. Double-duty gadgets: (a) a 3D puzzle serves both for entertainment and as an educational prop; (b) a beach ball serves also both for entertainment and as an educational prop (© 2012, photographed by A. Ursyn. Used with permission)

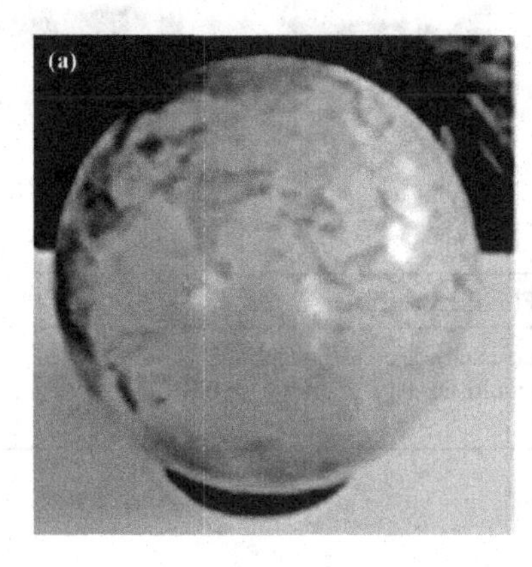

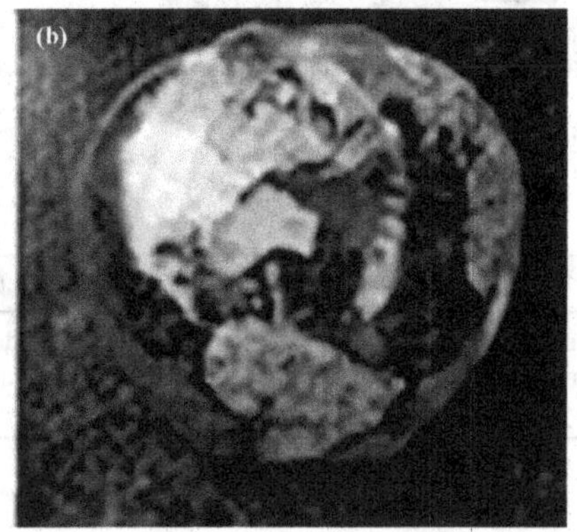

new places. There is a link from the iconic, well-known image to a hidden message of a pretender. Message design helps the user select a proper object. For example, it suggests the use of a bigger bar of soap for a bath rather than for washing hands, when a bar of bath soap is placed on a soap dish in a shape of a miniature bathtub.

In such cases pretenders enhance the culture of our living, especially when they help us recognize details and nuances. For example, a salt-and-pepper condiment, when shaped as porcelain eggs (Figure 12) indicates its usage at a breakfast table rather than during a dinner or supper. It enhances one's mood, entertains, and serves as a conversational theme. Thus, we may say such pretenders are addressed to a connoisseur, to somebody for whom a small change makes a big difference.

Double-Duty Gadgets and Multifunctional Tools

Some objects are designed for being able to perform various tasks (Figure 13). The aspect of functionality and the ergonomic design needs to be unified with the product's form and the users' expectations about the best performance. This leads us toward the concerns about product

Figure 14. A multi-tasking tool: a ten-function tool in a shape of a credit card, comprising a can opener, a knife edge, a screwdriver, a ruler, an bottle opener, three kinds of wrenches, a saw blade, and an orientation tool (© 2012, photographed by A. Ursyn. Used with permission)

Figure 15. A 59 cent tacos man (© 2012, photographed by A. Ursyn. Used with permission)

Table 3.

Your Visual Response: Finding Pretenders
List examples of pretenders of various kinds. Do you think that pretenders can be seen as metaphors? Draw a pretender, a misleader, an informer, a double-duty gadget, and a multi-tasker, showing your favorite tool: object you use every day, such as a memory stick, or a phone.

usability. A credit card that can be used as a set of mini tools (Figure 14), a Swiss Army knife (http://www.victorinox.com/stories) that has an additional duty as a memory stick, a screwdriver that is at the same time a beer bottle opener, and a porcelain egg, in which an egg yolk contains pepper and an egg white is filled with salt, can serve as examples of this kind of objects.

Sometimes people act as pretenders or informers, for example when they want to advertise something; a person wears a costume and advertises a product, for example pretending to be Mexican tacos (Figure 15).

See Table 3 for Your Visual Response.

PRETENDERS VS. CAMOUFLAGE

The notion of pretenders is somehow related to the concept of camouflage. It could be understood as the disguising forms, patterns, or coloring that enables to blend with the surroundings:

- Camouflage in the floral and animal world exists as a natural design of the outer surface; flounder can almost totally blend with small rocks at the bottom; patterns on a giraffe and zebra's skin, or the whiteness of polar bears in the winter months allow the animals to remain unnoticed or resemble something else (camouflage by mimesis). Some animals stop moving when being approached, to be perceived as a part of the background, for example pretending to be an inedible stick;
- Camouflage may take form of the ability to change color quickly as a means of survival, as it can be seen in chameleon, fish in coral reefs, foxes and hares in winter, and in many other cases;
- People can also to disguise their presence, mostly for military purposes by means of camouflage clothing, equipment, and installations;
- Camouflage may also mean dressing unusually and wearing masks to disguise one's identity or personality, especially during special holidays and festivities such as St Patrick Day parade, Halloween 'trick-or-treating activities, Venice carnivals in Italy with the costume and mask balls, New Orleans Mardi Grass celebrations, or the forty six days lasting Brazilian Carnival and parades;
- Costume design may transform a person into somebody different. A mask serves as camouflage and, at the same time a pretender – aimed to blend in or stick out. Many celebrities wear dark glasses or sun glasses;
- Toys are often designed as transformers and pretenders. Toys called Transformers (introduced in 1988) and the Ultra Pretenders featured a large exterior vehicle shell, then a secondary humanoid shell (which could also transform), and within that, the miniature interior robot. A puppet often hides another character within itself;
- Experiments in ship camouflage During World War I German submarines (called "U-boats") were dangerously successful in destroying Allied ships. It order to make these ships less visible, Norman Wilkinson, Everett L. Warner and other artists devised methods of camouflage. High-contrast, unrelated shapes were painted on a ship's surface, thus confusing the periscope view of the submarine gunner (Berens, 1999).

In visual arts, camouflage and pretending something else used to be applied mostly in painting. For example, the trompe l'oeil art technique created (first in the Greek and Roman times and then from the 16th century on) optical illusion of three-dimensionality with the use of exceptionally realistic imagery. One may find several other ways to pretend there is different actuality or reality disguised by camouflage, for example, the camouflage artwork of Bev Doolittle. Craig Reynolds (2011) designed an abstract computation model of the evolution of camouflage in nature, "The 2D model uses evolved textures for *prey*, a background texture representing the *environment*, and a visual *predator*. A human observer, acting as the predator, is shown a *cohort* of 10 evolved textures overlaid on the background texture. The observer clicks on the five most conspicuous prey to remove ("eat") them" (Reynolds, 2011, p. 123). One of the proponents of Op art Bridget Riley (2012) shows movement on still canvas. Considering that "Camouflage can be thought of as visual warfare," the Ohio State University Department of Theatre and the Advanced Computing Center for the Arts and Design produced in 2011 "The Camouflage Project" (http://camouflage.osu.edu/camouflage.html). Also, in his camouflage art an artist Liu Bolin (2010) turns himself into the invisible man. Inspired by how some animals can blend into their environment, Bolin uses camouflage principles to create his art.

There are many camouflage restaurants blended with the surroundings or pretending they are something else by applying unusual decor. Also, hundreds of hotels defy convention, hosting quests in the cars, aircraft, in a prison, a lighthouse, a treehouse, underground, underwater, on the top of cranes, and in other unusual places.

As a summary, when discussing pretenders and misleaders we may think about their look, functionality of the product and its effectiveness in terms of cost, maintenance, or ecology, messages it conveys, joy, fun, and emotions they evoke. We may ponder why the pretenders and misleaders are so ubiquitous and what kind of needs they satisfy. Maybe, they quench our inclination toward spontaneous playfulness. If so, we may seek advantages of applying them in communication, product design, and in instruction. We may use their ludic - relating to play or playfulness properties to entertain, evoke attention, emotion, and curiosity, and thus enhance the attractive qualities of a product or motivation of the learner. With cognitive approach to mental imagery, one can include pretenders as tools to visualize a concept and create image representation of it. Thus, we may make pretenders, misleaders, informers, and multitaskers a tool for instruction, as well as the product design and it's marketing; they convey both emotion and poetry, and make us aware of the artistic context of material culture of the products.

CONCLUSION

Meaningful messages may be conveyed in product design with the use of pretenders as the carriers of hidden messages; they refer to visual practices in design and visualization. Analysis of hidden messages conveyed by pretenders, misleaders, informers, double-duty gadgets, and multifunctional tools may be discussed as cognitive reflections of the real world, and also possible communication tools. The same applies to the importance of product design aesthetics in marketing, architecture, and visualization.

Camouflage, which could be understood as the disguising forms, patterns, or coloring that enables to blend with the surroundings may be seen somehow related to the concept of hiding messages through pretending. The existence of camouflage in art and social vents makes us aware of the artistic context of material culture of the products.

REFERENCES

Berens, R. R. (1999). The role of artists in ship camouflage during world war I. *Leonardo*, *32*(1), 53–59. doi:10.1162/002409499553000.

Bolin, L. (2010, November 20). Now you see me, now you don't: The artist who turns himself into the invisible man. *Mail Online*. Retrieved June 26, 2012, from http://www.dailymail.co.uk/news/article-1201398/Liu-Bolin-The-Chinese-artist-turns-Invisible-Man.html

Reynolds, C. (2011). Interactive evolution of camouflage. *Artificial Life*, *17*(2), 123–126. doi:10.1162/artl_a_00023 PMID:21370960.

Riley, B. (2012). *Brigdet Riley*. Retrieved July 13, 2012, from http://www.op-art.co.uk/bridget-riley/

Ursyn, A., & Lohr, L. (2010). Pretenders and misleaders in product design. *Design Principles and Practices: An International Journal*, *4*(3), 99–108.

S. Vihma (Ed.). (1992). *Objects and images: Studies in design and advertising*. Helsinki: University of Industrial Arts.

Chapter 8
Metaphorical Communication about Nature

ABSTRACT

Metaphors are present in our thoughts and make invisible concepts perceivable. The metaphorical way of perceptual imaging is discussed in this chapter, particularly the use of art and graphic metaphors for concept visualization. We may describe with metaphors the structure and the relations among several kinds of data. Metaphors may represent mathematical equations or geometrical curves and thus make abstract ideas visible. Most metaphors originate from biology-inspired thinking. Nature-derived metaphors support data visualization, information and knowledge visualization, data mining, Semantic Web, swarm computing, cloud computing, and serve as the enrichment of interdisciplinary models. This chapter examines examples of combining metaphorical visualization with artistic principles, and then describes the metaphorical way of learning and teaching with art and graphic metaphors aimed at improving one's power of conveying meaning, integrating art and science, and visualizing knowledge.

INTRODUCTION

Linking science and visual presentations may become easier with the use of metaphors that are suitable for creating science-inspired artistic projects, so these themes will return in chapters that follow. Metaphors present in our visual and verbal environment include iconic and symbolic images, representations inspired by the rules and phenomena observed in nature, as well as simulations and visualizations of concepts and events presented in metaphorical way. Metaphorical imaging of abstract

DOI: 10.4018/978-1-4666-4703-9.ch008

concepts includes metaphors based on natural objects, metaphors related to the physical senses, and conceptual metaphors that apply the known rules or phenomena as a way of translation of abstract concepts. Nature derived metaphors support data visualization, information and knowledge visualization, data mining, semantic web, and serve as the enrichment of interdisciplinary models. The further text examines examples of combining metaphorical visualization with artistic principles, and then describes metaphorical way of learning and teaching with art and graphic metaphors aimed at improving one's power of conveying meaning, integrating art and science, and visualizing knowledge.

TRANSLATION OF MEANING WITH VISUAL AND VERBAL METAPHORS

Few would challenge the assumption that people think for the most part in pictures. Communication proceeds with the use of language that is highly metaphorical, and many hold that there is no non-metaphorical thought. In many instances, art is metaphorical. A recurrent theme throughout this book is visual communication and visualization of ideas that is pictorial and linguistic at the same time, in both cases being metaphorical. Visualization has been generally seen as the presentation of pictures showing easy to recognize objects that are connected through some well-defined relations. The effectiveness of making the concepts or data comprehensible and visually appealing often depends on choosing a metaphor that is suitable to carry complex concepts and visual storytelling.

There is common agreement in opinion that a metaphor is a figure of speech in which a word or phrase that ordinarily designates one thing is used to specify another, thus making an implied mental comparison. Metaphor is a best known, persuasive rhetoric figure used to increase the effectiveness of a message, with an elaborate taxonomy of metaphorical tropes such as metonymy, simile, analogy, synecdoche, thought maps, and concept maps. Metaphor reflects cognitive operations; it makes us see one thing in terms of another and create a new meaning. Metaphors are not true or false. Thus, metaphors may involve mental models which otherwise wouldn't be easily grasped. A metaphor may indirectly suggest the meaning of something that is not easily understood, and transfer it from one thing to another without direct comparison, with the use of 'like' or 'as'. For example, we say 'the tip of the iceberg' to imply a small, visible part of a big problem; a symbol of 'heart' is a metaphor. We use interface metaphors everyday when we talk about online communication using familiar objects for organizing the corresponding elements related to the computer, a folder, files, and many other metaphors.

Metaphors often address our basic experience, for example, when our minds are conditioned by habit to visualize quantity: 'more' as something going up and 'less' as an object or a graph going down (Lakoff & Núñez, 2001). With inspiration coming from nature and/or mathematics, realities created with the use of metaphors refer to our imagination and experience based on physiological reality of the mind. One may say, the prevalent metaphorical imaging of abstract concepts includes natural metaphors of living organisms, often incorporating behavior such as motion and gesture, metaphors based on natural objects found in nature, visual, auditory, or other metaphors related to the physical senses (often in the use of symbols or icons), and conceptual metaphors that apply the known rules, phenomena, and mathematical ideas as a way of translation of abstract concepts (Lakoff & Núñez, 2001).

Metaphorical artwork does not necessarily present its content as representational depiction showing the physical appearance of objects or people. "The Swirl" (Figure 1) shows immaterial concepts translated into an abstract image. Viewers may want to create their own experience of the dynamic physical processes, find a counterpart in music, or think about the events occurring in cosmos. Somewhere in a space between the

Figure 1. Anna Ursyn, "The Swirl" (© 1987, A. Ursyn. Used with permission).

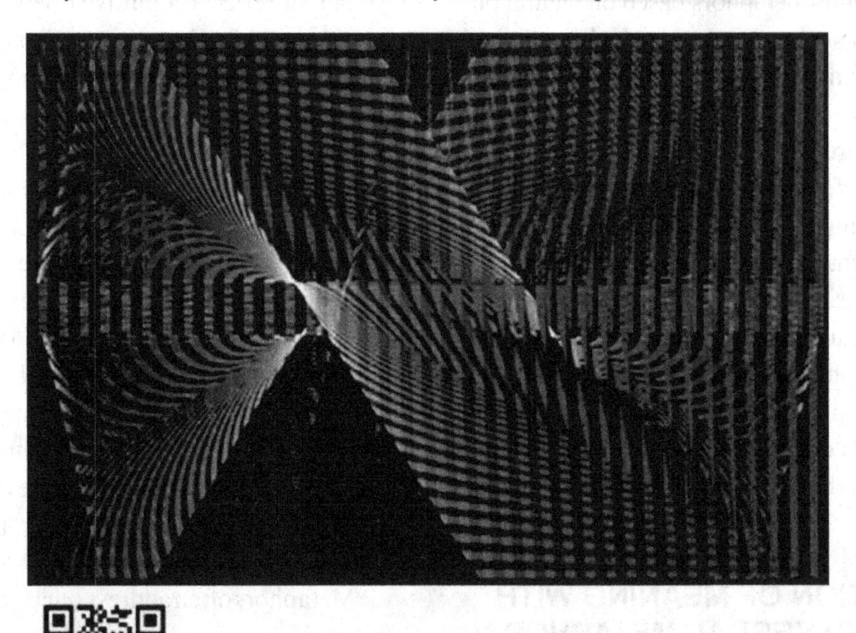

analog and the digital, generative art works result in precise images with perfect lines that follow intentional transformations.

Metaphors help us focus on an issue and facilitate understanding of the idea. Many times this happens through the use of a meta-language, as in case of tweets where people involved in a field of study communicate using specific terms and their metaphorical explanations. Metaphor may be seen as a crucial concept and a basic structure in our communication through the senses because any sensory signal: an image, melody, gesture, a figure of speech, or an action may serve for translation of an abstract concept, intangible idea, or a distant unreachable object into something accessible for our perception.

Metaphors are often visual. Images serve as metaphors, icons, and archetypes. A visual metaphor presents a picture of an easy-to-recognize object that represents another, so we may make a mental comparison of something that is not easily understood to something visible. Visual metaphors, which make undetectable concepts visible, are inherent in our thought, and thus enable visualization of abstract ideas. Visual metaphors enable us to translate abstract knowledge into a realm of familiar actualities that we can experience or see. With a metaphor as a basic structure for communicating messages we can select characteristics or qualities of a concept that are the most important for our purpose. A cognitive linguistics professor George P. Lakoff wrote that metaphors address cognitive abilities to abstract the essence of the idea. We do that by removing all features that are not crucial to the essence of the concept. Conceptual metaphors allow for understanding an abstract or unfamiliar domain in terms of another, more familiar concrete domain (Lakoff, 1990). In a similar way we compress a file for the web and compress audio or video, for example for distance learning.

Metaphor is not only an ornament to a language but also a tool of conceptual economy that is used in verbal communication, literature, for visualization, and for searching data on the web. Visual metaphors are omnipresent in commercial messages. According to Bertschi (2009), metaphor is a tool of conceptual economy but also a tool of discovery of structures within novel or unfamiliar situations. Metaphors are often applied to enrich literary productions, stimulate imagination and perception in works of art, music compositions, stage performances, as well as product design, marketing, and advertisements. It also happens often that a metaphor helps to convey meaning: a notion that has been defined and verbalized in our language is given quite different meaning when transferred into another language, alphabet, and culture. Application of the visual metaphor may eliminate such differences in the understanding of concepts. Bertoline, Wiebe, Hartman, & Ross (2010) stress the importance of the nonverbal method in communicating information as "descriptions of complex products or structures must be communicated with drawings. A visual image is formed in the mind, reviewed, modified, and ultimately communicated to someone else, all using visual or graphics processes" (Bertoline et al., p. 4). Metaphor is often seen an alternative way of expressing common sense, simplifying or stereotyping, but it can serve for capturing a cognitive content to achieve a new sense and valuable insights (Grey, 2000). Metaphors are a vital resource for the task of articulating novel insights into the human condition or refining old ones. George Lakoff, a UC Berkeley professor of linguistics and cognitive science discussed how Republicans could control House, Senate, etc., to some extent due to the effective use of metaphorical rhetoric. He argued that conservatives have spent decades defining their ideas, carefully choosing the language with which to present them, and building an infrastructure to communicate them (Powell, 2003). For Lakoff, the development of thought has been the process of developing better metaphors.

A computer graphics "Food for Thought" (Figure 2) presents metaphorical presentation of

Figure 2. Anna Ursyn, "Food for Thought"(© 2010, A. Ursyn. Used with permission).

recurring thoughts the driver may have during his long, often stressful travel. While he may be thinking about the meat and seafood tightly packed in his truck's cargo area, he may hope that living creatures will survive everything what happens to them. However, in spite of our efforts they often become listed as endangered species.

METAPHORS USING ICONIC AND SYMBOLIC IMAGES

In a visual metaphor one thing serves as a symbol representing another; for example, common symbols shown below (Figure 3) are metaphors that can be found everywhere, starting from a clip art included in writing software.

Images that have iconic properties or are used as generally accepted symbols are extensively used for visualization of the hard to explicate objects and notions. Interactive, real-time knowledge visualization methods use graphic templates, metaphors, mapping, diagrams, sketches, and other visual ways of explorations and narrations (Knowledge Communication.org, 2011). Signs, symbols, and metaphors in art, graphic design, and visual storytelling may support visual communication, organize and structure information graphically and convey human insight about the represented information through the key characteristics of the visualization metaphor. When we deal with several kinds of data, we may describe with metaphors the structure and the relations among data. Depictions of simple objects can often be seen in art works, serving as metaphorical statements, a container for cognitive constructs about the theme and the artist's thoughts on the theme. Common, primary components of our environment, such as raindrops, snowflakes, and simplified animal shapes may serve as a carrier of a notion. In cognitive terms, art works loaded with natural metaphors may thus be charged with meaning.

We may find a pervasive use of metaphors in literature, visual arts, and music. Some artists map the data into sensory experience. Auditory metaphors seem essential for music theory and appreciation. Zbikowski (1998, 2005) stressed how much cognitive science research on metaphor has to offer music theory, "There seems little doubt that musical analyses are not scientific explanations, but metaphorical ones. It remains to explore the nature and depth of these metaphors, and, in doing so, to come to a better appreciation of the processes through which we organize our understanding of music" (Zbikowski 1998). According to Desain and Honing (1992, 1996), metaphors in music theory inform and shape the ways we theorize about music: tempo is often compared to walking or moving, while the harmonic, melodic, rhythmic, and time related progression of music is characterized by flow or motion.

See Table 1 for Your Visual Answer.

Figure 3. Common clip art symbols serving as visual metaphors (public domain, free clipart published online)

Table 1.

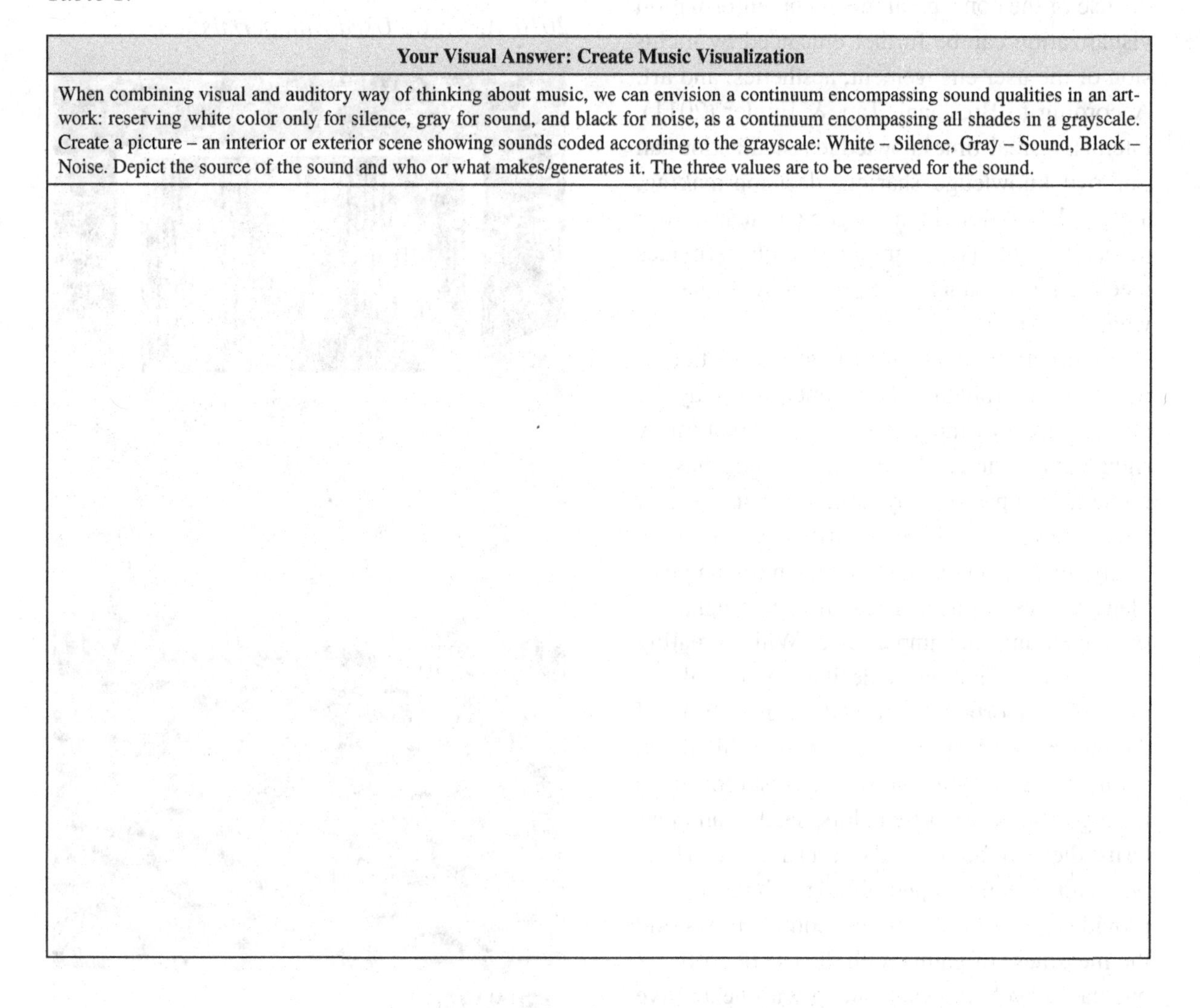

Your Visual Answer: Create Music Visualization
When combining visual and auditory way of thinking about music, we can envision a continuum encompassing sound qualities in an artwork: reserving white color only for silence, gray for sound, and black for noise, as a continuum encompassing all shades in a grayscale. Create a picture – an interior or exterior scene showing sounds coded according to the grayscale: White – Silence, Gray – Sound, Black – Noise. Depict the source of the sound and who or what makes/generates it. The three values are to be reserved for the sound.

INSPIRATION COMING FROM THE RULES AND PHENOMENA

Metaphors make invisible concepts perceivable; they are inherent in our thought and thus enable visualization of abstract concepts. Michelangelo supposedly visualized an intended sculpture inside a block of a marble; he posed that it needed to be uncovered by an artist. Metaphors are not necessarily based on familiar forms found in nature. We can assign visuals of objects and related events to non-physical numerical data (such as network system data or stock market values) and thus apply metaphors to process complex information about abstract data that lack a notion of position in space. Visual metaphors serve for representing the structure of and the relationships within abstract datasets; therefore, the use of suitable metaphors is crucial in computer program visualization (Vande Moere, 2008). As it was stated by Ittelson (2007), visualization helps to link immaterial concepts to images because "the ability to perceive objects and events that have no immediate material existence made possible the visualisation and creation of tools." Rhetorical images such as visual metaphors create meaning in infographics, evoke emotions, influence evaluations, and guide users' imagination (Lengler & Vande Moere, 2009). Due to the use of the conceptual metaphors information visualization can be further enhanced by inclusion of the user engagement, aesthetics, and art. According to Bresciani, Tan, & Eppler (2011), interactive, real-time knowledge visualization can improve knowledge sharing, decision-making, and collaboration in management teams dramatically. The use of graphic templates makes meeting participants more productive in sharing what they know.

Figure 4, "Intended Meaning" is about communication problems that sometimes may be alleviated with simple, iconic images that imply right connotations. A certain meaning has its home in one person only; others are unaware of our concepts, so we have to clarify what we mean. Many times a person involved in an exchange of ideas repeats words and sentences to emphasize their meaning and importance. While creating a visual presentation, a designer may enlarge the most important image and put it in front of the picture. In the process of creating this work, geometric and figurative images resulting from the algorithms comprise colors, shades, and patterns; they can be zoomed or transformed. Then, photosilkscreen and photolithograph techniques provide a new level of color combinations and the messiness of paint. With the use of software one can recycle drawings along with generative shapes and patterns. In a similar way we use the same words in different sentences. This can be seen in "Intended Meaning." The digital factor of the photosilkscreen and photolithograph techniques creates even more challenges for contemporary art.

Figure 4. Anna Ursyn, "Intended Meaning" (© 2010, A. Ursyn. Used with permission).

Metaphors derived from concepts in mathematics, cognitive science, and philosophy may facilitate assimilating more complicated topics, especially those related both to art and science, which can be negotiated easier as abstract concepts. Such conceptual metaphors may enable visualization of abstract concepts, for instance, represent mathematical equations or geometrical curves and thus make such concepts visible. One may say swarm computing, cloud computing, or the design of the linkage structure of the Web all originate from biology inspired metaphorical thinking. Those working on graph visualization or analyses of many kinds of topological objects may take advantage of applying the knot theory, a branch of mathematics that studies knots. Our everyday experience related to fastening hitches, shoelaces, or neckties may be inspiring for creating metaphors of mathematical knots, for example, knots closed without any open end to be untied, which have to be embedded in higher-dimensional space to be properly examined. Application of the knot theory allows graph visualization or analyses of many kinds of topological objects. Sculptors, architects, graphic artists, and designers draw inspiration from mathematical knots to create metaphorical art works. Helaman Ferguson aims at creating visual representations of mathematical objects, when he transfers the thought forms formulated in terms of topological mathematics to physical materials using a method of telecarving, where geometric forms drawn on a computer screen are translated into instructions on direct carving the stone. Ferguson claims that mathematics is an invisible art form, and computer graphics make mathematics visible; he writes (Ferguson & Ferguson, 1998), "Some math evokes art, some art evokes math." He created "A Bowline – a Twisted Knot," a small sculpture made from three large paper restaurant napkins. According to the artist, this sculpture is based on the 27,000-year old record of an algorithm. The impressions of cordage in ceramic materials were discovered and dated at Dolni Vestinice. Its algorithm is iterative and has three steps: twist, twist and counter twist. It is a very important algorithm, fundamental in all the cloths we are making of rope, cable, etc.

THE USE OF METAPHORS IN DATA, INFORMATION, AND KNOWLEDGE VISUALIZATION

Visualization is usually described as the presentation of pictures showing easy-to-recognize objects that are connected through some well-defined relations. Information visualization is the use of computer supported, interactive, sensory (mostly visual) representations of abstract data to reinforce cognition (Bederson and Shneiderman, 2003). Visualization is a part of art, science, technology, interaction, and delivery of data, because with visualization techniques computers transform data into information, and visualization converts information into picture form. At present, visualization means using a computer. Visualization creates graphic images directly from data, which let us comprehend data and make decisions. There are several approaches to the concept of visualization, for example data visualization, information visualization, and knowledge visualization. Scientific visualization is another approach to visualization, where physically based data are represented according to space coordinates, for example computer tomography data for medical use.

Visualizations are based on metaphors that are used to map from a program to an image. Meaningful images convey an insight about information contained in conceptual diagrams. In order to present information visually, one has to select a type of a metaphor. Some are two-dimensional, for example, a map or a graph metaphor, including thorough research on graph layout algorithms. Metaphors that are widely used in visualization as the representation of programs fulfill a dual function: they organize and structure information, but also convey an implicit insight through the key characteristics or associations of the metaphor (Eppler and Burkhard, 2007). Aesthetics of visualizations should become part of education and an agent in further developments in this field. Artists' cooperation could amplify imagery from which visualization metaphors may be selected. Some explore possible approaches from another perspective such as the use of avatars in visualizing knowledge or puppets used for explanation of a scientific concept.

See Table 2 for Your Visual Response.

Metaphors, in the words of Paton, Nwana, Shave, Bench-Sapon, & Hughes (1990) are "crucial to research as they provide alternative perspectives of seeing and understanding the world; indeed, they have been the basis for many scientific and technological creations." A success of any visualization project where images serve as metaphors, icons, and archetypes, may depend to a great extent on the use of right metaphor.

Table 2.

Your Visual Response: A Metaphor for Someone's Traits
You may now want to create a visual metaphor that shows someone's individual features. It could be thought provoking to draw a few sketches of some people that are familiar or in close friendship, and then draw a self-portrait (a quick sketch), trying to emphasize your traits you consider most characteristic of your personality. Devise metaphors for some features you are going to characterize in portraits you are going to draw, such as youth or old age, frailty, fidelity, sophistication, refinement, harshness in relation to others, etc. Create visual metaphors of these concepts by drawing simple symbols that characterize the person you are going to portray and include them into your portraits.

The use of suitable metaphor is crucial for successful program visualization. Metaphors that often serve visualization purposes include solar systems, video games, nested boxes, a city, house, parking lot, metro, library, street, and also facial expression. It seems clearly visible that the engineering-oriented research and inquiry have been focused mostly on the spatially referenced, time-dependent data. The choice of the type of metaphor used for data visualization depends on the kind of connections existing between the data, and defines the level of abstraction in mapping from a program model to an image. A great part of visualization techniques and tools are based on the graph metaphor. Graph theory supports also the extensive research on graph layout algorithms. Information can be looked at from various angles, so the user has an understanding of the hierarchy of the web site and its topology. When the data are organized into ranks with each level subordinate to the one above, a tree metaphor would be helpful, with a hierarchy of its limbs, branches and twigs.

When we can list many kinds of equally important data, we may need to describe the structure and the relations among these data, which means, explore their topology. A parking lot metaphor could be useful for this purpose. Since people are familiar with general concept of a parking lot or a skyscraper, this type of environment is often used to create metaphors for large web data organization. For instance, data are structured toward metaphorical representation of a building. Floors, rooms, elevators, or corridors are being used as visual interpretation of subsets of a data set. Thus by imaginative tour of a skyscraper, one may create associations and connotations about each subset and its relationship to the whole set of data and its links. Many authors (for example, Russo & Gros, 2002; Russo, Gross, Abel, Loisel, Trichaud, & Paris, 2000) developed multidimensional visualization tools using several kinds of metaphors: a system management viewed as a city metaphor with districts, blocks, houses, and buildings, topological views as cone trees, workstation views as a solar system, file system view or a site view as pyramids, geographic view as a landscape, and temporal view as a library. In a city, there are districts, residential blocks with houses, and financial blocks with tall buildings that have a height dimensionality, disks size, and status notation; there are also roads as a metaphor of connections.

A city metaphor that is widely used in data visualization presents static and dynamic information. Selecting this type of metaphor can reduce an effort, but also gives enjoyment when navigating through a city in 3D. Visual quality of data-, information-, and knowledge-visualization projects hinges on the choice of imagery. Artists' cooperation can amplify an aesthetic experience and enlarge possible imagery from which visualization metaphors may be selected. This requires that a visual language that presents information with images, symbols, and metaphors is created, or taken from the arts, both from the iconography of masterpieces and contemporary work of artists using technology. However, perception of digital, participatory, multisensory art has become even more transformed with the growing spectrum of domains included into the general notion of art. Blais and Ippolito (2006) confronted generally accepted definition of art by adding several fields of creative activity, such as computer code-based art, games, online autobiography, haktivism, computer virus making and preservation, and community building.

Figure 5, "Online Intelligence" conveys some concerns about networked communication. A simplified, programmed image of a man has been juxtaposed with the photos of apes and monkeys in their natural environment. With all advantages of intellectual development, humans are now subjected to some limitations due to the government and corporate intelligence measures:

Figure 5. Anna Ursyn, "Online Intelligence" (© 2010, A. Ursyn. Used with permission).

Apes developed a social warning system to communicate possible danger.
Humans do the same in different measures,
Even including art.
Artists' work can warn, challenge, and protect,
Against the social and cultural dangers.

Metaphors maybe considered an inherent element of the artificial life studies, systems, and simulations of life as it might be, as well as the virtual worlds such as Second Life. A postmodern literary critic in the field of electronic literature Katherine Hayles (2010) argues that narratives about and within the domain of artificial life constitute a multilayered system of metaphoric and material relays through which "life," "nature," and the "human" are being redefined; she regards the body itself as a congealed metaphor that is resonant with cultural meanings. In 2006 Ralph Lengler and Martin Eppler (2011) applied as a metaphor a grid-like, tabular form of displaying chemical elements, first devised in 1869 by the Russian chemist Dmitri Mendeleev and known as the Mendeleev's periodic table. They designed an impressive interactive presentation of one hundred visualization methods, "A Periodic Table of Visualization Methods" translating each element from the periodic table into particular meaning.

Data visualizations are visual representations of quantitative data in schematic form; they often take formats such as tables, pie or bar charts, histograms, area charts, line graphs, or Cartesian coordinates. Data visualization enables us to go from the abstract numbers in a computer program (ones and zeros) to visual interpretation of data. Text visualization means converting textual information into graphic representation, so we can see information without having to read particular sets of the data. Burkhard, Meier, Smis, Allemang, & Honish (2005) list types of visualization as sketches, diagrams, images, maps, objects, interactive visualizations, and stories. Virtual environment programs may also be constructed in a way resembling a video game.

Information Visualization, which provides multivariate display of data such as semantic networks or treemaps, uses interactive sensory (mostly visual) representations of data to amplify cognition (Shneiderman, 1996). Information Visualization supports exploring data (one-, two-, three-dimensional data, temporal and multidimensional data, and tree and network data). Animations help understand dynamic systems with interactive visualizations. Interactivity means that users are able to change the image as they work with the data. The data is transformed into an image and mapped to screen space. Edward Tufte presented in his books (1983, 1990, 1997) examples of visualizations from before the advent of computers.

Figure 6 "Ideas and Dogmas" shows how ideas, thoughts, and architectural insights got their almost perfect embodiments created by the masters in the city. Now we are filled with fear that dogma, although considered to be absolutely true, may destroy the work:

A fancy actually being in mind,
When the brain is weaker than the ego,
Seems to be absolutely true,
Unless we see disparate things differently.

Knowledge visualization uses visual representations to transfer knowledge, rather than data between people, to provide visual insight into the data. It helps us to identify, access, share, discuss, apply, and manage information, and understand the field as a problem solver, not only to learn a theory. Knowledge visualization focuses on the transfer of knowledge among people by sharing their insights, experiences, perspectives and predictions, in contrast with information visualization that concentrates on the use of computer-supported tools to explore large amount of data. Techniques used for this purpose are focused on the users, explanation (know-why, know-how), and presentation of knowledge in various visual formats.

Figure 6. Anna Ursyn, "Ideas and Dogmas" (© 2003, A. Ursyn. Used with permission).

NATURE DERIVED METAPHORS SERVING AS THE ENRICHMENT OF INTERDISCIPLINARY MODELS AND ARCHITECTURE

Natural metaphors organize and structure information in a meaningful way. They combine the creative imagery with the analytic rationality of conceptual diagrams. Most of metaphors we encounter or create are shaped upon natural forms. A spiral pattern, the helical curve, or a helicoidal shape are visible on a conical plane of a mollusk shell or an extinct fossil of an ammonite, while the concentric forms can be found on mussels; they may be described as geometrical curves or mathematical equations. These forms are familiar and because of that natural metaphors reside in thought as instinctively understandable concepts (Mateo & Sauter, 2007) and may help in visualization of complex structures such as the spiral chains of polymers characteristic of the nucleic acids (such as a double helix of the ribonucleic acid, RNA, the deoxyribonucleic acid, DNA) or other protein molecules with a helical form (displaying a helix – a curve drawn on a conical or cylindrical plane). We are fully aware that in reality configuration of molecules is ever changing and dynamic, but a double helix metaphor helps us grasp the essence of the protein structure. In contrast to photography, metaphorical visualization supports understanding of the invisible factors and not so much the real appearance of the object under scrutiny. Metaphors that draw from natural objects address our imagination in a great scope of formal, conceptual, geometrical and literary arrangements. Mateo and Sauter (2007) present diverse design approaches taken by the cross-disciplinary teams that address theoretical, material, and artistic challenges by finding correlations between nature and architectural design.

Figure 7. Anna Ursyn, "Green Architecture" (© 2008, A. Ursyn. Used with permission).

Using natural metaphors scientists and artists work at the interface between economics, biology and neuroscience to attain the enrichment of economic models aimed for instance, at environmental results and green architecture.

Figure 7, "Green Architecture" is about environmental architecture. We care about knitting together dwellings within a landscape, with roofs repeating the line of the hills, and slowly learn to draw natural resources from the power of sun, wind, and water.

METAPHORICAL WAY OF LEARNING AND TEACHING USING ART AND GRAPHIC METAPHORS FOR KNOWLEDGE VISUALIZATION

Developments in information and communication technologies strengthened the logical basis for the metaphorical approach to teaching and learning. One may also say, the improvements in visual literacy result in the growing importance of developing the way of thinking in terms of metaphorical presentation of information and knowledge. The Visual Literacy.org (2011) defines a skill of visual literacy as the ability to evaluate, apply, or create conceptual visual representations. Members of the International Visual Literacy Association (http://www.ivla.org/) consider this skill critical for business, communication, and engineering students. The use of natural metaphors offers the learners opportunity to sense the mathematical beauty while watching, for example, how a Nautilus shell obeys the rules of equations. Duality of mathematics as both the art form and the science is presented in nonverbal language in the art works, with many examples provided at Google Images. Eric Heller, professor of physics and chemistry (Harvard University) recorded a Rogue II http://www.ericjhellergallery.com/index.pl?page=image;iid=68 as a study of the effect of changing the speed of deep-water ocean waves, which is a simulation of the flow of wave energy as it negotiates a region of current eddies that is refracting waves. This scientific work is also perceived as artwork.

Developing abstract thinking abilities is essential both in countless areas of life and in education, as abstract thinking can be seen indispensable when one strives to be virtuous in mathematics, philosophy, poetry, or science. Non-physical, numerical data (such as network system data or stock market values) often refers to naturally occurring objects and related events. In such cases metaphors are often applied to process complex information. Interdisciplinary concepts, such as design cognition, user engagement, aesthetics, and art can contribute to knowledge comprehension as well as enhance information visualization (Vande Moere 2008). Moreover, in a study on wearable devices Fajardo and Vande Moere (2008) found that "a greater number of wearers appreciated the abstract metaphor consistently more than the overt one, in particular because of its aesthetic qualities, such as the use of expressive signs and symbols."

Visualization in educational psychology provides ability to create symbols that are understandable on the basis of convention, communicates knowledge and emotional reactions, changes the tacit knowledge into the explicit one as mental representations and images, and helps find meaningful patterns and structural relations in graphical displays of data. According to Edward Tufte (1983/2001, 1990, 1997), graphical display of information enhances density, complexity, dimensionality, and beauty of communication. Texts are linear and static, while pictorial and time-based constructions of net media are dynamic and often interactive. Visual learning projects aim to support learning process with the use of visual signs, symbols, icons, metaphors, visuals, photographs, and verbal coding. Visualizing knowledge in graphical form may contribute to combining both the precise and expressive way of thinking and embolden students to link creative imagery with the analytic rationality. Learning projects are aimed at encouraging non-linear thinking and enhancing image quality in student work.

Integration of the art and sciences related materials allows overlapping of the critical and visual thinking. Students learn about the data-related material in a visual way and create projects showing their understanding of the concept. Science-based topics inspire learners to create artistic, often metaphorical presentations. The hands-on instruction involves concrete operational rather than abstract thinking mode. Metaphorical imaging and abstract thinking prevail over hands-on instruction and mere memorization when it comes to learning higher-level thinking concepts and tasks such as writing programs and creating computer graphics. Learning through visualizing ideas in graphical form encourages learners to expand their visual literacy through artistic presentation, and present their findings with visual power. Projects are aimed to evoke a holistic, synaesthetic mode of learning engaging visual, verbal, and manual modes of action. Visual style of learning may reduce intrinsic cognitive load in structuring information, by shifting the explaining process from abstract to meaningful parts, which may be easier to learn and remember. Students draw sketches in order to capture the essence of the process under study, and control composition of their projects.

CONCLUSION

In this chapter, a way of perceptual imaging is discussed, with the use of art and graphic metaphors for concept visualization. An iconic or symbolic image, representation inspired by the rules and phenomena observed in nature, as well as simulation or visualization of concepts and events presented in metaphorical way may all be designated as a metaphor – a cognitive operation, which makes us see one thing in terms of another and create a new meaning. A metaphor indirectly suggests the meaning of something not easily understood, and transfers this meaning from one thing to another. Metaphors, especially biologically inspired metaphors are widely used in computing, data visualization, information visualization, data mining, and semantic web.

The evolving presence of information technologies and growing availability of networked communication media change our ambient environment and actions involving creation, research, and learning. The ubiquitous presence of metaphors used as a tool for translation of meaning has a crucial influence on the ways we are shaping our actions. Metaphors present in our visual and verbal environment include iconic and symbolic images, representations inspired by the rules and phenomena observed in nature, as well as simulations and visualizations of concepts and events presented in metaphorical way. Nature derived metaphors serve as the enrichment of interdisciplinary models and support data visualization, information and knowledge visualization, data mining, and semantic web. This caused a need for combining metaphorical visualization with artistic principles and working on metaphorical way of learning and teaching using art and graphic metaphors for the information and communication purposes. The use of metaphors is prevalently aimed at improving one's power of conveying meaning by merging information visualization with the principles of creative design, integrating art and science, and visualizing knowledge in graphical form. A set of reasons for focusing attention on metaphors as a way of communication include an impact of the online way of learning and teaching and the global dimension of the multicultural society of learners that differ in their cultural, linguistic, or experiential background but share resources and tools. Approaches to visual learning consist of visual presentation of scientific concepts, creating art by finding inspiration in a science-based topic, and learning by arranging data visually into a structured whole.

REFERENCES

Bederson, B., & Shneiderman, B. (2003). *The craft of information visualization: Readings and reflections*. Morgan Kaufmann Publishers.

Bertoline, G., Wiebe, E., Hartman, N., & Ross, W. (2010). *Fundamentals of graphics communication* (6th ed.). McGraw-Hill Science/Engineering/Math.

Bertschi, S. (2009). Knowledge visualization and business analysis: Meaning as media. In *Proceedings of the 13th International Conference Information Visualisation,* (pp. 480-485). IEEE Computer Society Press.

Blais, J., & Ippolito, J. (2006). *At the edge of art*. Thames & Hudson.

Bresciani, S., Tan, M., & Eppler, M. J. (2011). *Augmenting communication with visualization: Effects on emotional and cognitive response*. IADIS ICT, Society and Human Beings.

Burkhard, R., Meier, M., Smis, M., Allemang, J., & Honish, L. (2005). Beyond Excel and PowerPoint: Knowledge maps for the transfer and creation of knowledge in organizations. In *Proceedings of 9th International Conference on Information Visualisation (iV 05)*. IEEE.

Desain, P., & Honing, H. (1992). *Music, mind, and machine: Studies in computer music, music cognition, and artificial intelligence (kennistechnologie)*. Thesis Pub. ISBN 9051701497

Desain, P., & Honing, H. (1996). Physical motion as a metaphor for timing in music: the final ritard. In *Proceedings of the 1996 International Computer Music Conference*, (pp. 458-460). San Francisco, CA: ICMA.

Eppler, M. J., & Burkhard, R. A. (2007). Visual representations in knowledge management: Framework and cases. *Journal of Knowledge Management*, *4*(11), 112–122. doi:10.1108/13673270710762756.

Fajardo, N., & Vande Moere, A. (2008). ExternalEyes: Evaluating the visual abstraction of human emotion on a public wearable display device. In *Proceedings of the Conference of the Australian Computer-Human Interaction (OZCHI'08)*, (pp. 247-250). Cairns, Australia: OZCHI. Retrieved November 2, 2011, from http://web.arch.usyd.edu.au/~andrew/publications/ozchi08.pdf

Ferguson, H., & Ferguson, C. (1998). *Eightfold way: The sculpture*. Retrieved October 30, 2011, from http://library.msri.org/books/Book35/files/fergall.pdf

Grey, W. (2000). Metaphor and meaning. *Minerva – An Internet Journal of Philosophy, 4*. Retrieved February 6, 2012, from http://www.minerva.mic.ul.ie//vol4/metaphor.html

Gross, C., Gros, C. P., Abel, P., Loisel, D., Trichaud, N., & Paris, J. P. (2000). Mapping information onto 3D virtual worlds. In *Proceedings of the 4th Conference on Information Visualisation*. IEEE Computer Society Press.

Hayles, K. (2010). How we became posthuman: Ten years on. *Paragraph*, *33*(3), 318–323. doi:10.3366/para.2010.0202.

International Visual Literacy Association. (2011). Retrieved September 1, 2011, from http://www.ivla.org/

Ittelson, W. H. (2007). The perception of nonmaterial objects and events. *Leonardo*, *40*(3), 279–328. doi:10.1162/leon.2007.40.3.279.

Knowledge-communication.org. (2011). Retrieved September 1, 2011, from http://www.knowledge-communication.org/overview.html

Lakoff, G. (1990). The invariance hypothesis: Is abstract reason based on image-schemas? *Cognitive Linguistics*, *1*(1), 39–74. doi:10.1515/cogl.1990.1.1.39.

Lakoff, G., & Núñez, R. E. (2001). *Where mathematics comes from: How the embodied mind brings mathematics into being*. Basic Books.

Lengler, R., & Eppler, M. (2011). *A periodic table of visualization methods*. Retrieved November 2, 2011, from http://www.visual-literacy.org/periodic_table/periodic_table.html

Lengler, R., & Vande Moere, A. (2009). Guiding the viewer's imagination: How visual rhetorical figures create meaning in animated infographics. In *Proceedings of 13th International Conference on Information Visualisation*. IEEE. DOI 10.1109/IV.2009.102

Mateo, J. L., & Sauter, F. (Eds.). (2007). *Natural metaphor: Architectural papers III*. Zurich: ACTAR & ETH Zurich.

Paton, R., Nwana, H. S., Shave, M. J. R., Bench-Sapon, T. J. M., & Hughes, S. (1990). Transfer of natural metaphors to parallel problem solving applications. *Lecture Notes in Computer Science*, 496.

Powell, B. A. (2003, October 27). Framing the issues: UC Berkeley professor George Lakoff tells how conservatives use language to dominate politics. *UC Berkeley News*. Retrieved February 6, 2012, from http://berkeley.edu/news/media/releases/2003/10/27_lakoff.shtml

Russo Dos Santos, C., & Gros, C. P. (2002). Multiple views in 3D metaphoric information visualization. In *Proceedings of iV02, 6th International Conference on Information Visualisation*, (pp. 468-473). IEEE.

Shneiderman, B. (1996). The eyes have it: A task by data type taxonomy of information visualizations. In *Proceedings of IEEE Visual Languages*, (pp. 336-343). IEEE.

Tufte, E. R. (1990). *Envisioning information*. Cheshire, CT: Graphics Press.

Tufte, E. R. (1997). *Visual explanations: Images and quantities, evidence and narrative*. Graphics Press.

Tufte, E. R. (2001). *The visual display of quantitative information*. Cheshire, CT: Graphics Press. (Original work published 1983).

Vande Moere, A. (2008). Beyond the tyranny of the pixel: Exploring the physicality of information visualization. In *Proceedings of 12th International Conference on Information Visualisation*. IEEE.

Visual Literacy.org. (2011). Retrieved September 1, 2011 from http://www.visual-literacy.org/index.html

Zbikowski, L. M. (1998). Metaphor and music theory: Reflections from cognitive science. *Music Theory Online, 4*(1).

Zbikowski, L. M. (2005). *Conceptualizing music: Cognitive structure, theory, and analysis*. Oxford University Press.

Section 3

Tools for Translating Data into Meaningful Visuals

Chapter 9
Visual Approach to Translating Data

ABSTRACT

This chapter examines some of the tools that enable a visual approach to translating data, beginning with a comparison of the use of a computer versus pencil in visual communication. A short note follows, discussing the evolution of imaging with the use of computing: the history of computers and then some examples of graphic display and early computer-generated art works. This is followed by a discussion of the basic ways of graphical display of data and strategies for visual problem solving in the context of art and design. Thoughts on visual translation of data include an introduction to computer simulation. Examples of computer simulation and evolutionary computing conclude the chapter.

INTRODUCTION

The making of simple images may be done successfully at all levels of technical proficiency, without any introductory exercises, even before mastering the tools and without losing the novice's enthusiasm. Suitable tools are needed for completing meaningful projects. However, making a good data graphics requires developing diverse skills – the visual-artistic, mathematical, and statistical. An integrative approach to creating computer art graphics allows using the computer not only as a tool but also as a source of inspiration. Creation of art graphics depends not exclusively on a programming process or on the existence of human interface devices such as a mouse, keyboard, and joystick. It's good to keep in mind that the machine has the ability to accomplish the task precisely but without any evaluation; sometimes, it's good to learn not to ignore "happy accidents" so easily occurring while working with the computer.

It is usually accepted that the field of computer-based imaging belongs to the domain of computer science because graphics are used to process

DOI: 10.4018/978-1-4666-4703-9.ch009

digital images and manipulate numeral content. Also, computer graphics share many common techniques and applications with the field of visualization. Both two-dimensional and three-dimensional graphic presentations refer often to geometry, but also to the ways of representing, processing, manipulating, and projecting data in a static or animated way. Many educators oriented toward both visual and technological literacy integrate the imaging techniques, such as 3D, time-based, interactive, and other graphics including those with emphasis on computer animation, web design, video graphics, virtual techniques, robotics, games, web-based, data and knowledge visualization, with general curriculum to match the changing expectations of professionals and students. For example, a research-based organization Information Resources Management Association (IRMA) has collected vast materials concerning the ways to locate, comprehend, evaluate, and organize information using digital technology (Association, I. R., 2013). The University of Texas in Dallas launched out instruction through gaming and interplay of art, music, and narrative with the new media, and opened the Master of Arts and PhD programs as separate from the art department or the computer science department. Students in a section Art and Technology are required to learn programming; they develop projects aimed at interaction based environment such as gaming.

There is a saying "If it's not on the Internet, it does not exist". While this statement might seem accurate, more and more data gets lost or forgotten because it isn't of interest anymore, isn't novel, is expired, or there was a server error. While most of it is backed up and stored, some info gets lost. This may be one of the reasons for citing the Internet-based references as an URL followed by the date and time when the page was accessed.

In spite of all activities aimed at securing our achievements some of us may feel uncertain and anxious because of natural cataclysms and malevolent actions. Figure 1 "Timetable" tells about walking in the City with a regained confidence:

Long day of errands in the City,
Memories seen against the light,
Times and places in order again.

Figure 1. Anna Ursyn, "Timetable" (© 2002, A. Ursyn. Used with permission)

COMPUTER VS. PENCIL

We often compare a human brain to a computer, when we want to examine how it works and succeeds in doing various things such as art, for example. Visual thinking includes the perceptual and cognitive activities that involve mental imagery – an internal image representation that can be supported by external visualizations, the displaying of information either on a screen or paper, and saving the gained knowledge. Many claim that with digital imaging, where every step in a code execution clearly defines the next step, all worries about the creative process would disappear. However, there are many personal, psychological, cognitive, phenomenological, and aesthetical factors that result in the conflicting attitudes toward digital versus traditional image presentation. With all approaches and techniques adopted in digital imaging, some believe that freehand drawing can help in a search for finding, as the song tells, 'who you really are.' They ponder whether digital drawing techniques serve well for inspiration or attaining perfection, and feel more unrestrained when they spontaneously react to natural beauty by showing shapes, colors, and light. Somebody argued that it is hard to imagine an immense quantity of information and a great number of art works that can be created with just one pencil. Otherwise, one may transfer pencil drawings to the computer memory to make them a part of a larger whole and then use the computing techniques. Others believe that in the present state of digital technology there is no need to draw freehand because any photograph can be altered at will. Digital imaging adds an option of mimicking traditional media combinations, so the image may look like an artwork created in any medium: oil painting, a wet in wet watercolor, or a dry-brush painting; one can argue there is no need to learn and teach drawing. Visual representation of information, data, or knowledge called infographics, as well as conceptual or scientific information make good use of the precision coming up with a computer code or the 2D and 3D graphics transformations. Figure 2 shows a design for a t-shirt made by my student Leighanne Rayome, with a picture and a short verse about a pencil.

Figure 2. Leighanne Rayome, "Pencil," a t-shirt design (© 2013, L. Rayome. Used with permission)

Selecting meaningful content could be called abstracting the image. In such instances, freehand drawing and sketching is often a tool of choice because hand drawing may allow the artist choosing the essentials only. A British painter David Hockney, who argues that an artist has to be able to draw before anything else could be achieved (cited McCall Smith, 2010), draws and paints on the iPhone using the Brushes application. He suggests that the touchscreen applications such as iPhone or iPad would bring back drawing, and believes that the use of drawing as an instant communication tool would increase because children grow with computers.

If we would like to inspect processes of visualizing concepts, conveying knowledge, or creating art we may return to a simplest comparison: which activities can we perform better when we use the computer instead of a traditional tools such as a pencil. Below are students' opinions that took form of a collective statement:

One can find many circumstances in which one can better act with a computer. In animation we can do it by redrawing and pasting only changed portions of the image or using software capable of animating an image. Also, we have so many kinds of software capable of animating an image. We have ability to edit, without having to rewrite it from scratch, as changing or erasing is easier and cleaner, with the touch of a button or a mouse. We have the option of using different fonts, styles, colors and formats and we are able to copy and paste many times. Color mixing is faster and inexpensive, may be done with the use of additive or subtractive processes. 3D image transforming and manipulation is possible, by rotating, zooming, using proportional selection, distorting, placing in foreground or background, etc. Perspective drawing is quicker due to moving objects and changing relative sizes. Patterns can be done easier, for example, with the use of a lines and an airbrush tool. Magnifying a part of an image helps to work on the details.

Now for the advantages of using a pencil: it has been around for thousands of years and many believe that it will never become totally extinct. When we do the work with a pencil or other tools for drawing, we often feel more free and unrestrained when we use it for reacting to natural beauty by showing perfect shapes, vibrant colors, and light. Also, it is available to everyone; many times it is easier to find a pencil at hand, so we are ready to spontaneously scrabble and sketch. Possibly, tools for drawing on a computer screen serve us better for creating more conceptual, abstract works. Fortunately, we may transfer our pencil drawings to the computer memory using a scanner or a digital camera, and then make them a part of a larger whole by combining techniques that are available to us on a computer. And finally, a pencil will never become an outdated piece of hardware. All in all, everyone can select the best tool for the job and use the computer as a tool when working on a project that involves writing skills, simple artwork, or reworking the visual material by computing. The pencil may be a preferred choice when one wants to be more spontaneous using one's creative experience. Many times if the project requires a drawing, one would sketch it first before going to the computer. Maybe a combination of the two tools would be the best option.

See Table 1 for Your Visual Response.

Table 1.

Your Visual Response: Making a Drawing on a Screen and a Freehand Drawing
It may be interesting to apply pragmatic approach to this discussion and draw two pictures on the same theme: first to draw a scene on a screen and then as a freehand drawing. Possibly, drawing elements that would seem inconvenient while working in each of these frameworks may manifest themselves quite different than expected. You may draw your object on the screen of your phone or your graphic tablet, and then sketch a phone or a graphic tablet itself. You may find links for drawing and sketching apps for the portable devices on Google.

Tools for Cognitive Computation

One may ponder how the perceptual and cognitive activities related to visual thinking may depend on the digital or traditional tool selected for visual presentation. Colin Ware (2012a, 2012b) examined how visual presentation can help people solve cognitive problems. This process requires the use of visual thinking algorithms, dynamic queries, interactive design sketching, and reasoning involving social networks. Both the language of computer science and the language of vision research and cognitive psychology may describe cognitive processes, because some cognitive operations are carried out in a computer while others are carried out in a human brain. Ware describes the elements of the visual thinking as a part of perceptual and cognitive computations done in the brain of a person. These algorithms, according to Ware (2012a, 2012b) are using pseudo-code and contain (1) dynamic visual queries about patterns and attributes of perceived symbols and objects, transformed in order to accomplish a solution by means of a visual pattern search; it is constrained by visual pattern perception, as well as visual working memory capacity; (2) epistemic actions intended to seek information by the user, for example by eye movements or mouse movements aimed at navigating through a data space; (3) computer programs supporting epistemic actions by revealing and hiding information, changing scale, and adaptive highlighting; and (4) externalization as part of a thinking process – adding things to a display by the user, for example by circling a region to group objects or entering something into a computer.

Visual style of presentation may reduce the cognitive load in structuring information, by somewhat shifting the explanatory task from abstract to meaningful parts, which may be easier to understand and remember. Users of electronic devices interact with icons on the screen that carry information. The simple, less complicated icons that impose lesser cognitive load are understood easier, faster, and thus picked up first (Skogen, 2006). Lowering of the cognitive load may be achieved many ways: with the use of icons, symbols, and signs, or application of metaphors to particular concepts and principles. The use of multisensory ways of learning may also decrease the cognitive load; one can do it by making drawings and doodling about the content, listening and creating sounds and music, composing stories, mangas, and sending text messages. The physical act of drawing is considered beneficial in creating graphs on a computer, especially when one creates new images. According to Purchase, Plimmer, Baker, & Pilcher (2010, p. 80), drawing "the graph from scratch, rather than simply moving nodes in a pre-drawn graph – this removed any layout bias present in the original drawing.' Using a sketching tool with a stylus 'allowed for the physical action of creating the graph to be done as easily as if it were on paper, reducing any cognitive distance between the participant's desired drawing and what is presented. Using the interface of a formal graph drawing tool is a less natural way of drawing a graph than the free-form hand movements made possible by a sketching tool."

EVOLUTION OF IMAGING WITH THE USE OF COMPUTING

A story about digital tools for imaging through graphics, visualization, or online installations may begin with the evolution of computing itself, followed by the early ways of creating and displaying graphics, and computer art.

A Short Note about History of Computers

Evolution of computers came about in terms of their speed, accuracy, and memory, with a series of sequential generations discerned according to the technological progress. At first, computers were the accounting machines; they collected data from punched cards, performed calculations, reorganized them and then printed results. The first automatic calculator was the Mark I (1944) that used the binary number system. The binary number system is still the base of the present computer languages. The machine languages code electrical states in the computer as combinations of 0s and 1s. The binary number system uses the digits 0 and 1 and a base of 2, to convey data and instructions to the computer. The first electronic digital calculator called ENIAC was constructed in 1946 at the University of Pennsylvania. Eniac weighed 30 tons, covered 1,500 square feet of floor space, and consumed a lot of electricity with frequent electricity related failures. The developments in computer technology in the 1960s and 1970s brought about the evolution of hardware – the physical equipment of the computer, and then many types of software packages – programs that the hardware executes. Sequential advancements in computer technology were described as generations.

The first-generation computers used vacuum tubes. The UNIVAC 1 was produced in 1951. They were very complicated, heavy, expensive, large and unreliable, but faster than their predecessors. Punched cards still served for putting in instructions and data. The data, coded in a symbolic language had to be translated into machine language recorded on a magnetic drum, before being executed by the computer. Navy Commodore Grace Murray Hopper developed one of the first translation programs in 1952.

The second-generation computers (about 1956) used transistors; they were more reliable than vacuum tubes, smaller, and faster. Transistors, soldered to circuit boards, use less power and generate less heat. Computers had the magnetic core memory, so they had a larger storage capacity. They also used magnetic tapes and discs for storage and input/output. The first operating system was developed; the programming was done with the machine language and the assemble language; high-level languages were developed, such as COBOL and FORTRAN.

The third-generation computers (1964) used the integrated circuits etched on silicon chips (developed by Jack S. Kilby of Texas Instruments) to form the primary storage. Again, computers became smaller, faster, less expensive, and more reliable. This involved small-scale integration technology with only a few transistors, medium scale integration of integrated circuit chips with hundreds of transistors. The IBM's System/360 revolutionized the industry because of its main storage capacity in the central processing unit. First minicomputers were released, with the capabilities of the full-size system but smaller memory. Compatibility among computers became possible due to the introduction of software. Third generation computers were accessible by remote terminals.

The fourth generation computers (from 1970) have been using large-scale integrated circuits and microprocessors. They are characterized by increased speed, storage capacity, and versatility. Large-scale integrated (LSI) circuits progressed to the Very Large Scale Integration (VLSI) circuits that contain thousands of transistors densely packed in a single silicon chip. Microprocessor, built on a small silicon chip, is the central processing unit of a microcomputer; its small size and great versatility made possible producing home computer mainframes, minicomputers, and microcomputers. The next steps in the development of the fourth generation computers included the release of GUI (graphic user interface); Macintosh SE/30 in 1989, Macintosh Classic in 1990, Macintosh Classic II and Macintosh LC in 1990. Built-in math coprocessors speed up computing. In 1989 Tim Berners-Lee invented World Wide Web, a global information exchange. Browsers Netscape and Mosaic started in 1994. These computers have been used in the fields of simulation, virtual reality, multimedia, data communication because of their high accessing speed and storage capacity.

The label of a fifth generation of computers refers to the alternative developments in computer technology. A Japanese Fifth Generation Computer Systems (FGCS) project from the eighties was supposed to perform much calculation using a parallel processing system instead of a file system. The FGCS program ended in 1993. Computers from 1993-6, such as Intel Pentiums or a series of Macintosh are also called the fifth generation computers and then the sixth generation computers. Superconductors, dual processors, a release of Netscape Navigator 1.0 and 2.0 in the nineties, Netscape Communicator, Internet Explorer, and other web browsers, as well as the beginning of artificial intelligence characterize the fifth generation. DNA computers use DNA molecules and enzymes instead of silicon microchips. Researchers from Israel developed the first programmable molecular computing machine in 2004 that could diagnose cancer cells and release drugs.

Then, computer software took over the codes and interconnections securing instructions to run a specific processor and control the state of the computer. Many types of software applications and packages containing sets of computer programs and related data have been developed for computers to operate data processing systems. Generally speaking, software is considered a set of programs, procedures, algorithms, and its documentation. Issues pertaining to software quality refer to software architecture, documentation, libraries (collections of functions), standards related to programming languages, operating systems, and environments, the ways to execute the software, testing the software quality and reliability, and issues related to licenses and patents.

Presently in a networked environment specific kinds of software serve business (for example, accounting, tax calculation, word processing), home management (systems for inventory, security, and budgeting tasks; infrastructure for a digital home), education (geared to drill, tutor, or instruct, or for simulation), health (affordable technologies for personal health monitoring), and games (video games, adventure games, educational games). Computer-Assisted Instruction (CAI) direct interaction between a computer, acting as instructor, and a student. Interactive programs are being widely used for the classroom and online educational applications. Apple Computer, Inc. developed a multimedia educational workstation including text, sound graphics, animation and video, all controlled by Macintosh computers. The 'Multimedia in Education' consists of interactive video workstation (including a computer, videodisc player and a monitor), sound workstation (using a sound digitizer with sound sources and speakers), interactive imaging system consisting of a computer, a video digitizer and monitor, one or more of video sources (such as videodisc, VCR, camera) and scanner, or interactive presentation workstation, where videodisc-controlled computer program may be fed into an overhead projector with a projection pad. Tools for visual presentation of information and objects become present in our everyday environments due to the developments in ubiquitous computing and communication, which depending on the authors and context is described as pervasive computing and applications, physical computing, or ambient intelligence. Instruction is often delivered through handheld devices such as smart phones and tablets. Computing specialists design small gadgets, ubiquitous, and social robots – smart software or service providers within ambient intelligence environments integrated with semantic web, cloud computing and ubiquitous computing technologies.

Early Examples of Graphic Display

Graphics showing quantities or events going in time, such as data about weather, are called time-series charts and make the most frequently used form of graphic design. Information about these graphics comes from the books of Edward Tufte, a guru of graphic designers, computer scientists and visualization specialists. Time-series plots began in 1700s, however the oldest illustration of planetary orbits comes from the tenth century (Tufte, 1983, p.28). The use of abstract pictures to show numbers was introduced about 1750-1800. First data maps come from 17th century, 5,000 years after first maps on clay tablets. In the 1800s they appeared in scientific writings. Two inventors of modern graphical design are: Johann Heinrich Lambert (1728-1777) a Swiss-German scientist and mathematician who, for example, extended the existing visualizations of color mixing to three dimensions, and William Playfair (1759-1823), a Scottish political economist who contrasted his new graphical method with presentation of numbers with tables. The first graphic showing time (a time-series graphic), "Imports and Exports to and from England" was printed in 1786 (Tufte, 1983, p. 32). Étienne-Jules Marey pioneered graphical methods in physiology. He photographed man in black velvet in stick-figure images, explored the movement of a starfish turning itself over, and the advance of a gecko. Marey had also designed a time-series study of the tracks of the horse, by comparing an ordinary walk, long stride walk, quick walk, amble, jog-trot, and gallop (Tufte, 1983, p. 35-36). Marey became the time-series forerunner of Marcel Duchamp's *Nude Descending a Staircase*.

Figure 3, "Nothing to Undo" ponders about the reality of living in a networked community:

Reflection in water creates a virtual image of reality,
That somehow shows us,
How visualizations work in a similar way.
Everything is interconnected.
So, with all apps, connectivity, and interactivity of websites,
One has to think twice before making any change.

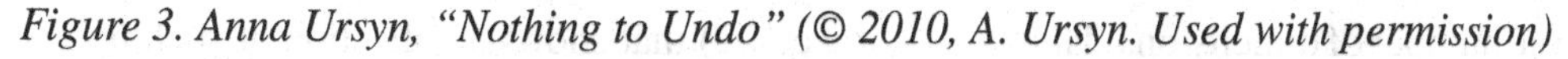
Figure 3. Anna Ursyn, "Nothing to Undo" (© 2010, A. Ursyn. Used with permission)

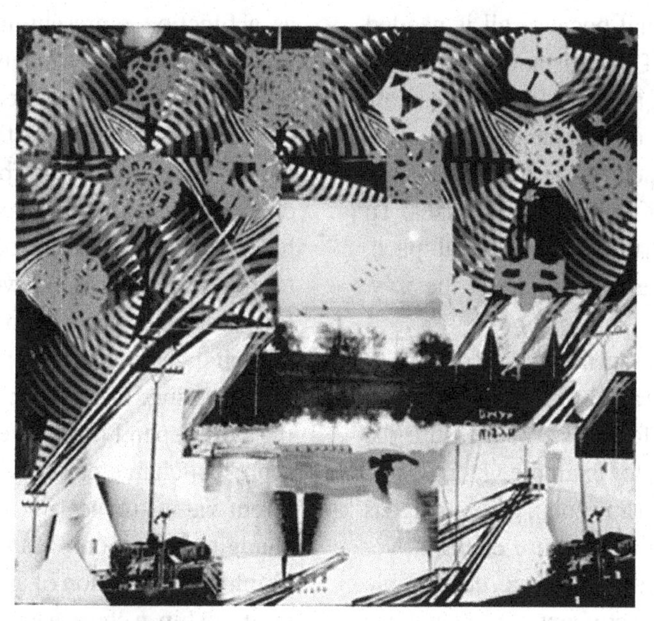

BASIC WAYS OF GRAPHICAL DISPLAY OF THE DATA

Dimensions, Variables, and Coordinate Systems

Basic concepts employed in visualization involve such notions as dimension and variable. According to Edward Tufte (1983), talking is a linear, non-reversible, one-dimensional sequence. Graphics overcome those restrictions and allow the viewers to reason about a multidimensional array of data at their own pace and in their own manner, communicate, document, and preserve knowledge. Basic concepts employed in visualizing higher dimensions include a mathematical definition of dimension as the measure of a distance of a single point from the origin of an axis (Cox, 2008). Visualizing higher dimensions requires specifying the position of a point. The number of measurements needed to determine the locus of a point depends on the number of dimensions. Thus, a line is one-dimensional because all is needed is one measurement of the distance between a point and the origin of an axis. In order to draw a two-dimensional plot on a plane, two measurements are needed to specify the distance of the point from the origin of the x and the y axes. To position a single point in three-dimensional space, three measurements are needed, along the x, y, and z axes.

The data are often described by sets of numbers. When we examine an object or a process, we determine some of its features. We measure the values of the variables, for example, geometrical dimensions of the object (how long, wide, or high is the object), its temperature, pressure, changes over time, etc. If our numbers do not change, the variable becomes a constant. The concept of dimension is related to the number of represented variables. A variable is a measurable whole to which we may ascribe a set of values. We use symbols of such wholes, for example, in the expression $a^2 + b^2 = c^2$, a, b, and c are variables.

Usually, the variables depend on each other. For example, if we have two variables, they are related in such a way that when we change the value of the first variable, we can determine the value of the second variable. The first variable is called the independent and the second the dependent variable. We use axes x and y to display the relationship of independent and dependent variables visually. We place independent variable (for example, time) at the horizontal axis (x) (we call point 0 the origin) and then we show values of the dependent variable (which changes in time, for example, distance) at the vertical axis (y). When we want to determine location of a point P on a two-dimensional plane in a Cartesian coordinate system, we talk about the x-coordinate or abscissa (it is positive if P is to the right of the y-axis) and the y-coordinate or ordinate (which is positive, negative, or zero if P is above, below, or on the x-axis). On our data graphics, the point P with coordinates (x, y) is represented as P (x, y). We may draw a graph (a straight line or a curve) that shows a collection of such points as a function representing relation between the independent and dependent variable.

When we want to describe a process, we have to look over characteristic traits of the object. When we want to describe how this object will change when conditions become different, we have to state what kinds of changes we are going to measure, that means, we have to list variables. For example, we may measure how a distance will change depending on the time (duration) of our traveling, or how the duration of our traveling will depend on its speed. Thus the variables represent characteristic traits, which take on different values (exact amounts or numbers) under changing conditions. Many times, when we draw a graphic presentation of an algebraic or geometric relationship, a chart, a graph, or a sketch, there are many variables dependent on each other and it is a real challenge to visualize them in one graph. When we want to visualize these variables, we can think about their dimensions.

See Table 2 for Your Visual Response.

Table 2.

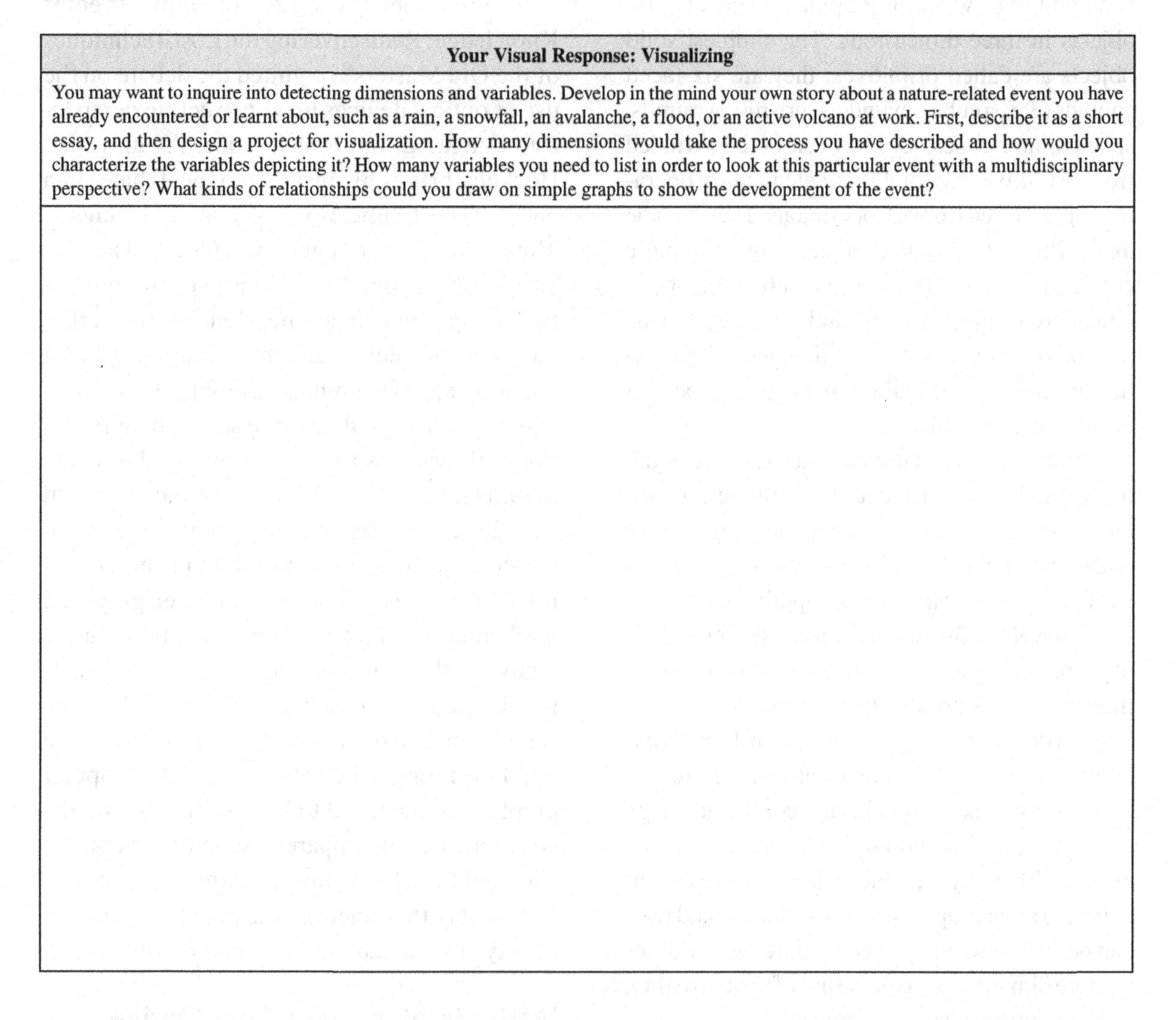

Your Visual Response: Visualizing
You may want to inquire into detecting dimensions and variables. Develop in the mind your own story about a nature-related event you have already encountered or learnt about, such as a rain, a snowfall, an avalanche, a flood, or an active volcano at work. First, describe it as a short essay, and then design a project for visualization. How many dimensions would take the process you have described and how would you characterize the variables depicting it? How many variables you need to list in order to look at this particular event with a multidisciplinary perspective? What kinds of relationships could you draw on simple graphs to show the development of the event?

Visualizing Higher Dimensions

Most of natural processes and events are complex and dynamic. In exploring high-dimensional space, we need measurements of many variables. For example, a three-dimensional location of the object and its conditions such as temperature, pressure or density may change in time. Thus, seven variables: x, y, z axes, time, temperature, pressure, and density describe the speed and state of this object in space. A two-dimensional plot or graph can only show one relationship between variables. However, most of natural processes and events are complex, dynamic, and multivariate – involving many variable quantities, and thus require multidimensional visualizations. Computers provide information in the form of billions of numbers providing the coordinates, numerical values resulting from the measurements of variables.

Graphics may show many variables in 2D, just telling stories about space and time and enhancing the explanatory power of the time-series display. One may model a solid object by approximating its form with a set of plane polygons as faces. There is a number of ways for graphic display of solid objects in three dimensions. The simplest solid objects are called primitives: they are six-faced cuboids (e.g., a cube), cylinders, prisms, pyramids, spheres, and cones. Objects can be constructed from primitives by combining elementary operations, most often Boolean operations. They include union (that mergers two objects, for example a cube and a sphere), difference (subtraction of one object from another one), and intersection (the portion that is common to both objects; it means the set that contains all elements of a cube that also belong to a sphere).

Other ways are to use parametric curve equations (mathematical functions defining the surfaces). Bezier curves are generated from control points on a plane. Bezier curves provide convenient method for interactive design applications.

To develop a 3-D model, one uses several kinds of commands: first geometric construction commands (such as points, lines, rectangles, circles, ellipses) in order to design a form, then the manipulation commands (such as rotate, move, mirror, copy, zoom), and the modifying commands (e.g., edit, erase, delete, redraw, scale, etc.). Complex events with many variables can be shown as 4 or 5 dimensional graphics. Stories of space and time can be even better displayed by showing variables in more dimensions, with the use of virtual reality, or in the immersive environment.

Construction methods build the solids from simpler shapes. One can sweep a two-dimensional pattern through some region of space, creating a volume. Solids with translational or rotational symmetry can be formed by sweeping a two-dimensional figure through a region of space (advanced modeling of complex surfaces). Another technique uses solid geometry methods to combine three-dimensional objects with a union operation which joints two objects to produce a single solid. Spline is a flexible strip used to produce a smooth curve through a set of plotted control points, and spline curves are drawn in this manner.

David Hockney's (2001) book "Secret Knowledge: Rediscovering the Lost Techniques of the Old Masters" reignited the debate on the use of optical devices for constructing perspective images in the Renaissance. As an artist, Hockney brings his insights to the debate. In a paper "The Implications of David Hockney's Thesis for 3D Computer Graphics" Theodor Wyeld (2011, pp. 409-413) argues that just as technology informed the Renaissance artist on ways of seeing and representing natural phenomena, 3D computer graphics today uses algorithms to simulate these same phenomena. For both, various techniques are used to make the images produced seem real or at least real enough. In the case of the Renaissance artist, painterly techniques were used to generate the illusion of clarity. For 3D computer graphics, mathematical algorithms are used to simulate many of the same effects. Striving for realism is a common theme. However, while the Renaissance artist never lost site of their role in interpreting what they see, 3D computer graphics is supposed to be underpinned by the certainties of its apparent scientific veracity. The author asks, is this certainty deserved or is it merely that science and art are intertwined in ways that mean one is reliant on the other?

Methods of Inquiry: Quantitative and Qualitative Research Design

Data graphics not only substitute statistical tables, but also are instruments for visual thinking about quantitative information. The way of thinking about the data we need to gather may determine our research and our inquiry strategies. Generally speaking, three types of research design are based on the data under consideration: quantitative research, qualitative research, and mixed methods research approach.

Quantitative data deal with measurable numbers and quantities such as length, height, area, volume, weight, temperature, radiation level, humidity, sound level, speed, acceleration, time, cost, or person's individual characteristics described by numbers. Quantitative numerical data are usually categorized, put in a rank order in units of measurement. Data are kept objectively separated from the subject matter and a selected research method.

We call our research quantitative when it examines the relationship between variables that are measured, (usually on instruments) and thus it tests objective theories in a deductive style based on logical analysis of the data. Quantitative research relates to the gathering, measuring quantitative data in numerical form, and then analyzing the data with the use of statistical procedures in order to answer the questions, test hypotheses, or solve a problem. The quantitative research design is aimed at preventing biases, controlling variables that may influence the result, and securing possibility to generalize and replicate the results. A structured form of reporting the results of quantitative research comprise a theoretical framework, review of literature, questions and hypotheses, data collection, methods, data analysis, results and interpretation, discussion and validation.

Qualitative data take form of both the words and images. They may result from observations that are done in particular, non-determined settings about colors, textures, tunes as note sequences, tastes, smells, appearances, styles, aesthetic perceptions, psychological experiences, and emotional feelings. Qualitative data are non-numerical, mostly verbal, and often translated by coding. They often serve the initial phases of a research study and are helpful in preparing a quantitative research design as its next phase. Research goes from particular data to general themes that evolve during the research and causes that further questions and procedures emerge as a consequence.

Qualitative research allows exploring what meaning the individual people or the whole groups ascribe to some notions, ideas, or problems, both individual and social. It assumes an inductive style of looking at the findings based on conclusions derived from the evidence and reasoning. We call our research qualitative when it is relating to quality of the subject of our study, examining what people think about other people, rules, projects, or behaviors. In a qualitative research the data are in the form of interviews, open-ended questionnaires, diary accounts, unstructured observations, videos, rather than numbers. Records on individual level may support studies of human direct experience and consciousness with a phenomenological approach; however, pursuits are often undertaken aside from the social context. Qualitative research seeks an understanding of human behavior and its reasons; data processing is often subjective in approach. A final report of the research has a flexible structure.

Mixed methods research incorporates, with a pragmatic worldview, methods of quantitative and qualitative research and combines in a sequence or associates some kinds of data such as closed questions on a questionnaire, structured interviews, or coded observations, which may produce both quantitative and qualitative information. The mixed methods research is useful when either the quantitative and qualitative research approach is inadequate to provide the best understanding of a research problem. They are taking into account that, in pragmatist framework, reality is not independent of the mind or within the mind (Creswell, 2008). This kind of research approaches the collected and analyzed data by introducing some philosophical assumptions. For example, a reason for mixing both research methods may be in a need for a postmodern study of social, historical, political, and other contexts that is reflective of social justice and political aims (Creswell, 2008).

Figure 4. Anna Ursyn, "Practice Makes Perfect" (© 2002, A. Ursyn. Used with permission)

"Practice Makes Perfect" (Figure 4) metaphorically pictures the works taken in pursuit of a goal:

Once around, ride again, up and down the mound,
To encircle barriers, grow a paragon,
To have no equal, be worthy to follow,
To become a nonesuch, a beau ideal.

Graphs, Data Graphics, Data Maps

To organize numbers well and facilitate comparisons, numbers can be arranged as a text-table, alphabetically, ordered by content, or data values (from small to big). A graph depicts a single relationship between the data or a whole network with the use of vertices (nodes) connected by edges (links). Graphics that present data, information or knowledge are called information graphics. Data graphics communicate information by combining points, lines, coordinate systems, numbers, shadings, color, words, pictures, and symbols, often in the form of well known, easy to grasp icons. Solutions for presentation include tables, small data sets, relationships among multiple variables, time-based displays, simulations for the multidimensional processes, and visualizations for both the abstract and the physically existing concepts. We use data graphics in everyday practice, often making them part of interactive design or multimedia projects, using them to visually communicate messages (such as road signs), advertise products (by creating logos or commercials), illustrate knowledge with diagrams, enhance performances (with scenery, posters, and programs), entertain with visual storytelling, manga, or comics, and provide information in newspapers, magazines, blogs, television and film documentaries.

In his classic books on theory and practice in the design of data graphics, Edward Tufte (http://www.edwardtufte.com/tufte/) provided analysis how to display data in graphics, charts, and tables. He advised to display numbers in tables when reporting small data sets of 20 numbers or less. Big and complicated sets of data, for example statistics about weather or some government issues are easier to grasp in graphic form than when expressed verbally or in numbers because graphics such as maps, charts, graphs, music and dance notations serve as explanatory tool for these data sets.

Data maps may contain much more data than we can arrange in a table. When a volume of data is big, for example taken from 3,056 counties in the United States data should be reported on data maps, which allows presenting and examining them in many ways: by region, by gender, and then making conclusions about the reasons of the events under study. Graphic presentation shows or makes clear how something works or explains the relationship between the parts of a whole (Tufte, 1983). The thematic maps combine maps with statistics, for example, maps showing reports of age-adjusted death coming from diseases (Tufte, 1983, p. 16-19) or the count of the galaxies, which places millions of bits of information on a single page before our eyes (Tufte, 1983, p. 26-27). According to Edward Tufte (1983), graphical displays should:

- Show the data;
- Induce thinking about the substance, not the techniques;
- Avoid distortion of the content;
- Present many numbers in a small space;
- Make large data sets logically connected;
- Encourage the eye to compare different pieces of data;
- Reveal levels of detail, from overview to the fine structure;
- Serve clear purposes: description, exploration, tabulation, or decoration; and
- Be closely integrated with statistical or verbal descriptions of a data set.

Visual language, which overlaps the art, science, and technology, uses hierarchy and composition to communicate visually what the alphabetic and character-based languages convey verbally. According to Manuel Lima (2011, p. 159), "visual elements and variations consider color, text, imagery, size, shape, contrast, transparency, position, orientation, layout, and configuration." Data maps can be enriched with shadings and colors. The use of color supports understanding of data graphics. Various elements of graphics can interact creating patterns and textures. With colors, we can show relationships, explain, add emphasis, use color-coding, and organize a graph. However, color looks different when placed in different surroundings (Tufte 1990, p. 92-3). The second color creates a new layer of information in a graphic. Joseph Albers (2010/1963) described the effect of color interaction as 1 + 1 = 3 or more. The use of several colors can be helpful in high-density maps. Tufte explains that the fundamental uses of color in information design are: to label (color as noun, used to discern things - Tufte 1983, p. 90), to measure (color as quantity, light-to-dark gradation - Tufte 1983, p. 91), to represent or imitate reality (color as representation), and to enliven or decorate (color as beauty) (Tufte 1983, p.81).

Interface

Communication between users and computers involves, as a necessary part, a concept of an interface, an interconnection between human beings and computer systems, devices and programs; it may be going in one or two directions. The concept of an interface may relate to theoretical problems and also to the message exchange systems between users and a product. It happens by creating and sharing over the Internet the multimedia-enabled, integrated documents with images, sounds, and text. For example, a sign providing useful information, a table of contents in a book, an information graphics, or a protocol for an input/output system used to communicate with other devices can be called an interface. Interface may also mean programs, tools, or equipment that enable separate parts or elements to interact in a coordinated way. For example, it may be a video card securing a supply of images to display, or a software application such as a web browser that enables using information resources on the web. Hardware or software interface couples parts of a system or provides access to its components, that is to say, selected resources. Physicists from the European Particle Physics Laboratory (CERN) created in early 90s HTML as an authoring language and distribution system. The Hypertext Markup Language (HTML) defines a set of rules, syntax of a document presentation by a browser. HTML is based on a hypertext system of links and is accessible on all computers and hand-held devices.

Interactive Websites

On an interactive website, data posted on the web may change their content, format, and even meaning because of the actions taken at the social networking. Facts, data, and opinions are posted on the web for further translation, discussion, additions and corrections, links exchange, enhancement with pictures, and emotional responses. Feedback from others may modify the object posted and its reception by viewers.

We may consider a website interactive when a message on the page relates to other ones and to relationships between them. It involves software that accepts and responds to input from the user's data and commands. It can be done using programs such as word processors or spreadsheet applications, and also social interfaces that facilitate responses from users during human-to-computer interaction and thus allow social interactions.

Video posted on the web may be linked to social networking services such as Facebook where it becomes available for over one billion of users, more than a half of them using it on a mobile device. As a social networking and microblogging (up to 140 characters) service, Tweeter handles over 340 million tweets daily and over 1.6 billion search queries a day; this service also stores the tweeter's location. Further services enable the visitor to utilize 'Explore' option to explore photos on Flickr by choosing a point in time; another Flickr feature called 'Interesting' allows the user choose most interesting content. The 'Tag' feature allows giving one's photos and videos up to 75 'tags' as keywords or category labels, which help finding things having something in common. The 'Bublication' option provides information on books, publication design, selections, prices, and shipping.

A text posted on the web may be translated into another language such as CSS, Java, Java Script, or other languages. A computer program allows a text to be organized by manipulating, modifying, and editing the text. The elements such as Java applets may be inserted depending on whether the intended shape of the web page. The web page content may be static – delivered to the user exactly like stored, often as a HTML document. The web page may become dynamic, such as a web site generated by the web application that can change due to interactions with the user scripting and the server-side scripting, for example in the bank–client exchange. It may be shared and interactive – allowing it's visitors to communicate with it and make changes in the web page content, for example by playing a game.

Each option requires using different language. Computer programming or coding is a process consisting of designing, writing, testing, debugging, and maintaining the source code of computer programs written in one or more of languages. For example, to produce dynamic web pages, a general-purpose server-side scripting language is PHP. The open source database management system MySQL (Structured Query Language) is used in Wikipedia, Google, Facebook, Twitter, Nokia, You Tube, and other large-scale web products. It was developed, distributed, and supported by Oracle Corporation, which runs as a server providing multi-user access to databases for use in web applications. A compiler may be needed, which is a (non-interactive) computer program that transforms the language of a source code into another (target) language to create an executable file (not just general data file) that can perform a task according to instructions.

Open source content management platforms such as Drupal serve for powering websites, building personal blogs, or enterprise applications. WordPress, a free open source blogging tool and dynamic content management system for websites is based on PHP and MySQL. They may support general coders writing software who are not specialists in a particular area as the computer programmers, web or software developers, and bridge their work with programmer analysts using specific language such as C, C++, Java, Lisp, Python, etc., who have also other software engineering skills. Thus the introduction of interactivity into users' communication may be done faster, easier, and more efficiently.

Web Presentation with Graphics or Numbers

Computer generated data graphics, which are omnipresent on the Web, serve to describe, explore, and summarize information in the form of the data. Information is usually presented on the web in numerical or graphic (or diagrammatic) form. When expressed as web graphics, factual information may be shown as drawings, plans, sketches, outlines, graphs, algebraic or geometric relationships, maps, photographs, designs and patterns, family trees, diagrams, architectural or engineering blueprints, bar charts and pie charts, typography, schematics, line art, flowcharts, and many other image forms. They are designed to represent the ideas, enhance, entertain, educate, or evoke emotion (W3C, Graphics, 2012).

Each year trillions (10^{12}) of images of statistical graphics are printed or displayed on the web. The amount and variety of data we may encounter online may be overwhelming, as there are currently billions of web pages available and this number is increasing by millions pages per day. When we look for data on the Internet, we get words, addresses, links, and other kinds of written information. A need for handling this amount of information resulted in the advent of the network culture that unifies, according to Manuel Lima (2011), the two rising disciplines – network science (that examines interconnections of natural or artificial systems) and information visualization (that translates data into meaningful information thus bridging data and knowledge). Surfing the web, data mining, and manipulating the data are made simpler when a collection of facts or data is shown with the use of information visualization and interactive techniques adding more dimensions to web presentation. The network-based design reflects diversity, decentralization, and nonlinearity of data. According to Lima (2011), the purpose of network visualization is to document (record the surveyed structure for posterior knowledge), clarify (explain in a simple, effective way important areas of the system, reveal (find a hidden pattern causality, relationships, and correlations), expand (for other uses, multidimensional behaviors, further explorations), and abstract (for hypothetical and metaphorical expression, depicting intangible concepts).

Figure 5. Anna Ursyn, "Legacy" (© 2003, A. Ursyn. Used with permission)

One may see this list as an obligation to look after continuous progression. Figure 5 illustrates this feeling:

Ancestors, predecessors from the past,
Handed down all this for us. What obligation!

COMPUTER SIMULATION

The word simulation means to take on the appearance or form of something, or to create a model; it indicates imitation and presents how the real-world processes or systems operate. It may also mean the act of giving a false appearance. The word model denotes a scale model or a mathematical model. A scale model is a copy of an object, constructed from smaller surfaces, that has been recreated either larger or smaller than the original by drawing a scale model and entering information into computer's memory. Changing shapes on a computer until a successful model is created can create an interactive model. A mathematical model is considered a set of mathematical equations that describe physical laws, which define the features or behavior of objects. One can design computer graphics to picture an individual as a scale model, while to describe how that person walks one has to make a mathematical model. Physical modeling can be a subtractive modeling, by carving in a block of material away everything that doesn't belong to an object, or an additive modeling (also called solid or constructive modeling) that involves building up a model from simple pieces.

In computer science, computer simulation translates real-world abstract models and physical systems to computer programs that create computer-generated environment or virtual environment. It helps to understand, explore, and control systems in many areas of science and engineering. Any phenomena that can be reduced to mathematical data can be simulated on a computer and used for designing a model of a problem or a course of events. For example, simulation of probabilities for market events may examine the effect of a price change on a market and the behavior of competitors and consumers in response to a price change. Computer simulation can refer to the process of imitating a real phenomenon with a set of mathematical formulas (for example, weather conditions, chemical reactions, or biological processes). A great many simulation languages exist, e.g., Simula. Simulation is a powerful tool for imaging concepts and structures as computational models involving programs and processes running on one or a set of computers. In technology, it is an artificial situation or environment. Simulation may refer to virtual reality techniques, which use an interactive computerized simulation or synthesis of an experience in several senses.

Computer simulation, visualization and databases are considered new cultural forms of the information society (Vande Moere, 2008). By creating a model, we can predict effects of the conditions of the thing being simulated. Computational simulations describing the space phenomena can be enhanced with the use of a method of mapping colors. In a false-color image close correspondence between subject color and image color is violated. Complex waveforms are converted to a composite color map with the use of such variables as amplitude, frequency, and phase. Colors represent measured intensities outside the visible part of the electromagnetic spectrum. This method allows the users immediate visual feedback; it proved to be an effective educational tool (Cox, 1988). Simulation may also refer to virtual reality techniques for learning and training, as it happens when animated runways and enemy flight forces are recreated in a room called a simulator. A real-time simulator also involves a headset, a chair, and other elements to ensure that it moves and sounds like the real thing. Ready simulation software packages are designed for specific kinds of computer simulation. Multimedia applications that combine text, high-quality sound, two- and three-dimensional graphics, animation, photo images, and full-motion video are often simulated.

Computer simulations that imitate real phenomena often use computer programs that simulate facts and events and show the effects of changes induced in a particular setting or a model under study. Mathematicians create simulations to work on the mathematics foundations, the formulation and analysis of the language, axioms, and logical methods on which mathematics rests. Simulation applications in object-oriented programming and in computer program (software) design support combining data and procedures (sequences of instructions). Mathematical models that imitate internal processes in the fields such as archeology, art, architecture history, and many other fields can predict the system behavior in the changed circumstances. Computer simulations of physical and mathematical systems use computational algorithms of different kinds, for example, Monte Carlo methods relying on repeated random sampling, to perform calculations. Abstract data such as those needed for financial models, textual analysis, transaction data, network traffic simulations or digital libraries, lack a natural notion of position in space (Vande Moere, 2008). They are often too complex to compute them with a deterministic algorithm and present as a graph or a matrix. Computer generated imagery simulations are intense graphical animated displays that show motion, changes in structures, and allow predicting events. Typical example is an aircraft flight simulator, equipment that represents real conditions in an aircraft or spacecraft used for the airline and fighter pilot training. Simulations model the dynamics of processes and events such as structural engineering, fluid flow, or aerodynamics, a study of gases in motion and of the forces acting on bodies moving through the gases. In physics, simulation has been applied to analysis of the dark matter, material that is believed to make up more than 90% of the mass of the universe but is not readily visible because it neither emits nor reflects electromagnetic radiation such as light or radio signals. In geology, simulated dynamic events provide understanding of the origin of rocks. Petrology, a branch of geology, is specifically concerned with the origin, composition, structure, and properties of rocks. Processes have been simulated, to clarify the events determined by the forces that change some kinds of igneous and metamorphic rocks into sedimentary ones. In medicine, Body Simulation for Anesthesia is an interactive PC-based multi-media software program for teaching and training in the field of anesthesia with user interface that represents the operating room environment. Users may explore scientific and clinical aspects as they run through life-like cases. Simulation has been applied in biology to study homology (the correspondence between structures and the similarities in function in different species). Motion sensing devices based on acceleration rather than on steady motion may simulate and control a virtual glider or create virtual driving or snowboarding. Electronic circuit simulators serve the experimental and instructional purposes. Real-time simulations allow estimating probabilities related to the market events, weather conditions, or chemical reactions. Models may serve for exploring biological processes; for example, they help to study how certain organs having different structure in various species relate to similar functions they perform. Some simulations are models of ongoing events such as military operations or human behavior; in such cases they are running on networked computers. Architects cooperate with computer graphic firms to create models for customers and builders that show possible relationship of a building with the environment and its surrounding buildings. Architectural simulation tools allow an architect to visualize a space and perform interactive demonstrations and explanations.

Many times simulation units take form of a room, such as CAVE (Cruz-Neira, Sandin, DeFanti, Kenyon, & Hart, 1992) or EON Reality ICUBE™ (www.eonreality.com/products_icube.html) immersive systems. In the CAVE immersive virtual reality environments projectors create and display a one-to-many visualization tool that utilizes large projection screens. Currently, immersive, interactive VR environments are used as a public display medium at universities, engineering companies, commercial industry, museums, galleries, conferences, and festivals.

EVOLUTIONARY COMPUTING

Evolutionary computing utilizes features of natural selection and genetic inheritance when applying computers as tools for automatic optimization and design. The study of biological evolution revealed mechanisms that enable life forms to adapt to their environments over successive generations: reproduction, mutation, and the Darwinian principle of survival of the fittest. Main approaches to evolutionary computing: genetic algorithms, evolution strategies, evolutionary programming, and genetic programming are dealing with a random or semi random set of possible solutions and a fitness function – a feedback system that measures how well each individual within the population performs the designated task or defines how close is a design solution from the designer's goal.

Genetic programming is often described as a biology-inspired method based on evolutionary algorithms, which is used for finding computer programs that perform a given task; it is considered a machine learning technique for optimizing a set of computer programs. A fitness function indicates how close a string of numbers (denoted as a chromosome) is to meeting the specification. In the field of genetic algorithms, designing a fitness function is performed to find workable optimization solutions. In analogical way as in biological systems, individuals that fit well have a greater probability of passing their genetic information onto the next generation. The operators that control the features of the offspring include reproduction (representation of individuals, ways of selecting the candidates for breeding); mutation (the dynamics of random changes occurring in genetic information); recombination (the exchange of genetic information with other successful individuals to form offspring); the choice of the representation and the choice of genetic operator in a particular evolutionary approach.

Bales and Kitzmann (2011) believe, "any model of communications that displays evolution (a selection process that acts on variation) is biologically inspired." Designers of communication and coordination models take inspiration from insects and many other species. Features such as complexity of animal's behavioral style, small versus large groups, methods of avoiding eavesdropping (such as specific color undetectable to predators, or behavioral tactics (for example, hiding spot-containing fin except during courtship) create challenges for bio-inspired models.

CONCLUSION

Suitable tools are needed for representing, processing, manipulating, and projecting data in a static or animated way. Tools for translation into visual or multisensory content may be as simple as a pencil or involve cutting edge computing technologies. However, an integrative approach to creating computer art graphics allows using the computer not only as a tool but also as a source of inspiration. A short description of evolution, first of computer related technologies, and then of graphical display of the data and methods of inquiry reflects the growing possibilities offered by the evolving thought processes and the advanced technologies.

REFERENCES

Albers, J. (2010). *Interaction of color*. Yale University Press. (Original work published 1963).

Association, I. R. (2013). *Digital literacy: Concepts, methodologies, tools, and applications* (3 vols.). doi: doi:10.4018/978-1-4666-1852-7.

Bales, K. L., & Kitzmann, C. D. (2011). Animal models for computing and communications. In *Bio-Inspired Computing and Networking*. CRS Press. doi:10.1201/b10781-3.

Cox, D. J. (2008). Using the supercomputer to visualize higher dimensions: An artist's contribution to scientific visualization. *Leonardo, 41*(4), 390–400. doi:10.1162/leon.2008.41.4.390.

Creswell, J. W. (2008). *Research design: Qualitative, quantitative, and mixed methods approaches*. Sage Publications, Inc..

Cruz-Neira, C., Sandin, D. J., DeFanti, T. A., Kenyon, R. V., & Hart, J. C. (1992). The CAVE: Audio visual experience automatic virtual environment. *Communications of the ACM, 35*(6), 64–72. doi:10.1145/129888.129892.

Hockney, D. (2001). *Secret knowledge: Rediscovering the lost techniques of the old masters*. London: Thames and Hudson.

Lima, M. (2011). *Visual complexity: Mapping patterns of information*. New York: Princeton Architectural Press.

McCall Smith, A. (2010). *The unbearable lightness of scones*. New York: Anchor.

Purchase, H. C., Plimmer, B., Baker, R., & Pilcher, C. (2010). Graph drawing aesthetics in user-sketched graph layouts. In *Conferences in Research and Practice in Information Technology (CRPIT)*. Brisbane, Australia: CRPIT.

Skogen, M. G. R. (2006). An investigation into the subjective experience of icons: A pilot study. In *Proceedings of the 10th International Conference on Information Visualization,* (pp. 368-373). IEEE.

Tufte, E. R. (1983). *The visual display of quantitative information*. Cheshire, CT: Graphics Press.

Tufte, E. R. (1990). *Envisioning information*. Cheshire, CT: Graphics Press.

Vande Moere, A. (2008). Beyond the tyranny of the pixel: Exploring the physicality of information visualization. In *Proceedings of the 12th International Conference Information Visualization*, (pp. 469-474). IEEE.

W3C. (2012). *Graphics*. Retrieved July 13, 2012, from http://www.w3.org/standards/webdesign/graphics#uses

Ware, C. (2012a). *Information visualization: Perception for design*. Morgan Kaufman.

Ware, C. (2012b). Visual thinking algorithms. In *Proceedings of the International Conference on Computer Graphics Theory and Applications and International Conference on Information Visualization Theory and Applications*. ISBN: 978-989-8565-02-0

Wyeld, T. (2011). The implications of David Hockney's thesis for 3D computer graphics. In *Proceedings of the Information Visualisation 15th International Conference*, (pp. 409-413). London: IEEE. ISBN 978-1-4577-0868-8

Chapter 10
Digital and Traditional Illustration

ABSTRACT

Traditional and computing-based illustrations make a great part of our everyday experience. This part of the book examines how traditional illustration types have found their continuation in computing-based media, even when the products mimic the old appearance. The next part includes several projects addressed to the reader and illustrated by student solutions, which refer to various fields of interests or areas of activities and apply selected illustration techniques.

INTRODUCTION

Both traditional and computing-based illustrations make a great part of our everyday experience including communication, learning, productive work, and artistic activities. It seems difficult to classify illustrations into particular types or groups because each categorization would depend on our interest: whether we would inspect techniques, products, or recipients. Some subcategorize illustration into the techniques such as: drawing, painting, printing and pasting ready images. These frames of reference would mingle, bind, and interlace with the objects of illustration, especially when accomplished in a digital, interactive, and shared environment. Great many of traditional illustration types have found their continuation in computing based media, even when the products mimic old appearance while obtained with different media. Interaction techniques for digital transfer from the old resources to a currently demanded destination start from the old cut, copy, and paste operations, often without any concern about the copyright issues.

DOI: 10.4018/978-1-4666-4703-9.ch010

Without doubt, the art of illustration is closely related to printing techniques, even when printing is a final step of the process. A timeline of printing techniques has been listed on Wikipedia (History of Printing, 2013) as follows: Woodblock printing (ancient but documented after 200), Movable type (1040), Printing press (1454), Etching (*ca.* 1500), Mezzotint (1642), Aquatint (1768), Lithography (1796), Chromolithography (1837), Rotary press (1843), Offset printing (1875), Hectograph (19th century), Hot metal typesetting (1886), Mimeograph (1890), Screen printing (1907), Spirit duplicator (1923), Dye-sublimation (1957), Phototypesetting (1960s), Dot matrix printer (1964), Laser printing (1969), Thermal printing (*ca.* 1972), Inkjet printing (1976), 3D printing (1986), and Digital press (1993).

The further text contains several illustration projects that are inspired by biology and other sciences, refer to various fields of our interests and areas of our activities, and apply selected illustration techniques.

BACKGROUND INFORMATION: TECHNIQUES

Traditional Illustration

Someone interested in traditional illustration may produce reproducible works in several ways going beyond creating a pencil-on-paper artwork. Traditional techniques may include the following.

Pen-and-ink illustration, which is still used frequently, applies ink according to the ancient traditions. Chinese ink was used since the 23rd century BC, the India ink since at least the 4th century BC, and so was in many other countries. For example, an old East-Asian type of brush painting is called Sumi-E ink wash painting. This technique also utilizes acrylic inks and uses brushes pens, or wooden sticks. Illustrations are made with black ink in various concentrations. The same ink is used in calligraphy. Currently, illustrators often replace the use of pen-and-ink with graphics software and page layout software. Calligraphy characters are placed on computer keyboards, specific for particular languages, for example those used in China, Japan, Korea, or Vietnam.

Figure 1. Anna Ursyn, "Rats," ink drawing (© 1982, A. Ursyn. Used with permission)

Figure 1, "Rats" results from observation of a family of the pet black hooded rats (Rattus rattus).

Calligraphy characters placed on computer keyboards are specific for particular languages. In China, it is believed a beautiful calligraphy is made by a beautiful person: one can tell a lot about a person's education, integrity, and talent when one can appreciate this person's calligraphy. Arabic, Chinese, Vietnamese, Korean, Japanese, Nepalese, Indian, Georgian (in three alphabets from Georgia), Persian (for many languages in Persia and contemporary Iran), Kufic (the old form in Iraq), Sini (Chinese Islamic), Tibetan, Mongolian, and Western calligraphies have their particular, specific features and rules. Western calligraphy in the medieval ages included the art of illumination of the first letter of the book's chapter. Fine penmanship was valued before the advent of typewriters and then computers, and so calligraphy was a required subject in schools. At present calligraphy is used for embellishing event invitations, announcements, memorial documents, and as a part of logo and other documents design on paper and on the web. In the past people developed calligraphic characters and fonts looking at nature; visual context had been gradually changing. Characters transpired emotions, which caused visual impact; characters became later more complicated and more abstract. For over thousand years emperors were establishing criteria and setting standards. People now make the best of their talents in calligraphy art working in several styles coming from ancient times and sharing online their experiences.

Printmaking techniques, described below, include relief (which produces inverted images), intaglio (involving metal-related techniques of engraving, etching, or stippling, as well as applying acid on glass), planographic printing such as lithography or photolithography, and stencil (silkscreen and serigraphy):

- **Engravings:** Attained by cutting grooves into a hard surface (such as silver, gold, steel, or glass and plexiglass) to incise a design. An engraved metal plate may result in intaglio prints as illustrations. In the intaglio technique, the incised image, called a matrix or a plate, holds the ink that is transferred on paper. Image may be also obtained by the etching, dry point, aquatint, mezzotint, or collography techniques;
- **Etching:** One of the intaglio techniques, along with engraving, drypoint, mezzotint, and aquatint, can be achieved with the use of chemicals (in older times it was strong acid or mordant – an adhesive compound for fixing a dye or stain) applied on a covered with a waxy, acid-resistant 'ground' material, eating away the uncovered surface of a metal plate.

Figure 2. Anna Ursyn, "Birds," etching (© 1982, A. Ursyn. Used with permission)

Drawing by scratching with an etching needle means not only creating an illustration but also selectively dissolving the surface of a printed circuit or a semiconductor, using a solvent, laser, or a stream of electrons. For example, precise etching of electronic materials, such as high electron mobility transistors, can be achieved with a two-step digital etch technique (consisting of an argon exposure followed by a surface treatment with boiling potassium hydroxide) (Keogh et al., 2006).

Figure 2, "Birds" was created before the emergence of fascination with the collective behavior of natural and artificial self-organized systems such as flock of birds, ant or bee colonies, or schools of fish:

- **Woodcut:** A relief printmaking technique; the non-printing parts of an image are carved into the surface and then removed with a knife, chisel, and gouge (a chisel with concave blade), while the original, raised surface level is covered with ink by using a brayer, an ink-covered roller. A sheet of paper placed on the block and pressed to print the image. When separate blocks of wood are used for different colors, a woodcut can be printed as multicolored. Illustrators cre-

Figure 3. Anna Ursyn, "Rain," linocut (© 1982, A. Ursyn. Used with permission)

Figure 4. (a) and (b) Anna Ursyn, "Sailing," linocut (© 1982, A. Ursyn. Used with permission)

ate single-leaf woodcuts or the series of woodcuts for book illustration. Woodcut-style illustrations, art works, postcards, or t-shirt design projects can be produced in a painting or graphic software such as Adobe Photoshop, Adobe Illustrator, Corel Painter, or other software. Some artists choose to create illustrations by combining woodcut and digital techniques;

- **Linocut:** A variant of woodcut where the relief surface is cut in linoleum. Design can be cut with a knife, a V-shaped chisel, or a gouge. The raised (un-carved) areas will be seen as mirror images. Either water or oil-based inks can be applied with a roller and then pressed to a fabric or paper. Figure 3 shows an example of linocut technique.

A series of linocuts showing boats (Figure 4) served later as a starting point for a computer graphic work "Clean Water Act" (Figure 5), which was inspired by the growing concern about our water resources (EPA, 2012):

Can we find many rivers, lakes, and bays safe enough for swimming, more than forty years after the Clean Water Act has set up water quality standards?

Figure 5. Anna Ursyn, "Clean Water Act," archival print, computer graphics (© 2007, A. Ursyn. Used with permission)

- **Lithography:** An image is drawn on a polished slab of lithographic limestone (or a prepared metal plate) by using an oil-based lithographic crayons or greasy ink, in the same way as with watercolors or crayon drawings. Then the stone is brushed with a chemical solution, such as plant resin mastic (also called Arabic gum), to have ink attracted to the image and repelled from the blank areas. After wiping the stone with a solvent the greasy parts repel ink and the drawing is bonded to the stone (MOMA, 2012).

  Figure 6, "A Dream" displays a soft texture and a fuzzy appearance of an image that is characteristic of many lithographs:
- **Screenprinting:** This produces illustrations as a kind of stenciling by using a woven mesh to support an ink-blocking stencil. Artistic screen prints are called 'serigraphs.' Andy Warhol created his screenprints from manipulated photo-

Figure 6. Anna Ursyn, "A Dream," lithography (© 1982, A. Ursyn. Used with permission)

Figure 7. Anna Ursyn, "City Neighborhood" (© 2000, A. Ursyn. Used with permission)

graphic images with the use of stencils (MOMA, 2012). Color separation and a silkscreen look can be created with software such as Adobe Photoshop.

Figure 7, "City Neighborhood" is a photosilkscreen after computer plot. It tells about connectedness of people, even without the use of networking technologies. The multiplicity of windows surrounding everybody returning home provides one with a feeling of the presence of others and gives an assurance coming from being a part of the big community.

Figures 8a and 8b present examples of printmaking techniques combined with digital technologies:

- **Digital Illustration:** A great variety of other techniques applied in digital illustration encompass the 2D, 3D, interactive, and many other types of illustration including smart graphics, intelligent and adaptive graphics often used for the production of mailing postcards, flyers, brochures, door hangers posters, business cards, magnetic plastic cards, presentation folders, stickers, labels, and also sketch-based interfaces and rendering techniques for the 3D models (Knödel, Hachet, & Guitton, 2009).

Figure 9, "Moon" was done with the use of airbrush, a pre-digital tool often used by illustrators and commercial artists. In this case the airbrushing was combined with application of the programmed images.

Figure 10a and 10b show the process of programming, and then manipulating a simplified image of a horse and rider. Images of the programmed horses served later for further transformation in other media.

Figure 8. (a) Anna Ursyn, "Battle," photolithography after computer program (© 2000, A. Ursyn. Used with permission) (b) Anna Ursyn, "Horses," silkscreen after computer program (© 2000, A. Ursyn. Used with permission)

Figure 9. Anna Ursyn, "Moon," airbrush (© 2000, A. Ursyn. Used with permission)

Figure 10. (a) Anna Ursyn, "A Horse" (© 1987, A. Ursyn. Used with permission). A 3D program in Fortran shown on the left corresponds the image of a horse (b) and (c) Anna Ursyn, "Horses" (after computer program) (© 1987, A. Ursyn. Used with permission)

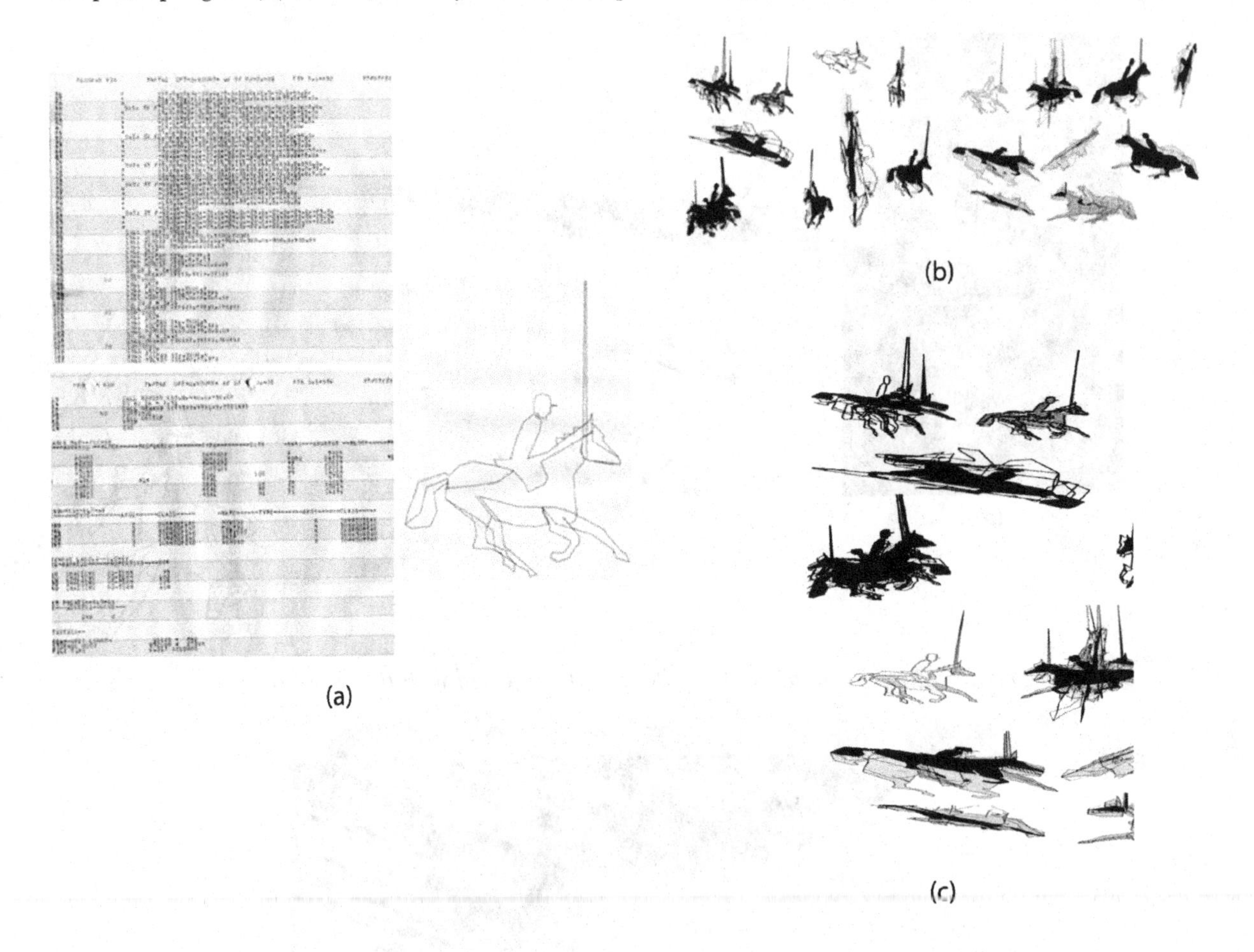

ILLUSTRATION: PROJECTS

Illustrators use selected types of illustration depending on their primary aims. The following set of projects is aimed at encouraging the reader to develop their own illustrations in any technique they prefer to choose. Thus, the selected illustration techniques, both traditional and digital, may serve the readers in visual responses to the discussed themes.

Project 1. A Cover for a Children's Book: Impossible Creatures

Designing a cover for a children's book provides an opportunity to instill fantasy and imagination into a reader's mind and at the same time introduce some nature- or science based information. One may create, for example, some not-known or non-existing characters. One way of imaging an impossible creature is to combine parts of different animals (such as a trunk of an elephant on a body of a ladybug); however, it may be seen a banal way of imaging an impossible creature. We may find a lot of sources for inspiration for creating this project. Inspiration may come from biological systems because our imagination comes for the most part from biology, predominantly from botany and animal world. It is easy to spot many differences in anatomy and functioning of organs in various species. For example, gills in fishes and some amphibians extract oxygen from water while bronchi and lungs provide oxygen supply into the blood of vertebrates other than fish, such as humans. The same variety can be seen in the types of locomotion (where legs do not resemble neither the butterfly's nor the bird's wings) and eating practices (by consuming, devouring, munching, feeding, and grazing) that often do not correlate with the animal size: whales eat plankton. A variety of the plant forms: trees, shrubs, herbs, grasses, ferns, and mosses may serve as an inspiration for creating impossible creatures. Inspiration can also come from inspecting the advantageous associations of two different organisms called symbiosis.

The interesting thing is that when we create a creature inspired with biology, we may end up with a notion that inhabitants of this planet become aliens. In his book titled "Bestiary" Nicholas Christopher (2008) discusses the fate of those animals that missed the Noah's Arc. According to the author's tale, an offspring of the individuals capable of exchanging genes became the harmful and quite unfriendly creatures such as a werewolf. Christopher created a glossary at the end of his book, listing each creature's name and explaining its origin and behavior. He also included in his index sketches and descriptions of some of them. In literature, we often see transformations described as unearthly creatures; sometimes, such as in a case of "Beowulf," (2001), the Anglo-Saxon epic created between the 7th and 10th centuries, the strange features may become exaggerated by conversion from old culture and language, for example, through translation of an old language into contemporary one.

Many times mysterious creatures are inhabitants of mythical worlds; for example in the ancient Greece myths and beliefs there were several kinds of mythological figures such as Minotaur; we may call to mind the earliest, primeval deities, Gods, Half-Gods, and their offspring: Titans, Giants, Nymphs, Muses, deities inhabiting seas, skies, and rustic areas. Each one had their powers and weaknesses mirroring the earthly, human, and unnatural forces. Supernatural entities were imagined in Sumerian and Akkadian myths from ancient Babylonia, first told and then written in cuneiforms. Babylonian epic tales preserved on clay and stone tablets introduced the god Enlil existing at the beginning of heaven and earth. Ancient myths of the Near East myths link the creation of humans with a mother goddess and Enki, the god of freshwater and practical inventions. Belisama was the Celtic goddess of fire, bodies of water, and metallurgy. Poseidon, brother of Zeus and the god of the seas and rivers, sent a sea monster against Troy (which was killed by Heracles), and then other monsters Scylla and Charybdis (his daughter) against Odysseus. The Maya people believed that life emerged from the lifeless, dark universe filled with water. The Aztec deity Tlaloc, called the Rain Sun, reportedly carried four magic jugs. Water from the first jug caused crops to grow, water from the second jug was killing crops, the third jug contained water that frosted plants, and the fourth jug destroyed everything (Mythology, 2008).

When we want to go beyond mere combination of different animal body parts in order to draw our creature, we may focus on animal features, functions, and habitats. In the past, after domestication of some of the animals, humans tried to improve animal functioning for their own purposes, so the animals would better serve them. They produced new breeds by cross breeding and then perfected the resulting hybrids over centuries. For example, a 5,000 years old Japanese breed of the dog Akita appeared perfect to watch over young children, protect the household, hunt bears in a pack of dogs, and be a good, trustful companion. Those dogs, which fall into the category of "working dogs," have a double coat that protects them from extreme cold. The bio-inspired research advances combining the best features found in different species. It also results in a progress of bioengineering: combining knowledge about molecular biology, biophysics, and biochemistry with mathematics and computing science to apply artificial tissues and organs into living organisms. As a result, we eat seedless oranges or grapes, celebrate holidays with blue carnations, or simply drink juices enhanced with vitamins.

This project may be seen akin to the bio-inspired way of thinking that change the way people develop new technologies in the fields of computing, software management, material science and material design, resource management, developments in computer technologies, and many other fields. The use of bio-inspired ways for bringing into being new forms results in designing new applications and devices. One of the solutions is translation of form to function, often discussed in terms of the 'form follows function' approach. For example, studying a symmetrical makeup of the wings might support developments in flying machines. Another approach is in translation of biological data to material science: creating materials that function in a similar way as in living beings. Also, inventors attempt to imitate structures that can be seen in nature and create materials that respond to external stimuli such as optical fibers, liquid crystals, or structures that scatter light. Yet another way is to combine biomaterials with artificial ones to create hybrid materials and technologies (National Research Council, 2008).

While working on this project you may want to construct different levels of departure from the earthly laws. People have always wanted to have skills and abilities possessed by animal creatures. One of the everlasting human dreams is a flying ability, which we envy the birds, bats, insects, and any other flying animals. The myth of Icarus or the flying machines designed by Leonardo da Vinci reveal this dream. (According to Ovid and Appollodorus, Icarus - the son of the master craftsman Daedalus took flight from imprisonment on Crete wearing wings that his father constructed from feathers and wax. In spite of his father's warning he flew too close to the sun. The sun melted his wings and caused him to fall into the sea and drown.

Figure 11 presents, with a view from two sides, a clay sculpture illustrating a story entitled "Man and Animal":

As for humans, there was always a want for flying. According to old myths, they tried to escape on wings as they tried to prove their ability. There is something to think about the fact that birds can fly. Fish can fly, perhaps humans also could fly. Not so well, yet they can.

And so, they say that who one keeps company of birds for a long enough time develops the power and endurance of a bird. Many joined the birds, in spite of all the differences.

The flock lived in a collective, hierarchical, and rigorous way. The most unique and important moments happened when the birds lifted from the ground. Those who belonged to mankind stayed aground following the birds with their eyes for a long time. For this reason, many decided to join penguins.

See Table 1 for Your Visual Response.

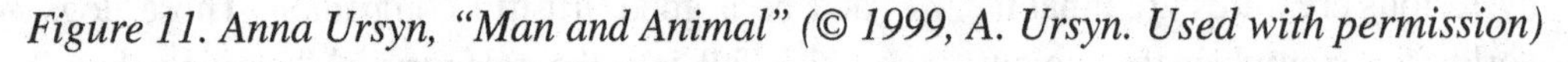

Figure 11. Anna Ursyn, "Man and Animal" (© 1999, A. Ursyn. Used with permission)

Table 1.

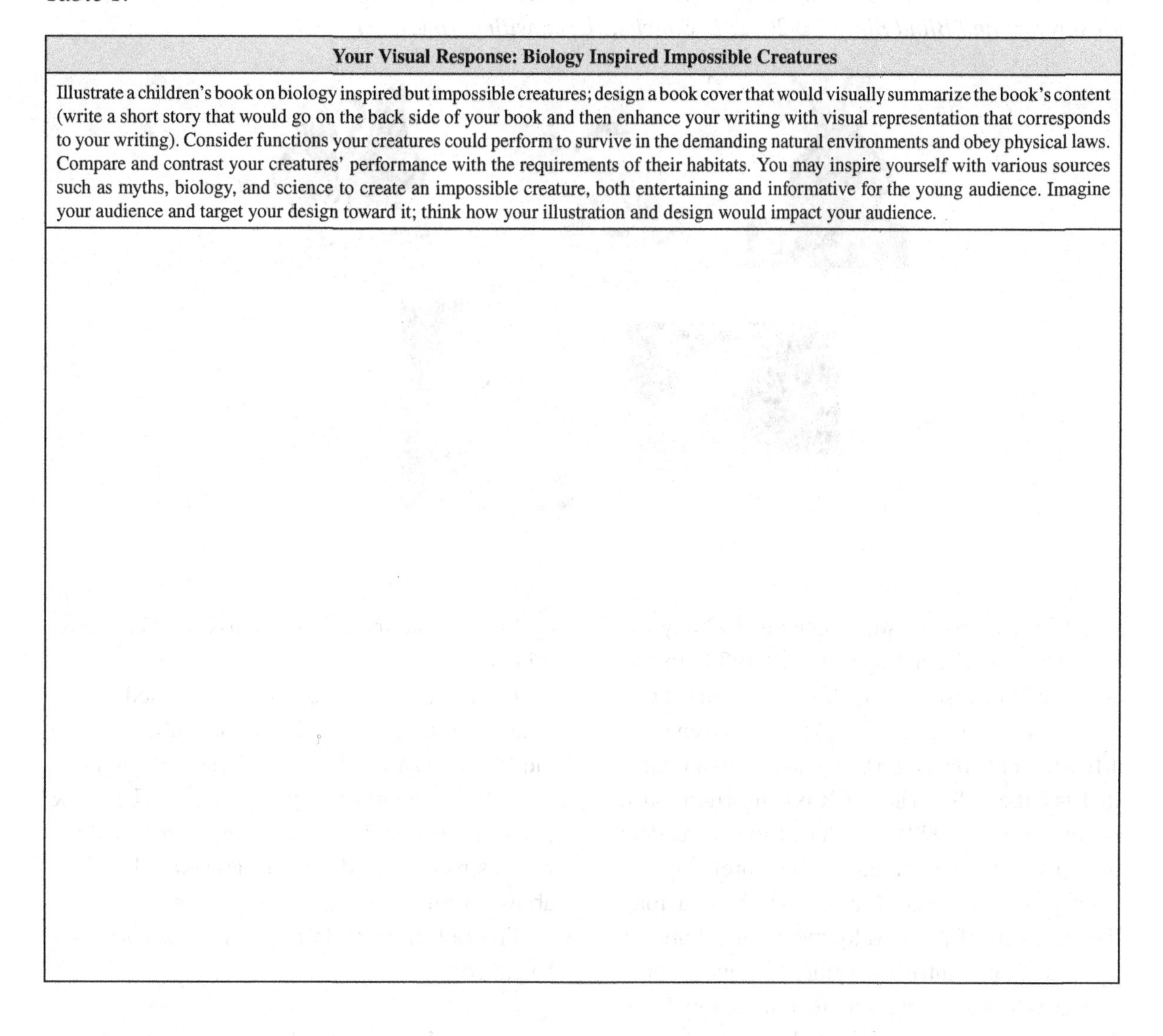

Your Visual Response: Biology Inspired Impossible Creatures
Illustrate a children's book on biology inspired but impossible creatures; design a book cover that would visually summarize the book's content (write a short story that would go on the back side of your book and then enhance your writing with visual representation that corresponds to your writing). Consider functions your creatures could perform to survive in the demanding natural environments and obey physical laws. Compare and contrast your creatures' performance with the requirements of their habitats. You may inspire yourself with various sources such as myths, biology, and science to create an impossible creature, both entertaining and informative for the young audience. Imagine your audience and target your design toward it; think how your illustration and design would impact your audience.

Figure 12 presents student's work on this theme created in the "Digital Illustration" class:

Explanation of the characters: The creatures show adaptations to environment and some specialized senses that enhance their survival skills. Modern Bat is flightless; it uses electric senses to manipulate electronics. It licks wing tips to make them more conductive and can squeeze them into tight spots. The Rock Eater is very slow, eats rocks after dissolving them with its special saliva. Cat Bird is like regular bird; special markings make it look like a cat when it puffs up. Glowing Mushroom glows to attract animals in the dark. Predators lie in wait, and the fungi get any leftovers. Spores make you sleepy. Blind Bird lives in caves; it has no color but has a special vibration-sensing beak that can be held against stone to feel what is moving nearby. It is very maneuverable; can fly blind by clicking its beak.

Apart from the nature inspired approach, you may want to compare this project to a robot design. A free of charge educational programming language named Karel after the Czech writer Karel Čapek may be helpful here.

Figure 12. Erik Sanchez, "Impossible Creatures: Modern Bat, The Rock Eater, Cat Bird, Glowing Mushroom, and Blind Bird" (© 2012, E. Sanchez. Used with permission)

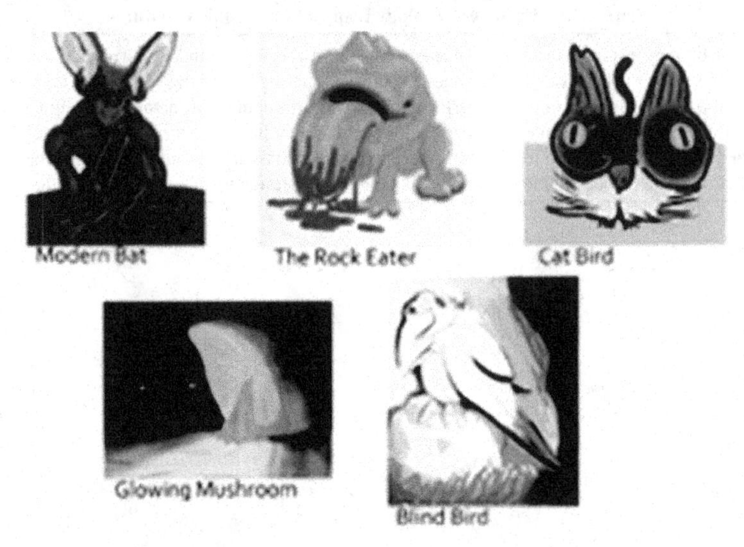

The word "robot" was coined in 1920 by the Czech writer Karel Čapek (1890-1938) in the course of conversation with his brother, the cubist painter and writer Josef Čapek; it was given to an artificial man that could be mistaken for humans. In 1941 the U.S. writer of Russian origin Isaac Asimov (1920-1992) introduced to the readers the term 'robotics' based on the Karel Čapek's concept of the robot. Many icons have a long life; in spite of the developments in robotics a picture of an anthropomorphic, humanoid robot still serves as an icon for a great many robots. In fact, the industrial, servicing robots designed for specific or general tasks, and bots (virtual software agents), all of them do not resemble humans.

Problems that are difficult for humans or robots may be solved by multi-agent systems consisting from multiple interacting intelligent agents, often software agents, which are present within their environment. Programs written for the weighted request and response, telling which features are very important, medium important, or unimportant, are often applied for computer simulations and models (Pérez et al., 2012).

See Table 2 for Your Visual Response.

Figure 13a and b show class projects about robots.

In Figure 13a, Royce Wood designed his interactive project entitled "Robot Maker" during the "Web Construction Design" course I co-taught with the Computer Science professor (Ursyn & Scott, 2006, 2007). Depending on the visitors' choices, robots with different features (as described above) would show up on the screen.

Figure 13b. Matt Hall presented a set of robots he called:

- **Pentabrach:** Intelligence 9, Speed 1, Weaponry 0; performs multiple tasks with great dexterity and intelligence, such as conducting an orchestra with all five arms, or delivering a baby with three arms and playing chess with the other two. His AI chip allows multitasking and problem solving, however it is utterly defenseless;
- **Autotank:** Intelligence 2, Speed 1, Weaponry 7; a massive fighting machine, completely auto-piloted and driver-free robotic tank. Armed with several cannons, a massive tank gun with two additional tur-

Table 2.

Your Visual Response: Design a Robot
You may want to write a program for a robot, which possess some features. When programming a robot, assign on the scale of 10, accepted as a maximum for all three factors, the numbers for each of three values: intelligence, speed and weaponry. When you decide how much would you distribute for intelligence and how much for speed (maybe power?), the rest goes to weaponry, but the total may not exceed 10. Would the one robot with higher intelligence win over the one with higher speed? Or the robot with better weaponry would take down the one with lower speed? You may also think about making a bot or an avatar, or making your bot the avatar. In Second Life avatars are designed to fulfill the players' dreams, hopes, and alleviate their deficiencies and complexes.

Figure 13. (a) Royce Wood, "Robot Maker" (© 2006, R. Wood. Used with permission). (b) Matt Hall, "Robots" (© 2013, M. Hall. Used with permission)

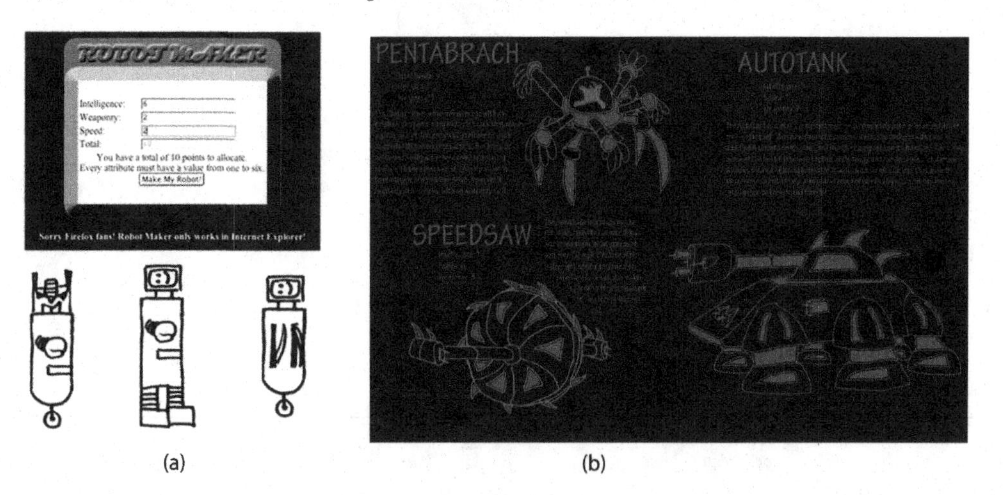

rets on its sides, and four laser gun ports on its front and sides, this is one of the best fighters on the battlefield today, and can do significant damage without the risk of losing human life. It walks with six robotic legs that allow it to traverse, albeit slowly, across any terrain of variable slopes, and it has enough intelligence to know friend from foe;

- **Speedsaw:** Intelligence 1, Speed 6, Weaponry 3; a buzzsaw-like robot designed to run over things and shred them at speed of well over 70 mph. It does not need a great deal of intelligence, which also makes it prone to forgetting whom its master is. Watch out.

See Table 3 for Your Visual Response.

Table 3.

Your Visual Response: Impossible Creatures as Robots
This time, draw an image for a children's book cover that would illustrate the book about robots. Write a short story introducing the concept of robotics and telling about operation and applications of robots, those resembling humans and also the big and miniature devices that take place of humans in inaccessible spaces. This time inspire yourself with the science and science fiction themes to create a robotic creature, both entertaining and informative for the audience. While picturing robots, you may want to consider a notion of permutations described in mathematics as small rearrangements of objects or their values in a combinatorial way. Place your story on the back cover of the book.

Figure 14. (a) Jennifer Funnel, "Robot" (© 2011, J. Funnell. Used with permission); (b) Jason Baird, "Robot" (© 2010, J. Baird. Used with permission). It is suggested that a robot could be a box we speak to. (c) Preston Stone, "Robot" (© 2012, P. Stone. Used with permission)

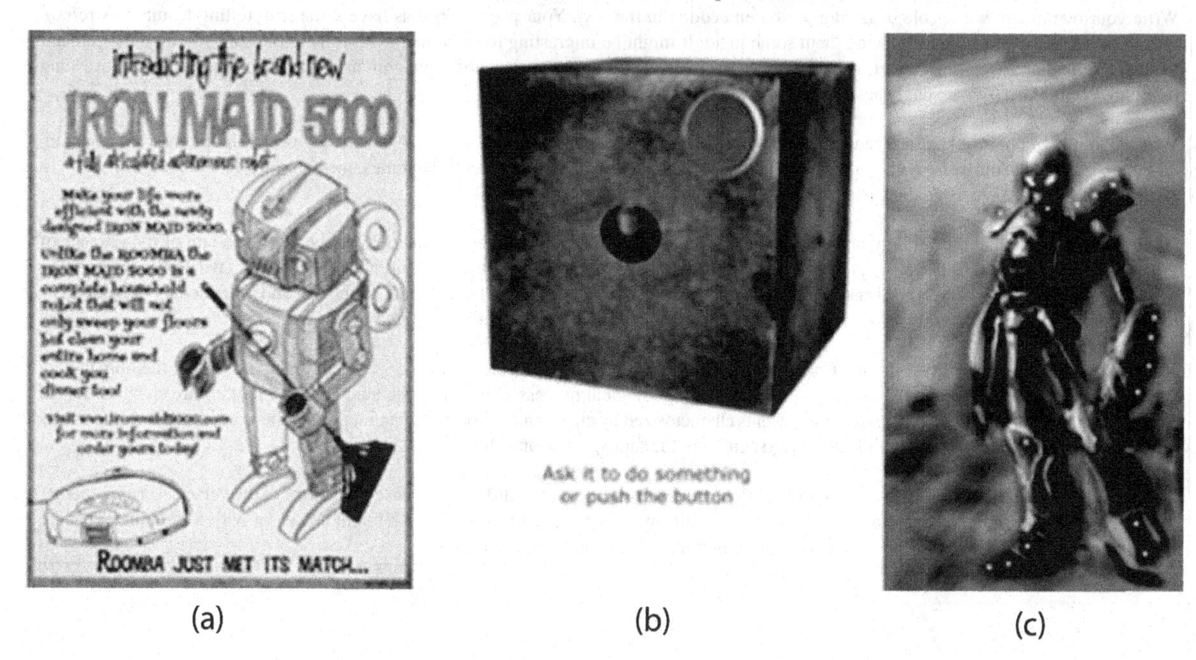

Figures 14a, b, c present three different student approaches to this theme. Jennifer Funnel presented her robot on a poster that advertises a friendly humanoid robot. Jason Baird designed his robot as an interactive cube box with a microphone on its side saying to the user, "Ask to do something or push the button: you tell an order and it is done." Preston Stone presented his robot in a painterly, heroic, and almost romantic way.

Project 2. Poem Illustration: Illustration to a Geology Inspired Poem, "Fate and Chance"

Many times a poet working with an illustrator become a successful team and create together several works that seem better than a text or an image alone. However, high visual literacy is a prerequisite for both a writer and an illustrator to achieve this goal. A book "Alice's Adventures in Wonderland" was written by Lewis Carroll (1898/1993) and illustrated by Sir John Tenniel. "Winnie the Pooh" (1926) and "The House at Pooh Corner" (1928) were written by Alan Alexander Milne (2009) and illustrated by Ernest H. Shepard.

In many cases poets illustrate their works themselves, attaining an undivided, complete quality distinctive to the work. For example, Antoine de Saint-Exupéry (1943/2000) illustrated his book "The Little Prince." Beatrix Potter (2006) wrote and illustrated "The Tale of Peter Rabbit" in 1893.

See Table 4 for Your Visual Response.

Illustration in Figure 15 relates to a story entitled "Fate and Chance" telling about eight generations of members of one family, with their works and their growing, more and more advanced achievements interrupted in a violent way by natural disasters resulting from geological processes.

Table 4.

Your Visual Response: Illustration to "Fate and Chance"
Write your own poem with geological information embedded in the text. Your poem will thus have some storytelling features. A refrain may tie described events together giving them some logic. It might be interesting to examine the refrains of the songs written by Leonard Cohen, a Canadian singer-songwriter, musician, poet, and novelist (http://www.leonardcohen.com/us/works/albums), and also those sung by Barry White (http://www.songlyrics.com/barry-white-lyrics/). Writing your own poem containing geological information means combining storytelling with scientific process and product information, all in poetical style. Your verse will contain a little bit of facts related to geological events, tectonics, meteorology, and archeology. Also, to share the feelings of persons in the verse, one has to think about their psychological traits. While writing a poem about geology and disasters often induced by geological processes one may think about the fate of the citizens of Pompeii and Herculaneum. There was, however, a pure chance-driven effect of the disaster deciding about what can we now piece together about the life in Pompeii before the disaster. What was totally covered in Pompeii and Herculanum by the volcano remained for archeologists to study, while the rest was sadly gone forever in fires, with people drown in the sea or suffocated by the fumes. Illustrate the refrain of a poem "Fate and Chance" provided below by showing geological processes. The verse tells about the human fate, which may result from the ordered, natural rules, in this case geological processes driven by the temperature and pressure changes. It also tells about a chance depending on the stochastic events characterized by chaos and entropy. We cannot predict the exact time when the events would happen and affect the mountains and valleys, creating the faults, folds, and shifts. You may want to focus on the refrain supported by a block of verse: each refrain is different because each block of the verse describes another process. Processes are linked; one forms another. A plate shift may create an earthquake, which might come out with a tsunami. What are the consequences? A fire, flood, strong wind, and falling trees coming as a domino effect.

Figure 15. Anna Ursyn, "Fate and Chance" (© 2012, A. Ursyn. Used with permission)

Fate and Chance

He woke up to plow his plot,
Plant the seeds, water them,
Wait a while to collect the goods,
And happily consume and share them year-round,
Exchange, or give away.

(Volcano)
The shift occurred,
The rock responded,
Its density changed and transformed.
The mud slid down,
The air was filled with sound of gas,
Plates covered the field he just plowed.

His son fed his hens early in the morning,
His horse, pigs, and cattle,
And his cat and dog.
He combed his sheep,
Moved his rabbits' cages,
So they could mow his lawn and get fed too.

(Fault)
The shift occurred,
The temperature rose,
The structure got changed.
The matter converted into something else.
It covered the lawn, the chicken, the ox,
And the ground rose higher despite his despair.

His grandson went to the mine at sunset.
He put his coat on the rack.
The coal was black, heavy, and shiny.
He carried it up toward the light.

(Fire)
The plate got shifted,
Shear, stress, and pressure,
Pushed it through and over.
The mine collapsed,
The coal all burnt.
The daylight was covered by dust.

His great-grandson finished his breakfast,
To work on the paperwork.
He calculated data,
Put it all together,
Compared the numbers,
And added it all,
Divided, multiplied,
Subtracted, and averaged.

(Flood)
When the plates shifted,
Rocks covered the voids.
The wells collapsed, the building sunk.
Water emerged and filled the holes.
The desk and cabinet broke,
The papers burned,
The pen got lost,
All then transformed into ash.

His great-great-great-grandson,
Got up at noon,
To perform on a stage with people watching.
The audience applauded,
The children laughed.
Director read those great reviews.

(Geiser)
Then the plates shifted,
The geyser erupted,
The surfaces tilted,
And rocks became soft.
Stage scenery went down,
The floor became a wall,
And the props started to dance.

His great-great-great-granddaughter,
Went into her office,
She hung her coat on the rack.
She downloaded some files,
For that important meeting,
And added some slides to her talk.

(Earthquake)
When planes shifted,
Wood sent out cracking sounds.
The colors faded, murmurs were heard.
The glass walls melted, iron structures bent,
The skyscraper went down,
Way under the ground,
Covered by another one.

His great-great-great-great-grandson,
Woke up on his ship,
He shared his captain's breakfast with passengers.

(Tsunami)
First Officer blew the horn,
To signal a tsunami.
The engines stopped,
The cruise ship went down,
Covered by the foam.

The actions of nature mixed it all together,
Shifted angles, elevated it all.
What once was a valley became an ocean,
What ocean once was became a mountain,
And the mountain became a plain.
The remnants were dug out,
With derricks and tools,
And sent to a refinery.

The grand-grand-grand-grand-grand-grandchildren refilled their cars,
When admiring the valleys, the ocean, the mountains.
They stopped for a picnic,
Thankful to their predecessors for oil,
Resulting from ashes and bones.

Project 3. Poster Art

The role of posters as an art form and as illustration of the society's cultural life seems to be evolving now into digitally conveyed forms of videos, vimeo – a video-sharing website, home pages for many kinds of websites, and visual announcements in television broadcasts.

Traditionally, posters were placed on the walls of buildings, on special stands, fences, or any surfaces suitable for gluing paper containing information, yet providing exposure. There was a special profession: a person covered a surface with a wheat-based glue, applied a poster, and the covered it again with this glue by sweeping it with a huge, broom-size brush. A cylindrical advertising column/pillar (also called a morris column) was an iconic object characteristic of the 19th and 20th century outdoor sidewalks in European cities. These columns had been densely covered with posters made by outstanding illustrators announcing the theater, opera, or operetta spectacles, philharmonic concerts, film projections, and artists' recitals, and circus performances. Posters were efficient as a method for spreading information and messages. They affected the passersby on several levels. People walking, riding a bus, a train, a car, or on bicycles had a chance to catch a glimpse of the unique characteristic of a poster. If such visual summary was catchy enough, they could come back to the poster (often in a different part of the city) to look at the image and read about the event. Thus the first imperative resulting from a visual power of the image often surpassed its verbal component.

Along with illustrations, which almost till the end of the 20th century were far more common than photographs, these informative posters comprised a spectacle title written in large lettering and particulars about the cast, place, and dates of performances. Posters of this kind, which were also installed inside the theater, opera, or philharmonic buildings, have been considered art works, displayed in galleries and museums, and treasured by the poster art collectors. There are international poster exhibitions, for instance at the Colorado State University, or at the Warsaw, Poland International Poster Biennale (http://www.postermuseum.pl/en/page/show/biennale-history) and poster galleries, for example, the Wilanów Poster Museum in Warsaw (http://www.postermuseum.pl/). A Poster Gallery in Krakow, Poland specializes from 1985 in Polish promotional and commercial posters, with a collection of tens of thousands film, opera, theater, exhibition, music, and circus posters, some of them very old. Currently, vintage poster art galleries and stores (one can find them on some campuses in California), often specialize in antique posters from many countries, and mostly sell poster art online.

Every day we come upon eye catching visual messages promoting or denoting some actions, attitudes, or decisions. Some of such affirmative posters are called motivational or inspirational posters. For example, we are encouraged to 'Reduce, Reuse, Recycle (water and energy)' and 'Donate Blood.' Some posters help organize our workplace or private life, often telling about detailed, specific actions: 'Be kind, rewind' (the VCR era is over, so it became history now) or even 'Sketch daily' (in accordance with this book's mantra). There are also demotivational posters that parody the motivational posters, for example, one may find them in "The Onion" magazine. There is a grey area where many things cannot be classified as good or bad (for example, during a war, an opponent is neither a hero or a bad guy).

See Table 5 for Your Visual Response.

Table 5.

Your Visual Response: A YES Poster
Select an important topic or an issue worth a discussion, promotion, and sharing. Create an image that provides encouragement for the viewers to actively support a cause.

Figure 16. (a) Adam Smith (b) Erik Sanchez, "Reduce, Reuse, Recycle" (© 2012, A. Smith, E. Sanchez. Used with permission)

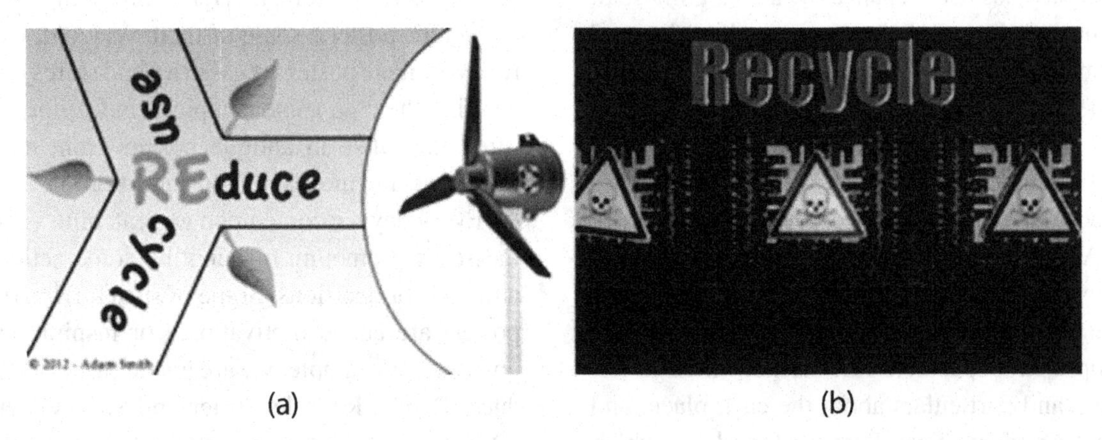

Figure 16a, and 16b show works done in my "Digital Illustration" class, both promoting the idea of recycling. Many times a "yes" poster makes the viewers aware of danger and thus motivates them to avoid detrimental effects or put a stop to bad practices. Work by Adam Smith provides easy to understand, symbolic design; Erik Sanchez tells about pollution caused by discarded electronic circuits.

Apart from being appealing and informative a great part of posters are designed with the purpose to protest, to oppose, or to fight against unwanted events and phenomena. This trend is apparent in relation to the health and environment related issues. It might be interesting to follow the developments in the social awareness of harmful events or processes and the ways of dealing with them. For example, posters that warn people about the danger of smoking have been changing in parallel with the developments in the social and medical research about the effects of the tobacco addiction. One may trace the approaches to this problem evolving in the sequential decades and view the examples of innumerable publications:

- In the 1950s people began to warn the society of a *danger* of smoking but the tobacco industry was against educating people about it. One could see posters stating that smoking is harmful;
- In the 1960s scientists found some association of smoking habits with the *lung cancer* induced by the products of combustion reaching the surface of the bronchi. The smokers and the tobacco producers have denied this connection in fervent discussions and in statements published in press (Fisher, 1958). Cigarettes with a filter made from cellulose acetate begin to dominate the market;
- In the 1970s people started to realize that smoking is a *disease* and is addictive. The tobacco industry contested the scientific results and founded their own analyses: it coordinated a scientific controversy with the aim of forestalling regulations of their products. Herbal, non- tobacco cigarettes are used not only in acting scenes but also as tobacco cessation aid;
- In the 1980s posters were telling about *second-hand smoking*: when somebody smokes, others suffer and get sick. The National Research Council published results about the environmental effects of the tobacco smoke and second-hand smoke.

Transdermal nicotine patches were introduced, for the nicotine replacement therapy;

- In the 1990s the genetic aspects of the tobacco *addiction* have been examined (for example, Madden et al., 1999). Lawsuits against cigarette manufacturers and industry organizations followed (The United States Department of Justice, 1999);
- The 2000s brought research on the effects of smoking on the female and male *infertility*, ongoing *pregnancy* (early rupture of membranes, placenta complications, pregnancy outside of the uterus, placental separation and the premature birth), and birth defects including future obesity (Nguyen, 2010). It was stated that both active and passive smoking might impact the fetal growth, both because of the carbon monoxide and the nicotine content (Fried, 2002). National Cancer Institute (2012) has been distributing free publications about cancer that begins in the lung, many times in connection with smoking. The tobacco industry coordinated a scientific controversy with the aim of forestalling regulation of their products (for example, Tong & Glantz, 2007). Smoking bans in many public areas begin;
- In the 2010s: Smoking in public places is *banned* in the United States and many other countries. People have to keep distance from the air intake, and are encouraged to smoke artificial cigarettes. Several social media websites, such as Smokefree.gov, Smokefree Women (http://women.smokefree.gov/), and Smokefree Teen (http://teen.smokefree.gov/) have been installed, containing information and support. It was supported with a production of the electronic cigarettes known as e-cigarettes – electronic inhalers vaporizing liquid solution into an aerosol mist, thus simulating the action of tobacco smoking; e-cigarettes might contain the same amount of nicotine as conventional cigarettes.

See Table 6 for Your Visual Response.

Table 6.

Your Visual Response: A NO Poster
Create a "Do not smoke" poster: "Adults do not start smoking; children do." Consider the 2020s: create a poster that would fit the expectations of the viewers living ten years ahead of us. Would it show drastic image of the lungs with problems caused by smoking? Will smoking be banned totally and become illegal? Will there be a public discussion of the issue similar to present discussions about marihuana?

Figure 17. (a) Jason Johnson, "Cancer Sticks," (b) Adam Smith, "Smoke is No Joke," (c) David Frisk, "Smoking Kills" (© 2012, J. Johnson, A. Smith, D. Frisk. Used with permission)

Three works of students from the "Digital Illustration" course are presented in Figure 17.

Project 4. Street Art vs. the Hyper Realistic Digital and Traditional Paintings

Quite another type of illustration can be seen in the form of street poster art, scraffiti (layers of tinted plaster) and street art called graffiti; these art forms, sometimes considered controversial, present artists' reaction to social, political, and cultural events. For example, Banksy (http://www.banksy.co.uk/), the street artist with an international reputation, is a painter, filmmaker (e.g., "Exit Through the Gift Shop" nominated for the Academy Award Best Documentary film), and political activist. His graffiti art, infused with humor and social commentaries, keeps abreast of current events and illustrates common immediate concerns. Banksy goes beyond graffiti, for example by cutting in half the London's iconic phone booth, re-welding it as a bent one, and placing it back on the street under the cover of the night. Street art has several facets, from art painted or stenciled on buildings, through objects hung from the buildings, to pictures created on trains, buses, cars, or underpasses. These types of activities are considered illegal, cause penalties, and create graffiti-cleaning jobs. There is a famous story about a man living on London's streets, who'd used to moist his own socks to create art on the buildings' walls. When caught and brought into justice, he made his way out of the court, and possibly jail, by stating he actually was cleaning the city by removing dirt with his own socks.

Photorealism masters are on the other side of a spectrum, producing hyper realistic pictures that look like photos (for example, http://www.boredpanda.com/photorealistic-paintings/). Trick photography and special effects are shared online, for example by Evan Sharboneau at http://trick-photographybook.com/?hop=14713 or at http://photoextremist.com/. Techniques of this type allow satirical approaches, such as those displayed by Pawel Kuczynski in satirical illustrations at the Bored Panda magazine website (http://www.

Table 7.

Your Visual Response: A Joke, a Cartoon and the Photorealism
Maybe you will now make an attempt to contrast the synthetic style of the compressive forms with the elaborate, narrative presentation. First draw a cartoon (like the cartoons in your newspaper or The New Yorker magazine), a joke, or comics using a few lines to convey the essence of your thought or statement. Then present it in a photorealistic way based on own photograph and possibly applying a live trace option in Adobe Illustrator.

boredpanda.com/satirical-illustrations-pawel-kuczynski/). Many enjoy aberrant art by Barry Kite (1997, 1994/2000) who recycles masterpieces from the Renaissance to Post-Modernism to create his photo-collages.

See Table 7 for Your Visual Response.

Project 5. Technical Illustration

Technical illustration can be a component of technical drawings and graphs, especially useful in communication with a nontechnical audience; according to Ivan Viola & Meister E. Gröller (2005), the aim of technical illustration is "to generate expressive images that effectively convey certain information via the visual channel to the human observer. Not only expressivity but also the overall visual harmony and aesthetics play a very important role in illustration." (Viola & Gröller, 2005, p. 7). Illustrative techniques derived from technical illustration are often applied in visualization; they employ high level of abstraction. Viola & Gröller (2005) described expressive visualization techniques that uncover important information through dynamic changes (cut-away views), deformations (ghosted views), or spatial modifications of the parts of the data (exploded views).

Figure 18. Jason Johnson, "A Spectrum" (© 2012, J. Johnson. Used with permission)

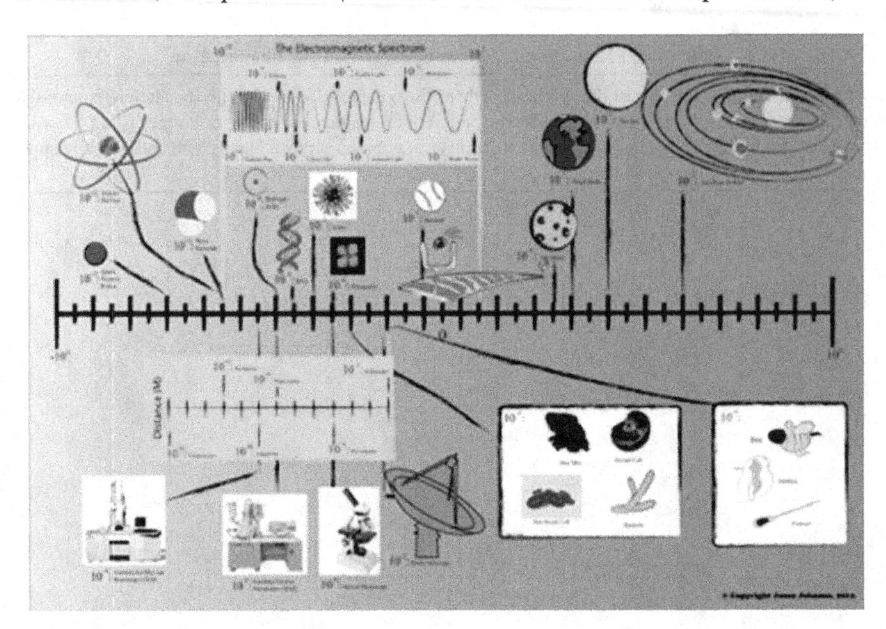

Figure 18 presents a student work showing a combination of the electromagnetic spectrum (from the shortest wavelength to the longest: gamma-rays, X-rays, ultraviolet, visible light, infrared, microwaves, and radio waves) and selected objects as examples of things existing in each range of wavelength. The task to be done was to add pictures to the graph about the wavelengths and thus depict the world and the tools used to visualize its electromagnetic properties.

PRODUCT DESIGN: MAKING A DRAWING FOR A DECK OF PLAYING CARDS

A deck of playing cards can be thought of as an iconic object; however, a great variety of styling, from the times of ancient China where playing cards were invented (Needham, 1962), to their modern design make them carriers of variety of symbols, meanings, and messages (for example, used for marketing). Designing the playing cards links the illustration related issues with the product semantics. Product design issues involve addressing the card design to user preferences, obeying some traditional demands, and taking into account card geometry related conditions. At variance with the chess design, which should comply with the rules of one game only, card design should conform to many different games such as bridge, poker or canasta, to say nothing about a variety of solitaire games.

Due to the many centuries lasting tradition several features are present in most of card designs. In the early 1400s, playing cards were created using block printing, which was the first use of prints in a sequenced and logical order. However, no examples of printed cards from before 1423 survived. Cultural and semantic traditions held in the countries from all continents make the playing cards varied; yet the cards are usually readable and playable due to their common basic design. Some decks have bar code markings on the edge of the face, so they can be machine sorted.

When creating an image for the both the face and the back of a card one has to make several decisions: about the style and aesthetics of images, the leading theme that often depends on the prospective users, and the games that will be

Figure 19. Fatma Alabdullaziz: the meaning of the images on the playing cards (© 2012, F. Alabdullaziz. Used with permission)

played with the cards. Playing cards are usually designed for a specific, sometimes broad group of people. Many times decks of cards serve some special occasions, are used as event related promotional materials distributed at conferences, or are designed with a specific leading theme in mind. For example, newspapers in the 2000s presented illustrations showing the most wanted Al-Qaida terrorists as particular cards in a deck, followed by announcements in newspapers informing for example, that a jack was caught). This metaphor puts the idea of a game, almost on the brink of hunting, often referred as a sport in relation to the game animals.

Other times cards are designed with an educational frame of mind, showing cultural or historical information, travel attractions, plants, emphasizing national values, or images and patterns characteristic of particular culture. Cards designed by some producers such as Piatnik (Wiener Spielkartenfabrik Ferdinand Piatnik & Söhne) have traditionally been especially valued.

Below are two designs of the decks of cards created by my students taking a "Digital Illustration" course. First project (Figure 19a, b, c) displays rich semiotic content by merging card design with the QR codes; it shows Islamic cultural icons presented as playing cards. The author of this project describes the meaning of the images.

In Figure 19a, calligraphy gives a visible form to the revealed word of the Qur'an and is therefore considered the most noble of the arts. It manages to combine a geometric discipline with a dynamic rhythm. Interestingly, none of its many styles, created in different places at different periods, has ever completely fallen into disuse. In the Islamic world it takes the place of iconography, being widely used in the decorative schemes of buildings.

In Figure 19b, Kufic is the oldest calligraphic form of the various Arabic scripts and consists of a modified form of the old Nabataean script. Its name is derived from the city of Kufa, Iraq, although it was known in Mesopotamia at least 100 years before the foundation of Kufa. At the time of the emergence of Islam, this type of script was already in use in various parts of the Arabian Peninsula. It was in this script that the first copies of the Qur'an were written.

Figure 19c shows Islamic calligraphy, also known as Arabic calligraphy, is the artistic practice of handwriting, or calligraphy, and by extension, of bookmaking, in the lands sharing a common Islamic cultural heritage. This art form is based on the Arabic script, which for a long time was used by all Muslims in their respective languages. They used it to represent God because they denied representing God with images. Calligraphy is especially revered among Islamic arts since it was the primary means for the preservation of the Qur'an. Suspicion of figurative art as idolatrous led to calligraphy and abstract depictions becoming a major form of artistic expression in Islamic cultures, especially in religious contexts. The work of calligraphers was collected and appreciated.

Figure 20. Fatma Alabdullaziz: selected cards for a deck of playing cards (© 2012, F. Alabdullaziz. Used with permission)

Figure 21. Structure of a QR code, highlighting functional elements (© 2012, Creative Commons Attribution-Share Alike License)

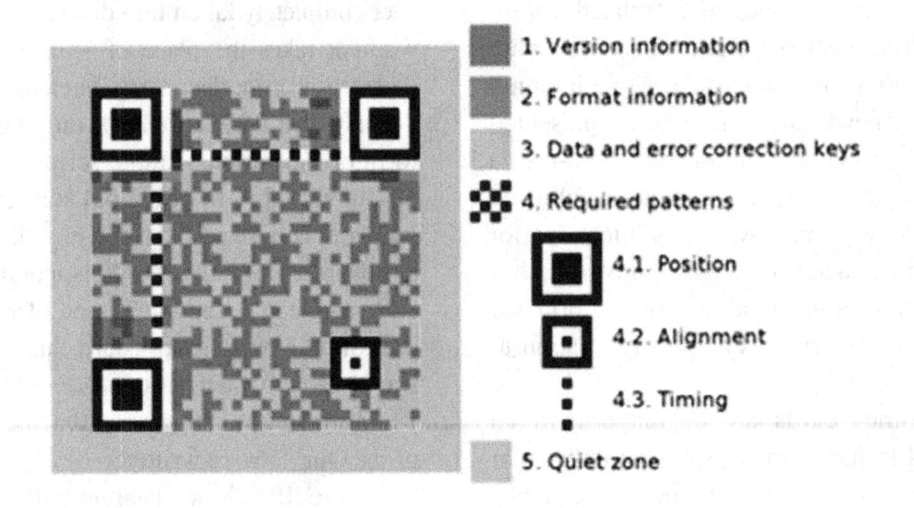

In a similar way as in Figure 19, the design for playing cards shown on Figure 20 combines traditional Arabic design with the QR code signs. It shows a variety of cultural icons from several cultures interwoven with the matrix barcodes called images of the quick response codes (QR codes – Figure 21). There is a strong impact in juxtaposing old cultural symbols with the contemporary QR code matrices designed to be detected as a 2-dimensional digital image by a semiconductor image sensor and then digitally analyzed by a programmed processor. The first known book on cards called *Yezi Gexi* was allegedly written by a woman from the Tang Dynasty (618-907 AD) (Needham, 1962), while QR codes have been used and printed on Chinese train tickets since late 2009.

Figure 21 presents a quick response code (QR code) – a matrix barcode consisting of black modules (square dots) arranged in a square pattern on a white background, with an explanation of its structure (Wikipedia, 2012). QR codes became a part of the playing card design presented in Figures 9 and 10.

See Table 8 for Your Visual Response.

Table 8.

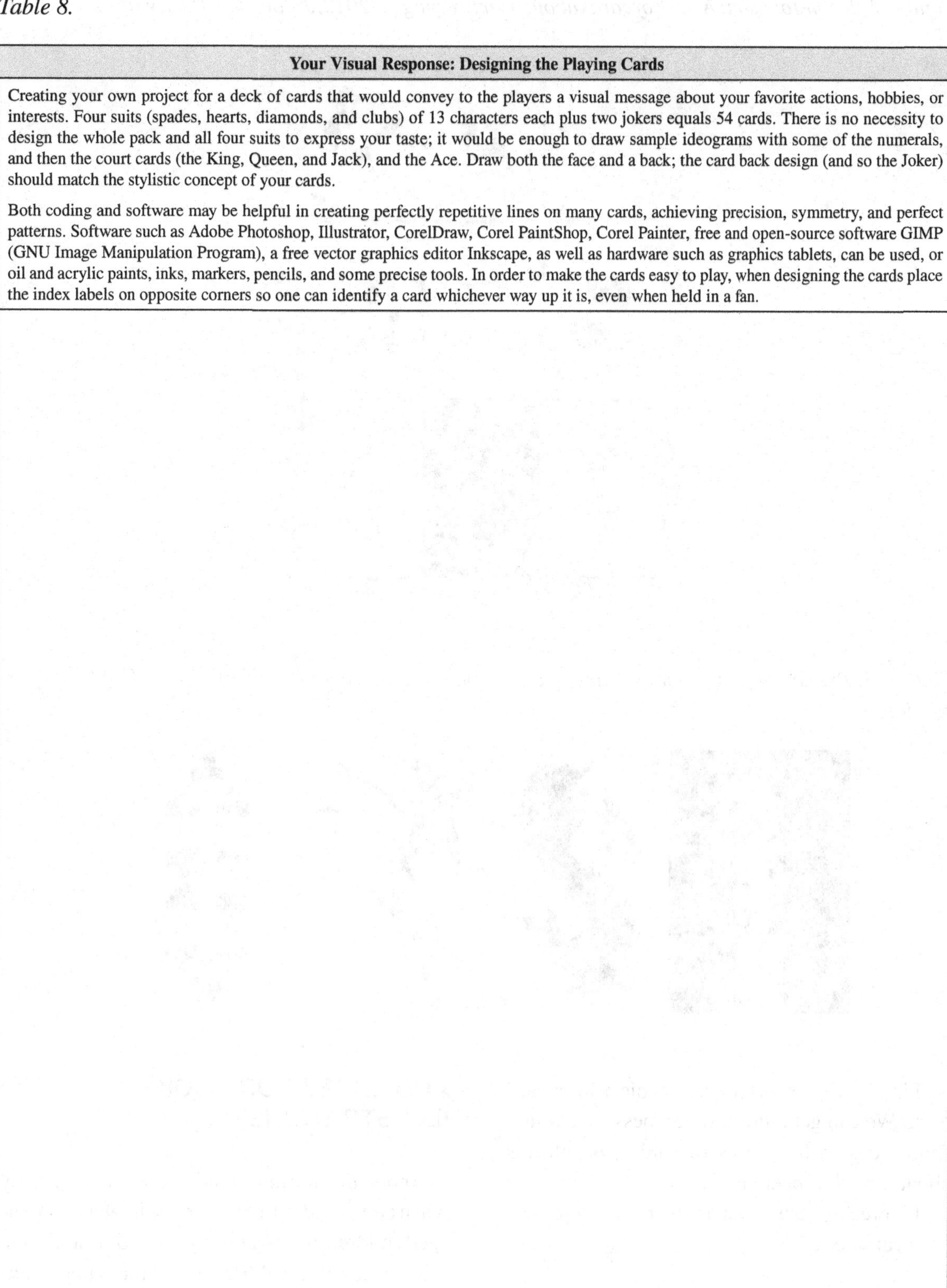

Your Visual Response: Designing the Playing Cards
Creating your own project for a deck of cards that would convey to the players a visual message about your favorite actions, hobbies, or interests. Four suits (spades, hearts, diamonds, and clubs) of 13 characters each plus two jokers equals 54 cards. There is no necessity to design the whole pack and all four suits to express your taste; it would be enough to draw sample ideograms with some of the numerals, and then the court cards (the King, Queen, and Jack), and the Ace. Draw both the face and a back; the card back design (and so the Joker) should match the stylistic concept of your cards. Both coding and software may be helpful in creating perfectly repetitive lines on many cards, achieving precision, symmetry, and perfect patterns. Software such as Adobe Photoshop, Illustrator, CorelDraw, Corel PaintShop, Corel Painter, free and open-source software GIMP (GNU Image Manipulation Program), a free vector graphics editor Inkscape, as well as hardware such as graphics tablets, can be used, or oil and acrylic paints, inks, markers, pencils, and some precise tools. In order to make the cards easy to play, when designing the cards place the index labels on opposite corners so one can identify a card whichever way up it is, even when held in a fan.

Figure 22. Jason Johnson: A deck of cards about beer brewing (© 2012, J. Johnson. Used with permission)

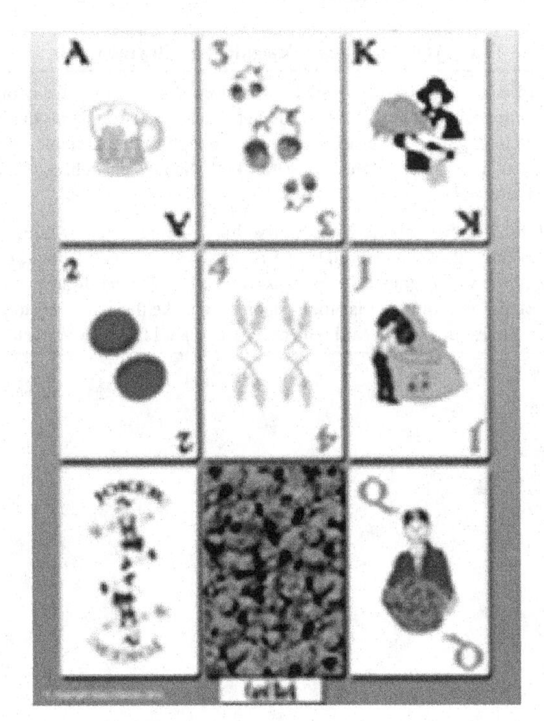

Figure 23. Preston Stone: A pack of cards based on fantasy imagery (© 2012, P. Stone. Used with permission)

Figure 22 presents another project for a card deck. We can get quite another message coming from a design for a deck of cards that informs about the art of beer brewing.

Figure 23 presents yet another student project for a card deck.

AUDIENCES FOR BOOK ILLUSTRATIONS

The book illustration would be created differently when addressed to toddlers, preschoolers, school-aged children, teenagers, or young adult audience. Thus, every type of illustration may assume different form depending on the prospective audience.

It will take yet another forms when images are created to help develop social skills or overcome learning disabilities. Book illustration for adults may accompany poems and verses, fiction, art, media, or sport related albums. It may be satirical, humorous, entertaining, with spiritual, religious, travel and leisure, or erotic content, and also may be socially or politically involved. Each area of human involvement determines somehow the illustrator's work. The same may relate to drawings present in multiple kinds of press, news coverage, and magazines, with the growing part placed by images in journalism and communication media.

CONCLUSION

Traditional and computing-based illustrations are present in our communication, learning, productive work, and artistic activities. Techniques typical of traditional illustration include pen-and-ink illustration, calligraphy, printmaking, engraving, etching, woodcut, linocut, lithography, screenprinting, and photosilkscreen. Digital illustration and infographics may often follow old traditions of illustration, for example, in children books illustration, poster art, street art, cartoons, comics, product design, or playing cards design. Several illustration projects are presented, which are inspired by biology and other sciences, refer to various fields of our interests and areas of our activities, and apply selected illustration techniques. The readers are invited to make their own illustrations.

REFERENCES

Carroll, L., & Tenniel, J. (1993). *Alice's adventures in wonderland*. Dover. (Original work published 1898).

Christopher, N. (2008). *The bestiary*. Dial Press Trade.

de Saint-Exupéry, A. (2000). *The little prince*. Mariner Books. (Original work published 1943).

EPA. (2012). Retrieved March 9, 2013, from http://cfpub.epa.gov/npdes/cwa.cfm?program_id=45

Fisher, R. A. (1958). Cancer and smoking. *Nature*, *182*(596). doi: doi:10.1038/182596a0 PMID:13577916.

Fried, P. A. (2002). Tobacco consumption during pregnancy and its impact on child development. In *Encyclopedia of Early Childhood Development*. Retrieved June 22, 2012, from http://www.child-encyclopedia.com/documents/FriedANGxp.pdf

Heaney, S. (2001). *Beowulf*. W. W. Norton & Company.

History of Printing. (2013). Retrieved March 9, 2013, from http://en.wikipedia.org/wiki/History_of_printing

Keogh, D., Asbeck, P., Chung, T., Dupuis, R. D., & Feng, M. (2006). Digital etching of III-N materials using a two-step Ar/KOH technique. *Journal of Electronic Materials*, *35*(4), 772-776. Retrieved June 18, 2012, from http://sigma.ucsd.edu/research/articles/2006/2006_9.pdf

Kite, B. (1994/2000). *Rude awakening at Arles: The aberrant art of Barry Kite: Postcard book by Barry Kite and Katie Burke*. Pomegranate Communications Inc..

Kite, B. (1997). *Sunday afternoon, looking for the car: The aberrant art of Barry Kite*. Pomegranate Communications.

Knödel, S., Hachet, M., & Guitton, P. (2009). Interactive generation and modification of cutaway illustrations for polygonal models. In *Proceedings of the 10th International Symposium on Smart Graphics*, (pp. 140-151). Berlin: Springer-Verlag. ISBN 978-3-642-02114-5

Madden, P. A. F., Heath, A. C., Pedersen, N. L., Kaprio, J., Koskenvuo, M. J., & Martin, N. G. (1999). The genetics of smoking persistence in men and women: A multicultural study. *Behavior Genetics, 29*(6), 423-431. doi 001-8244/00/1100-0423

Milne, A. A., & Sheppard, E. H. (2009). *Winnie the pooh*. Dutton Juvenile.

MOMA. (2012). *What is a print*? Retrieved June 18, 2012, from http://www.moma.org/interactives/projects/2001/whatisaprint/flash.html

Mythology. (2008). *National Geographic essential visual history of world mythology*. Berlin: Peter Delius Verlag GmbH & Co KG. ISBN 978-1-4262-0373-2

National Cancer Institute. (2012). *What you need to know about™ lung cancer*. Retrieved June 22, 2012, from http://cancer.gov/cancertopics/wyntk/lung

National Research Council of the National Academies. (2008). *Inspired by biology: From molecules to materials to machines*. Washington, DC: The National Academies Press.

Needham, J. (1962). *Science and civilisation in China*. Cambridge University Press.

Nguyen, L. (2010, April 27). Teen obesity linked to pre-birth tobacco exposure: Study. *Gazette*.

Pérez, J. B., Corchado Rodríguez, J. M., Ortega, A., María, N., Moreno, M. N., Navarro E., & Mathieu, P. (Eds.). (2012). Highlights on practical applications of agents and multi-agent systems. In *Proceedings of the 10th International Conference on Practical Applications of Agents*. Berlin: Springer. ISBN-10: 3642287611

Potter, B. (2006). *The complete tales*. Frederick Warne & Co. (Original work published 1902).

QR Code. (2012). Retrieved June 22, 2012, from http://en.wikipedia.org/wiki/QR_code

The United States Department of Justice. (1999). *Office of consumer protection litigation against tobacco companies*. Retrieved June 22, 2012, from http://www.justice.gov/civil/cases/tobacco2/index.htm

Tong, E. K., & Glantz, S. A. (2007). Tobacco industry efforts undermining evidence linking second hand smoke with cardiovascular disease. *Circulation, 116*, 1845–1854. doi:10.1161/CIRCULATIONAHA.107.715888 PMID:17938301.

Ursyn, A., & Scott, T. (2006). Web with art and computer science. In *Proceedings of the Consortium for Computing Sciences in Colleges (CCSC)*. CCSC.

Ursyn, A., & Scott, T. (2007). Web with art and computer science. In *Proceedings of the ACM SIGGRAPH Conference on Computer Graphics and Interactive Technics*. ACM. ISBN 978-1-59593-648-6

Viola, I., & Gröller, M. E. (2005). Smart visibility in visualization. In L. Neumann, M. Sbert, B. Gooch, & W. Purgathofer (Eds.), *Computational Aesthetics in Graphics, Visualization, and Imaging*. Retrieved June 18, 2012, from http://www.cg.tuwien.ac.at/research/publications/2005/Viola-05-Smart/Viola-05-Smart-Paper.pdf

Chapter 11
Making the Unseen Visible: The Art of Visualization

ABSTRACT

Themes and examples examined in this chapter discuss the fast growing field of visualization. First, basic terms: data, information, knowledge, dimensions, and variables are discussed before going into the visualization issues. The next part of the text overviews some of the basics in visualization techniques: data-, information-, and knowledge-visualization, and tells about tools and techniques used in visualization such as data mining, clusters and biclustering, concept mapping, knowledge maps, network visualization, Web-search result visualization, open source intelligence, visualization of the Semantic Web, visual analytics, and tag cloud visualization. This is followed by some remarks on music visualization. The next part of the chapter is about the meaning and the role of visualization in various kinds of presentations. Discussion relates to concept visualization in visual learning, visualization in education, collaborative visualization, professions that employ visualization skills, and well-known examples of visualization that progress science. Comments on cultural heritage knowledge visualization conclude the chapter.

1. INTRODUCTION: FROM DATA TO PICTURES AND THEN TO INSIGHT

The following text examines the language of visualization (often conceived as cross disciplinary), the interactive culture of knowledge visualization, and comprises some notes about the visual content analysis applied to data and knowledge. Art, graphic design, visual storytelling, and the use of signs support visual communication of human insight and understanding. As Rudolf Arnheim (n.d.) put it, "All perceiving is also thinking, all reasoning is also intuition, all observation is also invention." Visualization uses visual imagery and visual thinking to understand complex information, therefore it differs from a metaphor or

DOI: 10.4018/978-1-4666-4703-9.ch011

analogy. Through creating symbols, meaning can be conveyed and expressed by means of images. Graphs, diagrams, and animations can visualize messages as well. As productive thinking in whatever area of cognition takes place in the realm of imagery, visual perception should be considered a cognitive activity. In his structure of intellect model, Joy Paul Guilford (1968) designated a single factor as the cognition of visual figural systems. Visualization has also been considered a semiotic process because of the use of signs to present ideas. One may say visualization meant in traditional terms the intuitive use of a visual presentation of a concept; one of the most often cited examples is August Kekulé's vision of dancing atoms and molecules telling about a structure of benzene. In general non-technological terms, visualization means a mental image that is similar to a visual perception. It is also the technique of creating a mental image of a desired outcome, and repeatedly playing that image in the mind. It is sometimes used in conjunction with medical treatment, including cancer treatment.

There are several approaches to the concept of visualization and the ways it mediates between the user and the physical world. Data is seen as an essential abstract concept from which further levels can be derived: information, and then knowledge. The most important domains in visualization can be seen as data visualization, information visualization, and knowledge visualization. Scientific visualization is another approach to visualization, where physically based data are selected, transformed, and represented according to space coordinates, for example, visualizing computer tomography data for medical use. In broadcast media, visualization techniques are often used to explain a process that is important for action development; for example, in an American television medical drama House, M.D. dynamic visualizations show the interior of human organism.

Information is usually presented in numerical, graphic, or diagrammatic form. When expressed as visualization, information may be shown as a sketch, drawing, diagram, plan, outline, image, geometric relationship, map, music and dance notation, object, interactive installation, or a story. Diagrams visualize information in a pictorial yet abstract (rather than illustrative) way, as plots, line graphs and charts, or the engineers' or architects' blueprints. Big and complicated presentations of data organization and interpretation, for example governmental statistics, are easier to comprehend in a graphic than in a numerical form, when they serve as explanatory tools for the data sets. Data provide us a raw material that has no meaning if we do not process it, so it becomes useful for our own purpose. We may find online visual presentations, for example "The crisis of credit cards visualized" (www.crisisofcredit.com) created by Jonathan Jarvis or "Wind map" (file://localhost/Users/ursyn/Desktop/Wind%20Map.html), a personal art project interactively showing the actual tracery of winds.

Graphic presentation shows how something works or explains the relationship between the parts of a whole. However, surfing the web, data mining, and manipulating the data are easier when the data is shown with the use of information visualization and interactive techniques, with more dimensions in a presentation. Data-, information-, and knowledge-visualization have been present in different disciplines and in various modes since early days of civilization. Visualizations of many kinds are powerful cognitive tools useful in our everyday life; they take their form from several domains, and so there are no defined boundaries between the different disciplines of visualization.

Essential terms: data, information, and knowledge may need some consideration before discussing the data-, information-, and knowledge visualization issues.

1.1. Data: Analog and Discrete Data

Data in the form of incoming signals such as light, temperature, electrical, mechanical, pneumatic, hydraulic or other signals can be conveyed in an analog or digital format. An analog signal changes continuously over time; we can measure small fluctuations in its quantity. For example, a microphone diaphragm responds to sound (which means changes in air pressure) and causes changes in a voltage or a current in an electric circuit. Analog signals have higher than digital resolution because digital resolution goes in discrete, separate steps. A digital system uses discrete changes in information, so signals are conveyed as electronic or optical pulses with their amplitude presented as a logical 1 (pulse present) or a logical 0 (pulse absent). Digital audio or digital photography use such binary numeric system.

Data can be described as categorical, ordinal, or arbitrary. A categorical variable of the data (also called a nominal variable) can take on non-numeric values with two or more categories, for example colors. The ordinal, ranked values have an implied ordering (for instance, a mild, medium, or hot salsa). The arbitrary variables can take on any range of values without ordering (for example, geographical location in a country represented by that country's international telephone access code, or the make or model of a car). One cannot clearly arrange the data, for example by adding or subtracting them, as well as we cannot say, 'less than' or 'greater than' about the nominal numbers. For example, gender, marital status, race, religious or political affiliation, college major, and birthplace are examples of categorical data measured at a nominal level.

In ordinal measurement the ordered, numerical data are called the ordinal or rank data. They can take on numeric values that are binary (with values of 0 and 1 only), discrete (taking on integer values, sometimes being in a subset), or continuous (representing real values, for example, in the interval [0, 5]). The numbers called ordinals represent the rank order (1st, 2nd, 3rd etc.) of the data. Examples are data such as economic status of some people (low, medium and high), a mineral hardness scale, or the horse race results.

In terms of a scale of variables, the spacing between the values may not always be the same across the levels of the data. When the intervals between the values of data are equally spaced and the differences between pairs of measurements can be compared, the variable is called an interval variable. When we can determine a lowest value of the data (such as an absolute zero) we can apply mathematical operations of addition, subtraction, multiplication, division, and exponentiation (Ward, Grinstein, & Keim, 2010; UCLA Academic Technology Services, 2012; Wikiversity, 2012).

Coordinate systems (with the x and y axes) should contain clear labeling and a scale with regular intervals that tells about the amount of change (Tufte, 1983, p. 14). Graphs must not quote data out of context. Without axes and scales, a graph does not answer "compared to what?" and does not tell if the change is big, what happened before and after our measurements, is it a seasonal, temporary, or serious change, and how this change compares with events in other places (Tufte, 1983, p.74-75).

Time-series charts are the most frequently used form of graphic design for complex statistical material, e.g., thousands numbers about weather. Time series charts can explain big data sets where changes in recorded values go along to the regular rhythm of seconds, minutes, hours, days, weeks, months, years, centuries, or millennia. About 75% of all graphics published were the time-series. The problem with time-series is that the passage of time is not a good explanatory variable: descriptive chronology is not causal explanation (Tufte, 1992).

See Table 1 for Your Visual Response.

Table 1.

Your Visual Response: Hours of your Day
Draw a time-series graph showing the amount of the daylight hours in sequential days throughout the spring. Check your graph whether it is labeled correctly and what units are placed on the x and y-axes. Now, change the vertical x scale from the number of daylight hours to the 'units' of your work, tiredness and exhaustion, and label the horizontal y-axis as time in minutes spent in sequential days on workout, running, or your other physical activity. Draw a time-series graphic. You may choose to use charts (bar charts, plots, scattergrams, etc.) or standard presentation tools such as Excel or PowerPoint presentation to make informative labels, colors, etc. Add a small caption box on the chart (above or below). Make your graph visually pleasing, heartening, and inspiring.

Data are usually sensory (mostly visual), computer-supported, and interactive: users are able to change the image as they work with data. Data is usually presented in the form of a row of numbers, characters used to form words, or images produced by various devices. For computing purpose, the data takes form of a sequence of symbols, for example, in a binary alphabet comprising "0s" and "1s." A computer program is also a collection of data. Data is often shown as symbols. Visual thinkers are needed because most of the data, information, and knowledge can be found on the web is presented visually as pictures, movies, and graphs. As an example of the data collection, names of schools, number of students in particular schools, number of teachers, number of classes are considered the data (Table 2). If there are few numbers to present, using the table might work better then using a graph.

See Table 3 for Your Reaction and Visual Answer.

Table 2. An example of the data gathered about the schools in a school district N

Name of a School	Number of Students	Number of Teachers	No of Classes
Franklin	180	18	5
Martin Luther King	220	15	3
Meadows	326	15	8
Sunflower	187	16	4

Table 3.

Your Reaction and Visual Answer: Collect the Data of your Choice
You may want to make a data collection in the form of a table. You may choose to collect the kinds of jobs in your workplace, a number of glasses of water you have drunk every day in this month, names of planets in a Solar system, or temperature at noon during the last ten days. The observations or recordings you collect will not be significant by themselves. As a product of your project, you will obtain just the data: numbers or names. You will need to analyze them later to get information about the theme of your investigation.

1.2. Information

Information makes sense of the data and gives them meaning. To make sense of a row of data we have to abstract those that carry meaning useful for our purpose. We can locate information and explore the structure of any kind: from simple charts and graphs showing one dependent variable against one independent variable, to the 3D computer-generated virtual reality environments happening in real time, with human interaction possible through various input/output devices. The data becomes information when people or computers find the data patterns and characteristics that give them meaning. For example, a pamphlet describing characteristics of a school district, its schools' density, addresses, names and other features can be called information (Figure 1).

See Table 4 for Your Reaction and Visual Answer.

Figure 1. An example of information provided by an educator, based on the data: A fact sheet entitled "A State of School Distribution in District N, as for February 2011" by John Doe (© 2012, A. Ursyn. Used with permission)

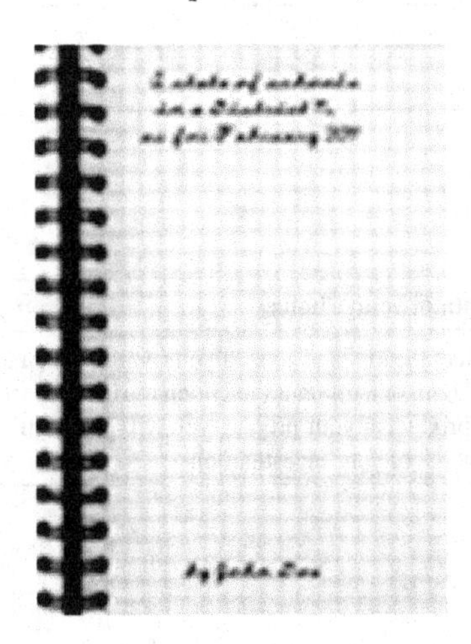

1.3. Knowledge

Information that we got by processing data can enhance our knowledge. We confirm as truthful the information obtained as the patterns in the data; we authorize it as verified, and check its ethical, political, or aesthetical values. For example, we may present our knowledge about the school district in a report containing practical information and conclusions about how to improve education in your district, or even the district's rankings, if you also collect the data and sort out information about other districts, to determine your district's position in a scale of a status (Figure 2).

See Table 5 for Your Reaction and Visual Answer.

Table 4.

Your Reaction and Visual Answer: Obtain Information from the Data
You may now want to use the data you have collected in previous activity and make them useful by determining their meaning and then writing or sketching a concept map about information you have gained from the data. For example, you may use the raw data about shoes in order to obtain information about the market needs in terms of sizes, you may use the data about how many glasses of water you used to drink every day in order to get information if it is similar to the average liquid consumption, you may write a handout about the Solar system, use the temperature data for future discussion of the climate change, or maybe you want inform your friends about your preferences. We all know that we have to collect much bigger samples of data to be serious about drawing meaningful information from the data. Thus, as a product of your project, you will draw information from the data you have collected before.

Figure 2. Example of knowledge based production: "Ways to Improve the State of Education in the School District N." A paper presented by John Doe at the Metro 2013 Educational Conference. (© 2013, A. Ursyn. Used with permission)

Table 5.

Your Reaction and Visual Answer: Construct Knowledge Base for your Theme
While you already have your data set and some meaningful information about the objects of your interest, you are ready to construct knowledge about the theme of your activity and then write or visually present your knowledge you have gained from information you gathered. You will need to perform some cognitive and analytical activity to make information useful for finding some relations, reasons, explanations, and solutions: how does the preferred fashion of shoes depend on activities of the owners, are your water drinking habits good for you, how the planets got their names, or is it really warmer now. The product of this project would be just knowledge constructed on the basis of information gained from the data you have collected.

2. VISUALIZATION

Matthew Ward, Georges Grinstein, and Daniel Keim (2010), p. 1) define visualization as "the communication of information using graphical representations." There is a wide range of visualization techniques that still grows along with the developments in computing and information technology. Visualization techniques help understand complex data, communicate, and navigate on the web (Marchese & Banissi, 2013). They support exploring data (such as one-, two-, three-dimensional data, temporal and multi-dimensional data, and tree and network data). Ward, Grinstein, & Keim

(2010) list techniques for visualization: visualization techniques for spatial data (in one, two, three dimensions and their combinations), techniques for geospatial data (point-, line-, area-based data, and their combinations), techniques for imaging multivariate data (point-, line-, region-based techniques, and their combinations), techniques for visualization of trees, graphs, and networks, text representations, and interaction techniques. When considering visualization, the word 'model' is used both as a copy of some object recreated either larger or smaller than the original (often with the use of an interactive modeling system), or as a mathematical model describing physical laws and the behavior of physical objects. Chen (2010) discusses information visualization in terms of structural modeling aimed "to detect, extract, and simplify underlying relationships" and graphical representation aimed "to transform an initial representation of a structure into a graphical one, so that the structure can be visually examined and interacted with" (Chen, 2010, p. 27). Structural modeling to describe the relationships may involve applications with the use of graphs, trees, or cones; detecting proximity and connectivity; clustering and classification using word search, multi dimensional-scaling, and network analysis; glyphs on charts and graphs; virtual structures; applying complex network theory, and network representations (Chen, 2010).

Visualization tools derive its form from many domains; moreover, there are no defined boundaries between disciplines of visualization, while definitions and theoretical approaches change with the advances in technologies. Masud, Valsecchi, Ciuccarelli, Ricci, and Caviglia (2010) identified the most important domains in visualization:

- **Data Visualization:** "Information which has been abstracted in some schematic form" (Friendly, 2009), to provide visual insights in sets of data. Data may be 1D linear, 2D, 3D, and multidimensional. Sabol (in Bertschi et al., 2011, p. 333) describes data as "sequences of numbers or characters representing qualitative or quantitative attributes of specific variables. To obtain information, data is processed and brought into a context within which it gains a specific meaning and becomes understandable to users;
- **Information Visualization:** Means the use of computer-supported, interactive, visual representations of abstract data to amplify cognition (Bederson and Shneiderman, 2003; Card, Mackinlay, & Shneiderman, 1996) and derive new insights. "Information visualization makes use of human visual perception capabilities for recognition of patterns and extraction of knowledge from raw data and information" (Sabol, in Bertschi et al., 2011, p. 333);
- **Knowledge Visualization:** Uses visual representation to transfer insights and create new knowledge, rather than data, between individuals; it concentrates on the recipients, other types of knowledge (know-why, know-how), and on the process of communicating different visual formats (Burkhard, Meier, Smis, Allemang, & Honish, 2005). Sabol (in Bertschi et al., 2011, p. 333) defines knowledge as "an acquired, established set of facts, recognized to be valid and valuable within a specific domain";
- **Scientific Visualization:** Established in 1985 at the National Science Foundation panel, deals with physically based data defined in reference to space coordinates, such as geographic data and computer tomography data of a body (Voigt, 2002). Biomolecular structures were first visualized as balls connected with sticks, and then as spheres with rods; advanced visualizers, the 3D graphics software generate images from an electron microscope, thus providing information visualization (Ward, Grinstein, & Keim, 2010, p. 24);

- **Information Aesthetics:** Forms a cross-disciplinary link between information visualization and visualization art;
- **Infographics:** Refers to tools and techniques involved in graphical representation of data, mostly in journalism, art, and storytelling.

Infographics are graphic visual representations of data, information, and knowledge. Visual presentation of concepts, content analysis, and visual transfer of knowledge are on demand due to the developments in communication media. Not only we talk, write, call, fax, text, page, e-mail, tweet and blog, enjoying high speed Internet; we also exchange images and videos through smart phones, Facebook, and Skype, to name just a few. Boundaries between the ways of delivering information (through the book, newspaper, TV, video, or web journalism) are being talked as vanishing (Flowing Data, 2012). It seems reasonable to assume that merging verbal and visual ways of communication makes the central part of many tasks and areas of interest, such as scientific visualization and simulation, visualization of big sets of data, virtual reality environments, web based environments, web graphics, game design, semantic web, and data mining. According to Martin Lindner (in Bertschi et al., 2011), Visual Thinking trend started in the 1990s; digital data, Apple-driven graphic engines, and the web resulted in a doodle revolution. A wave of making doodles and napkin sketches contributed to new visual languages and new cognitive styles. "Because they only require a small attention span to get their ideas and messages, these simple user-generated objects can be easily circulated in the cloud: in blogs, via flickr, SlideShare, or YouTube. This is part of a paradigm shift from 'published ideas' to 'circulating ideas (and) collaborative thinking enabled by the Web 2.0 ecosystem" (Lindner, in Bertschi et al., 2011). Sometime ago the founder and CEO of Netflix Reed Hastings described on Wikipedia the phases of the web, telling that Web 1.0 was dial-up, 50K average bandwidth, Web 2.0 has an average bandwidth of 1 megabit, and Web 3.0 would be 10 megabits of bandwidth, allowing making a full video web. And thus Web 1.0 was a "red-only organization of a site, Web 2.0 expanded into "read-write" web that involved users, while Web 3.0 could allow users to modify the site itself. Tim O'Reilly (2005) and John Battelle described the term 'Web 2.0' in 2004. The big type, massive fonts, glossy navigation bars, background gradients, rounded corners, and textures with subtle 3-D compositing effects are design trends typical of the Web 2.0 design style of the 2000s (Beaird, 2007, p. 88-9).

2.1. Visual Metaphors, Glyphs, and Icons as Visualization Tools

Visual metaphors in information or knowledge visualization make a basic structure for communicating messages, because they organize and structure information in a meaningful way and combine creative imagery with the analytic rationality of conceptual diagrams. A metaphor indicates one thing as representing another one, not so easy to grasp, thus making mental models and comparisons. For example, instead of developing a nomenclature specific for computing, we use everyday metaphors of desktop items when we use a computer: we open a new *window* or a *file* with a *mouse*, and then put them in a *folder* to organize files, place *icons* on a *desktop*, use *tools*, search *engines*, *canvas*, *mailbox*, *document*, *in* and *out boxes*, and a web *portal*, we *cut*, *copy*, and *paste*, thus applying names of familiar objects and actions for organizing computing-related items and activities). The desktop metaphor is now fading because cell phones and tablets, that are growing more powerful and cheaper, are replacing PCs as the main gateway to the Internet. Metaphors traditionally used in visualization projects to translate information about the qualities and quantities show programs or data as natural objects, such as a solar system, or man-made objects, for example an architectural design, a city, a house, parking lot, metro, library, street, but also are mapping the data to facial expressions, video games, or nested boxes.

Program visualization can enhance program understanding, software production, maintenance and cost related issues. Selecting options can lessen an effort, reduce the time for providing software maintenance, but also it can give fun when navigating through a software city in 3D. An example of the use of a city metaphor to support the program understanding is visualization of a program development (Panas et al., 2003). In their project, buildings denote program components, and the city itself represents a software package. The developers need to address both static and dynamic data. The static information includes the size of the buildings that shows the amount of lines of code, the density of buildings tells about the amount of coupling between components, while the buildings' structure shows the quality of system implementation. The dynamic data are shown as the speed and type of vehicles and the traffic density. Cars moving through the city indicate how the program runs. They leave traces in different colors and show the origin, destination, and density of communication between components. A cost-focused metaphor uses color-coding to show work distribution, recycling, hot execution spots, aspects, and high-cost components. The authors assume that everybody is already familiar with the structure of the city and it's basic features, such as the floor 4 is above floor 10. Corridors run horizontally, while elevators vertically, boats and cars can vary in sizes on their horizontal paths, while clouds move above the entire building. Large-scale virtual city 3D models transport huge amounts of different information, such as facade textures, aerial photographs, infrastructure models, and city furniture, thus creating information overload.

See Table 6 for Your Reaction and Visual Answer.

Table 6.

Your Reaction and Visual Answer: Create a Metaphor Visualization to Provide Visual Explanation
Create information that would be organized around the user's task. Perhaps you want to make it easy for the user to follow and perform. For this purpose, you may want to present visual instruction about a familiar situation. Choose your own metaphor and show it as simply as possible, preferably as an icon or a readily recognized object that directly reflects the thing you explain. For example, someone may want a visual explanation how to check pressure and inflate a tire, how to charge a cell phone, or how to make sushi. The best approach to this task is to pretend you are doing it for a character that doesn't speak in any human language. Your instructional message should be clear, so nobody would follow it in a wrong way. For example, the design of the slots and connectors in Apple computers leaves no doubt where each connector should go, so the users may feel safe they will not spoil anything, because they simply cannot connect it in a wrong way.

Most visualization techniques and tools are based on the graph metaphor, including the extensive research on graph layout algorithms. Metaphor-based animated environments can map the abstract programming concepts to concrete metaphors. For example, virtual environment may be based on a garden metaphor for visualization of complex systems, with programs constructed in a way resembling a video game. In a similar way as the game players, data miners rapidly become familiar with the topic. Visualization created in the immersive virtual reality is an example of interactive visualization; people can explore in real time information perceived as a close and real (not abstract) thing, and receive feedback about their actions. Many times interactive visualization becomes collaborative, when people who are physically distant communicate and control visualization using networked computers, mobiles, and video conferencing, to focus on particular approaches to information visualization such as theory, techniques, usability, and applications. Information visualization may serve scientific theories and fields including applications of graph theory, geometric modeling and imaging, interaction design for information visualization, visual analytics, virtual environments, geo analytics, biomedical informatics, biomedical visualization, web visualization, cultural heritage knowledge visualization, aesthetics, education, visualization in software engineering, architecture, visualization in built and rural environments, and many other fields. It serves for conveying information in online journalism, business management, technical writing, social networks, and in education. For instance, the architecture of a software system can be understood easier when a visual query, constructed on the display, helps us to find a pattern of highly connected components in a node link diagram. Girot and Truniger (2006) proposed replacing maps, perspectives and photographic stills with dynamic video presentation as a tool for visualizing landscape. They introduced an analytical grid for the evaluation of video as a means of visualizing landscape perception from the vantage point of the slowest traveler, the pedestrian.

Visual framework of the visualization science provides a way of dealing with massive amounts of data. Visual presentation of large data sets is in demand because web became the main carrier of information. When we use the search engines, too much data must be scrolled on the screen, so new browsers or new ways of visualization, such as cloud visualization, are necessary to present information. It can be done with the use of information visualization, data mining, and semantic web. The semantic web means the use of visualizations providing metaphors for web navigation and communication. A structured semantic network above information resources bridges knowledge representation and information management. Data mining discovers meaningful patterns in large data. It is often used in business, e.g., for assessing client risk in a bank loan. It is also used against possible terrorist acts. Results can be visualized using virtual environments based on a metaphor.

Web-search result visualization can be done in a glyph form. Glyphs are single graphical units, which are able to portray many variables by adapting their properties, just helping to overview, examine details, and abstract information about the very large, multivariate data sets. An early example was developed by Chernoff (1973) who represented multi-variable data through face expressions as a way of displaying interactive content. Further works involving face expression as information visualization tool included works by Loizides and Slater (Loizides, 2012; Loizides & Slater, 2001 and 2002). By observing data points represented on a glyph, such as a sphere or a "bubble," one can determine the quality and quantity of the links within a presentation, because they are displayed as five variables, by the position (x, y, and z), size, shape, and color of the sphere. The 3-D aspect of presentation enhances understanding of data flow. Web-search result visualization with glyphs can be seen in linked views: data about the keywords search terms, the domain-name, URL, text snippets of the pages; also, the size of the HTML file, number of links (anchors) on a page, media type (html, text, images, movies, animation, sound), last-modified-date, and page structure. With a system of intersecting circular and linear muscles, the human face is capable of motions varying in intensity and velocity. Davies et al. (2012) described factors defining the identity and expression parameters of human faces presented as a 3D Morphable Model, where the identity parameters are set in the initial frame and the expression parameters are adjusted in subsequent frames. Othman, El Ghoul, and Jemni (2012) developed algorithm based on the topology of 3D face serves for segmenting a 3D head and extracting vertices of feature points; they assessed it on full 3D heads with different surfaces, gender and ethnicity.

Richard Brath (2009) described multiple shape attributes that can be used within information visualizations. Prior art from many fields and experiments inform what the attributes of shape are and the potential ways that we may effectively utilize shapes to represent multiple data values within information visualization. The independent attributes of shape include closure, curvature, corner angle and type, edge and end type, notch, whiskers, holes, intersection, and local warp (Brath, 2009). They can be used separately or together to convey data attributes, as opposed to icons, numbers, common symbols, or compound glyphs. Scientific visualization represents physical phenomena and is therefore restrained to a spatial context, while information visualization often use shape to represent only a single data attribute.

Scientific visualization often uses curvature-based parametric shapes, glyphs with curvature and twist, and blobs. Information visualization utilizes shapes, such as Chernoff faces, physical objects, star coordinates, sticks, radar plots, 'growth' visualizations, and other organically inspired visualizations. The use of multiple shape attributes increases the expressive range and the information density of visualizations. The experiments show the potential to convey ten or more different data attributes within a glyph based on shape attributes.

See Table 7 for Your Reaction and Visual Answer.

When we deal with several kinds of data, we may use a metaphor to describe the structure and the relations among data. Visual metaphors fulfill a dual function: first, they graphically organize and structure information, and second, they convey an insight about the represented information through the key characteristics of the metaphor that is employed (Lengler & Eppler, 2006). We may use graphics or show virtual environments, often shaped by artist's fantasy. The success and quality of any visualization depend on imagination, the retrieval of necessary data for visualization, the choice of a suitable metaphor, and the delivery method: whether to apply animation, interconnection, or interaction.

For example, in an assignment "Day and Night" students from my Computer Graphics course presented visually their daily schedule of activities using exclusively visual means of communication, such as metaphors, icons, signs, or symbols, and avoiding inserting any verbal content (even ZZZZZ for sleeping was not used). The work of Tiffany Mulford titled "Home" presents an interactive use of icons to show daily actions on her cellphone (Figure 3).

Visualization is not only making the unseen visible – it is building a meaningful net of associations and connotations. We create connotations, analogies, signs, icons, acronyms, idioms, symbols, synonyms, metaphors, and paraphrases. The reader may want to perform in that order, imaging the essence of a physical law, visualizing processes, events, and relations, and then finding metaphors for invisible forces.

Yet, we should remember that a metaphor might have a different meaning in other cultural or social environments, so some viewers may react in unforeseen way.

See Table 8 for Your Reaction and Visual Answer.

Table 7.

Your Reaction and Visual Answer: Creating a Metaphor for a Set of Factors to Make a Visualization
Choose a theme that interests you. It may be a project you are just working on or a group portrait of people you are working with. First, you may want to visualize the theme in two ways, journalistic and artistic: writing a description of your project or a group profile of people (verbal approach) or graphically presenting a group of people of your choice (visual approach). Depending on the kind of visualization you will work on, whether it be a personal, cultural, social, political, or psychological depiction, you will cope with a different set of variables you will have to take into account. For this reason, you will have to structure your data toward different metaphorical representations. Design a visualization using a metaphor showing the organization of your concept. It would be useful to create a metaphor for a set of characteristics that make up a whole group profile and show individual features. Explore and communicate your concepts and their relevance to one another. Decide if would be a hierarchical order or an assembly without ranks that would characterize your theme better and would be helpful in exploring and communicating relations between individuals. For example, for a hierarchical presentation of your concept, you may use a description of the animal kingdom, the primary division in biological classification with its phyla (such as Chordata), classes (such as Mammalia), orders (such as Carnivora), families (such as Canidae), genera (such as Canis - Latin 'dog'), and species (such as doglike species - dogs, wolves, jackals, coyotes, and foxes). If the connections among concepts you describe are not hierarchical, you may draw a garden and arrange various plants, shrubs, and trees in specified relations. For this purpose, you may want to use stamps, a collection of downloadable brushes, or a clipart.

Figure 3. Tiffany Mulford, "Home" (© 2010, T. Mulford. Used with permission)

Table 8.

Your Reaction and Visual Answer: Visualizing What You Know or Think
You may now want to devise simple visualization with the use of visual metaphors and analogies. It may be done on a computer, a tablet computer such as an iPad, or on a smart phone. You may choose to draw a nature- or knowledge-derived visualization; however, the process of creating visualization has its inherent tendency to apply abstract thinking; you may prefer to find a way of depicting an abstract concept by assigning visuals to abstract ideas and forces, such as love, infinity, or an expanse, thus making the unseen visible. A large number of scientists perceive the visualization techniques as a chance to show what is invisible – translation of mental, abstract, formal concepts into images: pictures or graphs. In these cases, visualization is not only making the unseen visible – it is building a meaningful net of associations and connotations. You will thus create connotations, analogies, signs, icons, acronyms, idioms, symbols, synonyms, metaphors, and paraphrases. For this purpose use a semiotic way of applying signs, symbols, and iconic images; also, employ some tools of visual thinking such as abstracting details that are not crucial for conveying the core of your message. Apply the principles that control the basic elements in art, such as line or color, and then convey the essence of your response to a selected slice of reality. Make your visualization simple and convincing. A grasp of visualization techniques is a requisite for designing effective visual marketing. Sellers extract superior features of their products and present them as quite distinct images in a way that everybody can apprehend the product's quality and buy the marketed article. Choose a simple theme for your visualization; for example, you may want to present your daily activities using simple symbols, make a visual explanation of how your video camera is working, visualize in an explanatory way a structure of a plant cell, visually present the overall organization of your work place, or picture your family tree. Be careful to avoid simplifications and steer clear of clichés. To be understood and able to convince others, try to find the very essence of the problem you discuss and show it in a simple way.

In the Manuel Lima's book (2011), Lev Manovich indicates the important features that make information visualization unique: projects are visually dense, with more data; they show relations between data; in aesthetical terms, they show complexity (chaos theory, emergent complexity theory) rather than reduction (breaking down into the simplest elements). According to Lima, visualizations fall into all three categories of science, design, and art; they are used as a tool for understanding data – i.e. discovering patterns, connections, and structure. As a scientific tool, information visualization serves for discovery of new knowledge; as a design tool it facilitates the perception of patterns and evokes emotions in the viewers; as art it is a technique to produce something non-utilitarian and aesthetically interesting (Lima, 2011, p. 12).

Vande Moere and Boltzmann (2009) propose taxonomy of information display based on context, in terms of the data it represents and the environment it is located in, with three categories: visualization as translation (with an impartial display without context or contextual relationship with the data, location, and the viewer), visualization as augmentation (a display within context or specifically designed within the functionality of a specific physical context, where both object and display provide cues how to understand the displayed content), and visualization as embodiment (where a display is context, for instance, pixel sculptures, physical installations with matrices of repeated objects to simulate text and objects; they address the physical medium of the display itself rather than communicate specific content, and often the display medium overwhelms the meaning of the content, and the display cannot exist without the information). The authors point out that the emerging field of data sculptures addresses the materialization of information into three-dimensional form, where metaphorical distance between sign (the information) and object (the materialization of the information) plays an important role in how the information is interpreted.

Applications for mobile devices such as handhelds and smart phones make easy to use them as desktop applications. Usage of information visualization techniques reduces limitation caused by a small visual area for analysis, and enhances the presentation, search, and analysis of data. When applying information visualization techniques for mobile devices, for the geo-referenced data in the maps visualization techniques include, for example, the scatterplot graphs, while visualization of hierarchical structures involves the treemaps (Clayton, Pinheiro, Meiguins, Simões, Meiguins, & Almeida, 2008).

In order to explore the visual way of learning and create an opportunity for students to study visually, I asked them to work on a visual learning project lasting the whole semester. Students were asked to study visually for another course they were enrolled concurrently (they considered the most difficult in the semester) by applying knowledge gained in my Visual Learning course, where they learned about data-, information-, and knowledge visualization, and techniques related to visual display of quantitative information introduced in the books of Edward Tufte. Students presented in a visual way all the material required for another course. The progressive stages of their study culminated in a final project taking the form of an illustrated graph or a set of graphs, visualization, a game, and several other forms of display. Some examples of student work are presented below.

In the data visualization project, students used their knowledge to summarize time-based processes within a specific field. This approach utilized design skills and knowledge that students developed over the semester. In the "Aesthetic explanation of motion picture making" the author provided, in a visual manner instruction concerning the use of a video camera. Corwin Bell, the author of this project (who titled it also "Cinematic Mass Option Syndrome") set down all factors on a surface of the graphics, so viewers could examine relations between artistic and technical aspects (Figure 4). In summary, this graph could possibly serve as a basis for writing a book on the subject of motion picture making because it juxtaposes concepts with objects, and then shows the relationships and dependencies between them

See Table 9 for Your Reaction and Visual Answer.

The website http://prezi.com/8_i4pmtb6t_t/timeline/ shows a timeline of major events in art history presented in space, time, and context. Student created it using Prezi in order to show how major artworks were set in time and geographically, with attention placed on events shaping them.

Figure 4. Corwin Bell, "Aesthetic Explanation of Motion Picture Making" (©2004. C. Bell. Used with permission)

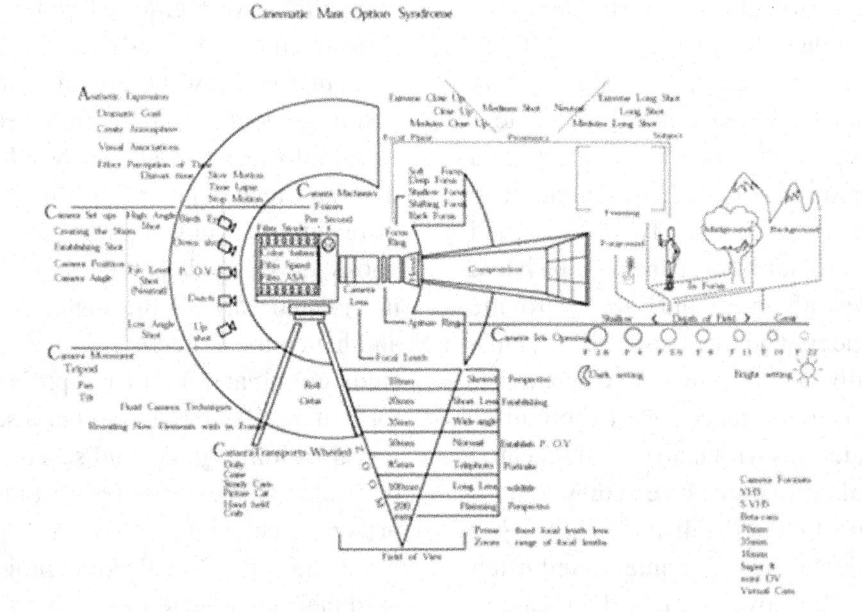

Table 9.

Your Reaction and Visual Answer: Create Information Visualization
Gather the data from your earlier activities and then make an information visualization project using the data you have collected. Try to present information visually, so it will be self explanatory without words. You may want to use data related to your work and/or your interests. Or, returning to previous text, you may draw a shoe for each size of items you collected and then put shoes of a specific size into a box that would fit the number of shoes you counted. You may assign colorful circles to specific days of your search, with the size of a circle telling about the number of glasses of water you drunk that day (information you have gathered earlier). Maybe you prefer to draw images of muses, avatars, or other individuals to make the names of planets easy to learn and remember; or present visually the temperature changes as related to people behavior, or clothes they wear any specific day.

2.2. Data Visualization

Possibly, visualization is the best way of learning, teaching, or sharing the data, information, and knowledge. Many agree that visualization outperforms text-based sources and increases our ability to think and communicate. Data visualization enables us to go from the abstract numbers in a computer program (ones and zeros) to visual interpretation of data. Text visualization means converting textual information into graphic representation, so we can see information without having to read the data. For example, designers may choose to apply tables, pie or bar charts, histograms, or Cartesian coordinates to present their data. Abstract or model-based scientific visualizations present real objects in a digital way directly from the scientific data. They may enrich learning by presenting art-science cooperative projects, and making knowledge accessible and comprehensible to a wide audience. With visualizations, you can avoid providing dry context that is difficult to absorb. We can apply visualization when we design a presentation of abstract data in a visual, often interactive way. For this purpose, we use easy-to-recognize objects that are connected through some well-defined relations. At present, visualization means using the computer. Computers transform data into information, and then visualization converts information into picture forms and creates graphic images and symbols to convey and express meaning; this let us comprehend data and make decisions. Visualization is a kind of storytelling comprises interactive graphics, animated graphics, multimedia features, interactive narratives, and explanation graphics (Burmester, Mast, Tille, & Weber, 2010). Analysts, decision makers, engineers, or emergency response teams depend on the ability to analyze information contained in the data. Communication is more efficient when it is connected with knowledge of the audience. For example, a search conducted in such companies as Amazon.com, Netflix, or eBay results in suggestions about products that users would enjoy most. It is based on an averaging of how users navigate the database.

Visualization means the ability to locate information and explore its structure; it results from creating and manipulating graphic images directly from the data. This way, visualization is an efficient tool that assists graphic designers, artists, and other practitioners in visual ways of thinking about creative projects, strategies, and performance. Visualization serves a wide range of professionals performing scientific research and presentations, creating communication media art and installations, drafting architectural projects, designing newspapers and magazines, working on website design, or originating animated video or film projects. Different types of users apply visualizations to understand how data analyses and queries relate to each other. According to Fox and Hendler (2011), visualization should serve not only as an end product of scientific analysis, but also as "an exploration tool that scientists can use throughout the research life cycle. New database technologies, coupled with emerging Web-based technologies, may hold the key to lowering the cost of visualization generation and allow it to become a more integral part of the scientific process."

From simple charts and data graphics to the 3D computer generated virtual reality environments happening in real time with human interaction possible through input/output devices, visualizations let us fly around the organized data, comprehend, and make decisions. Francis Marchese (2011) indicated that tables of the past acted as visualization modalities. Many kinds of tables that have been preserved since ancient times, such as the Near East Akkadian clay tablets, Sumerian accounting tables, Aztec calendars, or the Egiptian stele Rosetta Stone, as well as the medieval chronicles, canon tables, and calendars are representations of some of early genres in information visualization. Analysis of these tables demonstrates the constant need to visualize information, which had transformed the ways in which it has been communicated and used.

Figure 5. An example of data visualization: A color wheel (© 2012, A. Ursyn. Used with permission)

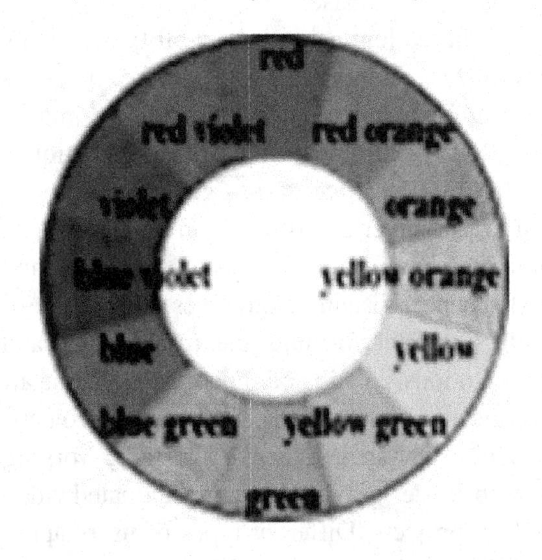

According to Hartman and Bertoline (2005), computer graphics learning environment takes advantage of a learner's ability to quickly process and remember visual information. Technological world uses graphics "to plan, produce, market, and maintain goods and services", because about 80% of sensory input comes from our visual system (Hartman and Bertoline 2005: 992-997). Producing good graphics requires knowledge based on color theory, projection theory, cognitive visualization, and geometry, used to communicate and store information, solve problems, and affect through the senses the human experience. Therefore, Hartman and Bertoline postulated that a body of knowledge called Visual Science – defined as the study of the processes that produce real images or images in the mind – should be studied, practiced, and scientifically verified as a discipline. The authors listed three major categories of visual science: geometry, spatial perception, and imaging. It would have its own knowledge base, research base, history, and public perception, resulting in considerable improvement from the time when Edward Tufte complained that sophisticated graphics on slides or PowerPoint presentations would not communicate effectively (Tufte, 2003). Below is an example of the use of data visualization in a classroom: an art teacher presents the data about color, and provides the students with data visualization in the form of a color wheel annotated with the color pigment names (Figure 5).

See Table 10 for Your Reaction and Visual Answer.

2.3. Information Visualization

Information visualization, considered novel language for visual communication, helps to explore data, understand its complexity, communicate and navigate, for instance, on the web. With computer programs, abstract data is identified and gathered, selected and transformed into pictorial form that makes less difficult human interactive exploration and understanding. Then textual labels and related information are combined with visuals to be published online. Human language and thought are metaphorical; to communicate knowledge, visualizations provide two-dimensional and three-dimensional metaphors, familiar and understandable in social and cultural terms. They are used for navigation and communication in the dynamic, interactive way, often in real-time.

Information visualization is often characterized as representation plus interaction. The data and information visualization techniques include, interaction techniques, which allow modifying what and how the users see the data space that may encompass a screen, data value and structure, attribute, object, or visualization structure. According to Ward, Grinstein, & Keim, (2010, p. 315, 333), classes of interaction techniques include navigation; selection of objects; filtering and thus reducing the size of the data; reconfiguring the data; encoding by changing the graphical attributes and reveal selected features; connecting to show relations; abstracting/elaborating to modify the level of detail; and hybrid combinations of the above techniques. Interactive visualizations help understand dynamic systems that form the

Table 10.

Your Reaction and Visual Answer: Create Data Visualization
Drawing basic shapes like squares, triangles, and circles connected by lines and arrows, and then inserting simple drawings inside of these shapes make us creating visualization of our concepts. Visualization is an effective tool for conveying information. If we intend to make our visualization an effective communication tool as well, we can make it interactive, so people we are talking to could add their drawings to the picture. Even those who think they can't draw may cooperate by adding meaningful ideas and uncovering the hidden relationship. To design data visualization, you may want to present your data graphically: as a list, a circular pie chart, a bar chart, or a diagram showing a number of cases with a feature you have chosen to collect. For example, you may choose to color-code for the sizes of shoes, use textures to tell apart the glasses of water drunk each day, represent planets as amusing drawings, or visually exaggerate temperature changes. You may prefer to visualize data in the form of art; in that case picture in a synthetic way the essence of the data using symbols or finding metaphorical way of data presentation in relation to the background conditions.

objects of our scrutiny: libraries, cirques, theaters, stadiums, etc. For example, visual representation of geographic data allows spatial analysis of the number of schools in various school districts and then focus on our own district. Designers often use diagrams of many kinds to make understanding easier: they may apply timelines, Venn diagrams (collections of circles showing logical relations between data sets), data maps, flow charts, or they may design a semantic network (described later). Bederson and Shneiderman (2003) defined information visualization as the use of interactive visual representations of *abstract* data to amplify cognition. Ben Shneiderman (1996) included theoretical approaches and discussion about visual literacy as a scientific discipline. He proposed the Visual Information Seeking Mantra and also the task-by-data-type taxonomy. According to Craft and Cairns (2005, pp. 110-118), Mantra serves as a methodological guidance to practitioners who seek to design novel systems and many authors reference to Mantra as a holistic approach to visualization design. It offers many implementations, such as information visualization systems, document analysis visualization tools, software visualization tools, or design implementations, methodologies, evaluations, and taxonomies Information visualization application should support seven high level tasks: overview, zoom, filter, detail-on-demand, relate, history, and extract; they were later elaborated in prescriptive way by Craft and Cairns (2005, pp. 110-118):

- **Overview of the Entire Collection:** For the file system – the size of files, for the newspaper – the number and size of articles. 'Overview' can provide assistance in understanding the information that is encoded, a picture of a whole dataset and data relationships, so the user can filter the extraneous information and exclude unimportant aspects of the representation;
- **Zoom on Items of Interest:** 'Zooming and filtering' involve reducing the complexity of the data representation. Zooming is often used as an expression of scalar changes of space, i.e., vantage points rather than changes of discrete screen objects such as text or icons. Zooming involves two cognitive tasks. Zooming-in enlarges smaller data elements of interest, and removes from the visual field or reduces the size of larger data that are not of interest. Zooming-out reveals hidden, often contextual information and integrates the close inspection into a larger understanding;
- **Filter Out Uninteresting Items:** Filtering diminishes complexity in the display but it disturbs the general context. The adjustment of widgets in the interface allows for control of which data points are visible, so that information can be simplified to aid cognition. The dynamic filters allow users quickly see how the changed variable affects the data representation;
- **Details-on-Demand on a Selected Item, with the Level-of-Detail Design:** 'Details-on-demand' (a simple action such as a mouse-over or selection) can limit visual complexity without requiring a change of view or the data context, on a point-by-point basis. Data items can number from dozens to millions. Specific data elements can be identified amongst many or relating attributes of some data points;
- **Relate:** Viewing relationships between the data items with hierarchical parent-child relationships of data, so the user can separate nodes from leaves, and make comparisons among the characteristics of different data objects in the display;
- **History of Actions:** To support undo, replay, and progressive refinement, to allow the user to delete, create, and move files and directories. 'History' means an ability to return to a previous state, recover from a mistake, and replay a sequence of changes in refining the data;
- **Extract Sub-Collections and Query Parameters:** To save the current state of visualization. 'Extraction' provides a means of selecting and applying important findings for use in other computing systems and saving work exploration when the use of the information visualization tools involves lengthy and complex operations (Craft and Cairns, 2005).

According to Ward, Grinstein, & Keim (2010, p. 25), the process of visualization involves mapping from the data to the display. The entered, presented, monitored, analyzed, and computed data are thus translated into more visual and intuitive formats for users For this reason data values and their attributes are used to define graphical objects. With user interaction and collaboration, with the analysis, computational, and synchronization tools, the raw data (information) may be translated into data tables, then visual structures, and finally visualization (multiple views of visual things (Ward, Grinstein, & Keim, 2010, p. 129). Models of information management address the data presentation, mapping, and issues related to the temporal dimension. Card, Mackinlay, & Shneiderman (1999) discussed activities necessary to create visualizations that included transforming raw data into structured data as data tables, and then further converting the data for calculations of meta-data attributes. Then goes visual mapping of the structures essential to the data into the abstract visual structures that can be interactively transformed on a screen by the users as changes in shape, color, size, location, etc. According to Graham (2005), challenges for IV and tools development for IV are identified with respect to: import data; combining visual representations with textual labels; seeing related information; viewing large volumes of data; integrating data mining; collaboration with others, and; achieving universal. It became a norm that large projects with several partners should provide a website publicizing their project. Selecting, transforming and representing abstract data in computer programs facilitate human understanding, interaction, and exploration, often resulting in an interactive, dynamic way of the visual representation. Visualization shows spatially referenced, time-dependent data that can be then modified in real-time, enabling perception of patterns and structural relations in the abstract data. Craft and Cairns (2008) described methods for designing information visualization as:

- Design examples that provide solutions;
- Taxonomies that categorize and list artifacts;
- Guidelines that recommend best practices;
- Reference models that describe how visualization systems work as a whole.

Information visualization helps to explore data, understand their complexity, communicate and navigate, for instance, on the web. With computer programs, most often utilizing Java, abstract data are gathered, selected, and transformed into pictorial form that makes human exploration and understanding easier.

Human language, thought and communication use metaphors that are familiar and understandable in social and cultural terms. To communicate and share knowledge, visualizations provide two-dimensional (2D), tree-dimensional (3D), interactive techniques, and interaction metaphors, which are used for web navigation and communication. Along with the advances in 3D computer graphics, data are often displayed in an interactive 3D virtual space (Ware, 2000). The use of virtual reality allows constructing shared, multi-user virtual environments providing information visualizations (Chen, 2010).

2.4. Knowledge Visualization

Knowledge visualization uses visual representations to transfer knowledge between people, rather than to show only data or information. Professionals, who make knowledge visible and understand patterns of communication, hope to live in a knowledge-based society (with people used to identify, access, share, discuss, apply, and manage information, and thus understand the field as problem solvers) rather than information-based society (of people behaving only as passive learners). Knowledge visualization provides us with visual insight into the data and shares our experiences, perspectives and predictions. It contrasts with information visualization that concentrates on the use of computer-supported tools to explore large amount of data. Knowledge visualization techniques are focused on the users, explanation, and presentation of knowledge in various visual formats. We can see knowledge visualization as a tool to enhance cognitive processes and reduce cognitive load for working memory, and also as a communication tool. Knowledge visualization specialists use computer-based (and also non-computer-based) graphic representation techniques, such as information graphics, sketches, diagrams, images, concept maps, animations, interactive visuals, or storyboards to produce information design solutions concerning readability, simplification, and effectiveness of visual presentations for a wide spectrum of users. Designers co-work with communication science specialists for social network users (such as cell phone users, e-mail archives, criminal networks, or underage audience sensitive messages).

See Table 11 for Your Reaction and Visual Answer.

Martin Eppler describes the domain of knowledge visualization as a "discipline that focuses on the collaborative use of interactive graphics to create, integrate and apply knowledge – particularly in the management context. This emerging approach nevertheless builds on decades of research on using images collaboratively for sense making and knowledge sharing" (Eppler, 2011, pp. 349-353). Visual representations allow organize information and concepts, convey knowledge, amplify cognition, and enhance communication. Examples include conceptual diagrams, knowledge maps, visual metaphors, interactive visualizations, information visualization applications, stories as pictorial visualizations, objects, and

Table 11.

Your Reaction and Visual Answer: Reduce Cognitive Load for One's Working Memory
Pretend you have to prepare a colleague for a trip to far-away place where your colleague may approach communication problems. Create your instruction about a region you are familiar with. Design a visual aid how to manage without knowledge of local language. Your instruction must incur as low cognitive load as possible. Think about creating some simple signs, symbols, icons, and metaphors for the building blocks of your instruction, for example, pictures that correspond with typical actions or situations. This way, your colleague will cope with a reduced cognitive load, just combining ready to use chunks to solve a specific problem, formulate a question, or convey a message. This little booklet may resemble some existing travel books assigning names to objects. For example, a book by Dieter Graf (2009) "Point it" is a collection of photographs presenting typical objects useful in every day situations. It contains no words, can be shown in any place, for instance to ask for directions, places, services, or points of interest. Develop your own aid along another strategy or idea.

sketches. According to Eppler & Burkhard (2007; 2008), knowledge visualization designates all (interactive) graphic means that can be used to develop or convey insights, experiences, methods, or skills. Eppler reviews concepts from different disciplines that help to explain how visualizations can effectively act as collaboration catalysts and knowledge integrators. His review "makes it apparent that many different labels and conceptions exist in very different domains to explain the same phenomenon: the integrative power of visuals for knowledge-intensive collaboration processes. These concepts can be used to compile a list of the requirements of an effective knowledge visualization."

Remo Burkhard (2005) describes knowledge visualization as a field that investigates the use of visual representations to transfer knowledge between at least two people, and suggested applying it as a problem solver, not only a meta-science. Agent-based crowd simulation tools have been primarily used in architecture and urban planning for analytical purposes, such as the simulation of pedestrians or fire escape scenarios in buildings. Because of the high cost-benefit ratio, they were only rarely used for communication purposes, for example for marketing purposes. To bridge the fields of architecture and commercial crowd simulation, Burkhard, Bischof, & Herzog (2008) discuss crowd simulations for analytical purposes and case studies and consider this relevant for architects, urban designers, communication and PR experts, and for researchers in the fields of architecture, knowledge visualization, communication science, and agent-based simulations. Martin Eppler put a stress on the importance of professional knowledge communication as a key activity for today' s specialized workforce. "Knowledge communication thus designates the successful transfer of know-how (e.g., how to accomplish a task), know- why (e.g., the cause-effect relationships of a complex phenomenon), know-what (e.g., the results of a test), and know-who (e.g., the experiences with others) through face-to-face (co-located) or media-based (virtual) interactions" (Eppler, 2005, p. 317). Eppler and Burkhard (2008) identify concepts related to knowledge visualization in a multidisciplinary context and the structures of scientific domains:

Knowledge maps – cartographic depictions of knowledge sources, structures, assets, and development or applications steps, which do not directly represent knowledge, but reference it for easier identification and assessment; knowledge animations – interactive applications that consist of interactive mechanisms that foster the reconstruction of knowledge or the generation of new insights; visual metaphors – graphic depictions of seemingly unrelated graphic shapes that are used to convey an abstract idea by relating it to a concrete phenomenon; heuristic sketches – ad-hoc drawings that are used to assist the group reflection and communication process by making unstable knowledge explicit and debatable; and conceptual diagrams – schematic depictions of abstract ideas with the help of standardized shapes such as arrows, circles, pyramids, matrices, etc.

Computational solutions for knowledge domains require availability of adequate tools, with visualization techniques belonging to the most important ones because of the complexity of data, information, and knowledge within many application areas. Visualization improves communication because knowledge needs to be seen and without successful and sustainable transfer knowledge is meaningless; in the words of Stefan Bertschi (Bertschi et al., 2011), "the act of visualizing is more important than the image itself: medium>message." Visualization enhances visual communication through display of information with the use of combination of letters and numerals, art, signs, and application software.

Stefan Bertschi (Bertschi et al., 2011) states that strategic and operational processes rely on operation and interaction, planning, implementation, project and change processes (Burkhard, 2005). "Knowledge Visualization aims to understand how the sender's intended meaning can be transferred in such a way that it is not distorted in the recipient's perception, therefore allowing effective and efficient communication to take place. ... For most people we may state that complex dependencies and interactions can more easily be understood when illustrated: an intelligent process flow chart makes more sense than a numbered list describing the same process in words." The author stresses that visual thinking helps to make best use of the understanding of others, and advises, "Draw and sketch in front of a live audience, or even better, sketch collaboratively. ... Speaking and listening with your eyes also means making full use of the available methods" (see Lengler & Eppler, 2007, Periodic Table of Visualization Methods). According to Bertschi, visualization improves communication, in particular the interaction around cognitive processes. "Visuals stimulate discussion, and discussion creates knowledge" (Bertschi et al., 2011, pp. 334-335).

Bertschi and Bubenhofer (2005) analyzed metaphorical face of knowledge visualization by examining interrelation of metaphor, linguistic learning, and mediation of knowledge through visualization. Communication calls for language, imaginary or articulated: pictures need a caption. Because language is metaphorical, our thoughts are mostly metaphorical. Thus our communication through visualization is at the same time pictorial and linguistic. Also, it is socially and culturally conditioned, based on frequently used linguistic patterns, for example a 'pie chart' metaphor of market shares uses a metaphor about cutting a pie, and a 'starry night' metaphor can show it in 3D. Thus, according to Bertschi and Bubenhofer (2005), metaphor is a tool of conceptual economy but also a tool of discovery of structures within novel or unfamiliar situations. People think in pictures, so knowledge must be recreated in the mind of the receiver. Modes of thought can be adequately understood if their social origins are familiar and the pictures are defined by language used in a specific culture at a specific time. Visual metaphors combine the creative leap of sketches with the analytic rationality of conceptual diagrams and organize information meaningfully. They can either be natural objects or phenomena (e. g., mountains, tornados) or artificial, man-made objects (e. g., a bridge, a temple), activities (climbing), or concepts (war, family). They fulfill a dual function: organize and structure information, but also convey an implicit insight through the key characteristics or associations of the metaphor (Eppler and Burkhard, 2012, http://www.knowledge-communication.org).

The core of knowledge visualization, which is recognized as an independent discipline (Burkhard, 2005), is all about communication, with steps such as gathering, interpreting, developing an understanding, organizing, designing, and communicating the information; it is a recursive loop rather than a linear process (Crawford, in Bertschi et al., 2011). Knowledge maps is a subset of knowledge visualization. Andrew Vande Moere states:

In order for information to transform into knowledge, one must share some context, some meaning, in order to become encoded and connected to preexisting experience. In that sense, Knowledge Visualization can be considered as data visualization 'in context' with the "emerging popularity of data visualization in current online media, expectations will inevitably shift from simply delivering information to conveying the causally influencing factors that drive the events in our world today (Vande Moere, in Bertschi et al., 2011, pp. 330-331).

According to Sabrina Bresciani (in Bertschi et al., 2011), knowledge visualization "means mapping concepts graphically, by structuring text and visuals in a meaningful way. Examples include conceptual diagrams, knowledge maps, visual metaphors and sketches. A mind map drawn with pen and paper is a common example of Knowledge Visualization which is not computer generated." Bresciani indicates three main paths for developments in knowledge visualization as:

- Studying and measuring the impact of visualization, especially in emerging forms of collaborative interactions, including visual groupware, group support systems, and social media;
- The diffusion of input devices such as (multi-) touch screens that provides fluid forms of interaction;
- Testing knowledge visualization in new domains such as intercultural communication that can be useful to overcome linguistic and cultural barriers.

Martin Eppler (2011) reviews seminal concepts related to knowledge visualization coming from different domains, and then compiles a list of the requirements of effective knowledge visualization, showing the theoretical and practical implications derived from five principles:

- **Visual Variety:** Visual vocabulary to express ideas through various ways;
- **Visual Unfreezing:** Re-elaborating the captured and frozen visuals to be changed again;
- **Visual Discovery:** Connecting elements in new ways and detecting new patterns;
- **Visual Playfulness:** Inviting to change perspectives, assume new roles, immerse in the collaborative effort, and reframe issues creatively;
- **Visual Guidance:** Providing a clear 'road-map' of how it should be iteratively populated or completed.

George P. Lakoff, a cognitive linguistics professor, says that conceptual metaphors address cognitive abilities to abstract the essence of the idea as they allow understanding an abstract or unfamiliar domain in terms of another, more concrete and familiar domain (Lakoff, 1990). Abstracting would mean removing non-crucial information, in a similar way as in an optimized file or in a compressed video where the computer outlines the shapes and fills them in with a sample picture allowing the computer to send only an outline and a picture to a destination.

See Table 12 for Your Reaction and Visual Answer.

3. TOOLS AND TECHNIQUES USED IN VISUALIZATION

The pace of growing the number, range, and quality of tools and techniques used in visualization quickens because of advances in technology. Below, data mining, concept maps, knowledge maps, open source intelligence, visual analytics, tag clouds, network visualization, network visualization, web search result visualization, visualization of the semantic web, and tag cloud visualization will be mentioned briefly.

3.1. Data Mining

Data mining is a search for new, implicit, nontrivial, previously unknown, and potentially useful information within data. A data set with M items has 2^M subsets, any one of which may be the one we really want. Our fantastic pattern recognition ability can cut swaths and also extract insights from the visual patterns (Inselberg, 2009). Spatial data mining discovers patterns from large data that may be spatial (such as location, shape, geometric and topological properties) or non-spatial (human age or activity). It allows for analysis of large quantities of data such as stock market, telecommunication, or scientific data in order to discover meaningful patterns and search for new information. It applies visualization techniques for discovery and the transfer of knowledge from a database and often includes interactive techniques and the use of graphical tools.

Data mining allows making inferences, predictions, and decisions made on the basis of reasoning about the data. Data mining, as the process of selecting, preprocessing, and mapping the

Table 12.

Your Reaction and Visual Answer: Create Knowledge Visualization
Make knowledge visualization using data and information gathered till now for your previous activities. First, you may want to expand your knowledge to provide a context for your visualization. Then, create a metaphor that shows the data in an easy way because metaphor is considered a structure of choice for communicating messages. For example, you may pretend to be a visual analyst for a glove company and you are expected to analyze the market from the users' point of view. You may develop a metaphor for client expectations, demands for gloves with particular length and width, store networks database, and other data sources. Draw a funnel where all these data will enter from the top, while results of the analysis will go out from the bottom opening of the funnel: cost allocation, storing capacity planning, reports on planning, resources, workflow for creative development, marketing, visibility, cooperation, etc.
For presenting your knowledge-based statement about future solutions in developing instruction in your school district, design a knowledge map showing a colorful map of your district as an island in the sea of various needs, with geographically presented information about teaching, learning, tutoring, and your conclusions.
For the analysis of the daily amounts of fluids you drink, you may choose a tree metaphor. Your data may be organized into ranks with each level secondary to the one above, for example, as a hierarchy of the tree limbs, branches and twigs. The length and thickness of branches characterize various factors that determine your present state. They may show the variety of drinking options you can choose from, what is their content and how healthy they are, cost analysis, etc., while the roots below may represent your conclusions and decisions about future structure of your liquid diet.
For knowledge visualization about planets you may want to create a still life consisting of a table with a vase with different kind of flowers.
To visualize temperature related climate issues you may create a knowledge map in the form of a city street, where the house sizes, number of stores, displays in store windows, as well as the car colors, speeds and directions will inform the viewers about factors involved in temperature changes, reasons for future drifts, and possible ways to control them.

data, such as facts or cases, to make them more compact, abstract, and useful, makes a first step in a knowledge discovery process. Data mining methods are aimed at determining patterns in the observed big data sets to extract information that is significant but not directly expressed and otherwise unknown, and also at searching for patterns in the large data with spatial attributes (such as location, shape, geometric and topological properties). Both with the statistical approach for random data and logical models for deterministic systems with predictable outcomes, 3D computer graphics and interactive visualization became important tools in developing data mining and data analysis applications. Interactive data mining developers choose appropriate web browser based applications, such as the server-based rendering with Flash and Java to detect relationships or anomalies in the data, classify, cluster, find functions for models, and visualizations to represent the findings. The primary applications are in science, for example for extracting sky-survey images, medicine and health policy, in communication media, marketing, for business, industrial, manufacturing and investment companies, detecting and monitoring fraud, and organizing the large sets of data in information-rich environment of big companies, associations, and institutions. Data mining strategies are also applied in game theory and studies on games and endgames.

3.2. Clustering, Biclustering

Clustering technique (clusters are subsets of observations) is used in data mining for statistical analysis, pattern recognition, and bioinformatics. Biclustering serves for finding subsets in a dataset, with biclusters of different types displayed as the rows and columns in a matrix. For example, in gene expression analysis simultaneous clustering shows both genes and conditions, by displaying rows as corresponding to genes and columns displaying conditions. This technique allows better representation of genes with many functions and many factors that regulate these functions. It can be done with the open source software tools for retrieving and visualizing information (e.g., Gonçalves, Madeira, & Oliveira, 2009). Visualization includes color matrices, expression evolution charts, pattern charts, dendrograms derived from the results, and ontology graphs highlighting biological terms in the Gene Ontology for specific organisms.

Visual analytics methodology presented by Santamaria, Theron, & Quintales (2008) permits biology experts study biclustering results. In the field of bioinformatics, biclustering has been in many instances used for microarray analysis, in particular the microarray data classification. Visualization of interactive representations of biclusters can complement biological analyses. Biclustering provides simultaneous grouping of genes and conditions, overlapping, and determine the transcript of an organism's genes. Visualizing the biclusters (for example, where nodes represent genes or conditions, and edges join nodes that are grouped by one or more biclusters) allows to extract interesting features of the biclustering results, and reduce the time needed for interpreting results. For example, visualization might present biclusters as flexible overlapped groups of genes and/or conditions.

3.3. Concept Mapping

Concept mapping is a knowledge visualization tool for representing the structure of information visually and building intelligible knowledge models in many fields of science, that is useful for many purposes: strategic planning, product development, market analysis, decision making, and measurement development. As described by William Trochim (2006), "concept mapping is a structured process, focused on a topic or construct of interest, involving input from one or more participants, that produces an interpretable pictorial view (concept map) of their ideas and concepts and how these are interrelated." Concept maps reduce cognitive load in learning and improve recall of information. They support the learning about new concepts by actively constructing knowledge. Six steps in designing a concept map include preparing project, generating ideas, structuring ideas (sorting and rating), computing maps (using multidimensional scaling and cluster analysis), interpreting maps, and utilizing maps (Trochim, 2006). The lines or arcs (that can be one-way, two-way, or non-directional) connect nodes representing concepts (displayed usually in boxes or circles) and denote relationships between concepts. Concept maps integrate information and knowledge visualization, thus giving the overview about knowledge and holistic understanding of the relationship between concepts. They enable browsing information resources. The map can serve for spatial presentation for imaging non-spatial relationships. A map metaphor is widely used to visualize non-geographic knowledge domains.

As Bertschi and Bubenhofer (2005) recalled, according to the Wittgenstein's constructivist concept of learning, viewers give meaning to the concept through the way they use it. They gain the first-order knowledge as an awareness of object, and then meta-knowledge as knowledge about things, meanings, associations, and possible improvements. Third-order knowledge involves comprehension or conclusion about meaning: how it is and why it is that way. Such mental processes are always socially or culturally mediated and context-specific. Internal constructions evolve, as culture enters our personal understanding. A source of mediation can be an artifact, a system of symbols, behavior of others, but mostly language (signs and symbols), with meaning making resulting from people's active participation in social and cultural contexts and settings. Knowledge visualization has to be impressive, work identically in all recipients' heads, and has to be conformable to all existing linguistic patterns. Knowledge visualization acts as a mirror of linguistic (or idiomatic) coinage, which in turn is a mirror image of culture and context (Bertschi and Bubenhofer, 2005).

3.4. Knowledge Maps

Knowledge maps is a subset of knowledge visualization that helps to find the knowledge and build assessment of particular sets of information. Thus, according to Burkhard et al. (2005), frameworks for knowledge maps involve questions about their function type (what?), recipient type (whom?), and map type (how?) that may include an experience based heuristic, diagrammatic, metaphoric, geographic, 3D, interactive, or mental map. A search for information usually consists of browsing visually by visiting web pages with their images, and querying – entering individual queries into the web search engines. Knowledge maps often provide features such as houses, stores, or bridges, and their connections, for example by roads, forests, or rivers, and use cartographic symbols. They show changes, interrelationships, and help to design strategies.

A project "My Virtual Town: Communication Links" presents visual presentation of communication media and related concepts and processes existing in an everyday schedule of the project author's activities. After creating this map, the author – the University of Northern Colorado student discovered that he felt like practically living within the Internet (Figure 6).

3.5. Network Visualization, Web-Search Result Visualization

Keyword search is not efficient any more because the size of the data causes information overload. Search results have been mostly presented in rank-ordered lists spread over multiple pages. The user could only view the top 10 ranks in a window and spend more time on searching, surfing, and shifting data than on using them. Visual search machines present information with the use of information visualization, data mining, and semantic web.

Figure 6. Nathan Lowell, "My Virtual Town: Communication Links" (©2004, N. Lowell. Used with permission)

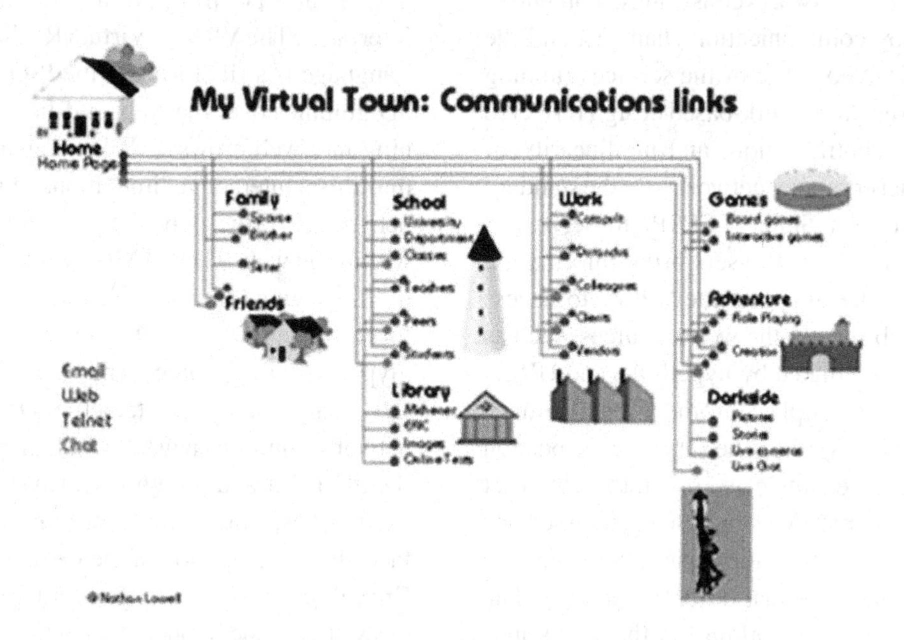

With several formats and programming languages, the semantic web is an enhancement of the World Wide Web that allows finding and analyzing data on the web. In 2007 Feigenbaum, Herman, Hongsermeier, Neumann, & Stephens described the semantic web in Scientific American following an earlier article on this theme written for the Scientific American by Berners-Lee, Hendler and Lassila (2001). According to the authors, "the data format, ontologies, and reasoning software would operate like one big application on the World Wide Web, analyzing all the raw data stored in online databases as well as all the data about the text, images, video and communications the Web contained." Ontologies have been meant here as sets of statements translating information from different databases into common terms and rules that allow software agents to reason about the information described in those terms.

A website must comply with many features that are required or necessary: it must open fast, should be visually appealing, contain all information needed, and must be easy to find on the web. A great part of visited pages are revisited pages, but it is not so easy to find them again, unless a good navigation system is developed. Developing visual ways of navigation on the web became one of the fastest growing information visualization techniques. To make web page hunting easy, pages should be clustered according to the themes, previous visits, interactions with other previously visited pages, then visually organized, many times, with visually catchy metaphors. The web developers and analysts are designing web architecture using many kinds of search machines, managing a large database of documents, and creating web interfaces keeping up with the concept of the semantic web. Some concepts might be useful when surfing the web, such as visualizing and manipulating data in multiple dimensions, Java, interaction techniques, 2D and 3D interaction metaphors, and data mining.

As Lima (2011) states, a tree metaphor serves to convey a variety of topics such as theological events or an encyclopedia's table of contents, and a tree metaphor has been widely used as a classification system. Later on, with the Internet – a global system of networks consisting of computers connected by communication channels, and the World Wide Web – one of the services running on the Internet, a network-based design reflected diversity, decentralization, and nonlinearity of data. Internet computer networks use the standard Internet Protocol Suite (TCP/IP) to serve billions of users. Web browsers bring information resources to the users, as they allow to access from the web servers the web resources, such as text documents linked by hyperlinks and URLs. Browsers (for example, Chrome, Firefox, Safari, or other browsers) settle the server-name portion of the URL (for example, w3.org) into an Internet protocol IP address. Web browsers, put in service by the users, must have the capability to view the modeling languages that describe the data. The web developers are working on the web's universality – the ability to publish regardless of the software, the computer, the language, the wired or wireless system, and the sensory or interaction mode (W3C, 2012), which would enable accessing the web from any kind of hardware that can connect to the Internet – stationary or mobile, small or large. W3C facilitates this blending via the international W3C standards that define an Open Web Platform for web design and web applications developments (such as HTML, CSS, SVG, or Ajax). According to Lima (2011), thinking about and drawing network visualization requires a scientific background, and pragmatic, utilitarian approach, so guiding principles are needed to improve existing methods and techniques.

For the Web-search result visualization, visual exploratory techniques for representations of the Web-search results go beyond charts and graphs and encourage the user interaction and the use of current tools and applications. Many times, visualization over the web requires creating collaborative groups organized to work together, all in different Internet locations, involving the data providers, the visualization service providers, and of course, the users. Search engines on the web can be constructed using the HTML forms, VRML, and a CGI script (the Common Gateway Interface). The VRML, Virtual Reality Modeling Language is a file format – the ISO standard for 3D graphics over the web, and Java applets running in a web browser. This language describes multi-user interactive simulations – virtual worlds networked via the global Internet and hyper linked within the web. The HTML language may serve for displaying information as Image Maps in the form of the 2D and 3D graphics, for example, glyphs (small graphic symbols). 3D animated GIF images or JPEG interactive graphics on the network require a navigation system that combines the 3D input and a high performance rendering capabilities. For example, web-based 3D interface allows navigation of the German Brockhaus Encyclopedia through a geospatial metaphor by browsing of encyclopedia content by geographical context.

Web-search result visualization can also be done in a glyph form. A glyph is a graphical unit that portrays many variables by adapting its many properties; just helping to overview, examine details, and abstract information about the very large, multi-variable data sets. This information-rich way of graphical representation allows displaying more data in a small space. H. Chernoff posed that humans interpret information encoded into facial features and developed in 1973 an early glyph application; he represented data through face expressions: different values could be shown by changes in the face shape, length of the eyes, nose, or mouth, the angle of the eyebrows, etc. Later on, several methods have been developed of automatic mapping from the data to emotional features of the face expression, so values could be shown by changes in the face characteristics. By observing data points represented as a glyph, such as a sphere or a "bubble," one can determine the quality and quantity of the links within a website, because they are displayed in five dimensions: by a position (x, y, and z), size, shape, and color.

3.6. Open Source Intelligence

Another concept that deserves our attention is an Open Source approach to designing, developing, and distributing software, which allows a peer production of a source code for an open-source software that is available for public collaboration. A user has access to a software source code, and may introduce legally, with relaxed or non-existing copyright restrictions, small changes in an already existing code, adapt the program to one's own needs, and thus use, change, or improve the software. Through the Internet, it provides access to various production models, communication paths, and interactive communities. A number of posting services offer opportunities for building a website without knowing a specific software or having expertise in the field; other applications allow even more sophisticated content management, which otherwise would require technical thinking.

Open source intelligence comes from publicly available sources that are not covert or classified. It refers to finding, selecting, and analyzing information (as opposed to covert or classified sources). The raw data have different type (web pages, claims, crime reports), source (Internet, Intranet - a private computer network accessible only to authorized persons, database, etc.), protocol (for example, HTTP - Hypertext Transfer Protocol, HTTPS - Hypertext Transfer Protocol over Secure Socket Layer, or FTP - File Transfer Protocol) and language used.

3.7. Visualization of the Semantic Web

Semantics (from the Greek semantikos, "significant,") defines the nature of meaning in language and its role in a sentence. The interconnection of topics to search for information goes through the associations (relations between topics) and roles linking the topics. Semantics make this network meaningful through making definitions of types for the different object, defined by the topics and used by these roles, topics, associations, and occurrences. Visual semantics is used in design and in presentation of three-dimensional objects, for example, in experimenting with the design products: how to apply in practice semiotics of utility products when describing purpose, function, and qualities of this product. In the '80s the process of developing the product semantics changed the concepts about the optimization, marketing, and aesthetics of commercial design.

Semantic networks are large structures comprising knowledge about interconnected categories, for example, taxonomies of knowledge about animals and plants. An extensive research is conducted in semiotic terms in relation to communication with the use of multimedia. Topic maps are being designed, which bridge knowledge representation and information management by building a structured semantic network above information resources. Topic maps help navigate on the web. The Data Web and the W3C semantic web are aimed to grow as a universal database, a medium for data, information, and knowledge exchange. A Web3D Consortium is working on transforming the web into a series of 3D spaces, realized also by the Second Life – a free 3D virtual world where users can socialize, connect and create using free voice and text chat (http://secondlife.com/).

As stated by Yu (2011, p.1), "Data integration on the Web refers to the process of combining and aggregating information resources on the Web so they could be collectively useful to us." The semantic web has been described by the W3C director Tim Berners-Lee as an evolving extension of the World Wide Web, where information can be expressed in a format that is readable and usable not only for humans but also for the software applications, so they become capable of finding, sharing, combining, and analyzing all the data on the web (Berners-Lee, 2000). Definition placed on the W3C Semantic Web site (2011) states, "Semantic Web provides a common framework that allows data to be shared and reused across application, enterprise, and community boundaries. It is a collaborative effort led by W3C with participation from a large number of researchers and industrial partners. It is based on the Resource Description Framework (RDF)," which is a standard model for defining information and the data interchange on the web. Pieces of data and links that connect them are identified as Universal Resource Identifiers (called URIs). The common Web addresses URLs are special forms of URIs.

In a semantic space, the meanings of words may be represented numerically depending on the frequency distributions of the words. The semantic web analyzes the data on the web – the content, links, and transactions between people and computers, and provides common metadata languages (such as Resource Description Framework RDF) that describe information about web resources to publish information. The Semantic web came to being due to the development of the XML standard. With the semantic web it is possible to share via the Internet large, formalized collections of facts or data derived from study, experience, or instruction. The grammar of the XML derives from the semantics and involves definitions of types for different objects, topics and their occurrences, links between topics, relations and interconnections between topics.

The content of the web can be conveyed not only in languages written by people; it also can be understood and used by software agents that act for users or other programs. It means the content of the web can be understood and used not only by people but also by software, so the semantic web analyzes the content, links, and interactions between people and computers. Soon web will provide an access to a semantic web integrated across huge data resources. It is a universal medium for data, information, and knowledge exchange comprising text documents (mostly in HTML language), images (mostly graphics), audio and video files, animations, and other materials such as data, applications, web pages, e-services, or archived e-mail messages. Thus, the semantic web comprises the philosophy, design principles, collaborative working groups, and technology, advancing the current web progress and allowing users to easily find, share, and combine information. In a semantic space, the meanings of words are represented numerically depending on the frequency distributions of the words; the relations are examined, in a network of concepts using words and symbols as labels for data. Information management of the data and the web itself is transformed into semantic web that bridges knowledge representation and information management. The semantic web is a web visualization tool that builds a virtual network and makes possible sharing data and knowledge by providing metaphors for web navigation and communication through the Internet and hyperlinks. Semantic web builds relationships based on a specific topic, for example on a scientific study, while social networking attracts friends through social connections.

Semantic web and personalized information management employ ontologies – terms and data that individuals or groups use and the relations among those items. Ontologies represent the semantic context of the domain and its classes (Katifori, Torou, Halatsis, Lepouras, & Vassilakis, 2006). Ontology is a format description of a domain that contains entities (classes). Thus, a formal description of a domain can be done with the ontology engineering environment tools for semantic web services. Thus ontologies operate above RDF and software programs called the inference engines operate above ontologies (Feigenbaum et al., 2007). Several ontology visualizations are available through the existing ontology management tools. Protégé and the Protégé–OWL editor are tools that support the web ontology language OWL that is endorsed by the World Wide Web Consortium (W3C) – an organization composed of hundreds of cooperating companies and universities. Protégé uses several kinds of visualization plug-ins for the representation of the ontology. Plug-ins for web browsers can render into a browser window the Scalable Vector Graphics files (SVG, an XML-based language for describing geometric objects).

In philosophy, ontology is a discipline that studies theories about the nature of existence. In information science, ontology is a basic part of the semantic web, a document or file that formally defines terms and relations among terms. Ontologies collect information about objects, their classes, properties, relations, and changes in properties and relations, restrictions, and rules, and thus help the users understand, exchange, analyze or share knowledge of a specific domain. Users can thus understand, exchange, analyze, or share knowledge of a specific domain. Visualization of the structure of ontology systems represents the real world objects. For example, curricular structures and instances are needed to represent individual college courses.

A user may feed a semantic web application with huge collections of data coming from various sources: collections of text documents (composed mostly of HTML), music and other audio files, photos and other images (mostly graphics), web pages, bookmarks, blog posts, library catalogs and worldwide directories, syndication and aggregation of news, software, animations, data, applications, and e-mails or archived e-mail messages. The semantic web tool distills core concepts from these files and then finds the relevant new information. Semantic web involves examining internal content of the relations in a network of concepts, and uses words and symbols as labels for data. Thus, the semantic web comprises philosophy, design principles, collaborative working groups, and technology (W3C Semantic Web Activity, 2011).

3.8. Visual Analytics

Visual analytics focuses on handling massive, dynamic volumes of information through application of related research areas including visualization, data mining, and statistics. Visual analytics combines the use of abstract visual metaphors, mathematical deduction, and human intuitive interaction to detect patterns within dynamic information resources and thus gain knowledge and insight. Huge data sets have millions of records that come from various different sources and provide a dynamic, many times inconsistent and incompatible data. Analytical reasoning allows finding significant, often unpredicted patterns in databases.

Visual analytics is a combination of computational and visual methods in exploration process. Visual and presentational exploration is shifting to visual analytic enquiries that aim at broadening our awareness of knowledge. Researchers developing techniques beyond visualization to analyze data visually, simplify the complexities, reveal uncertainties, and complete incompleteness, advocate a new framework that emerges both from information-rich disciplines like humanities, psychology, sociology, and business and the science-rich disciplines (Banissi, 2011). Integration of interactive visualization tools, data mining, and statistics enable analytical reasoning and collaboration. Visual metaphors, mathematical deduction, and human intuitive analysis detect patterns, and gain knowledge and insight. Visualization of large amounts of financial data for investments can support decision making for investors on the financial market.

As Jern and Franzén (2006) stated, visual analytics requires interdisciplinary science, going beyond traditional scientific and information visualization to include statistics, mathematics, knowledge representation, management and discovery technologies, cognitive and perceptual sciences, decision sciences, and more. By applying visual analytics people derive insight from data, ant then gain and communicate assessments for action. For example, visual analytics may be used for a study of the entire genome of an organism, where data and visual representations are analyzed at many abstraction levels, formats, and scales: molecules, gene networks, signaling networks, and cells, organisms, and ecosystems of different types (Wong and Thomas, 2004). Multidimensional statistical and semantic representations of data are required for national security. Temporal analytics of visual and interactive data enable us to compare the new data to huge digital libraries of information and discover what might have changed and why.

According to Wolfgang Kienreich (in Bertschi et al., 2011), visual analytics is the science of analytical reasoning facilitated by interactive visual interfaces; it combines automated analysis, visual representations, and user interaction in a close loop intended to provide users with new insights (Thomas & Cook, 2006). "Visual Analytics builds upon information visualization to facilitate analytical reasoning by combining automated discovery and interactive visualization" (Sabol, in Bertschi et al., 2011). According to Kienreich (2011), visual analytics emphasizes the use of visual abstractions to represent aggregated information and facilitate the formulation and validation of hypothesis by expert analysts, while knowledge visualization emphasizes the use of visual metaphors, facilitates collaborative dissemination and decision-making by domain experts. Personal digital universes (photo collection, etc.) require multimedia search and retrieval techniques. Human interaction with visualization solutions supports further reasoning about the data. Personal social network requires methods of social network analysis. As stated by Jusufi, Dingjie, and Kerren (2010), networks are used in modeling relational data, often comprised of thousands of nodes and edges, with many attributes to visualize. It is hard to avoid clutter in network elements in social network analysis, software engineering, or biochemistry if using traditional node-link diagrams. Jusufi et al. (2010) apply a tool called Network Lens to visualize such attributes in the context of the underlying network; users can interactively build and combine various lenses by specifying attributes and selecting suitable visual representations.

Kienreich proposes that "Visual Analytics and Knowledge Visualization join forces for analyzing, evaluating and, ultimately, utilizing the wealth of knowledge thus created" with multi-touch surfaces as a driving technological factor for a closer integration. Visual analytics could contribute techniques for the automated analysis of large amounts of information and the closed loop approach with analysis, visualization, and interaction, e.g., with user feedback for consumer sentiment and product quality on a personal level. Knowledge Visualization could contribute design- and user-specific visual representations, e.g., with a map, metro, and aquarium metaphor for users with limited visual literacy (Kienreich, in Bertschi et al., 2011, pp. 332-333), as well as a solar system, or flower metaphor (Van Tonder & Wesson, 2008). Further developments in visual analytics result in significant part from the needs for this type of insight in biology, medicine, environmental research, and national security.

3.9. Tag Cloud Visualization

Tag cloud visualization applications became popular in the social and collaborative software applications. Tag clouds are text-based visual representations of a set of tags; the tag importance is usually depicted by font size, highlighted by a font color, a background color, or tags are arranged semantically, according to their similarity. Interactive tag clouds and tag maps often serve for multidimensional explorations of large datasets, where tags are used to label digital content, for example in photographs (Flickr, www.flickr.com), videoclips (YouTube, www.youtube.com), WWW bookmarks (www.del.icio.us), and Instagram (a free photo and video sharing application for iPhone or Android apps, with web interfaces). Interactive tag maps, tag cloud techniques, and the rapid prototyping methods show geographical space by providing overviews and filtering by text and geography (Slingsby, Dykes, Wood, & Clarke,

2007). Recent trends in social and collaborative software have greatly increased the popularity of this type of visualization. Common approaches to the tag cloud visualization design give consideration to the large whitespaces, overlapping tags, and restriction to specific boundaries. The tag cloud layouts computed by algorithms that address these issues (Seifert, Kump, Kienreich, Granitzer, & Granitzer, 2008) are compact and clear, have small whitespaces, may feature arbitrary convex polygons as boundaries, and thus are useful for many application scenarios.

4. MUSIC VISUALIZATION

Visualization, in accordance with a name, usually relates to a visual form, however there may be other sensory representations. Examples of the non-visual creations are multimodal interactive data presentations (such as user interfaces that can be realized in visual, auditory, or tactile domains), sonification (such as sonification of atmospheric events or human motion), and haptic/touch interfaces (for example, pressure sensitive interfaces). For example, Noritaka Osawa (2004) proposed an auditory method that generates sound passages called "sound glyphs" (depicted in the form of notes) for nodes in hierarchical relationships or constraints. Different nodes at the same depth in this hierarchy are distinguished by melodies. We may find a pervasive use of metaphors in literature, visual arts, and music.

For example, one can envision a continuum encompassing sound qualities: silence, sound, and noise, represented as a grayscale, where white stands for silence, grey for a sound and black for noise. In a project that was a warm-up for the information visualization class projects students were asked to reserve color white for silence, grey for sound, and black for noise. One example of this can be seen in Figure 7.

Figure 7. Anna Melkumian, "The Drop" is a visual response to a sound (© 2006, A. Melkumian. Used with permission)

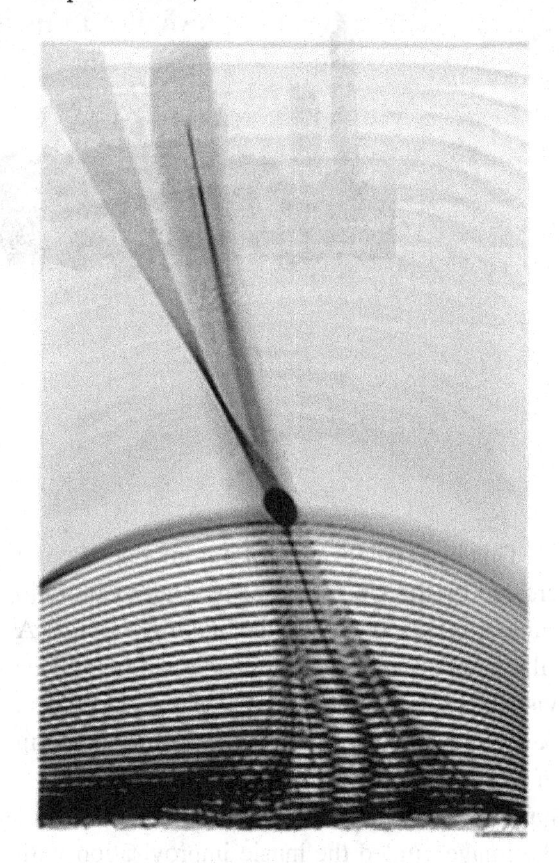

Many believe that metaphors in music theory inform and shape the ways we think about music so musical analyses are not scientific but metaphorical explanations. When one talks about rhythm, timing and tempo, often an analogy with physical motion (like walking or moving) is made.

Student artwork entitled "Two Directions for Two Trombones" is another example of visualization with the use of the visual thinking technique. It is a visual guide for playing music: it visualizes directions for playing a trombone by two people, visually presenting pitch and volume in this performance. The two directions are written on the graph as: change pitch and change volume (Figure 8).

Figure 8. Matthew Tolzmann, "Two Directions for Two Trombones" (© 1996, M. Tolzmann. Used with permission)

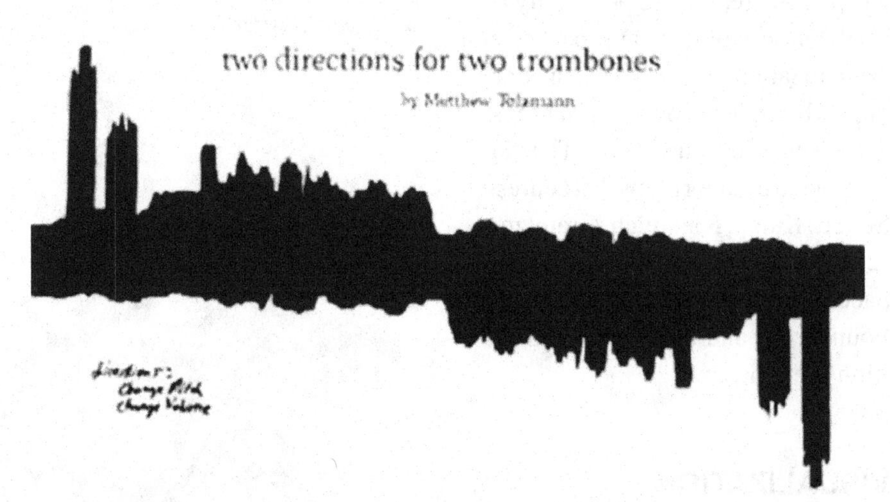

This is a visualization of a trombone playing created by the University of Northern Colorado student taking my "Visual Thinking" course. A silhouette of the Denver downtown served as visualization for interactive music performance. A scanned, transformed, and mirrored drawing of a skyline of the Denver downtown served as a guide for playing two trombones. The outline of the image guided the music improvisation with two directions (hence the title "Two directions for two trombones): (1) to change pitch (upper outline) and (2) to change volume (lower outline). An outline of a Denver, Colorado cityscape was thus used as a metaphor for music.

A visual guide for an orchestra conductor may serve as an example of knowledge visualization in music. Students were asked to visually present material they were studying for another course. The University of Northern Colorado student taking my "Visual Thinking, Visual Images" course created this visualization, instead of making notes in traditional way, when he was preparing himself for examination in conducting orchestra. He achieved clear delineation of time each trombone was played, and a detection of silent periods. Instead of using traditional staves and music notation, he color-coded instruments, and then indicated time, tempo, and the silent periods notations. Thus, he used the composition as a visual guide for himself as an orchestra conductor. He used it for his conducting exam, which he passed with excellent scores (Figure 9).

Graphic representation of four compositions shows:

1. Orchestration of forms composed for flutes, oboes, clarinets, sax, cornets, horns, trombones, euphonium, and tuba presented along the time axis (measures);
2. Visualization of composition dynamics and architecture; and
3. Formal analysis of compositions with graphic presentation of their parts.

5. THE MEANING AND THE ROLE OF VISUALIZATION IN VARIOUS KINDS OF PRESENTATION

Visualization of a complex information structure in a small area poses a conflict between showing a high-level general context and presenting at the same time the low-level details. According Chen (2010), the three general approaches to

Figure 9. Paul Grimes, "Visual Guide for an Orchestra Conductor" (© 1996, P. Grimes. Used with permission)

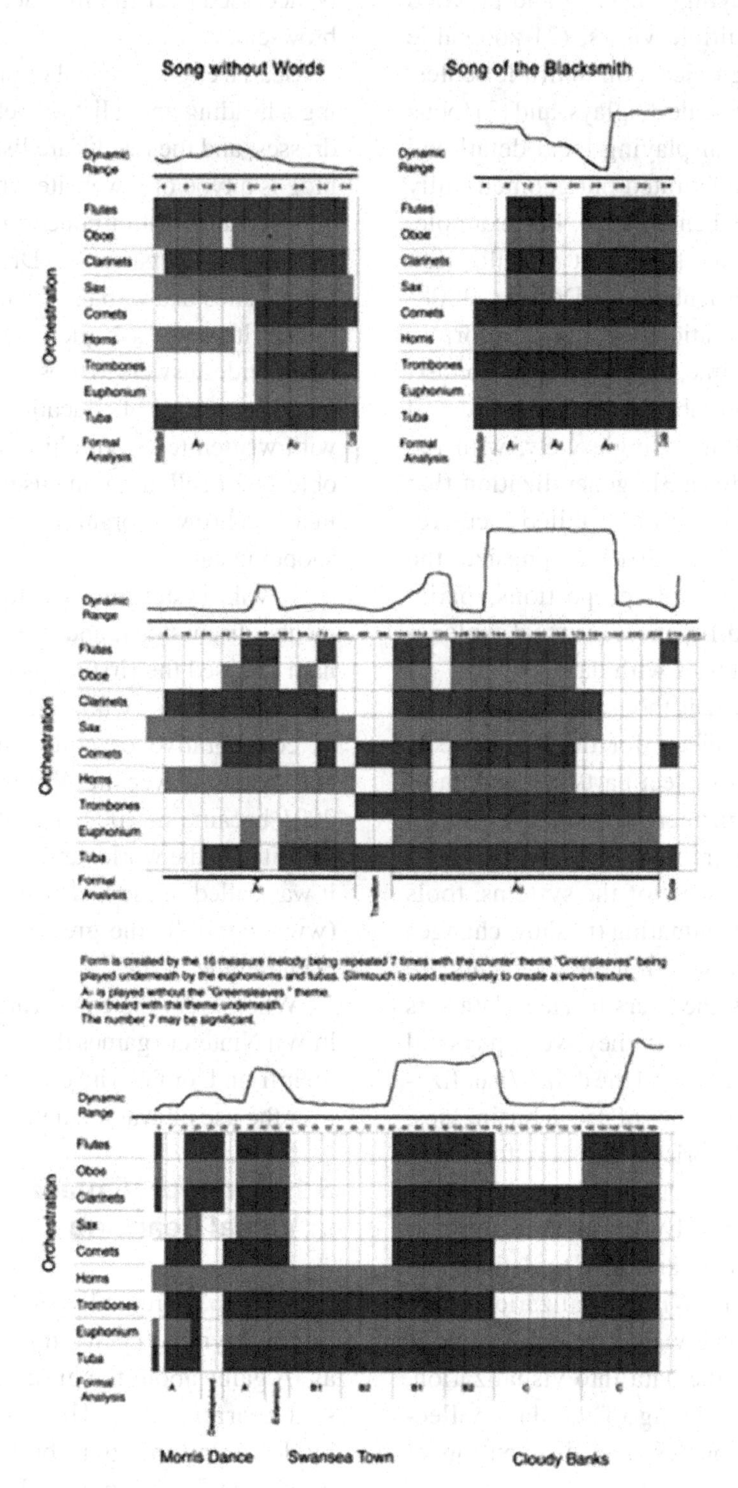

ease the tension are "(1) the overview + detail views – displaying overview and detailed information in multiple views, (2) zoomable views – displaying objects on multiple scales, also known as multiscale displays, and (3) focus + context views – displaying local detail and global context in integrated but geometrically distorted views" (Chen, p. 118). For example, the interactive focus + context visualization (Trapp, Glander, Buchholz, & Döllner, 2008) facilitates the exploration of complex information spaces. At runtime, the generalized context representation is combined with a full-detail representation within a single view, with the use of the lenses for a 3D generalization that direct users attention to the detailed focus region. Nathan Yau (2011, 2013) emphasizes the role of visualizing patterns, proportions, spatial relationships, and differences in visual explanations and telling stories with data.

Gigabytes of data and thousands of web pages that have been produced for the big projects must be analyzed to select parts of the data or artifacts that contain interesting features, and to support different users of the measured systems. Because of the big scale of the systems, tools for visualizing and animating (to show changes over time) the data sets in network events are produced, to allow the users to view data sets in three dimensions as if they were physical objects and to move around the data. Visualizations of different kinds are of use in businesses, companies, and enterprises, and serve for media development and applications; they address the needs of designers involved in software production and management, and help to deliver and manage information. With visualization tools, network performance visualizations are made through rendering the data into visualization, abstraction, and modeling of the data collection, mining the sources, and distribution of information.

A great number of web applications that can be accessed over the Internet serve various web browsers.

Search engines reveal information by displaying a heading and a list of websites and their addresses, and the results are listed as keywords. A blog is a type of a website where one can reveal personal information concerning one's private life from her/his perspective. Discussion boards are textual sites in a web browser. Users register and then write posts, ask question, and then other users provide answers to those questions, often used in various forms of education. Ecommerce sites, with written texts, graphics, and pictures of the objects for sell allow the users to see a webpage in a web browser organized by software called a shopping cart.

A wiki system is used for annotation, comments, discussions, and search. Ward Cunningham released the first software for wiki in 1995. Users can access wiki, add to, or modify, often as collaborative or community websites. The collaborative website Wikipedia, launched in 2001 became in 2007 one of ten most popular websites in the world, and on its 10th anniversary it was called in the German newspaper Die Zeit (www.zeit.de/) "the greatest work of mankind" (Zeit Online, 2012).

When everyone can create characters for use in Wii Nintendo games that look like themselves, their friend, or favorite celebrity, one can ponder over the use of avatars in visualizing knowledge.

5.1. Concept Visualization in Visual Learning

Craig Howie created visual explanation of the internal symmetry in simple or infinite groups as "A game about the group theory" for the Visual Learning class. He made learning playful by designing things to be learned in the form of a game. This project related to the Abstract Algebra class. It is a presentation of the basic concepts of the group theory, a mathematical

Figure 10. Craig Howie, "The Group Theory," work of the University of Northern Colorado student (©2004, C. Howie. Used with permission)

study of internal symmetry in structures called simple or infinite groups. Group theory is the abstraction of ideas common to many areas, so it has many applications in mathematics, physics, and chemistry. For this board game, the author created symbols for the concepts, for example, an algebraic operation as a sort of input-output machine. Data were presented interactively, as the learner might play with item configurations to single out a chunk of data. The basic idea of a Group is laid in the middle and all examples of the Group abstraction were built from there. Color was used to show properties of symmetry in a group Cayley table showing a structure of the group in a square table (Figure 10).

Figures 11, 12, and 13 present another example of learning visually, this time for computer science. It is a student project about the basic structure of a programming language C++ (figures 11, 12, and 13). It was one of projects aimed at learning visually subjects from different disciplines that were learnt by the students at the same time in other courses. The program was written to explain the concept of programming visually to the classmates in "Visual Thinking" class at the University of Northern Colorado. After Ben's presentation many students admitted that programming did not seem that difficult or confusing and they began learning programming actively.

To show a visual learner how the basic structure of C++ works, Ben Hobgood connected the user-defined elements of a simple C^{++} program with visual symbols. Using these visual representations he presented the topic of computer programming to a diverse group of students and received a positive reaction. He used pictures of a cow to describe some simple aspects of the C++ class as an abstract data type. The class is used in a program that asks for the pounds of food that one wishes to feed a cow named Betsy, and it is telling how many steaks can be made from her. This graphical representation of C++ has a source code and an executable to back it up; someone who understands C++ may present it to an audience that has little or no programming experience.

The names of the files are in *italics*.

5.1.1. Example of a C++ Header File or Class Declaration

This is a *declaration* of the *class* called cow or more formally known as a *class declaration*. Defined as an ordinary every day cow. Nothing fancy, just some of the basic stuff about a cow:

cow.h (the declaration of the cow class)

Figure 11. Ben Hobgood, "Example of a C++ Header File or Class Declaration" (©2004, B. Hobgood. Used with permission)

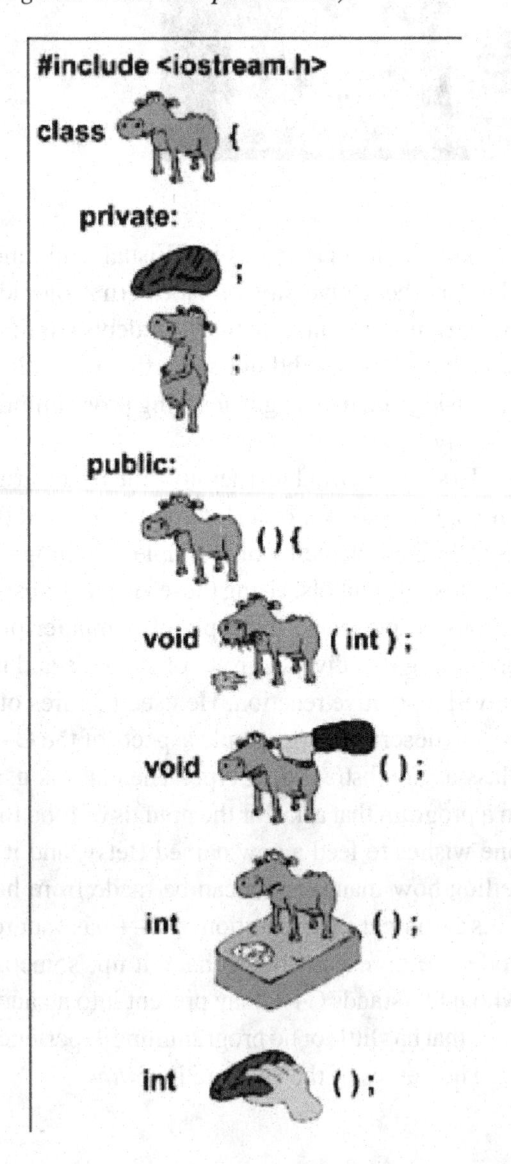

5.1.2. Example of a C++ Class Definition

This is a *definition* of the *class* called cow that was declared above in the cow.h file. More formally known as a *class definition* this file describes exactly how the *abstract data type*, cow, works:

cow.cpp (The definition of the cow class)

Figure 12. Ben Hobgood, "Example of a C++ Class Definition" (©2004. B. Hobgood. Used with permission)

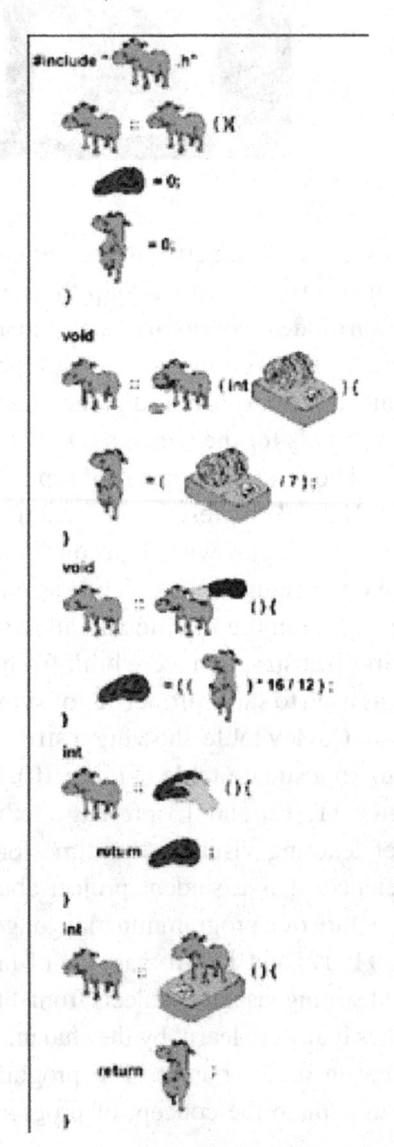

5.1.3. Example of a C++ File that Uses the Cow Data Type

The abstract data type cow that we declared in *cow.h,* and defined in *cow.cpp* is used as a data type to collect information about an *instance* of the cow class called Betsy (black and white cow with a little more character):

betsy.cpp (Implementing the cow class in a program)

Figure 13. Ben Hobgood, "Example of a C++ File that Uses the Cow Data Type" (©2004. B. Hobgood. Used with permission)

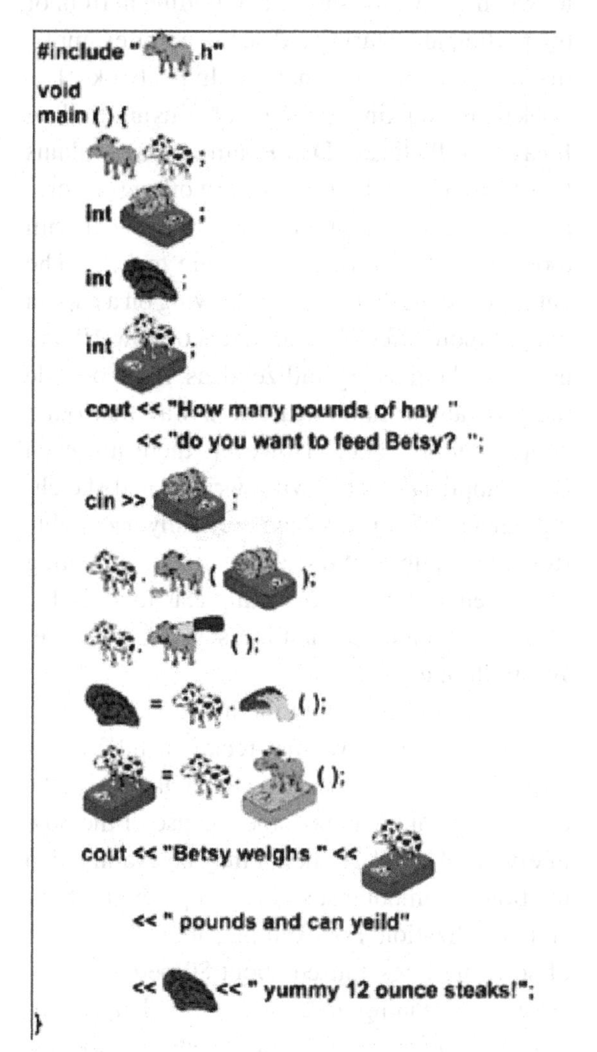

5.2. Visualization in Education

In educational psychology visualization means the ability to create symbols that allow communicating knowledge, conveying meaning, and changing tacit knowledge into explicit one by expressing it as mental representations and images. Visualization helps the learners solve problems by finding patterns and structural relations in graphical displays of data, which is especially valuable when students learn about abstract concepts and ideas, or when it is difficult to see the object because it is too small or too big.

Educational visualization is one of significant applications of knowledge domain visualization, which is often completed using computer-generated simulations to model natural systems and show the concepts, processes, and events to be taught. Phenomena that can be reduced to mathematical data can be simulated on a computer to represent the real world by computer-generated environment. Educational simulations often use a virtual reality environment or interactive computerized simulation. Educational visualization is especially valuable in teaching about small, otherwise invisible structures such as atoms, molecules, or nano technology applications; it is also conducive to making abstract concepts comprehensible. Present-time workplaces need a broad range of skills in the visualization domain. Such skills are highly desired and support a lot of job opportunities. Designers cooperate with the new media artists, electronic artists, visual designers, and digital design managers in designing educational interfaces that interact with learners. They also cooperate with creative directors (or art directors) for the advertising, media, or entertainment industry on integrating educational content with the design, electronic artifacts, and interactive installations to enhance the process of learning and instruction. Instructional designers cooperate with interaction designers, media exhibit designers, and multimedia architecture engineers to make innovative instructional tools and products, by designing novel interfaces that allow the students interact with physical environment where they learn through playing.

Interactive information graphics are considered a tool of choice as an aid in teaching and learning. They combine text, photos, audios, videos, charts, maps, graphs, and illustrations, with such features as multimodality, interactivity, and hypertext (point-and–click method to link topics and graphics), and this makes the main difference between a print graphic and online representation. Interactive graphics focus on a story or topic, and online reader may choose their own navigation path through the information graphics, their own pace, and order. Students interact with icons, the small graphic elements on the screen of the computer that carry information. The simple, less complicated icons that impose lesser cognitive load are understood easier, faster, and thus picked up first (Lang, 2006). The active ways of learning go through picturing various kinds of data in a quick, simple way.

Interactive information graphics make an essential part of journalism, information visualization, and information design. They are important in conveying information in education, online journalism, business management, and technical writing. Visualization and semantic analysis supports understanding of a social network, for example it combines text-based and image-based data analysis with the occurrences search. Visual instruction strategies and visual learning environments lower cognitive load and transform traditional courses to interactive modes. Instructors visualize on-site and online curricular content for e-learning, assist in staff development, work as digital technology developers and consultants, educational game designers, instructional strategists (using data mining), developers and editors of motion graphics and visuals. Visualization techniques help them integrate technology into curriculum and create web-based multimedia instructional materials, and maximize efficiency and appeal of learning and teaching.

Visualization supports learning abstract dynamic concepts, amplifies understanding and retention, and also helps us generate ideas and improve metacognitive skills that enhance awareness and understanding of our own thought processes. Choosing adequate visualization tools helps us to select efficient strategies for instruction or learning. Visualization may also increase our problem solving ability. It was demonstrated in many instances; the most cited example is the renowned story of Dr. Snow who found the source of a cholera epidemic in London (see description below, in section 5.5).

Mutual understanding and interest go even better when simple pictures are unfolding in front of the participants during a discourse, for example, made on the back of a napkin. In his book "The Back of the Napkin: Solving Problems and Selling Ideas with Pictures" Dan Roam (2008) explains how to use visual thinking while working on complex business ideas and provides the reader with tools and rules to facilitate picture making. The author believes that a simple drawing on a napkin may be more effective than Excel or PowerPoint, as it "can help us crystallize ideas, think outside the box, and communicate in a way that other people simply 'get'." However, there are even better applications allowing writing and sketching our visual notes, which are easily accessible from any equipment due to the cloud computing service environment (for example, iCloud (2012) storage and cloud computing service, with over 100 million users).

When we think about the instruction in visualization strategies, we may recall the 80-20 rule, also called a Pareto principle, which states that 20% of specific data may become a cause of the 80% of effects. The 80-20 model may be examined in relation to computer science, computer graphics, and visualization. For example, about 20 percent of software bugs caused about 80 percent of all errors, so cleaning up 20% of errors helped Microsoft improve reliability of products (Rooney, 2002). For ray tracing – a computer graphics technique used in 3D image generation – in partitioning the dynamic scenes a simple approach is often sufficient according to the 80/20 rule: 20 percent of rays intersect 80% of graphic geometry (Slusallek, 2009). When applying metaphorical visualization to transfer knowledge, we employ only the key characteristics of the metaphor, and then the familiar, easy to remember images foster the retrieval of necessary data for teaching or learning the topic. Another 80-20 approach, not clearly related to the Pareto principle, is often taken in the textbook publishing practice, when new materials, methods, or techniques find their place in up to 20 percent of the text, with the remaining content of the textbook remaining traditional in the content and form.

Another trend in art education is creating and keeping art journals (McDonald, 2011) by students and the use of art journals as a tool for teaching students. Art journals created by students contain both their writings and visual notes. Therefore, they are considered not only tools for learning about the creative process but also artistic objects. Art journals found their place on phones and tablets. In some way, a combination of photos, pictures, and descriptive texts may be seen a part of networking.

5.3. Examples of Collaborative Visualization of Interdisciplinary Projects

Researchers using semantic web tools analyze databases serving for designing consumer applications in many areas including health and disease, genetics, industry, among many other applications. Visualization supports understanding of a microarray of data technology used in many biological experiments. For example it helps neurobiologists analyze and understand data about activity of genes (called as gene expression or gene regulation). Interactive exploration of gene combinations bears more information and can lead to new hypotheses about the data (Tominski and Schumann, 2008).

Visualization techniques allow comparison of metabolic networks of bacterial or larger organisms. Large-scale networks such as gene networks, protein–protein interaction networks, or metabolic networks, comprise sets of interconnected reactions that can be modeled as small graphs called metabolic pathways and clusters. Visual mining tools allow mining the newly discovered metabolic networks and comparing them to already known ones using pathways. By defining color-coded drawing algorithms and navigation methods going through levels, Bourqui and Jourdan (2008) visualized the networks and their hierarchies.

One may see many advantages of showing action in pictures. Eriksson, Johansson, & Björndal (2011) point out that even small products are "accompanied with big instructions in several languages that may weigh more than the product itself. The substitute for written instructions is visual instructions…One challenge for the global market is to overcome communication problems of different kinds. The largest communication problem is language; people speak different languages and have limited skills in other languages. This problem is central in manuals and instructions for assembly and installations." One hopeful solution is that pictures can replace verbal instructions by showing action in drawings. However, replacing language with pictures is complex, since language and pictures belong to different symbol systems, and therefore demand different strategies from the designers of instructions. The goal is to overcome differences in cultural traditions; for example, engineering drawings that come from French encyclopedists (Jean le Rond d'Alembert) and James Diderot (Barthes, 1964) are difficult for people speaking Chinese because they use instrumental pictures coming from the 13th century (Eriksson et al., 2011). The EON Reality Corporation (2012) applies virtual reality solutions for simulation-based learning and safety training; an

EON IPortal Tablet provides real time 3D guidance and information about products that support users' knowledge and decision-making. For example they provide a virtual display of steps that need to be performed in order to fix a copier (or copying machine) when one touches a part of the machine that causes problem. EON Coliseum is an online multiuser technology platform where people can present their ideas, communicate complex concepts and collaborate in 3D using rich media such as 3D worlds, slideshows, videos, avatars, voice, chat and interactive 3D objects virtually from anywhere in the world. It is a web conferencing tool that allows you to virtually meet online rather than in a conference room.

Information visualization is often displayed on a physical surface. Models employed in computing include (Tomitsch, Grechenig, Vande Moere, & Renan, 2008):

- Mediatecture (the interface between virtual and physical spaces, such as large-screen displays in public spaces for display of visualizations of data, for example weather, or stock data);
- Ubiquitous computing technology (user-driven applications allowing interaction over a distance);
- Ambient display acting at the periphery of human attention that can be perceived at a glance without causing cognitive load for working memory.

Within an architectural environment, the ceiling may serve as a new physical surface for information visualization. Tomitsch et al. (2008) present a notion of 'information sky' based on the metaphor of the natural sky, historical examples of ceiling art (such as paintings in Altamira cave or Medieval and Renaissance frescos), and recent computing paradigms. The authors define three categories for the information sky. Contextual information, based on the concept of ambient display, provides awareness of information such as weather conditions. Navigation and guidance within architectural environment using navigational cues on the ceiling provides a sense of orientation and directions toward exits. Storytelling – short dialogues informing people about historical events or conversations that happened in that space.

The use of visualization is considered a superior strategy in the communication of business tactics. In an experimental study of Kernbach and Eppler (2010), managers who were exposed to visualization paid significantly more attention to the strategy, agreed more with the strategy, and recalled the strategy better than did subjects who saw text in the form of PowerPoint. According to Bresciani, Eppler, Kaul, and Ylinen (2011, pp. 365-370), "knowledge visualization has the power to increase the effectiveness of the message compared to text, and these benefits replicate across different cultures…Visualizing knowledge means mapping concepts graphically, by structuring text and visuals in a meaningful way." Visual presentations enhance both the cognitive and emotional response to the presented content by creating involvement and engaging employees. As the authors state, "in recent years we are witnessing a growing interest and use of knowledge visualization for communicating ideas and insights. Companies are deploying diagrams and knowledge maps to convey crucial business concepts. Scholars are reporting successful company cases, theorizing on the topic and compiling classifications and best practices." Bresciani et al (2011) examined the effects of visualizations on the attitude toward its content and assessed if visualization effectiveness is universal or culturally bounded. The authors state, "images have an impact also on the emotional attitude of the user, by providing engagement and motivation. Visualizing knowledge is useful for collaborative work: mapping the group dialogue can facilitate the integration of knowledge." The authors conducted an experimental study, comparing text and two knowledge visualization types: a timeline for the Westerners (a sequential and abstract diagram) and the mountain trail for Asians (a holistic and metaphorical representation). The results of 231 subjects in Europe and India demonstrate that knowledge visualization has the power to increase the effectives of the message compared to text, and that these benefits replicate across different cultures. Communicating a business strategy with knowledge visualization, compared to text, has a significant positive impact on the cognitive and on the emotional response of the subjects.

Companies of many profiles use information visualization for aesthetically rendered projects, which provide solutions, taxonomies (that categorize data), guidelines (that recommend best practices) and reference models (that describe how visualization systems work as a whole). Designers work for the industry by cooperating with usability engineers, user interface developers, human factors design engineers, and usability consultants. They specialize in designing and evaluating novel human-computer interfaces, being knowledgeable about how users perceive, learn, and use new products.

An important field in data visualization is product visualization made with the use of many kinds of visualization software that provide vivid, often photorealistic images of products to be sold and marketed. Product visualization serves for designing, producing, marketing, and sales of manufactured articles. This is another implementation of visualization techniques, which involves visualization of software technology and management. Technical visualization of models is useful in product development, 3-D modeling, simulation, and rapid prototyping, among other needs.

Another kind of visualization is the immersive 3D interaction experience. In the immersive multi-wall virtual reality and interactive visualization environment participants are completely surrounded by virtual imagery and 3D sound, and the world comes to them in depth when smart interactive 3D content appears to float in space in and outside of the screen without any glasses; large transparent display screens let you see objects and manipulate them by simply pointing with bare hands from a distance or by voice interaction. In a stereo digital showroom one can get a complete visual feature demonstration, seeing objects on a large 20-foot 3D immersive holographic display, then again on the web. Software for simulation technology creating and publishing interactive 3D rich media is easy to use as the high-end standalone application on a computer or at the Internet. The portable stereoscopic 3D visualization can be set up in minutes for event marketing, sales presentations, and product animations for easy publishing on the web or wireless. Internet users have 3D objects available on the web. Hartman and Bertoline (2006) examined how instruction in a stereoscopic, immersive environment would cast light on assessment of 3D constructs and abilities, as compared to the use of traditional paper-based assessments.

5.4. Professions that Employ Visualization Skills

Present-time workplaces need a broad range of skills in the visualization domain. Such skills are highly desired and support a lot of job opportunities. Interactive information graphics make an essential part of journalism, information visualization, and information design. They are important in conveying information in education, online journalism, business management, and technical writing. Visualization and semantic analysis supports understanding of social networks, for example it combines text-based and image-based data analysis with the occurrences search.

There are numerous areas of work where knowledge about visualization techniques is supportive or required. Knowledge visualization specialists integrate methods from different fields, and work in numerous professions, such as:

- Designers use information visualization as a creative concept for aesthetically rendered projects, which provide solutions, taxonomies (that categorize data), guidelines (that recommend best practices) and reference models (that describe how visualization systems work as a whole);
- Usability engineers, user interface developers, user researchers, human factors design engineers, and usability consultants specialize in designing and evaluating novel human-computer interfaces. They are knowledgeable about how users perceive, learn, and use new electronic products;
- New media artists in the field interactive art, electronic artists, visual designers, and digital design managers design interfaces that interact with electronic appliances and applications;
- Creative directors (art directors) work for the advertising, media, or entertainment industry on integrating the design of special effects, electronic artifacts and interactive installations to enhance the communication and the awareness of a brand;
- Interaction designers, media exhibit designers, technical art directors, and multimedia architecture engineers make innovative products and systems interactive, by designing novel interfaces that allow the users interact with physical environment where they live, work, and play;
- Designers for services visualization, for example, services such as waste removal companies, home video and videogame rental chains, corporations offering video streaming over the Internet, as well as banking services need visualizations informing about customer activities;
- Knowledge visualization specialists use computer-based (and also non-computer-based) graphic representation techniques, such as information graphics, sketches, diagrams, images, concept maps, animations, interactive visuals, or storyboards and produce information design solutions concerning readability, simplification, and effectiveness of visual presentations for a wide spectrum of users;
- Communication science specialists work, for example, for social network users (such as cell phone users, e-mail archives, criminal networks, or underage audience sensitive messages);
- Specialists in mining the web are tracking customer preferences, or analyzing the web content, including text, table, image, audio and video. Data mining can show new business opportunities, for example, assessing client risk when giving a bank loan. Its use in relation to possible terrorist acts has been vividly discussed in the U.S. Senate;
- Data mining analysts represent real-world features such as atmosphere, transportation networks, or hydrology, landscape formation, population flows, or urban development;
- Producers of the Global Positioning System (GPS) military and civilian applications use visualizations fostering data gathering from the earth, earth-circling solar-powered radio transmitters above the earth surface, which support ground-based stations, receivers, and the users. Visualized information is applied to organize and access information in digital and physical worlds;
- Cognitive psychologists create instructional, consulting, or therapeutic visual materials concerning decision making, problem solving, attention, and memory, while working in the academic, business, governmental, or private consulting fields;
- Graphic data designers create visualizations for magazines, business reports, merchandise specialists, marketing companies to present projects over the worldwide web;

- Multimedia production specialists/artists working on creative development and production of visual presentations for college, company, or other programs;
- Interactive web designers and web developers create websites and build application user interfaces to create and maintain company, brand, or school identity;
- Visual mining and data analysts find and integrate the data they need to analyze, integrate educational tools, scientific data interfaces, databases, data graphics, and other visualization tools to create user interface design, cross-platform browsers, software, and interactive web-based applications. An analyst can present and share gained insights online using interactive visualizations;
- Consultants for the humanities and social sciences work, for example, as behavior analysts and convey information visualized in laymen's terms when converse with clients effectively;
- Visual perception and visual communication experts develop visual signage for clients, consult client as portfolio analysts, and conceptualize user interfaces;
- Knowledge management professionals apply visualization techniques to work on business development (insurance and financial services, investment and production companies and corporations);
- Enterprise marketing specialists design visual communication essential in the competitive enterprise-marketing environment, train and coach in applying visual communication tools to improve productivity;
- Visual analysts develop interactive visualization tools, often designed from the users' point of view, to support digital workflow for creative development, marketing, and improve speed, visibility, and cooperation;
- Designers for services visualization. For example, services such as home video and videogame rental chains, corporations offering video streaming over the Internet, waste removal companies, as well as banking services need visualizations informing about customer activities;
- Instructional designers maximize efficiency and appeal of learning and teaching by creating visual materials that lower cognitive load and transform traditional courses to interactive modes. They visualize onsite and online curricular content for e-learning, assist in staff development, work as digital technology developers and consultants, educational game designers, instructional strategists (using data mining), developers and editors of motion graphics and visuals. Visualization techniques help them integrate technology into curriculum and create web-based multimedia instructional materials.

5.5. Well-Known Examples of Visualization that Helped Make Progress in Science

Events and abstract ideas have been visualized from the very beginning of human history, starting from cave paintings and aborigine sand painting. Ancient maps of the sky, earth, and oceans visualized knowledge of the day. A digital artist and information designer Martin Wattenberg (2005) indicated that some of the biggest scientific discoveries have hinged on the turning data into pictures and collected some examples of early visualizations:

- The visual proof of the Pythagorean theorem is one of the oldest visual explanations in mathematics (e.g., http://en.wikipedia.org/wiki/Pythagorean_theorem);
- In1869 Russian chemist Dmitri Mendeleev applied a grid-like, tabular display of chemical elements. He thus created a simple metaphor of known data that made

gaps in current knowledge visually noticeable, so scientists could know where to search for new elements. Ralph Lengler and Martin J. Eppler (2006) designed an impressive interactive presentation of one hundred visualization methods, titled "A Periodic Table of Visualization Methods," where the data are presented on several levels. The general design of this visualization follows the construct of the periodic table of chemical elements; different types of visualization pop up upon touching the screen of the Visual-Literacy.org: http://www.visual-literacy.org/periodic_table/periodic_table.html;

- A Swiss psychologist and psychiatrist Carl Jung (1875-1965) constructed a round house in a low populated area to create a space where he could work without distractions. He created visualization of his scientific ideas on the walls of his building. As a Japanese writer Haruki Murakami (2011, p. 871) described it, "He created paintings himself on the wall. These were suggestive of the development and split in individual consciousness. The whole house functioned as a sort of three-dimensional mandala. It took him twelve years to complete the entire work." For Jungian researchers, it's an extremely intriguing building;
- Santiago Ramon y Cajal was a doctor, also trained as an artist, who won the Nobel Prize in 1906. Using a new staining technique to examine the brain tissue, Ramon y Cajal created precise drawings that are used even today in neuroscience courses. Building on this visual presentation, he discovered that the brain was made of discrete neurons (e.g., http://retina.umh.es/webvision/imageswv/drwCajal.jpeg);
- A map used by Dr. Snow to chart the patterns of an epidemic in London is famous early data visualization. It shows a relationship between the disease and the water sources. The map made possible to locate the source of cholera that had taken more than 500 lives in 1854 (this case was described by Edward Tufte, 1983, 1992, p. 24). At this time people were unfamiliar with a connection between cholera and water. Dr. Snow suspected the cholera outbreak was connected with water, so he plotted the addresses of people who died because of epidemic on a map of London and then added wells on the map. Visualization of two sets of data: deaths and wells laid one over another as a display on a map, provided Dr. Snow with a powerful visual tool that was explanatory because it showed the causal link. This visual tool enabled Dr. Snow to point a well on the Broad Street as a source of the epidemic (http://en.wikipedia.org/wiki/File:Snow-cholera-map-1.jpg). Then the local city council removed the pump handle from the Broad Street well, thus suppressing the epidemic. This kind of graphic visualization is currently used by geographic information systems;
- In 1861 Charles Joseph Minard (1781-1870) designed perhaps the most cited information visualization. He created a map of Napoleon's invasion of Russia in 1812 (http://www.edwardtufte.com/tufte/posters; or http://mappery.com/Napoleon's-Invasion-of-Russia-Map). Minard visualized the Napoleon's campaign using several variables: the size of the army, its location on a 2-D surface (along with the names of rivers they crossed), time and direction of movement of the French army in both directions, temperature on various dates, and also geographical information such as rivers. One copy of this graph might serve as an adequate resource for writing a book about the event.

We are surrounded by visualizations in our everyday routines. On a TV screen, visualizations often advertise new products; visualizations of such abstract qualities as the beauty, novelty, or the prestige that would come from using a product persuade us to buy this product, even if we did not plan to do it. Everyday encounters with visualizations may include, for example histograms of test scores showing average score and score frequency; or cycle diagrams helping us to identify and assess a specific task, identify and review drawbacks, design solutions, and communicate the results.

We learn through visualizations about the atomic structure of matter, as we cannot see anything on the molecular, nanoscale level. Researchers are developing nanoscale visualization tools because we cannot see "images" coming from the scanning, transmission, and scanning tunneling electron microscopes, polarizing, 3D imaging fluorescence confocal microscopes, or any other tools for imaging in nano scale. For example, DNA chains have diameter about 2 nm (nanometers, $2x10^{-9}$ m), while human eye can see wavelengths from about 390 nm to 750 nm. One of the most popular visualizations of the biological data on the molecular, nanoscale level is a spiral, rotating picture of a DNA (deoxyribonucleic acid) molecule. Visualization tools for research on nanostructures are usually developed in cooperation between biology and computer science specialists.

The Cosmos, which is on the other side of the micro-macro range of dimensions, can also be "seen" almost exclusively through the use of visualization tools. Visible spectrum of light constitutes only a small part of the radiation wavelength available for observation. Most of telescopes used in astronomy do not form an image in the same way as telescopes at visible wavelengths. Experts in particular fields of study in astronomy develop visualization methods to be able to communicate their results. In books and online pages we can often also view artists' impressions of the technical imaging. Astronomical space observatories use space telescopes that operate in selected frequency ranges of the electromagnetic spectrum. Therefore, astronomers are specializing in radio (radar) astronomy, sub-milimeter astronomy, far-infrared and infrared astronomy, visible –light astronomy, ultraviolet, X-ray, and gamma ray astronomy. For example, radio telescopes can record wavelengths 1 mm and more, ultraviolet telescopes operate in the spectrum 10 nm – 400 nm, X-ray telescopes work in 0.01 nm – 10 nm frequency range, while gamma ray telescopes operate in less than 0.01 nm wavelengths. As these are not wavelengths available for observation, visualization is a tool of choice to learn and teach about Cosmos.

6. CULTURAL HERITAGE KNOWLEDGE VISUALIZATION

As David Brooks (2011) wrote, "We inherit a great river of knowledge, a flow of patterns coming from many sources. The information that comes from deep in the evolutionary past we call genetics. The information passed along from hundreds of years ago we call culture. The information passed along from decades ago we call family, and the information offered months ago we call education. But it is all information that flows through us. "Our cultural and heritage resources are explored with interdisciplinary research projects involving visual, sonic and algorithmic techniques for interactive visualization (Kenderdine, 2010). The practice of visualization of the digital cultural heritage is well established; it is aimed at developing a digital heritage treasury of archival media, and also reflects contemporary artistic practices, culture, and heritage traditions. Material culture is presented as digital content with the use of augmented visualization offered by mobile technologies, game engines, and interactive environments. Research on interpretative heritage comprises studies of previously lived cultures and recently also the users' role in the formation of cultural knowledge. Pumpa & Wyeld (2006) used a multi-dimensional database used in a 3D game environment as a particularly efficient method to represent the types of narratives used by Australian

Aboriginal people to tell their stories about their traditional landscapes and knowledge practices. They found this tool complementing rather than supplanting direct experience of these traditional knowledge practices.

7. CONCLUSION

Interactive knowledge visualization, and the visual content analysis applied to data and knowledge can be considered significant characteristics of current culture of information system. This chapter discusses the use of visualization techniques in various kinds of presentations applied in visual learning and education, and then the collaborative visualization, professions that employ visualization skills, and provides examples of visualization that helped making a progress in science.

REFERENCES

W3C Semantic Web Activity. (2011). Retrieved June 17, 2012, from http://www.w3.org/2001/sw/

Arnheim, R. (n.d.). *BrainyQuote.com*. Retrieved March 20, 2012, from http://www.brainyquote.com/quotes/quotes/r/rudolfarnh374327.html

Banissi, E. (2011). Preface. In *Proceedings of the Information Visualisation 15th International Conference*. London: IEEE. ISBN 978-1-4577-0868-8. Retrieved June 17, 2012 from http://ieeexplore.ieee.org/xpl/mostRecentIssue.jsp?punumber=6003377

Barthes, R. (1964). *Image, reason, déraison*. Paris: Les Libraires Assoiés.

Beaird, J. (2007). *The principles of beautiful design: Design beautiful web sites using this simple step-by-step guide*. SitePoint Pty. Ltd.

Bederson, B., & Shneiderman, B. (2003). *The craft of information visualization: Readings and reflections*. San Francisco, CA: Morgan Kaufmann Publishers.

Berners-Lee, T. (2000). *Weaving the web: The original design and ultimate destiny of the world wide web*. New York: HarperBusiness.

Bertschi, S., Bresciani, S., Crawford, T., Goebel, R., Kienreich, W., Lindner, M., et al. (2011). What is knowledge visualization? Perspectives on an emerging discipline. In *Proceedings of the Information Visualisation 15th International Conference*, (pp. 329-336). London: IEEE. ISBN 978-1-4577-0868-8

Bertschi, S., & Bubenhofer, N. (2005). Linguistic learning: A new conceptual focus in knowledge visualization. In *Proceedings of the Information Visualisation 9th International Conference*, (pp. 383-389). IEEE.

Bourqui, R., & Jourdan, F. (2008). Revealing subnetwork roles using contextual visualization: Comparison of metabolic networks. In *Proceedings of iV, 12th International Conference on Information Visualisation*. IEEE.

Brath, R. (2009). The many dimensions of shape. In *Proceedings of IV'09, International Conference on Information Visualization*. Barcelona: IEEE.

Bresciani, S., Eppler, M., Kaul, A., & Ylinen, R. (2011). The effectiveness of knowledge visualization for organizational communication in Europe and India. In *Proceedings of the 15th International Conference on Information Visualization*. IEEE. ISBN 978-1-4577-0868-8

Brooks, D. (2011, January 17). Social animal: How the new sciences of human nature can help make sense of a life. *New Yorker*.

Burkhard, R. (2005). *Knowledge visualization: The use of complementary visual representations for the transfer of knowledge – A model, a framework, and four new approaches*. (D.Sc. thesis). Swiss Federal Institute of Technology (ETH Zurich), Zurich, Switzerland.

Burkhard, R., Bischof, S., & Herzog, A. (2008). The potential of crowd simulations for communication purposes in architecture. In *Proceedings of iV, 12th International Conference on Information Visualisation*. IEEE Computer Society Press.

Burkhard, R., Meier, M., Smis, J. M., Allemang, J., & Honish, J. (2005). Beyond Excel and PowerPoint: Knowledge maps for the transfer and creation of knowledge in organizations. In *Proceedings of iV, International Conference on Information Visualisation*. IEEE.

Burmester, M., Mast, M., Tille, R., & Weber, W. (2010), How users perceive and use interactive information graphics: An exploratory study. In *Proceedings of iV, 14th International Conference on Information Visualisation*, (pp. 361-368). IEEE.

Card, S. K., Mackinlay, J. D., & Shneiderman, B. (Eds.). (1999). *Readings in information visualization: Using vision to think*. San Francisco, CA: Morgan Kaufman.

Chen, C. (2010). *Information visualization: Beyond the horizon*. Springer.

Chernoff, H. (1973). The use of faces to represent points in k-dimensional space graphically. *Journal of the American Statistical Association, 68*, 361–368. doi:10.1080/01621459.1973.10482434.

Clayton, S., Pinheiro, V., Meiguins, B. S., Simões, A., Meiguins, G., & Almeida, H. L. (2008). A Tourism information analysis tool for mobile devices. In *Proceedings of iV, 12th International Conference on Information Visualisation*. IEEE Computer Society Press.

Craft, B., & Cairns, P. (2005). Beyond guidelines: What can we learn from the visual information seeking mantra? In *Proceedings of the 9th International Conference on Information Visualization*, (pp. 110-118). IEEE.

Craft, B., & Cairns, P. (2008). Directions for methodological research in information visualization. In *Proceedings of the 12th International Conference on Information Visualisation*. IEEE Computer Society Press.

Davies, A., Henrik, C., Dalton, C., & Campbell, N. (2012). Generating 3D morphable model parameters for facial tracking: Factorising identity and expression. In *Proceedings of the International Conference on Computer Graphics Theory and Applications and International Conference on Information Visualization Theory and Applications*, (pp. 309-318). IEEE. ISBN: 978-989-8565-02-0

EON Reality Corporation. (2011). Retrieved June 17, 2012, from http://www.eonreality.com/products_iportal.html

Eppler, M. J. (2005). Knowledge communication. In *Encyclopedia of Knowledge Management*. Hershey, PA: IGI Global. doi:10.4018/978-1-59140-573-3.ch042.

Eppler, M. J. (2011). What is an effective knowledge visualization? Insights from a review of seminal concepts. In *Proceedings of the 15th International Conference on Information Visualization*. IEEE. ISBN 978-1-4577-0868-8

Eppler, M. J., & Burkhard, R. A. (2007). Visual representations in knowledge management: framework and cases. *Journal of Knowledge Management, 4*(11), 112–122. doi:10.1108/13673270710762756.

Eppler, M. J., & Burkhard, R. A. (2012). *Knowledge communication*. Retrieved June 17, 2012, from http://www.knowledge-communication.org

Eriksson, Y., Johansson, P., & Björndal, P. (2011). Showing action in pictures. In *Proceedings of the 15th International Conference on Information Visualization,* (pp. 403-408). IEEE. ISBN 978-1-4577-0868-8

Feigenbaum, L., Herman, I., Hongsermeier, T., Neumann, E., & Stephens, S. (2007). The semantic web in action. *Scientific American, 297,* 90–97. doi:10.1038/scientificamerican1207-90 PMID:17894177.

Flowing Data. (2012). Retrieved March 24, 2012, from http://flowingdata.com/category/visualization/infographics/

Fox, P., & Hendler, J. (2011). Changing the equation on scientific data visualization. *Science, 331,* 705–708. doi:10.1126/science.1197654 PMID:21311008.

Friendly, M. (2009). *Milestones in the history of thematic cartography, statistical graphics, and data visualization.* Retrieved March 8, 2012, from http://www.math.yorku.ca/SCS/Gallery/milestone/milestone.pdf

Girot, C., & Truniger, F. (2006). The walker's perspective: Strategies for conveying landscape perception using audiovisual media. In *Proceedings of the 10th International Conference on Information Visualization.* IEEE Computer Society Press.

Gonçalves, J. P., Madeira, S. C., & Oliveira, A. L. (2009). BiGGEsTS: Integrated environment for biclustering analysis of time series gene expression data. *BMC Research Notes, 2,* 124. doi:10.1186/1756-0500-2-124 PMID:19583847.

Graf, D. (2009). *Point it: Traveller's language kit - The original picture dictionary - Bigger and better.* Graf Ed.

Graham, D. (2005). Information visualization theory and practice. In *Proceedings of iV, International Conference on Information Visualisation,* (pp. 599-603). IEEE.

Guilford, J. P. (1968). *Intelligence, creativity and their educational implications.* San Diego, CA: Robert Knapp, Publ..

Hartman, N. W., & Bertoline, G. R. (2005). Spatial abilities and virtual technologies: Examining the computer graphics learning environment. In *Proceedings of 9th International Conference on Information Visualisation.* IEEE.

Hartman, N. W., & Bertoline, G. R. (2006). *Virtual reality-based spatial skills assessment and its role in computer graphics education.* Retrieved March 19, 2012 from http://www.siggraph.org/s2006/main.php?f=conference&p=edu&s=44

iCloud. (2012). Retrieved June 17, 2012 from https://www.icloud.com/

Inselberg, A. (2009). *Parallel coordinates: Visual multidimensional geometry and its applications.* New York: Springer.

Jern, M., & Franzen, J. (2006). GeoAnalytics - Exploring spatio-temporal and multivariate data. In *Proceedings of the Information Visualisation (IV) 10th International Conference,* (pp. 25 – 31). IEEE.

Jusufi, I., Dingjie, Y., & Kerren, A. (2010). The network lens: Interactive exploration of multivariate networks using visual filtering. In *Proceedings of the Information Visualisation (IV) 14th International Conference,* (pp. 35-42). IEEE.

Katifori, A., Torou, E., Halatsis, C., Lepouras, G., & Vassilakis, C. (2006). A comparative study of four ontology visualization techniques in protégé: Experiment setup and preliminary results. In *Proceedings of the Information Visualisation (IV) 10th International Conference,* (pp. 417-423). IEEE.

Kenderdine, S. (2010). Immersive visualization architectures and situated embodiments of culture and heritage. In *Proceedings of the 14th International Conference on Information Visualization,* (pp. 408-414). IEEE.

Kernbach, S., & Eppler, M. J. (2010). The use of visualization in the context of business strategies: An experimental evaluation. In *Proceedings of the Information Visualisation (IV) 14th International Conference*, (pp. 349-354). IEEE. ISBN 978-1-4244-7846-0

Lakoff, G. (1990). The invariance hypothesis: Is abstract reason based on image-schemas? *Cognitive Linguistics*, *1*(1), 39–74. doi:10.1515/cogl.1990.1.1.39.

Lang, S. B. (2006). Merging knowledge from different disciplines in search of potential design axioms. In *Proceedings of the 10th International Conference on Information Visualization,* (pp. 183-188). IEEE.

Lengler, R., & Eppler, M. (2006). *A periodic table of visualization methods*. Retrieved March 26, 2012, from http://www.visual-literacy.org/periodic_table/periodic_table.html

Lengler, R., & Eppler, M. (2007). *Towards a periodic table of visualization methods for management*. Retrieved March 26, 2012, from http://www.visual-literacy.org/periodic_table/periodic_table.pdf

Lima, M. (2011). *Visual complexity: Mapping patterns of information*. New York: Princeton Architectural Press. ISBN 978 1 56898 936 5

Loizides, A. (2012). *Andreas Loizides research home page*. Retrieved February 12, 2012, from http://www.cs.ucl.ac.uk/staff/a.loizides/research.html

Loizides, A., & Slater, M. (2001). The empathic visualisation algorithm (EVA), Chernoff faces re-visited. In *Conference Abstracts and Applications, Technical Sketch SIGGRAPH 2001*. IEEE.

Loizides, A., & Slater, M. (2002). The empathic visualisation algorithm (EVA) - An automatic mapping from abstract data to naturalistic visual structure. In *Proceedings of iV02, 6th International Conference on Information Visualisation.* IEEE.

F. M. Marchese, & E. Banissi (Eds.). (2013). *Knowledge visualization currents: From text to art to culture*. London: Springer-Verlag. doi:10.1007/978-1-4471-4303-1.

Marchese, F. T. (2011). Exploring the origins of tables for information visualization. In *Proceedings of the Information Visualisation 15th International Conference,* (pp. 395-402). London. ISBN 978-1-4577-0868-8

Masud, L., Valsecchi, F., Ciuccarelli, P., Ricci, D., & Caviglia, G. (2010), From data to knowledge - Visualizations as transformation processes within the data-information-knowledge continuum. In *Proceedings of the Information Visualisation 14th International Conference* (pp. 445-449). IEEE.

McDonald, Q. (2011). *Raw art journaling*. Cincinnati, OH: North Light Books.

Murakami, H. (2011). *1Q84*. New York: Knopf.

O'Reilly, T. (2005). *Design patterns and business models for the next generation of software*. Retrieved February 2, 2013, from http://oreilly.com/web2/archive/what-is-web-20.html

Osawa, N. (2004). Visual and sound glyphs for representing constraints. *YLEM, Artists Using Science & Technology, 24*(2).

Othman, A., El Ghoul, O., & Jemni, M. (2012). An automatic approach for facial feature points extraction from 3d head. In *Proceedings of the International Conference on Computer Graphics Theory and Applications and International Conference on Information Visualization Theory and Applications* (pp. 369-372). IEEE. ISBN: 978-989-8565-02-0

Panas, T., Berrigan, R., & Grundy, J. C. (2003). A 3D metaphor for software production visualization. In *Proceedings of the 7th International Conference on Information Visualization* (pp. 314-319). IEEE.

Pumpa, M., & Wyeld, T. G. (2006). Database and narratological representation of Australian aboriginal knowledge as information visualisation using a game engine. In *Proceedings of the iV, 10th International Conference on Information Visualization*. IEEE Computer Society Press.

Roam, D. (2008). *The back of the napkin: Solving problems and selling ideas with pictures*. New York: The Penguin Group.

Rooney, P. (2002). *Microsoft's CEO: 80-20 rule applies to bugs, not just features*. Retrieved March 27, 2011 from http://www.crn.com/

Santamaría, R., Therón, R., & Quintales, L. (2008). A visual analytics approach for understanding biclustering results from microarray data. *BMC Bioinformatics*, *9*, 247. doi:10.1186/1471-2105-9-247 PMID:18505552.

Seifert, C., Kump, B., Kienreich, W., Granitzer, G., & Granitzer, M. (2008). On the beauty and usability of tag clouds. In *Proceedings of 12th International Conference Information Visualisation*, (pp. 17-25). IEEE.

Shneiderman, B. (1996). The eyes have it: A task by data type taxonomy for information visualizations. In *Proceedings of the 1996 IEEE Conference on Visual Languages*, (pp. 336-343). IEEE.

Slingsby, A., Dykes, J., Wood, J., & Clarke, K. (2007). Interactive tag maps and tag clouds for the multiscale exploration of large spatio-temporal datasets. In *Proceedings of iV07, 11th International Conference Information Visualisation*, (pp. 497-504). IEEE.

Slusallek, P. (2009). *Computer graphics, ray tracing III*. Retrieved March 23, 2012 from http://graphics.cs.uni-saarland.de/fileadmin/cguds/courses/ws0809/cg/slides/CG04-RT-III.pdf

Thomas, J. J., & Cook, K. A. (2006). A visual analytics agenda. *IEEE Computer Graphics and Applications*, *26*(1), 10–13. doi:10.1109/MCG.2006.5 PMID:16463473.

Tominski, C., & Schumann, H. (2008). Visualization of gene combinations. In *Proceedings of iV, 12th International Conference on Information Visualisation*. IEEE.

Tomitsch, M., Grechenig, T., Vande Moere, A., & Renan, S. (2008). Information sky: Exploring the visualization of information on architectural ceilings. In *Proceedings of iV, 12th International Conference on Information Visualisation*. IEEE.

Trapp, M., Glander, T., Buchholz, H., & Döllner, J. (2008). 3D generalization lenses for interactive focus + context visualization of virtual city models. In *Proceedings of iV, 12th International Conference Information Visualisation*, (pp. 356-361). IEEE Computer Society Press. ISBN: 9780769532684. DOI: 10.1109/IV.2008.18

Trochim, W. M. (2006). *Concept mapping*. Retrieved March 24, 2012, from http://www.socialresearchmethods.net/kb/conmap.htm

Tufte, E. R. (1983/2001). *The visual display of quantitative information*. Cheshire, CT: Graphics Press.

Tufte, E. R. (1992/2005). *Envisioning information*. Cheshire, CT: Graphics Press.

Tufte, E. R. (2003). *The cognitive style of powerpoint*. Cheshire, CT: Graphics Press.

UCLA Academic Technology Services. (2012). *Stat computing*. Retrieved June 17, 2012, from http://www.ats.ucla.edu/stat/mult_pkg/whatstat/nominal_ordinal_interval.htm

Van Tonder, B., & Wesson, J. (2008). Visualization of personal communication patterns using mobile phones. In *Engineering Interactive Systems: EIS 2007 Joint Working Conferences EHCI*. Springer.

Vande Moere, A., & Boltzmann, L. (2009). Beyond ambient display: A contextual taxonomy of alternative information display. *International Journal of Ambient Computing and Intelligence, 1*(2), 39–46. doi:10.4018/jaci.2009040105.

Visual-Literacy.org. (2012). Retrieved March 27, 2012, from http://www.visual-literacy.org/pages/documents.htm

Voigt, R. (2002). *An extended scatterplot matrix and case studies in information visualization*. (Unpublished Masters thesis). Virtual Reality and Visualization Research Center, Vienna, Austria.

Ward, M., Grinstein, G. G., & Keim, D. (2010). Interactive data visualization: Foundations, techniques, and applications. Natick, MA: A K Peters Ltd. ISBN 1568814739

Ware, C. (2000). *Information visualization: Perception for design (interactive technologies)*. Morgan Kaufmann.

Wattenberg, M. (2005). *From data to pictures to insight*. Retrieved March 18, 2006, from http://www.alphaworks.ibm.com/contentnr/introvisualization

Wikiversity. (2012). *Level of measurement*. Retrieved March 18, 2012, from http://en.wikiversity.org/wiki/Level_of_measurement

Wong, P. C., & Thomas, J. (2004). Visual analytics. *IEEE Computer Graphics and Applications, 24*(5), 20–21. doi:10.1109/MCG.2004.39 PMID:15628096.

Yau, N. (2011). *Visualize this: The flowing guide to design, visualization, and statistics*. Wiley.

Yau, N. (2013). *Data points: Visualization that means something*. Wiley.

Yu, L. (2011). *A developer's guide to the semantic web*. Berlin: Springer-Verlag. doi:10.1007/978-3-642-15970-1.

Zeit Online. (2012). Retrieved April 6, 2012, from http://twitpic.com/3pesre

Chapter 12
The Intelligent Agents: Interactive and Virtual Environments

ABSTRACT

Tools available for enhancing and sharing knowledge include intelligent agents, Augmented Reality (AR), and Virtual Reality (VR), among other solutions and paradigms. Collaborative computing became possible due to the advances in social networking, collaborative virtual environments, multi-touch screen-based technologies, as well as ambient, ubiquitous, and wearable computing. Examples of simulations in various domains include virtual computing machines, transient public displays of the data, mining for patterns in data, and visualizations of past events with the use of immersive technologies, virtual reality, and augmented reality. Further discussion relates to the tools for creating and publishing interactive 3D media and the Second Life culture.

INTRODUCTION

Intelligent agents can recognize changes in their environment and then comprehend and react accordingly to amplify their chances of success. Interacting intelligent agents such as software, computers or robots may form intelligent systems. Applications of multi agent systems involve multiple venues such as managing complex structures, e.g., transportation, traffic, or parking; detection and solving faults in industrial systems; and managing machine-to-machine systems, among many other implementations (Demazeau et al., 2012).

Studies on artificial intelligence, especially machine learning is resulting in constructing systems that can learn from the data they search for and receive. Biology inspired computational intelligence studies pertain to adaptive mechanisms that enable or facilitate intelligent behavior in complex and changing environment (Engelbrecht,

DOI: 10.4018/978-1-4666-4703-9.ch012

2003). Paradigms and methods developed in the field of computational intelligence involve the developments in artificial neural networks, fuzzy systems, and evolutionary computing, for example, building algorithms based on swarm intelligence or artificial immune systems. Researchers make the mind models and map them to the corresponding brain parts, thus attempting to unify natural and artificial intelligence (Weng, 2012).

People can create virtual environment in their own minds without any technology. Picture books, theatre, television, and movies extract abstract information of the story into concrete visual scenes to enhance experience and evoke emotions. Readers and listeners convey the words into virtual environment in their own minds. One can achieve simulation with or without immersion in the simulated world. In the desktop-type 3D virtual environment, the real world is not blocked out from the user who can see through a window and communicate by mouse and keyboard. Virtual objects that are not presented in life size do not create the illusion of immersion.

Developing virtual reality, ambient intelligence, multimedia, and robotics becomes crucial for technological development for both corporations and universities. Apart from data graphics displayed in various graphical ways, tools used for work and entertainment include augmented reality and virtual reality; they link science, engineering, technology, and art in service of real-time, immersive, and 3-dimensional interaction with the collaborative and intelligent environments, where people can interact with each other and with artificial agents.

AUGMENTED AND VIRTUAL REALITY

On the spectrum between virtual reality and the real world, augmented reality is closer to the real world. The hardware and software is designed as a desktop type or an immersive one that visually and physically isolates users from the environment and awareness of reality. Avatars, characters created in artificial environment represent the users who may control them with head-mounted displays and gloves. Without a need of any glasses smart interactive 3D content seems to float in space in and outside of the screen. Visual displays, body and head tracking interfaces, aural (acoustic) and haptic (force and touch) feedback, and peripherals such as acoustic and haptic displays provide the illusion of immersion. Data communication goes through the wired, wireless, stand-alone or networked channels.

Augmented reality adds graphics, sounds, haptics (force and touch), and smell to the natural world. Integrated solutions combine the single-sense display types or provide virtual stimuli to several sensory modalities; visual, audio, haptic, or, less frequently, smell and taste (Coquillart, 2012). Augmented reality interfaces build applications with an audio-visual augmentation, realistic object augmentation (e.g., with augmented shadows), image augmentation, textual annotations, and audio augmentation (Liarokapis, White, & Lister, 2004). Gimeno, Morillo, Ordu, & Fernández (2012) developed the software framework, an easy-to-use augmented reality authoring tool for non-programming users, to develop the AR prototypes for industrial applications; the time needed for developing the prototypes was much lower than with computer graphics programming.

Virtual reality links technology and art in service of real-time, immersive, and 3-dimensional interaction with computer-generated environments. Augmented, virtual reality, and Second Life play an increasing role in lives of participants. Projects may refer to various configurations and visual appearances, involving the use of light, sound, such as music and voice including songs, haptic experiences, touch, and gesture. Particular solutions may be also attained with the use of avatars, telecasting, TV, groupware implementations, social networking, You Tube, and any other ways of social activities. We may think about sharing knowledge as a performance act. It may pertain to the performing arts, acting for a theatre, dancing, performing sport activities, filmmaking and recording on video, may involve costume design, as well as creating comics and animations. We may also consider artificial life (A-Life), which models living systems, ecosystems and life in general.

Working environment can be enhanced by possibilities of talking to a computer or touching a screen instead of typing, or even marking hand movements by a virtuoso to attain master performance on a digital musical instrument. One can imagine no more instances of a carpal tunnel syndrome; the success of the performance would depend on the brain–hand coordination of the musician, without application of strength. One may also contemplate playing chamber music with a small ensemble of musicians who actually reside in far away places. Some scientists approach the virtual worlds as a combination of both physical and electronic environments that focus on processes and actions rather than on objects and perceptual signs. Models created by physical modeling or computer simulation allow for experimenting in fictional scenery. Computer models serve to simulate a physical phenomenon, for instance atmospheric conditions such as snow, rain, or wind. Most of the major feature animation studios and the film industry incorporate simulation in their production process.

Virtual Music

Virtual music has been described as acoustic or psychoacoustic phenomenon occurring in the inner ear and the brain. According to the music theory professor Robert C. Ehle (2012), the 17th century violinist Giuseppe Tartini (who in 1753-4 wrote Trattato di Música) discovered the existence of the difference tones whose frequencies, resulting from the additive or subtractive interference between the pairs of sound waves, produced complex chords. Periodicity pitches, the tones that result from the linear mixing of acoustic waves, are phantom pitches that have no energy at their frequencies, but many writers claim that we can hear them. The less known kinds of virtual music are psycho-acoustical phenomena such as virtual pitches (where one's brain interprets tones in music that don't actually exist, in contrast to a spectral pitch, which is a tone that physically exists), subjective tones (two single-frequency tones present in the air at the same time, which interfere with each other and produce a beat frequency, and thus other tones), missing or phantom fundamentals (when the brain perceives the same pitch even if the fundamental frequency is missing from a tone), Schouten's residue pitches (evoked by harmonic complex tones and corresponding to the fundamental frequency of which their component frequencies are all integer multiples), and others. Experiments with the use of the imaging techniques such as fMRI (functional magnetic resonance imaging), PET (positron emission tomography), MEG (magnetoencephalography neuroimaging), and CAT (computerized axial tomography) have shown that the amygdala, the brainstem, the cerebellum, and other lower parts of the brain light up in scans in response to virtual music. These phenomena are evidence of emotional response in the centers of emotion in the brain, often subliminally, subconsciously, and unknowingly to musicians. Ehle suggests there is a wide range of acoustical and psychoacoustical experiences that modulate emotion; the emotional component in the music of some composers results from these unwritten and unaccounted for pitch phenomena.

Virtual Reality Immersive Systems

In a computer-generated artificial environment people can be immersed, and they can interact. One has the complete sense of presence in an immersive visualization environment, where everything moves and sounds like the real thing. Virtual reality techniques provide an interactive experience in several senses at once. Interactive fiction can offer an enjoyable reading experience. In the immersive virtual reality system, the user is visually and physically isolated from the real environment and feels blocked from real life due to the use of the visual display, HMD (head mounted device), and such peripherals as body and head tracking interface, 3D acoustic display, and haptic (force and touch) feedback (Xu & Taylor, 2000). Digital showroom provides a large 20-foot 3D immersive holographic display with connection on the web.

Intelligent environments and virtual reality can be linked to the human computer interaction domain through user interfaces. Easy to use software for creating and publishing interactive 3D rich media is available online. Helmets and gloves, along with speech and gesture recognition support older kinds of interface such as a keyboard and a mice or a light pen. Intelligent environments with multimodal interfaces allow controlling appliances. Thus, one may control with hand gestures without words movements of robot arms, or use one's voice or a pointer to operate a projector or a computer. One can interact with the use of animations providing a lifelike behavior of characters in virtual environment. In some intelligent environments appliances may be controlled by interfaces, for example, a life-size wooden puppet serves as an interface, robot arms are controlled by hand gestures, or computer functions are controlled by voice and pointer. In an immersive virtual reality, the user is represented in an artificial environment by an avatar character controlled by the user with a head-mounted display and a glove.

Examples of the Virtual Reality Immersive Simulations

Virtual reality simulations can be applied in every field of science and engineering, for research or learning and training. For example, a pilot training program uses a model of real conditions in an aircraft or spacecraft. People learning to fly are completely surrounded by virtual imagery and 3D sound in an aircraft flight simulator equipped with a pilot's chair, a headset, control panels, and a sight of animated runway. Real actions performed by pilots in extreme conditions are recreated for purpose of training.

Many times simulation units take form of a room, such as CAVE (Cruz-Neira, Sandin, DeFanti, Kenyon, & Hart, 1992) and also EON Reality ICUBE™ (www.eonreality.com/products_icube.html) immersive systems. In both cases a large theatre is sited within a large room. Immersive virtual reality environments projectors are directed to three, four, five or six walls made up of rear-projection screens in a room-sized cube, to create and display a one-to-many visualization tool that utilizes large projection screens, often via mirrors. Users wear special glasses to see 3D graphics. With the use of electromagnetic sensors visitors can see objects apparently floating in the air and can walk around them, getting a proper view like in reality. CAVE was invented by Thomas A. DeFanti and Daniel J. Sandin, and developed in 1992 at the University of Illinois, Chicago. The first CAVE was open to the public in 1996, when the City of Linz opened a Museum of the Future dedicated to art and technology. According to the CAVE creators, the name is also a reference to the allegory of the cave in Plato's Republic where a philosopher contemplates perception, reality and illusion.

Digital museums in a 3D virtual environment provide visitors with information about museum content and context, services, and personal recommendations. The current framework of content management system is integrated in the museum

3D virtual environment (Sookhanaphibarn, K., & Thawonmas, R. (2010). Content management systems – computer programs allowing publishing, editing, modifying, storing content of system components, websites, documents related to commercial companies, and managing workflow interactively with the users. A unified content strategy identifies content requirements and manages it for reuse, thus reducing costs of creating, managing, and distributing content. Content is no longer managed as document; it is designed now as intelligent content by tagging and structuring it, so it is structurally rich and semantically categorized. One can automatically discover, reuse, reconfigure, and adapt the intelligent content. A unified content strategy may serve the enterprise tasks on web, publishing, production, and publishing (Rockley & Cooper, 2012).

Currently, virtual environments are developed and used in a variety of fields at universities, engineering companies, and commercial industry. Immersive, interactive VR systems are used as a public display medium in common spaces such as museums, galleries, conferences, and festivals. The projection-based VR brings forth artistic and educational experiences in entertainment and museum settings and serve the research community. Immersive virtual environments serve for the live-action telepresence videoconferencing where interactive narrative and virtual environment can be interfaced with its physical surrounding. The expressive and experiential features of VR environments are used for presentations in environments such as art, research, industry, architecture, or medicine. The visitors can fly across galaxies, go inside the cells and molecules, or experience turbulent gas flows. For example, in the seven-wall digital dome for 440 attendees at the American Museum of Natural History Hayden Planetarium projectors throw 70 feet hemisphere where you can go on a tour from earth, to the Milky Way, to the Virgo Cluster, and beyond into the large-scale structure of the universe. Artists, technologists, and scientists develop visualization exhibits, such as "The Passport to the Universe" (http://www.amnh.org/rose/passport.html) or the "Big Bang" (http://www.amnh.org/rose/hayden-bigbang.html) at the Hayden Planetarium, New York. Other projection environments provide a full-dome spherical projection surface or a large cylindrical 360-degree screen where the viewer interactively moves a projected image from a motorized platform in the center. Josephine Anstey, a virtual reality dramatist and video artist, develops interactive installations, art videos, audio documentary, web and prose virtual fiction experiences where the user is the main character immersed in an intimate setting physically and emotionally, many times being alone, without sharing experiences and interactions with others.

Temporary Art Zone Manifesto, a social event that came about in the early nineties on the premises of the ACM/Siggraph Conference could be seen in some respects as a prototype of a social networking idea. The participants present in the zone received sheets of paper containing a short introductory text followed by two statements about art, science, and computing, along with two images. Many of the statements were quotations of various authors, while images were the details of computer art graphics or photographs. Participants reacted to the verbal statements by modifying them, and changed the printed images adding their own input. Leaflets were then exchanged a few times, so next individuals could contribute to their content. This cooperative work was tentatively resulting in a final collective manifesto in the form of texts and images. The introductory text announced, "In the spirit of the pirate Utopias of the 18th century, we are setting up a Temporary Art Zone. What is needed to create truly contemporary, truly interactive artwork that could be identified as an innovative masterpiece? The leaflets circulating in the zone display a variety of verbal and visual approaches to this question. Feel free to use whatever you see on any of these leaflets as prompts or stimuli to your own responses."

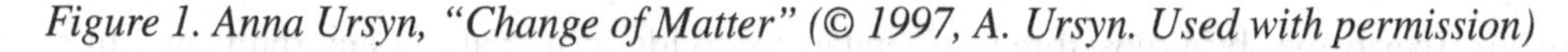

Figure 1. Anna Ursyn, "Change of Matter" (© 1997, A. Ursyn. Used with permission)

Figure 1, "Change of Matter" tells metaphorically about transitions experienced by someone entering a new community, as compared to the phase transition from one state of matter to another, such as transformation of a solid ice to liquid water, and then to gaseous air. Phase transitions may thus serve as subjects for study as well as metaphors for human experience. As we change various types of communities, we change our perspective and routines. While surviving snow and rain, ice and sunshine, liquid and solid phases, we may utilize computer networks to alleviate the impact caused by changes and differences perceived. In the same way as we may use networked digital art as an exquisite medium to convey the order and regularity of material forms (physical matter) in landscape, we may picture and share with this medium our experiences in the new surroundings.

Collaborative Biological Simulations

In neural network or neural computing, computer architecture is modeled upon the human brain's interconnected system of neurons. Through simulation, neural networks imitate the brain's ability to sort out patterns and extract the relationships that underlie the data. Below are two examples of current simulation projects resulting from collaboration in the fields of neuroscience, medicine, and computing.

The Blue Brain Project EPFL (École Polytechnique Fédérale de Lausanne, http://bluebrain.epfl.ch/) is focused on exploring the healthy brain and understanding neurological diseases. Scientists translate their observations into mathematics to develop algorithms that allow reconstruct the brain by building virtual brain in a supercomputer. The Human Brain Project (http://www.humanbrainproject.eu/index.html) performed by almost three hundred experts in neuroscience, medicine, and computing at an approximate cost of 1,190 euro, is a continuation of this approach that is carried out by scientists from European countries to simulate the actual working of the brain. According to the Human Brain Project Report (HBP, 2012), the project has four major goals: generate the data, advance theory by identifying mathematical principles that underlie brain organization; offer services to neuroscientists and clinicians; and develop applications of first draft models and prototype technologies. Research areas include neuroscience: integrative principles of cognition; medicine: understanding, diagnosing, and treating brain disease; and advancing computing technologies: interactive supercomputing for brain simulation. Six platforms for integrative action comprise neuroinformatics, brain simulation, high performance computing, medical informatics, neuromorphic computing, and neurorobotics.

Scientists at Stanford University and the J. Craig Venter Institute have developed the first software simulation of an entire organism and modeled an entire organism in terms of its molecular components and their interactions (Karr et al., 2012). Modeling of 525 genes and the simulation of the complete life cycle of a single-cell bacterium that lives in the human genital and respiratory tracts is a step toward developing computerized laboratories that could carry out many-thousands-of-factors problems involved in researching gene functions, metabolism, and individual cell processes (Markoff, 2012). According to Karr et al. (2012), "The simulation, which runs on a cluster of 128 computers, models the complete life span of the human pathogen *Mycoplasma genitalium* at the molecular level, charting the interactions of 28 categories of molecules – including DNA, RNA, proteins and small molecules known as metabolites, which are generated by cell processes. … Currently it takes about 9 to 10 hours of computer time to simulate a single division of the smallest cell – about the same time the cell takes to divide in its natural environment." In a few years the researchers will hopefully bring this to a bigger organism, like E. coli, yeast or even eventually a human cell.

Figure 2, "Micro Macro" examines the role of magnification. What seems blurry or insignificant from a distance, can transfer us on another level regarding the data details or a conceptual framework. Be it a telescope or a binocular, a microscope or an electron microscope, lenses and digital approaches allow us examining common truths in a variety of disciplines:

We zoom in on an ancient sailboat but we engage with life of tiny creatures that we can't see even through a loupe.

This picture contains images of flat pieces of wood that remained after completing an educational 3D model of an animal. The openings on a wooden surface retained the outlines, patterns, programs, and the meaning of objects that had been taken away. It is like a form without contents, which may one think about the imprinting occurring in early phases of learning, and the effects of the time and the others gradually washing away its content. We may insert memory sticks with software for marketing, architecture, or art creating, instead of the initial animal forms making a model. We may also acquire our knowledge about, for example software, marketing, architecture, or art images and insert it all into the imprinted framework in our minds.

Computing Based on Nanosize Structures in Living Cells

The increasingly overlapping areas of biology, technology, and art help us focus on the form, structure, and function of living or life-like things. Studies on nanosize structures in living cells serve the computer scientists for developing models and creating biological computing devices. For example, a model of a dynamical transport network

Figure 2. Anna Ursyn, "Micro Macro" (© 2008, A. Ursyn. Used with permission)

formation and evolution of the synthetic virtual plasmodium based on a single-cell organism *Physarum polycephalum* serves as a virtual computing material for designing distributed unconventional computing devices (Jones, 2011). Adamatzky, De Lacy Costello, Holley, Gorecki, & Bull (2011) imitated arrangements of vesicles as Voronoi automata – finite state machines defined on a planar Voronoi diagram. According to Wolfram (2012), a particularly notable use of a Voronoi diagram was the analysis of the 1854 cholera epidemic in London, in which physician John Snow determined a strong correlation of deaths with proximity to a particular (and infected) water pump on Broad Street).

Voronoi diagrams are applied in many fields of science and engineering. Voronoi diagram is generated from Voronoi polygons where each polygon contains exactly one generating point and every point in a given polygon is closer to its generating point than to any other (Wolfram, 2012). Andrew Adamatzky (2010) experimentally demonstrated that plasmodium *Physarum polycephalum* approximates a planar Voronoi diagram, and then developed a Physarum construction from Voronoi diagram. One can see how such diagrams are formed on stones colonized by lichens. A Physarum machine is a biological computing device, a programmable biological computer implemented in the true slime mould *Physarum polycephalum*. It comprises an amorphous yellowish mass with networks of protoplasmic veins, programmed by spatial configurations of attracting and repelling gradients that computes by propagating diffusive or excitation wave fronts. The Physarum machine is a green, environmentally friendly unconventional computer that is cheap and easy to maintain, functions on many substrates and in many environmental conditions.

Scientists studying human genome architecture look for the logic of genes or proteins interactions. As stated by Ricard Solé (2008, p. 253) "Scientists find a familiar pattern: the way molecules interact within cells is not very different from the way the Internet is organized." In cells, scientists are able to achieve single molecule detection by applying electrophoretical molecule driving. Single-stranded DNA (ssDNA) and RNA molecules can be driven through a pore-forming protein and detected by their effect on the ionic current through this nanopore (Branton et al., 2008).

INFORMATION SYSTEMS AND THE DESIGN SCIENCE

People used to organize their operations and management to interact and supervise processes, data, activities, and technology. Information system (IS) is according to O'Brien & Marakas (2009) an organized combination of people (especially information technology specialists), physical devices (hardware such as computer peripherals and servers), information processing instructions and procedures (software, such as system software, application software, and procedures), communication channels (networks made of communication media and network support), and stored data (data resources providing knowledge and databases). An information systems framework includes business applications, management challenges, information technologies, foundation concepts, and development processes (O'Brien & Marakas, 2010).

Historically, visual design and information systems have been philosophically and functionally independent from each other. However, people exhibit a fundamental preference for all things beautiful. As Liu put it (2003a, p. 1274), "The forgoing insights call for a guiding framework to help visual designers create systems that better serve user requirements." But aesthetics is not a hard science discipline because it lacks scientific and theoretical foundation or framework to organize, communicate, and explain related ideas and concepts" (Liu, 2003a, p. 1274). At the 2012 Decision Science Institute, Southwest Region Conference, Liu (2012) and also Peak, Prybutok, Gibson, & Xu (2012a,b) proposed that information systems (IS) can serve as a reference discipline for visual design, and vice versa, visual design can reciprocate as a reference discipline for IS. With the increasing number of visually sophisticated, design-savvy users it seems imprudent to overlook users' informed preferences. The time has come for Information Systems to become a reference discipline offering mature IS research and systems methodologies.

Design science involves a growing amount of computation and is becoming part of the information systems discipline. The design science research is getting increasing attention (March & Storey, 2008). According to the authors, a design science research contribution requires:

1. Identification and clear description of a relevant organizational IT problem;
2. Demonstration that no adequate solutions exist in the extant IT knowledge-base;
3. Development and presentation of a novel IT artifact (constructs, models, methods or instantiations) that addresses the problem;
4. Rigorous evaluation of the IT artifact enabling the assessment of its utility;
5. Articulation of the value added to the IT knowledge-base and to practice; and
6. Explanation of the implications for IT management and practice (March & Storey, 2008, p. 726).

Visual appeal of the design products (Liu, 2003b) is an important factor in business and marketing. The users' first visual impressions are made immediately and persistent. They are instantaneous, affective, precede cognitive process, and are long lasting; users give more weight to negative visual attributes than to positive ones. Aesthetic beauty of design influences consumers' perception of product features. Liu (2012b) offers a framework of visual systems design (VSD), where the visual design discipline utilizes information systems and systems development. The visual design discipline is part of the aesthetic paradigm and the IS discipline is contained in the positivist paradigm. Liu proposes a reference discipline for visual designers with access to IS knowledge systems and resources.

SOCIAL NETWORKING AND COLLABORATIVE VIRTUAL ENVIRONMENTS

There is an interplay and convergence between information visualization and virtual environments. Media spaces involve hybrid spaces combining real and virtual worlds. According to Chen (2004), virtual environment not necessarily need to have the presence of 2D or 3D spatial model; it may be text-based but a spatial-visual interface makes a lot of difference. Information visualization plays an important role in construction of a virtual, collaborative environment when people interact with visualized information in such environments (Chen, 2004, p. 211). As stated by Chen (2004, p. 216), "An important insight from the social navigation trends is that like-minded people may form a group or even a crowd as they are attracted by the content of a web page. In social navigation, people gather together because they are all interested in a particular topic. This distinguishes the concept from traditional chat rooms where a considerable number of people come, just for the sake of talking to someone."

Web Science: A study of the web as a vast information network of people and communities has transformed into a domain of the Web Science, which involves analysis and design of web architecture and applications, as well as studies of the people, organizations, and policies that shape and are shaped by the Web. "An understanding of human behavior and social interaction can contribute to our understanding of the Web, and data obtained from the Web can contribute to our understanding of human behavior and social interaction" (Web Science, 2012). Web Science integrates computer and information sciences, communication, linguistics, sociology, psychology, economics, law, political science, and other disciplines.

Social hierarchies that are common in animal societies control access to resources according to the fitness of the individual. Conversational expressions such as 'pecking order,' 'leader of the pack,' and 'alpha male' describe such societies or groups of individuals. This concept serves in developing the three different models of an adaptive social hierarchy among nodes in a homogeneous wireless network, in terms of their energy and connectivity (Markham, 2011).

Chen (2004) discerns general multi-user virtual environments, distributed virtual environments, and collaborative virtual environments with built-in facilities for various social activities. Virtual reality-based multiuser environments include Online Community (derived from blaxxun interactive, originally named "Black Sun Interactive," a company that developed a 3D community platform for the Internet environment), Community Place (Sony), and Active Worlds (The Circle of Fire Studio). Collaborative virtual environments, for example DIVE (which distributed interactive virtual environments) or MASSIVE, have applications such as distance learning, digital libraries, and online communities. The MASSIVE (Multiple Agent Simulation System in Virtual Environment) software is an example of the use of artificial intelligence, for example for film animation. Applications for computer animation displayed combinations of techniques depending on demands.

Countless options of social networking stimulate artists to create online works of different type, for example, interactive art where both the form and content of the artwork are affected by the behavior of the audience, or augmented reality art where the observer is experiencing and responding to events as if they were real. Multi-touch screens allow multi-touch collaboration of participants. For example, touching and waving at the screen brought up displays of zooming graphics on the 8ft x 3 ft large monitor called the Multi-Touch Collaboration Wall, which could be positioned at will around the screen (Farhi, 2008). Solutions of that kind promote the videoconferencing and Skype (a voice-over-Internet service and software application) contacts. Within this approach we may find examples of defeating the semiotic divide between fictional and material characters, or between virtual and realistic images. Users are represented as avatars, the 3D graphical icons. Communication through the use of avatars in a virtual social space often applies the action panels designed to animate avatars in a conventional way using ready expressions, gestures, and actions. Websites, e-mails, and game spaces are also often designed in a similar conventional way. According to Heinrich (2010, p. 9):

The digital iconic avatar seems to undermine the Western epistemic distinction between the human subject and pictorial representation, questioning the notion of the body as a mainly biologically defined entity. ... It is in and through the player's actions that the avatar portrait emerges as a prototype, thereby dissolving the distinction between subject and its representation and between material and virtual reality.

Pervasive applications are usually based on the shared virtual environment. Collaborative virtual environments involve interacting participants (often over large distances) for such actions as distributed simulations, 3D multiplayer games, or collaborative engineering software. Collaborative virtual environments include intelligent agents as an essential part. An intelligent agent, which is an object of study and design in the field of artificial intelligence, is a system that perceives its environment through sensors and takes actions to achieve goals. It may be as simple as a thermostat or as complex as a human being (an individual or a whole community). To attain goals, an intelligent agent may learn or use knowledge.

Experts show amazingly high perceptual skills. Experience-induced improvement is called perceptual learning and the accompanying neural changes are called neural plasticity (Eysel, 2002). Perceptual learning and neural plasticity have been studied in all sensory modalities including vision, hearing, and touch perception. Attention of researchers has been focused on developing a visual information processing model and defining a link between perceptual learning and attention (Yotsumoto & Watanabe, 2008).

Wahl, Winiwarter, & Quirchmayr (2011) evaluated several toolkits for natural language processing (NLP), and then designed the framework architecture for the intelligent integrated computer-assisted language learning (iiCALL). The use of artificial intelligence technologies such as natural language processing (NLP) allows developing computer-assisted language learning (CALL) environment. Integrated e-learning environment is a web-based language learning with the use of common working environments, for instance web browsers or e-mail clients.

The pervasiveness of mobile communication includes the text and multimedia messaging, along with the virtual publication of data in physical places called spatial messaging. The spatial messaging services, including gaming, location-based social networking, or advertising services, require providing the trust and security mechanisms that inform about honesty and truthfulness of the messages. With this approach Mayrhofer, Sommer, & Saral (2012) presented an Air-Writing, a globally scalable spatial messaging platform for private group messaging systems that preserves user privacy in anonymous message retrieval and client caching and filtering, as well as randomized queries for obscuring traces. Wei, Liu, Zhao, McFarlane, & Clapworthy (2011) proposed a Web-based 3D visualization for biomedical applications, which accepts two data resources as an input (local and remote) and copes with two types of algorithms (built-in and remote). Web technology makes it possible to use software maintained on a remote server. This approach, which can be used in many other application areas, provides a virtual client environment, in which users can employ remotely installed software interactively using any standard browser.

Concepts, units, and phenomena or real word objects such as documents have unique resource identifiers that serve as a name and locator identifiers. Publicly available media content produced by users reduces collaboration barriers and reflects media production through accessible and affordable technologies, such as open source, free software, and flexible licensing/related agreements. Categories of the user-generated content include crowd sourcing (e.g., Wikipedia) and expert sourcing, volunteered geographic information, feedback on reference mapping, among other options. They help create new, meaningful content for searching machines.

Electrolibrary – a book reading technique (Wegrzyn, 2012) is a device existing between a tangible book and an interactive digital content features sensors and flexible circuits printed onto paper; when inserted into a book and connected to a computer they inform what page the reader is on and also allow the reader to browse a companion website and gain additional information.

EXAMPLES OF INTERACTIVE VISUALIZATION TECHNIQUES

Hartmann, Pinto, Runkler, & Sousa (2011) adopt a strategy of trading optimality for efficiency in real world applications where a problem is constantly changing: a slightly suboptimal solution is much more desirable than small gains in quality in an optimal solution. In problem optimization, basic families of social insects: ants, wasps, bees, and termites were used as an inspiration source for designing challenging computational models. For example in ants, changes of the pheromone (a secreted substance that triggers social response) concentration on a path of ants would guide them in following the trail. The principles of swarm intelligence, where the colony agents cooperate to achieve a common goal without a higher supervision, apply to the mould building capability of termites. Amount of saliva deposited in the chewed earth-and-saliva pellets bias the termites to move in the direction of a new mould. Mathematical models of real colonies serve the authors in developing optimization algorithms. Optimization logistic process has also been studied in the environment-to wasp interaction, wasp-to-wasp hierarchical interaction within the nest, and in the bee colony where bees directly communicate by performing the waggle dance.

A Display with Cloudlets and Dynamic Virtual Computer

Satyanarayanan, Bahl, Caceres, & Davies (2009) provided a transient display model with cloudlets and dynamic virtual computer synthesis. This vision of mobile computing breaks free of the constraints resulting from the resource poverty of mobile hardware. Mobile devices are resource-poor and energy-limited comparing to static client and server hardware. Wide Area Network (WAN, for example the Internet) creates delays in cloud computing. Cloudlets form a decentralized and widely dispersed Internet infrastructure. Virtual machine (VM) technology provides customized service software on a nearby cloudlet, and then uses that service over a wireless LAN. A virtual machine-based cloudlet is a self-managed data center in a box shared by few users at a time, in contrast to a centralized cloud ownership by Amazon, Yahoo! etc. Thus, a cloudlet means a compute cluster, plus wireless access point, plus wired Internet access with no battery limitations (Satyanarayanan, 2010). Rather than relying on a distant "cloud," mobile users seamlessly utilize nearby computers to obtain the resource benefits of cloud computing. With the real-time two-way translation on mobile devices, the high-resolution image or HD video is sent to the cloudlet and applied to one of the virtual machines running in the cloudlet.

Display of Privacy-Sensitive Information

Human computer interaction and embedded technologies exist in everyday life as wearable mobile electronics. However, the ability to be constantly online and connected can have a negative impact on the user's privacy issues. Scientists are working on designing the optimal scenario where one can keep the privacy-sensitive part of information on one's own mobile device (for example, a smart cellphone, a tablet computer, a miniature wearable computer such as watch or wearable glove) and at the same time interactively project a visual part of information at large resolution to a large display screen. Clinch, Harkes, Friday, Davies, & Satyanarayanan (2012) envision a dynamic, interaction-rich transient display model whereby users walk up to a display and temporarily use it to augment their mobile device. User devices remain the root of identity, trust, customization and interaction. Likewise, privacy-sensitive information only appears on the mobile device. For example, Dr Jones, a physician who is outside of a hospital can interpret a pathology slide that must be interpreted while surgery is in progress. "Walking up to a large display in the lobby she views the slide at full resolution over the Internet. Using her smartphone for control, she zooms, pans and rotates the slide as if at a lab microscope. Privacy-sensitive clinical information displays on her smartphone. Dr. Jones interprets the slide, telephones the surgeon, and returns to dinner" (Clinch et al., 2012, p.1). Other scenarios involving transient display may involve personalized tourist information or a visiting expert consultation realized through the described above dynamic VM synthesis supported by cloudlets. As the authors put it, transient public display by mobile users is now possible due to:

- The available display hardware for affordable deployment;
- Virtual machines for transient customization;
- Smart phones for actuation and access control;
- Wireless networks for untethered communication; and
- Cloud offloading for resource-intensive tasks (Clinch et al., 2012, p.1).

Visual Search for Patterns in Categorical Data Sets

Interactive visualization techniques support search, exploration, and summarization of multiple sets of temporal (with built-in time aspects) categorical data. Temporal Pattern Search (TPS) is an algorithm for visual search and exploration of temporal patterns of events in personal histories (Wang, Deshpande, & Shneiderman, 2010). Temporal patterns discernible in large databases such as medical records, web server logs, legal, academic, or criminal records can lead to the discovery of cause-and-effect phenomena and support higher-level tasks such as hypothesis generation (Wang, 2010). Software Lifelines2 (http://www.cs.umd.edu/hcil/lifelines2/) is an interactive visualization tool for visualizing temporal categorical data

Figure 3. Anna Ursyn, "Human Nature Takes Over" (© 1992, A. Ursyn. Used with permission)

across multiple electronic health record systems. Lifelines2 comprises three operators Align, Rank, and Filter; it improves the user performance speed in understanding of relative temporal relationships across records such as complaints, diagnoses, treatments, etc. Wang, Wongsuphasawat, Plaisant, & Shneiderman (2010) examined the usage data and user comments on Lifelines2.

Several visual methods have been proposed to overcome the complexity and scale of making medical decisions based on electronic health records. To support physicians and clinical researchers in interacting with the electronic health record systems, Rind, Wang, Aigner, Miksch, Wongsuphasawat, Plaisant, & Shneiderman (2010) compared twelve information visualization systems for exploring and querying electronic health records in terms of the data types, multivariate analysis support, number of patients records, and user intents. The authors concluded that most systems handle numerical or categorical data but not both, are designed for looking at a single patient or multiple patients but not both, utilize horizontal time lines to represent time, and not always have good support for Filter or address specific user intents.

Figure 3, "Human Nature Takes Over" tells about the persistence of human actions in assuming control of harmful microorganisms, plants, insects, or seaweed. Computer-derived forms such as programmed plots have been combined with images of nature presented as a photosilkscreen and a photolithograph. The concomitance of disparate media develops the depth of the work and provides inspiration for further reworking of the artwork.

Interactive Visualization of Objects from the Past

Interactive visualization techniques have been used to recreate the images of and bring into being physical or virtual reconstructions of historically significant places. Simulation of works relating to objects from the past supported reconstruction of medieval mosaics, the mural works of Rome, of Saint Mark's Church and Torcello at Venice, treasures of architecture in Palermo, Monreale, and Cefalù in Sicily, and the medieval marble and mosaic floors with geometric patterns. Simulated reconstruction of the ruined monastery at Georgenthal, in Thuringia, Germany provides the viewer with an insight how the monastery looked like in the past (Interactive Storytelling, 2008).

Alan Price (2011) created several virtual interactive and immersive reconstructions of historical places that are no longer accessible to the public, to install them permanently in a museum setting. The real time interactive simulation shows two Marlborough Apartments from the 1930s where Etta and Claribel Cone amassed a collection of over 3,000 works of early 20th century French art by Matisse, Gauguin, Picasso, Cezanne, Van Gogh, Renoir, and others. A touch plasma screen version was developed to give viewers an immersive experience; interactive floor plan of the building allowed moving to a specific room, and interactive polygonal models of sculptures with detailed mesh and texturing could be opened and their contents explored (Virtual Tour, 2005). The Marlborough Apartment building no longer exists as it did during the sisters' lives. Paintings and furniture were donated to the Baltimore Museum of Art, which in 2010 organized a travel exhibition to the Jewish Museum in New York and the Vancouver Art Gallery. The immersive stereoscopic experience was based on research, rendering, creating models, and digital reproductions (LoPiccolo, 2003).

A real time simulation The Sun Dagger Explorer (Price, 2011) is an interactive computer model based on laser scanning, reconstructed in collaboration with an archeoastronomer Anna Sofaer, of the ancient calendar site in Chaco Canyon, New Mexico. This is a part of an exhibition on cultural astronomy at the Adler Planetarium and Museum in Chicago, and then a Space Frontiers exhibit at the New Mexico Museum of Natural History and Science, Albuquerque. It reconstructs the solar and lunar calendar created in stone approximately one thousand years ago by ancient Chacoans or Anasazi – the ancestral Pueblo people. The assembly of three nine foot stone slabs lines up sunlight into patterns of light and shadow onto a spiral petroglyph in the cliff wall (petroglyphs are carvings or line drawings on a rock, many of them made by prehistoric people). The patterns mark the year's solstices and equinoxes and are believed to track the 18.6-year cycle of the moon. Solstices are astronomical events that happen twice each year when the sun is at its greatest distance from the celestial equator. The summer solstice is the longest day of the year when the sun is over the tropic of Cancer (about June 21). The winter solstice occurs when the sun is over the tropic of Capricorn (about December 21) and the day is the shortest. Equinoxes take place two times during a year, when the sun crosses the celestial equator and when the length of day and night is approximately equal: the vernal equinox or the autumnal equinox.

UBIQUITOUS COMPUTING

There is no single definition of ubiquitous computing and not all possible properties are supported in particular cases. Scenarios for ubiquitous models require working on several system designs at once, for example, communication, data storage, sensors controllers, and processors. They involve human computer interaction and physical objects interaction. Constructing pervasive applications requires the design tools that can be used in an interdisciplinary setting (Forsyth & Martin, 2012). Mobile devices embedded in everyday objects and surroundings or implanted into humans can communicate and interact with the users and with each other. Active software agents organize and arrange such activities making them flexible and adaptive to changes in an unsupervised, self-governed way. As a result, ambient computing becomes second nature.

Ubiquitous computing (ubicomp) stems from the advancements in human-computer interaction (HCI). Users are not always aware that they are using devices and systems where information processing is integrated into everyday objects and activities. Pervasive computing, ambient intelligence, physical or haptic computing, all represent the new way of thinking about computing; small and powerful computing devices that are worn, carried, or embedded in our environment, enable us to place the work, not the tool (in the form of a desktop or portable computer) in the focus of our attention. Developers of ambient intelligence aim at supplying physical environment with a network of devices: sensors, actuators, and computational resources, in order to help the users in completing their tasks (Chong & Mastrogiovanni, 2011). With ambient intelligence, processing of information is interactively embedded in our ordinary activities and objects, often without our awareness. This model, which relates to pervasive computing and ubiquitous computing, is considered a future paradigm, and an advanced stage in comparison with the desktop paradigm, because of its capability to record and oversee our surroundings and support our interactions with other people or objects. Calm computing, developed in the late 1990s, in which the computer did not require full attention to operate, empowered peripheral (sensory) processing in our brains by switching between the center and the periphery of our locus of attention, bringing more details and the location awareness into periphery.

Mark Weiser (1991) of the Xerox Palo Alto Research Center (Xerox PARC) coined the phrase "ubiquitous computing" around 1988 and described this post-desktop model in 1991, calling it the third wave in computing. Research on ubiquitous computing has been advanced in the USA and in several other countries. MIT started in Massachusetts the Things That Think consortium (http://ttt.media.mit.edu/). Georgia Tech, New York University UC Irvine, Microsoft Research, and Intel Research and other places of research contribute to the progress. Claytronics Project from the late 2000s (http://www.cs.cmu.edu/~claytronics/ "combines modular robotics, systems nanotechnology, and computer science to create the dynamic, 3-Dimensional display of electronic information known as claytronics" (Goldstein et al., 2009). Ubiquitous Songdo City was built in 2000 in South Korea not far from Seoul. The city and a theme park Robotland provide the users with remote control of appliances, video connections between residents, schools, offices, stores, and the security network. The European Union launched the Disappearing Computer initiative in a way of the Ubicomp infrastructure. In Japan, the Ubiquitous Networking Laboratory (UNL) in Tokyo has been led by Ken Sakamura. Several conferences such as Pervasive, ACM Ubicomp, IEEE Percom, IEEE ICPS, and IEEE SUPE, and a number of academic journals are dedicated to pervasive computing.

The core feature of the ubiquitous computing is the availability of computer-based services everywhere, with devices interacting with humans in a hidden, non-intrusive way. Pervasive devices are often small (and are becoming even smaller), embedded, interconnected, hidden in an unobtrusive way, supporting intuitive solutions, and aware of environmental conditions such as location, time, and user activity, in order to be able to optimize their interaction with human and physical environment. They are usually networked, distributed for the coordinated use for tasks or storage, and transparently accessible. A user can model their key properties; one person may apply many computers at a time for several purposes such as: interacting with others; processing and managing personal visual and written materials; computing and storing professional data; contacting food, services, transport, and communication related facilities; benefiting from the ability to sense and control smart devices and interconnected environments, make the most of the devices' healing possibilities, among many other options.

Smart devices interact with environment by tagging, annotating, sensing and monitoring, filtering, adapting, or controlling surroundings by assembling and regulating (Poslad, 2009). However, computers can be self-governed and function autonomously without human intervention, performing several interactive tasks, and often using artificial intelligence to make decisions and self-organize their actions. Ambient intelligence is another name assigned to electronic environments that are sensitive and responsive to the presence of humans, integrated into our environment so the technology is hidden and only the user interface is visible. Actually or in a near future devices are co-working to help people in their work, everyday activities, and entertainment; they support communication, computing, telecommunication, and consumer electronics. Digital real-time streams are replacing actions based on a particular location. These technological trends change our social behavior as people show willingness to share. While making transactions, social engineering is replacing algorithm engineering, so analogue transactions convert into real-time feeds of data and services. Poslad (2009) summarizes main features of ubiquitous computing devices and environments as distributed, iHCI, context-awareness, autonomy, and artificial intelligence. Implicit human-computer interaction (iHCI) does not act on a computer system but rather provides an input to a computer, for example when setting time for the first time in the device that uses absolute time for setting actions, while explicit human-computer interaction occurs when a user configures for the first time the controls of a timer interface. Context aware systems may be aware of a person, environment, context, or physical, spatial or temporal features; they may be adaptive when tailored to an individual user or type of user. Autonomous systems display self-configuring, self-healing, self-optimizing, and self-protecting behavior. Individual intelligent systems display a large number of properties including reactive, reasoning, several goals-oriented, adaptive behaviors that may be cooperative in case of multiple agents and competitive, even malevolent when referred to individual agents (Poslad, 2009).

Ubiquitous system devices are low-cost, networked processing devices that control elements of the physical world spontaneously and effortlessly. Some of them, originally proposed by Weiser, have visual output displays that take form of wearable tabs (the size of a centimeter), hand-held pads (about a decimeter), and boards, for interactive display (the size of a meter). Three forms of Ubiquitous Computing devices have been extended later to include the ICT devices, some of them without visual output displays (ICT is an acronym for Information Communications Technology signifying products that serve for the storage, retrieval, manipulation, transmission or receipt of information in a digital form. Examples of ITC devices are phones, cameras and game consoles, automatic teller machines (ATMs) vehicle control systems such as antilock brakes, smart phones, electronic calculators, household appliances, and computer peripherals). The three additional forms for ubiquitous devices comprised (a) miniature dust, called also smart dust devices (with a nanometer to millimeter size); examples are Micro Electro-Mechanical Systems (MEMS) inter-communication devices, that are often solar powered, that can be spread in millions like pigment in the wall paint; (b) skin devices – fabrics from light emitting and conductive polymers that could serve as flexible display surfaces, or could be painted with MEMS to act as networked surfaces; (c) clay devices – three-dimensional artifacts comprising MEMS, for example, tangible interfaces. As Poslad (2009) describes, “Micro fabrication and integration of low-cost sensors, actuators, and computer controllers, MEMS enable devices or motes to be small enough to be sprayed or scattered untethered into the air, to become embedded throughout a digital environment, creating a digital skin that senses a variety of physical and chemical phenomena of interest" (Poslad, 2009, p. 47). Information from digital skin can be localized, current, and directly accessible by the end-users and applications. Smart dust motes contain micro sensors, an optical receiver, passive and active optical transmitters, signal-processing and control circuitry, and a thick film battery power source. MEMS are based on integrated circuit (IC) silicon chips and fabricated in millions with the use of photolithography. MEMS attached to a substrate (paint, gel, in air or in water) form smart surfaces or smart structures that can reorganize. "Smart paint coating on a wall can sense vibrations, monitor the premises for intruders, and cancel noise" (Poslad, 2009, p. 197, after Abelson, 2000).

Goldstein et al. (2009) describe Claytronics, as programmable matter made out of millions of sub-millimeter sized spherical robots. Their goal is to create ensembles of cooperating sub-millimeter robots, which work together to form dynamic 3D physical objects, for example, in telepresence, to mimic, with high-fidelity and in 3-dimensional solid form, the look, feel, and motion of the person at the other end of the telephone call. Claytronics can be used to implement pario, a new media type, which would render physical 3-dimensional objects that one can see, touch, and even hold in hands, and thus may change how we communicate with others and interact with the world around us. According to Kirby et al. (2005), modular robots called Catoms can move relative to one another without moving parts, which allows manufacturing at smaller and smaller physical scales using high-volume, low-unit-cost techniques such as batch photolithography, multi-material submicron 3D lithographic processing, and self assembly, radically altering the relationship between computation, humans, and the physical world. "Claytronics envisions multi-million-module robot ensembles able to form into three-dimensional scenes, eventually with sufficient fidelity so as to convince a human observer the scenes are real" (Kirby et al., 2005).

According to the Ubirobots workshop (2012) organizers, "ubiquitous robots as cognitive entities have been able to add value to services compared to traditional systems. They are able to coordinate their activities with other physical or logical entities, move around, sense and explore the environment, and decide, act or react to the situations they may face anywhere and anytime." At the UbiComp (2012) the 14th ACM International Conference on Ubiquitous Computing, digital designers and researchers are working on current advancements in creating location-based social networks; sensing, interpretation, and integration of events, behaviors and environmental states; using mobile devices as instruments to collect data and conduct studies; testing and using subliminal stimuli and information below aware perception; capturing and interacting with information on connected objects and devices, as well as capturing, processing, and sharing data on events; the implications of pervasive eye tracking for context-aware computing to assist users in their daily activities, among other issues.

Some hold that ubiquitous computing creates Synthetic Reality present in everyday life and directed toward commonplace targets without any sensory augmentation, in contrast with virtual reality or augmented reality. For example, ubiquitous computing devices may control environmental conditions such as light and heating according to the personal biometric monitors painted or woven into clothing; refrigerators might inform users about the amount and the state of the tagged food inside and plan a menu.

Moonlit Manifestation (Figure 4) reflects upon the ways the networked technologies and ubiquitous computing make the new city environments familiar and easy to absorb and accept:

Figure 4. Anna Ursyn, "Moonlit Manifestation" (© 1999, A. Ursyn. Used with permission)

Moon enters the city,
Its converging structures belong to other town interiors,
Left behind with inattention,
Showing the way to new encounters.
Wooden moon soaks familiar sandboxes, parks, and factories,
Well-retained in the memory,
A city, never visited before,
Welcomes the guest.

SECOND LIFE

Virtual worlds offer visualization and promote discussion among visitors located across the globe. Previously, there were virtual worlds developed for education, which attempted to simulate classroom teaching such as lectures, demonstrations, and group tasks. From 2006, Second Life (SecondLife_1_13_3_2.dmg) developed by Linden Lab became the popular virtual world, with millions of users. The downloadable client program allows the users called residents to enjoy communication, visualization, interaction, and simulation. Users represented as avatars visible to others in any shape and size (which provides a level of anonymity) can interact, manipulate objects, and communicate with other residents, socialize, participate in individual and group activities, and buy virtual property and services from one another (they can own land). Second Life has its virtual currency known as Linden Dollars that is exchangeable for US Dollars. Many companies and education centers own land at the Second Life to perform research, commerce, and teaching. Second Life has easy to use interface and global media coverage. It allows users building objects and developing scripts to run within them. It makes possible to create virtual lecture theatres with streamed media, interactive models, and virtual presentations.

Second Life provides the interactivity, storage, and access to a database such as Amazon or YouTube, and is linked to websites allowing users to expand the learning experience. Second Life suggests a new format for e-learning because of the relation of the flow experience to the Second Life's educational environment. Pessoa Forte et al (2011) indicate that there is flow in Second Life's e-learning environment, with interactive speed, exploratory behavior, and telepresence as the most significant constructs detected. Second Life is not a game because it does not have points, scores, winners, or losers, levels, or any characteristics of games. People may earn real money on Second Life by designing and selling virtual objects. Art shows are going at the Second life environments. Second Life became one of the virtual classrooms for major colleges and universities, including Harvard, Pepperdine, Ball State, and New York University. However, frequent visiting Second Life often results in addiction and procrastination of high-priority duties in the real life, as some residents find an escape there from frustration and the failure to fulfill goals in real life. Bastiaan Vanacker & Don Heider (2012) suggest that ethical harm can most likely occur in virtual communities when players see their avatars as extensions of themselves while other players do not have strong notions of wrong and right behavior, and a norm violating behavior and ethically relevant avatar harm may occur at the Second Life.

THE INTERNET OF THINGS

Many proclaim that Internet of Things is a new form of the Internet. The Internet of people has been connecting and interchanging electronic data through computers and the Internet and then through the various types of social media. It is now being transformed into the machine-to-machine technology that makes things intelligent and connected. Concepts pertaining to the evolving field of the Internet of Things, such as digital objects' memories, have been recently studied and theoretically analyzed (e.g., UbiComp, Pervasive, and MediaCity Conferences, 2012). Casaleggio (2011) describes the evolution of the Internet of Things as a process evolving in five stages:

Stage 1. The World is the Index: Technologies such as augmented reality, geotagging, and GPS enabled describing all things on the Internet, so the world became indexed. However, there is no direct interaction with the object. Short message service (SMS) text messages, Quick Response Codes (QR) – 2D matrix barcodes that may be read by a cell phone, and RSS feeds (web feed formats to publish temporary entries such as blogs, news, audio or video works in standardized format) are forms of geospatial metadata. Geotagging is based on position of things and provides geospatial, location-specific identification to various media and objects, for example, to photographs, video clips, and websites. We can organize photos according to location using Google Earth (http://www.google.com/earth/index.html);

Stage 2. Take the World Online: Technologies such as automatic identification and tracking e.g., RFID, visual recognition, barcode, and near field communication (NFC), all caused that information can shadow objects online (e.g., tracking packages and other moving objects identified by a code). Again, there is no direct interaction with the object. RFID tags attached to objects utilize wireless non-contact systems to transfer data, using radio-frequency electromagnetic fields. For example, runners can be timed with a chronometer and see their pace online, while the RFID collar on a pet can be recognized by doors that can open to let them in (Casaleggio, 2011). NFC is a set of standards for smartphones to establish radio communication with each other by touching them together or bringing them a few cm close for contactless transactions, data exchange, and simplified setup of Wi-Fi, often used as a synonym for WLAN – wireless local area network, technology that allows a device to exchange data using radio waves over a high-speed Internet);

Stage 3. Take Control of the World: Technologies such as remote control make possible that objects are connected to the internet and interact with people: they communicate, take orders, and state information about themselves. For example, lost or stolen objects such as keys or cars can tell us where they are, and windows open at certain temperature;

Stage 4. Let the Things Talk to Each Other: Technologies such as machine-to-machine (M2M) wirelessly connect devices into a network allow objects to communicate with each other and take action in certain conditions. For example, they monitor patient heartbeats, enable plants alert sprinkler systems when they are dry, and business cards swap data or download them to a computer;

Stage 5. Let Things Become Intelligent: Objects that interact with the Internet can create value that interlinks them. Technologies such as Object Generated Content (OGC) and Device to grid cause that objects communicate with the Internet using sensors and counters. They provide information that can be elaborated and used as a new knowledge. For example, Nike places sensors in shoes; the alarm can ring earlier in case of traffic or bad weather; intelligent cups of medicine bottles signal with light, sound, or a telephone call a time to take a pill and send a monthly report to the doctor (Casaleggio, 2011).

STARTUPS

Startups are becoming a way to communicate with society. Startup company, establishing a business by bringing to existence some ideas saved on napkins or Photoshop files, takes advantage of the existence of open source databases and cloud apps that make the access to customers globally available. Startup companies are getting great visibility and may have an impact on other people around the world because one may use for free the experiences gathered by previous entrepreneurs, both successful and failed. One can run a distributed team due to the cloud apps. Staring a new startup company involves using strategically, with little capital, a variety of customer acquisition channels, public relations, inbound marketing, search engine marketing, in-person events, platform distribution, direct sales, affiliate programs, and information that currently exists. A new class of investors creates software, uses new technologies, platforms, and even new tools for startups that become available, and makes a living.

CONCLUSION

This chapter contains discussion of technologies involving augmented and virtual reality that are becoming crucial for technological development for both corporations and universities. Augmented, virtual reality, and Second Life play an increasing role in lives of participants. The desktop-type 3D virtual environment, where the real world is not blocked out from the user who can see through a window and communicate by mouse and keyboard, is especially promising for learning and education. Examples of the virtual reality immersive simulations are related to the immersive virtual reality environments used for pilot training programs, digital museums, galleries, conferences, and festivals.

REFERENCES

Abelson, H., Allen, D., Coore, D., Hanson, C., Homsy, G., & Knight, T. Jr et al. (2000). Amorphous computing. *Communications of the ACM*, *43*(4), 74–82. doi:10.1145/332833.332842.

Adamatzky, A. (2010). *Physarum machines: Computers from slime mould.* World Scientific Publishing Company.

Adamatzky, A., De Lacy Costello, B., Holley, J., Gorecki, J., & Bull, L. (2011). Vesicle computers: Approximating Voronoi diagram on Voronoi automata. *Chaos Solitons and Fractals, 44*, 480-489. Retrieved August 28, 2012, from http://arxiv.org/pdf/1104.1707.pdf

Branton, D., Deamer, D. W., Marziali, A., Bayley, H., Benner, S. A., & Butler, T. … Schloss, J. A. (2008). The potential and challenges of nanopore sequencing. *Nat Biotechnol., 26*(10), 1146–1153. doi 10.1038/nbt.1495. Retrieved August 28, 2012, from http://www.ncbi.nlm.nih.gov/pmc/articles/PMC2683588/

Casaleggio Associati. (2011). *From internet of people to internet of things*. Retrieved July 17, 2012, from http://www.casaleggio.it/pubblicazioni/Focus_internet_of_things_v1.81%20-%20eng.pdf

Chen, C. (2004). *Information visualization: Beyond the horizon*. Springer-Verlag.

Chong, N.-Y., & Mastrogiovanni, F. (2011). *Handbook of research on ambient intelligence and smart environments: Trends and perspectives*. Hershey, PA: IGI Global. doi:10.4018/978-1-61692-857-5.

Clinch, S., Harkes, J., Friday, A., Davies, N., & Satyanarayanan, M. (2012). *How close is close enough? Understanding the role of cloudlets in supporting display appropriation by mobile users.* Retrieved from http://www.cs.cmu.edu/~satya/docdir/clinch-percom-2012-CAMERA-READY.pdf

Coquillart, S. (2012). First-person visuo-haptic environment: From research to applications. In *Proceedings of the International Conference on Computer Graphics Theory and Applications and International Conference on Information Visualization Theory and Applications*. ISBN 978-989-8565-02-0

Cruz-Neira, C., Sandin, D. J., DeFanti, T. A., Kenyon, R. V., & Hart, J. C. (1992). The CAVE: Audio visual experience automatic virtual environment. *Communications of the ACM*, *35*(6), 64–72. doi:10.1145/129888.129892.

Demazeau, Y., Müller, J., Corchado Rodriguez, J. M., & Bajo, Pérez, J. (Eds.). (2012). *Advances on practical applications of agents and multi-agent systems: 10th international conference on practical applications of agents*. Springer. ISBN 3642287859

Ehle, R. (2012). *Virtual music.* University of Northern Colorado. Unpublished.

Engelbrecht, A. P. (2003). *Computational intelligence: An introduction*. Wiley.

Eysel, U. T. (2002) Plasticity of receptive fields in early stages of the adult visual system. In *Perceptual learning*, (pp. 43-66). A Bradford Book. ISBN 0262062216

Farhi, P. (2008, February 5). CNN hits the wall for the election. *The Washington Post*. Retrieved September 22, 2012, from http://www.washingtonpost.com/wp-dyn/content/article/2008/02/04/AR2008020402796.html

Forsyth, J. B., & Martin, T., L. (2012). Tools for interdisciplinary design of pervasive computing. *International Journal of Pervasive Computing and Communications*, *8*(2), 112–132. doi:10.1108/17427371211245355.

Gimeno, J., Morillo, P., Ordu, J. M., & Fernández, M. (2012). An occlusion-aware AR authoring tool for assembly and repair tasks. In *Proceedings of the International Conference on Computer Graphics Theory and Applications and International Conference on Information Visualization Theory and Applications*, (pp. 377-386). ISBN: 978-989-8565-02-0

Goldstein, S. C., Mowry, T. C., Campbell, J. D., Ashley-Rollman, M. P., De Rosa, M., & Funiak, S. et al. (2009). Beyond audio and video: Using claytronics to enable pario. *AI Magazine*, *30*(2).

Hartmann, S. A., Pinto, P. C., Runkler, T. A., & Sousa, J. M. C. (2011). Social insect societies for optimization of dynamic NP-hard problems. In *Bio-Inspired Computing and Networking*, (pp. 43-68). CRC Press, Taylor & Francis Group. ISBN 1420080326

HBP. (2012). *The human brain project: A report to the european commission*. Retrieved July 21, 2012, from http://www.humanbrainproject.eu/files/HBP_flagship.pdf

Heinrich, F. (2010). On the behalf of avatars: What on earth have the aesthetics of Byzantine icons to do with avatar in social technologies? *Digital Creativity, 21*(1), 4-10. Retrieved October 26, 2010 from http://www.informaworld.com/smpp/title~db=all~content=g922551031

Imaging Research Center & Baltimore Museum of Art. (2005). *Virtual tour: Cone sisters' apartments*. Retrieved July 15, 2012, from www.irc.umbc.edu/2005/10/01/cone-sisters/

Interactive Storytelling. (2008). *First joint international conference on interactive digital storytelling*. Erfurt, Germany: Springer. ISBN 3540894241

Jones, J. (2011). Influences on the formation and evolution of physarum polycephalum inspired emergent transport networks. *Natural Computing*, *10*(4), 1345–1369. doi:10.1007/s11047-010-9223-z.

Karr, J. R., Sanghvi, J. C., Macklin, D. N., Gutschow, M. V., Jacobs, J. M., & Bolival, B. ... Covert, M. W. (2012). The dawn of virtual cell biology. *Cell, 150*(2), 389-401. doi 10.1016/j.cell.2012.05.044. Retrieved July 21, 2012, from http://www.cell.com/abstract/S0092-8674%2812%2900776-3

Kirby, B., Campbell, J., Aksak, B., Pillai, P., Hoburg, J., Mowry, T., & Goldstein, S. C. (2005). Catoms: Moving robots without moving parts. In *Proceedings of AAAI (Robot Exhibition)*, (pp. 1730-1731). AAAI.

Liarokapis, F., White, M., & Lister, P. (2004). Augmented reality interface toolkit. In *Proceedings of 8th International Conference on Information Visualization*, (pp. 761-767). IEEE.

Liu, L. (2003a). Engineering aesthetics and aesthetic ergonomics: Theoretical foundations and a dual-process research methodology. *Ergononics*, *46*(13/14), 1273–1292. doi:10.1080/0014013031 0001610829 PMID:14612319.

Liu, L. (2003b). The aesthetic and the ethic dimensions of human factors and design. *Ergononics*, *46*(13/14), 1293–1305. doi:10.1080/001401303 10001610838 PMID:14612320.

Liu, L. (2012). *Information systems as a reference discipline for visual design*. Retrieved September 27, 2012, from www.swdsi.org/swdsi2012/proceedings_2012/papers/.../PA137.pdf

LoPiccolo, P. (2003). A virtual exhibit transports museum-goers back in time to view a famous art collection in its original setting. *Computer Graphics World, 26*(3). Retrieved July 14, 2012, from http://www.cgw.com/Publications/CGW/2003/Volume-26-Issue-3-March-2003-/Backdrop-3-03.aspx

March, S., & Storey, V. (2008). Design science in the information systems discipline: An introduction to the special issue on design science research. *Management Information Systems Quarterly, 32*(4), 725–730. Retrieved from misq.org/misq/downloads/download/editorial/152/.

Markham, A. (2011). Adaptive social hierarchies: From nature to networks. In *Bio-Inspired Computing and Networking*, (pp. 305-350). CRC Press, Taylor & Francis Group. ISBN 1420080326

Markoff, J. (2012, July 20). In first, software emulates lifespan of entire organism. *The New York Times*. Retrieved July 21, 2012, from http://www.nytimes.com/2012/07/21/science/in-a-first-an-entire-organism-is-simulated-by-software.html?_r=1&partner=rss&emc=rss

Mayrhofer, R., Sommer, A., & Saral, S. (2012). Airwriting: A platform for scalable, privacy-preserving, spatial group messaging. *International Journal of Pervasive Computing and Communications, 8*(1), 53–78. doi:10.1108/17427371211221081.

O'Brien, J., & Marakas, G. (2009). *Introduction to information systems*. McGraw Hill Higher Education.

O'Brien, J., & Marakas, G. (2010). *Management information systems*. McGraw-Hill/Irwin.

Peak, D. A., Prybutok, V., Gibson, M., & Xu, C. (2012a). Information systems as a reference discipline for visual design. In *Proceedings of the 2012 SWDSI Conference*. Retrieved September 27, 2012, from http://www.swdsi.org/swdsi2012/proceedings_2012/papers/Papers/PA137.pdf

Peak, D. A., Prybutok, V., Gibson, M., & Xu, C. (2012b). Information systems as a reference discipline for visual design. *International Journal of Art, Culture and Design Technologies, 2*(2), 57–71. doi:10.4018/ijacdt.2012070105.

Pessoa Forte, J. A., Arruda, D., Gomes, C. A., Nogueira, G., & Cavalcante de Almeida, C. F. (2011). Educational services in second life: A study based on flow theory. *International Journal of Web-Based Learning and Teaching Technologies, 6*(2), 1–17. doi:10.4018/jwltt.2011040101.

Poslad, S. (2009). *Ubiquitous computing: Smart devices, environments and interactions*. Wiley. doi:10.1002/9780470779446.

Price, A. (2011). *The sundagger explorer*. Retrieved July 14, 2012, from http://accad.osu.edu/%7Eaprice/works/sundagger/index.html

Rind, A., Wang, T., Aigner, W., Miksch, S., Wongsuphasawat, K., Plaisant, C., & Shneiderman, B. (2010). Interactive information visualization for exploring and querying electronic health records: A systematic review. *Human-Computer Interaction Lab*. Tech Report HCIL-2010-19.

Rockley, A., & Cooper, C. (2012). *Managing enterprise content: A unified content strategy* (2nd ed.). New Riders.

Satyanarayanan, M. (2010). The role of cloudlets in mobile computing. *Microsoft Networking Research Summit*. Retrieved July 13, 2012, from research.microsoft.com/en-us/events/mcs2010/satya.ppt

Satyanarayanan, M., Bahl, V., Caceres, R., & Davies, N. (2009). The case for VM-based cloudlets in mobile computing. *IEEE Pervasive Computing, 99*(1). doi:10.1109/MPRV.2009.64.

Solé, R. (2008). On networks and monsters: The possible and the actual in complex systems. *Leonardo, 41*(3), 253–258. doi:10.1162/leon.2008.41.3.253.

Sookhanaphibarn, K., & Thawonmas, R. (2010). Digital museums in 3D virtual environment. In *Handbook of Research on Methods and Techniques for Studying Virtual Communities: Paradigms and Phenomena*, (pp. 713-730). Academic Press.

UbiComp. (2012). *14th ACM international conference on ubiquitous computing*. Retrieved August 10, 2012, from http://www.ubicomp.org/ubicomp2012/

Ubirobots. (2012). *ACM international workshop, smart gadgets meet ubiquitous and social robots on the web*. Retrieved August 10, 2012, from https://sites.google.com/site/ubirobots2012/

Vanacker, B., & Heider, D. (2012). Ethical harm in virtual communities. *Convergence, 18*(1), 71–84. doi: doi:10.1177/1354856511419916.

Wahl, H., Winiwarter, W., & Quirchmayr, G. (2011). Towards an intelligent integrated language learning environment. *International Journal of Pervasive Computing and Communications, 7*(3), 220 – 239. doi 10.1108/17427371111173013. Retrieved July 15, 2012, from http://www.emeraldinsight.com/journals.htm?issn=1742-7371&volume=7&issue=3&

Wang, T., Deshpande, A., & Shneiderman, B. (2010). A temporal pattern search algorithm for personal history event visualization. *IEEE Transactions on Knowledge and Data Engineering, 99*. doi: 10.1109/TKDE.2010.257. HCIL-2009-14

Wang, T., Wongsuphasawat, K., Plaisant, C., & Shneiderman, B. (2010). Visual information seeking in multiple electronic health records: Design recommendations and a process model. In *Proceedings of the 1st ACM International Informatics Symposium* (IHI '10), (pp. 46-55). ACM.

Wang, T. D. (2010). *Interactive visualization techniques for searching temporal categorical data*. (PhD dissertation). University of Maryland, University Park, MD. Retrieved July 12, 2012, from http://hcil2.cs.umd.edu/trs/2010-15/2010-15.pdf

Web Science 2012 Conference. (2012). Retrieved July 15, 2012, from http://www.websci12.org/program

Wegrzyn, W. (2012). *Plug-in book offers digital content when the page is turned*. Retrieved November 22, 2012, from http://www.springwise.com/media_publishing/plug-in-book-offers-digital-content-page-turned/

Wei, H., Liu, E., Zhao, X., McFarlane, N. J. B., & Clapworthy, G. J. (2011). Article. In *Proceedings of the 15th International Conference on Information Visualization*, (pp. 632-637). IEEE.

Weiser, M. (1991). The computer for the 21st century. *Scientific American*. Retrieved January 17, 2012, from http://classes.dma.ucla.edu/Winter06/256/text/Weiser-21stCentury.pdf

Weng, J. (2012). *Natural and artificial intelligence: Introduction to computational brain-mind*. BMI Press.

Wolfram MathWorld. (2012). Retrieved August 28, 2012, from http://mathworld.wolfram.com/VoronoiDiagram.html

Xu, Z., & Taylor, D. (2000). Using motion platform as a haptic display for virtual inertia simulation. In *Proceedings of the Information Visualization Conference*, (pp. 498-504). IEEE.

Yotsumoto, Y., & Watanabe, T. (2008). Defining a link between perceptual learning and attention. *PLoS Biology*, *6*(8), e221. doi:10.1371/journal.pbio.0060221 PMID:18752357.

Conclusion

This book is based on a focal premise and a conviction that multisensory perception is becoming an important factor in shaping the current lifestyle, technology, and reasoning. Multisensory perception is a decisive factor in a growing number of biologically inspired technological solutions. Our knowledge about living organisms, which communicate in ways not resembling the traits of human senses, enables us to develop biologically inspired theories, applications, and devices. Investigating multisensory perception is also important because social communication is becoming multisensory, interactive, interdisciplinary, and technology-augmented. Researchers are often solving problems according to social behaviors and the heuristic ways the social societies of insects, fish, or birds solve their difficult situations. This book discusses the background material that would be useful in working on projects involving the reader's input. Text and images are of service to assist the readers in enhancing their solutions with explanatory visuals, and hopefully finding joy in these tasks.

Further Reading

Boden, M. A. (2006). *Mind as Machine. A History of Cognitive Science*. Oxford: Clarendon Press.

Boden, M. A. (2010). *Creativity and Art: Three Roads to Surprise*. Oxford University Press.

Damasio, A. (2010). Self Comes to Mind. New York, New York: Pantheon. Reprint: Vintage. ISBN 030747495X.

DeLanda, M. (2006). A New Philosophy of Society: Assemblage Theory and Social Complexity. Continuum; 1 edition. ISBN 0826491693.

DeLanda, M. (2011). *Philosophy and Simulation: The Emergence of Synthetic Reason*. Continuum; 1 edition ISBN1441170286.

Eco, U., & McEwen, A. (Translator, author) (2004). History of Beauty. Rizzoli. ISBN 0847826465.

Eco, U., & McEwen, A. (Translator, author) (2007). On Ugliness. Rizzoli. ISBN 0847829863.

Editors of Phaidon (Author). (2005). The Art Book. Phaidon Press. Midi ed. ISBN 071484487X / 9780714844879 / 0-7148-4487-X (pocket edition).

Editors of Phaidon (Author). (2012). *The Art Book (Hardcover)*. Phaidon Press.

Gardner, H. (1993c). *Art, Mind, and Brain: A Cognitive Approach to Creativity*. New York: Basic Books, A Division of Harper Collins Publishers.

Gardner, H. (1997). *Extraordinary minds. Portraits of exceptional individuals and an examination of our extraordinariness*. Basic Books, Harper Collins Publishers.

Kandinsky, W. (2011). *Concerning the Spiritual in Art*. Empire Books. (Original work published 1911).

Klanten, R. (Author, Ed., Bourquin N. (Ed.), Ehman, S. (Ed.), & Tissot, T. (Ed.). (2010). Data Flow 2: Visualizing Information in Graphic Design. Die Gestalten Verlag. ISBN 3899552782.

R. Klanten, N. Bourquin, S. Ehmann, F. van Heerden, & T. Tissot (Eds.). (2008). *Data Flow: Visualising Information in Graphic Design*. Die Gestalten Verlag.

Lima, M. (2011). Visual Complexity: Mapping Patterns of Information. New York: Princeton Architectural Press. ISBN 978 1 56898 936 5.

Murakami, H. (2011). 1Q84. Knopf. ISBN 0307593312.

Pearson, M. (2011). Generative Art. Manning Publications; Pap/Psc edition. ISBN 1935182625.

Tufte, E. R. (1983/2001). *The Visual Display of Quantitative Information* (2nd ed.). Cheshire, CT: Graphics Press.

Tufte, E. R. (1992/2005). *Envisioning Information*. Cheshire, CT: Graphics Press. Third printing with revision.

Tufte, E. R. (1997). *Visual and Statistical Thinking: Displays of Evidence for Making Decisions*. Cheshire, Connecticut: Graphics Press.

Yau, N. (2011). *Visualize this: The flowing data guide to design visualization and statistics*. Wiley.

Zeki, S. (2009). Splendors and Miseries of the Brain: Love, Creativity, and the Quest for Human Happiness. Wiley-Blackwell; 5th edition, ISBN 1405185570.

Compilation of References

Abbott Abbott, E. A. (1994/2008). *Flatland: A romance of many dimensions* (Oxford World's Classics). Oxford University Press.

Abelson, H., Allen, D., Coore, D., Hanson, C., Homsy, G., & Knight, T. Jr et al. (2000). Amorphous computing. *Communications of the ACM, 43*(4), 74–82. doi:10.1145/332833.332842.

About.com. (2012). *How many emails are sent every day?* Retrieved May 30, 2012, from http://email.about.com/od/emailtrivia/f/emails_per_day.htm

Adamatzky, A., De Lacy Costello, B., Holley, J., Gorecki, J., & Bull, L. (2011). Vesicle computers: Approximating Voronoi diagram on Voronoi automata. *Chaos Solitons and Fractals, 44*, 480-489. Retrieved August 28, 2012, from http://arxiv.org/pdf/1104.1707.pdf

Adamatzky, A. (2010). *Physarum machines: Computers from slime mould.* World Scientific Publishing Company.

Albers, J. (1969). Homage to the square: Soft spoken. In *Heilbrunn Timeline of Art History.* New York: The Metropolitan Museum of Art. Retrieved January 23, 2012, from http://www.metmuseum.org/toah/works-of-art/1972.40.7

Albers, J. (2010). *Interaction of color.* Yale University Press. (Original work published 1963).

al-Rifaie, M. M., Aber, A., & Bishop, J. M. (2012). Cooperation of nature and physiologically inspired mechanisms in visualisation. In *Biologically-Inspired Computing for the Arts: Scientific Data through Graphics.* Hershey, PA: IGI Global. doi:10.4018/978-1-4666-0942-6.ch003.

Amabile, T. M. (1996). *Creativity in context: Update to the social psychology of creativity.* Boulder, CO: Westview Press.

Ansburg, P. I., & Hill, K. (2002). Creative and analytic thinkers differ in their use of attentional resources. *Personality and Individual Differences, 34*(7), 1141–1152. doi:10.1016/S0191-8869(02)00104-6.

Apple in Education. (2013). Retrieved March 30, 2013, from http://www.apple.com/education/apps/

Aristotle (2013). *Poetics.* Oxford University Press.

Arlin, P. K. (1974). *Problem finding: The relation between selective cognitive process variables and problem-finding performance.* (Unpublished doctoral dissertation). University of Chicago, Chicago, IL.

Arlin, P. K. (1984). Adolescent and adult thought: A structural interpretation. In M. L. Commons, F. A. Richards, & C. Armon (Eds.), *Beyond formal operations: Late adolescent and adult cognitive development.* New York: Praeger.

Arnheim, R. (n.d.). *BrainyQuote.com.* Retrieved March 20, 2012, from http://www.brainyquote.com/quotes/quotes/r/rudolfarnh374327.html

Arnheim, R. (1969). *Visual thinking.* Berkeley, CA: University of California Press.

Arnheim, R. (1969/2004). *Visual thinking: Thirty-fifth anniversary printing.* University of California Press.

Arnheim, R. (1974). *Art and visual perception.* Berkeley, CA: University of California Press.

Arnheim, R. (1988). *The power of the center - A study of composition in the visual arts.* Berkeley, CA: University of California Press.

Arnheim, R. (1990). Language and the early cinema. *Leonardo*, 3–4.

Aron, E. N. (2006). The clinical implications of Jung's concept of sensitiveness. *Journal of Jungian Theory and Practice, 8*(2), 11-43. Retrieved January 6, 2012, from http://www.junginstitute.org/pdf_files/JungV8N2p11-44.pdf

Artlyst, London Art Network. (2012). Retrieved March 30, 2013, from http://www.artlyst.com/articles/art-students-to-be-given-free-ipads

Ascher, M., & Ascher, R. (1980). *Code of the Quipu: A study in media, mathematics, and culture*. Ann Arbor, MI: University of Michigan Press.

Ascher, M., & Ascher, R. (1997). *Mathematics of the Incas: Code of the Quipu*. Dover Publications.

Association, I. R. (2013). *Digital literacy: Concepts, methodologies, tools, and applications* (3 vols.). doi: doi:10.4018/978-1-4666-1852-7.

Auson, K. S. (2012). 0h!m1gas: A biomimetic stridulation environment. In *Biologically-Inspired Computing for the Arts: Scientific Data through Graphics*. Hershey, PA: IGI Global Publishing. doi:10.4018/978-1-4666-0942-6.ch004.

Baddeley, A. (2000). The episodic buffer: A new component of working memory? *Trends in Cognitive Sciences, 4*(11), 417–423. doi:10.1016/S1364-6613(00)01538-2 PMID:11058819.

Baddeley, A. D., & Hitch, G. J. L. (1974). *Working memory*. Academic Press.

Baddeley, A. D., Thompson, N., & Buchanan, M. (1975). Word length and the structure of memory. *Journal of Verbal Learning and Verbal Behavior, 1*, 575–589. doi:10.1016/S0022-5371(75)80045-4.

Baddeley, A., Eysenck, M. W., & Anderson, M. C. (2009). *Memory*. New York: Psychology Press.

Bales, K. L., & Kitzmann, C. D. (2011). Animal models for computing and communications: Past approaches and future challenges. In X. Yang (Ed.), *Bio-Inspired Computing and Networking*. CRC Press, Taylor & Francis Group. ISBN 1420080326

Bales, K. L., & Kitzmann, C. D. (2011). Animal models for computing and communications. In *Bio-Inspired Computing and Networking*. CRS Press. doi:10.1201/b10781-3.

Bancroft, J., & Wang, Y. (2011). A computational simulation of the cognitive process of children knowledge acquisition and memory development. *International Journal of Cognitive Informatics and Natural Intelligence, 5*(2), 17–36. doi:10.4018/jcini.2011040102.

Banissi, E. (2011). Preface. In *Proceedings of the Information Visualisation 15th International Conference*. London: IEEE. ISBN 978-1-4577-0868-8. Retrieved June 17, 2012 from http://ieeexplore.ieee.org/xpl/mostRecentIssue.jsp?punumber=6003377

Barbieri, M. (2010). On the origin of language: A bridge between biolinguistics and biosemiotics. *Biosemiotics, 3*, 201-223. doi 10 1007/s12304-010-9088-7

Bar-David, E., Compton, E., Drennan, L., Finder, B., Grogan, K., & Leonard, J. (2009). Nonverbal number knowledge in preschool-age children. *Mind Matters: The Wesleyan Journal of Psychology, 4*, 51–64.

Barthes, R. (1964). *Image, reason, déraison*. Paris: Les Libraires Assoiés.

Battail, G. (2009). Living versus inanimate: The information border. *Biosemiotics, 2*, 321–341. doi 10.1007/s12304-009-9059-z. Retrieved December 17, 2011, from http://www.springerlink.com/content/r376x-87u5mk68732/fulltext.pdf

Battail, G. (2011). An answer to Schrödinger's what is life. *Biosemiotics, 4*, 55–67. doi:10.1007/s12304-010-9102-0.

Baum, L. F., Hearn, M. P., & Denslow, W. W. (2000). *The annotated wizard of oz* (Centennial Ed.). New York: W. W. Norton & Company. ISBN 0393049922

Beaird, J. (2007). *The principles of beautiful design: Design beautiful web sites using this simple step-by-step guide*. SitePoint Pty. Ltd.

Beardon, C. & Malmborg, L. (Eds.). (2010). *Digital creativity: A reader (innovations in art and design)*. Routledge.

Bederson, B., & Shneiderman, B. (2003). *The craft of information visualization: Readings and reflections*. San Francisco, CA: Morgan Kaufmann Publishers.

Beltz, E. (2009). *Frogs: Inside their remarkable world*. Firefly Books.

Berens, R. R. (1999). The role of artists in ship camouflage during world war I. *Leonardo, 32*(1), 53–59. doi:10.1162/002409499553000.

Berker, T., Hartmann, M., Punie, Y., & Ward, K. J. (2006). *Domestication of media and technology*. McGraw-Hill International.

Berlyne, D. E. (1955). The arousal and satiation of perceptual curiosity in the rat. *Journal of Comparative and Physiological Psychology, 48*(4), 238–246. doi:10.1037/h0042968 PMID:13252149.

Berlyne, D. E. (1960). *Conflict, arousal, and curiosity*. New York: McGraw-Hill Book Company. doi:10.1037/11164-000.

Berners-Lee, T., & Cailliau, R. (1990). *WorldWideWeb: Proposal for a hypertexts project*. Retrieved May 1, 2012, from http://www.w3.org/Proposal.html

Berners-Lee, T. (2000). *Weaving the web: The original design and ultimate destiny of the world wide web*. New York: HarperBusiness.

Bernstein, K. (2005). Expression in the form of our own making. In *Proceedings of the Special Year in Art & Mathematics, A+M=X International Conference*. Univ. of Colorado.

Bertoline, G., Wiebe, E., Hartman, N., & Ross, W. (2010). *Fundamentals of graphics communication* (6th ed.). McGraw-Hill Science/Engineering/Math.

Bertschi, S. (2009). Knowledge visualization and business analysis: Meaning as media. In *Proceedings of the 13th International Conference Information Visualisation*, (pp. 480-485). IEEE Computer Society Press.

Bertschi, S., & Bubenhofer, N. (2005). Linguistic learning: A new conceptual focus in knowledge visualization. In *Proceedings of the Information Visualisation 9th International Conference*, (pp. 383-389). IEEE.

Bertschi, S., Bresciani, S., Crawford, T., Goebel, R., Kienreich, W., Lindner, M., et al. (2011). What is knowledge visualization? Perspectives on an emerging discipline. In *Proceedings of the Information Visualisation 15th International Conference*, (pp. 329-336). London: IEEE. ISBN 978-1-4577-0868-8

Bettencourt, A. (1989). *What is constructivism and why are they all talking about it?*. Journal Announcement: RIEMAR91. (ERIC Document Reproduction Service No. ED325402).

Binkley, T. (1990). Digital dilemmas. *Leonardo*, 13–19.

Binkley, T. (1996). Personalities at the salon of digits. *Leonardo, 29*(5), 337–338. doi:10.2307/1576396.

Bisbort, A., & Kite, B. (1997). *Sunday afternoon, looking for the car: The aberrant art of Barry Kite*. Pomegranate Communications.

Blais, J., & Ippolito, J. (2006). *At the edge of art*. Thames & Hudson.

Bloom, B. S. (1956). *Taxonomy of educational objectives*. Boston, MA: Allyn and Bacon.

Bloom, B. S., Hastings, J. T., & Madaus, G. F. (1971). *Handbook on formative and summative evaluation of student learning*. New York: McGraw-Hill Book Company.

Boden, M. (2007). *How creativity works*. Retrieved April 23, 2012 from cii.dmu.ac.uk/resources/maggie/Boden.pdf

Boden, M. (2007). Creativity in a nutshell. *Think, 5*(15), 83–96. doi:10.1017/S147717560000230X.

Boden, M. A. (1998). Creativity and artificial intelligence. *Artificial Intelligence, 103*, 347–356. doi:10.1016/S0004-3702(98)00055-1.

Boden, M. A. (2006). *Mind as machine: A history of cognitive science*. Oxford, UK: Clarendon Press.

Boden, M. A. (2009). Computer models of creativity. *AI Magazine, 30*(3).

Boden, M. A. (2010). *Creativity and art: Three roads to surprise*. Oxford University Press.

Boden, M. A., & Edmonds, E. A. (2009). What is generative art? *Digital Creativity*, *20*(1-2), 21–46. doi:10.1080/14626260902867915.

Bolin, L. (2010, November 20). Now you see me, now you don't: The artist who turns himself into the invisible man. *Mail Online*. Retrieved June 26, 2012, from http://www.dailymail.co.uk/news/article-1201398/Liu-Bolin-The-Chinese-artist-turns-Invisible-Man.html

Bolter, J. D. (1987). Text and technology: Reading and writing in the electronic age. *Library Resources & Technical Services*, *31*(1), 12–23.

Borgia, G. (1995). Complex male display and female choice in the spotted bowerbird: Specialized functions for different bower decorations. *Animal Behaviour*, *49*, 1291–1301. doi:10.1006/anbe.1995.0161.

Borst, G., Ganis, G., Thompson, W. L., & Kosslyn, S. M. (2011). *Representations in mental imagery and working memory: Evidence from different types of visual masks*. DOI 10.3758/s13421-011-0143-7. Retrieved May 6, 2012, from http://isites.harvard.edu/fs/docs/icb.topic561942.files/Borst_Ganis_Thompson_Kosslyn_2011.pdf

Borst, G., & Kosslyn, S. M. (2008). Visual mental imagery and visual perception: Structural equivalence revealed by scanning processes. *Memory & Cognition*, *36*(4), 849–862. doi:10.3758/MC.36.4.849 PMID:18604966.

Bossomaier, T., & Snyder, A. (2004). Absolute pitch accessible to everyone by turning off part of the brain? *Organised Sound*, *9*(2), 181-189. DOI 10.1017/S1355771804000263. Retrieved December 13, 2011, from www.centreforthemind.com/publications/absolutepitch.pdf

Boström, N. (2005). A history of transhumanist thought. *Journal of Evolution and Technology*, *14*(1). Retrieved January 31, 2012, from http://www.nickbostrom.com/papers/history.pdf

Bourqui, R., & Jourdan, F. (2008). Revealing subnetwork roles using contextual visualization: Comparison of metabolic networks. In *Proceedings of iV, 12th International Conference on Information Visualisation*. IEEE.

Boyd, R. B. (1989). *Identifying and meeting the needs of students functioning at the concrete operations stage of cognitive development in the general chemistry classroom*. RIEJAN90 (ERIC Document Reproduction Service No. ED309984).

Brannon, E. M. (2005). The independence of language and mathematical reasoning. *Proceedings of the National Academy of Sciences of the United States of America*, *102*(9), 3177–3178. doi:10.1073/pnas.0500328102 PMID:15728346.

Brannon, E., & Terrace, H. (1998). Ordering of the numerosities 1 to 9 by monkeys. *Science*, *282*, 746–749. doi:10.1126/science.282.5389.746 PMID:9784133.

Branton, D., Deamer, D. W., Marziali, A., Bayley, H., Benner, S. A., & Butler, T. … Schloss, J. A. (2008). The potential and challenges of nanopore sequencing. *Nat Biotechnol.*, *26*(10), 1146–1153. doi 10.1038/nbt.1495. Retrieved August 28, 2012, from http://www.ncbi.nlm.nih.gov/pmc/articles/PMC2683588/

Brath, R. (2009). The many dimensions of shape. In *Proceedings of the International Conference on Information Visualization*. IEEE.

Brath, R. (2010). Multiple shape attributes in information visualization: Guidance from prior art and experiments. In *Proceedings of the 2010 14th International Conference Information Visualisation*, (pp. 433-438). IEEE.

Bresciani, S., & Eppler, M. J. (2007). Usability of diagrams for group knowledge work: Toward an analytic description. In *Proceeding I-KNOW 07*. Graz, Austria: I-KNOW.

Bresciani, S., & Eppler, M. J. (2008). Do visualizations foster experience sharing and retention in groups? Towards an experimental validation. In *Proceedings I-KNOW 08*. Graz, Austria: I-KNOW.

Bresciani, S., & Eppler, M. J. (2010). Choosing knowledge visualizations to augment cognition: The managers' view. In *Proceedings of the 14th International Conference on Information Visualisation*, (pp. 355-360). IEEE.

Bresciani, S., Blackwell, A. F., & Eppler, M. J. (2008). A collaborative dimensions framework: Understanding the mediating role of conceptual visualizations in collaborative knowledge work. In *Proceedings of the Hawaii International Conference on System Sciences*. IEEE.

Bresciani, S., Eppler, M., Kaul, A., & Ylinen, R. (2011). The effectiveness of knowledge visualization for organizational communication in Europe and India. In *Proceedings of the 15th International Conference on Information Visualization*. IEEE. ISBN 978-1-4577-0868-8

Bresciani, S., Tan, M., & Eppler, M. J. (2011). *Augmenting communication with visualization: Effects on emotional and cognitive response*. IADIS ICT, Society and Human Beings.

Bridges. (2010). *Mathematical connections in art, music, and science*. Retrieved October 1, 2012 from http://www.bridgesmathart.org/

Brockman, J. (Ed.). (2011). *Is the internet changing the way you think? The net's impact on our minds and future*. New York: Harper Perennial.

Brooks, D. (2011, January 17). Social animal: How the new sciences of human nature can help make sense of a life. *New Yorker*.

Broudy, H. S. (1987). *The role of imagery in learning*. Malibu, CA: The Getty Center for Education in the Arts.

Broudy, H. S. (1991). Reflections on a decision. *Journal of Aesthetic Education*, *25*(4), 31–34. doi:10.2307/3332900.

Bruner, J. S. (1962). The connotation of creativity. In *Contemporary Approaches to Creative Thinking*. New York: Atherton Press. doi:10.1037/13117-001.

Bucik, V., & Neubauer, A. C. (1996). Bimodality in the Berlin model of intelligence structure (BIS), a replication study. *Personality and Individual Differences*, *21*(6), 987–1005. doi:10.1016/S0191-8869(96)00129-8.

Burke, K., & Kite, B. (2000). *Rude awakening at arles: The aberrant art of Barry Kite: Postcard book*. Pomegranate Communications Inc.(Original work published 1996).

Burkhard, R. (2005). *Knowledge visualization: The use of complementary visual representations for the transfer of knowledge – A model, a framework, and four new approaches*. (D.Sc. thesis). Swiss Federal Institute of Technology (ETH Zurich), Zurich, Switzerland.

Burkhard, R., Bischof, S., & Herzog, A. (2008). The potential of crowd simulations for communication purposes in architecture. In *Proceedings of iV, 12th International Conference on Information Visualisation*. IEEE Computer Society Press.

Burkhard, R., Meier, M., Smis, J. M., Allemang, J., & Honish, J. (2005). Beyond Excel and PowerPoint: Knowledge maps for the transfer and creation of knowledge in organizations. In *Proceedings of iV, International Conference on Information Visualisation*. IEEE.

Burmester, M., Mast, M., Tille, R., & Weber, W. (2010), How users perceive and use interactive information graphics: An exploratory study. In *Proceedings of iV, 14th International Conference on Information Visualisation*, (pp. 361-368). IEEE.

Burt, C. L. (2009). *Mental and scholastic tests*. Cornell University Library. (Original work published 1922).

Butterworth, B., Reeve, R., Reynolds, F., & Lloyd, D. (2008). Numerical thought with and without words: Evidence from indigenous Australian children. *Proceedings of the National Academy of Sciences of the United States of America*, *105*(35), 13179–13184. doi:10.1073/pnas.0806045105 PMID:18757729.

Camus, A. (1951/1992). *The rebel: An essay on man in revolt*. Vintage.

Card, S. K., Mackinlay, J. D., & Shneiderman, B. (Eds.). (1999). *Readings in information visualization: Using vision to think*. San Francisco, CA: Morgan Kaufman.

Carey, S. (2004). Bootstrapping & the origin of concepts. *Daedalus*, 59–68. doi:10.1162/001152604772746701.

Carroll, L., & Tenniel, J. (1993). *Alice's adventures in wonderland*. Dover. (Original work published 1898).

Casaleggio Associati. (2011). *From internet of people to internet of things*. Retrieved July 17, 2012, from http://www.casaleggio.it/pubblicazioni/Focus_internet_of_things_v1.81%20-%20eng.pdf

Chandler, D. (2001). *Semiotic for beginners*. Retrieved June 1, 2012, from http://www.aber.ac.uk/media/Documents/S4B/sem08a.html

Chandler, D. (2013). *Semiotics for beginners: Articulation*. Retrieved March 3, 2013, from http://users.aber.ac.uk/dgc/Documents/S4B/sem08a.html

Chen, C. (2004). *Information visualization: Beyond the horizon*. Springer-Verlag.

Chernoff, H. (1971). *The use of faces to represent points in n-dimensional space graphically*. Palo Alto, CA: Stanford University.

Chernoff, H. (1973). The use of faces to represent points in k-dimensional space graphically. *Journal of the American Statistical Association*, *68*, 361–368. doi:10.1080/01621459.1973.10482434.

Chi, R. P., & Snyder, A. W. (2011). Facilitate insight by non-invasive brain stimulation. *PLoS ONE*, *6*(2), e16655. doi:10.1371/journal.pone.0016655 PMID:21311746.

Chomsky, N. (1959). A review of B. F. Skinner's verbal behavior. *Language*, *35*(1), 26–58. doi:10.2307/411334.

Chong, N.-Y., & Mastrogiovanni, F. (2011). *Handbook of research on ambient intelligence and smart environments: Trends and perspectives*. Hershey, PA: IGI Global. doi:10.4018/978-1-61692-857-5.

Christopher, N. (2008). *The bestiary*. Dial Press Trade.

Chung, D. N. (2012). *Language arts*. Western Washington University. Retrieved May 6, 2012, from http://faculty.wwu.edu/auer/Resources/Hayakawa-Abstraction-Ladder.pdf

Cichy, R. M., Heinzle, J., & Haynes, J.-D. (2011). Imagery and perception share cortical representations of content and location. *Cerebral Cortex*, *22*(2), 372–380. doi:10.1093/cercor/bhr106 PMID:21666128.

Clayton, S., Pinheiro, V., Meiguins, B. S., Simões, A., Meiguins, G., & Almeida, H. L. (2008). A Tourism information analysis tool for mobile devices. In *Proceedings of iV, 12th International Conference on Information Visualisation*. IEEE Computer Society Press.

Clinch, S., Harkes, J., Friday, A., Davies, N., & Satyanarayanan, M. (2012). *How close is close enough? Understanding the role of cloudlets in supporting display appropriation by mobile users*. Retrieved from http://www.cs.cmu.edu/~satya/docdir/clinch-percom-2012-CAMERA-READY.pdf

Coie, J., Costanzo, P., & Farnill, D. (1973). Specific transitions in the development of spatial perspective taking ability. *Developmental Psychology*, *9*(2), 166–177. doi:10.1037/h0035062.

Collin, H. (2012). Biological translation: Virtual code, form, and interactivity. In A. Ursyn (Ed.), *Biologically-Inspired Computing for the Arts: Scientific Data through Graphics*. Hershey, PA: IGI Global Publishing.

Committee on Forefronts of Science at the Interface of Physical and Life Sciences. National Research. (2010). *Research at the intersection of the physical and life sciences*. Retrieved December 28, 2011, from http://www.nap.edu/catalog.php?record_id=12809

Copeland, J. (2012, June 19). Alan Turing: The codebreaker who saved millions of lives. *BBC News Technology*. Retrieved June 23, 2012, from http://www.bbc.co.uk/news/technology-18419691

Coquillart, S. (2012). First-person visuo-haptic environment: From research to applications. In *Proceedings of the International Conference on Computer Graphics Theory and Applications and International Conference on Information Visualization Theory and Applications*. ISBN 978-989-8565-02-0

Correa, P. N., & Correa, A. N. (2004). *Nanometric functions of bioenergy*. Akronos Publishing.

Counsel, J. (2003). Pointing the finger: A role for hybrid representations in VR and video. In *Proceedings of the Seventh International Conference on Information Visualization*. IEEE. ISBN 0-7695-1988-1

Courtois, C., Verdegem, P., & De Marez, L. (2012). The triple articulation of media technologies in audiovisual media consumption. *Television & New Media, 13*(2). Retrieved May 31, 2012, from http://tvn.sagepub.com/content/early/2012/04/11/1527476412439106.abstract

Cox, D. J. (2008). Using the supercomputer to visualize higher dimensions: An artist's contribution to scientific visualization. *Leonardo*, *41*(4), 390–400. doi:10.1162/leon.2008.41.4.390.

Craft, B., & Cairns, P. (2005). Beyond guidelines: What can we learn from the visual information seeking mantra? In *Proceedings of the 9th International Conference on Information Visualization,* (pp. 110-118). IEEE.

Craft, B., & Cairns, P. (2008). Directions for methodological research in information visualization. In *Proceedings of the 12th International Conference on Information Visualisation*. IEEE Computer Society Press.

Creswell, J. W. (2008). *Research design: Qualitative, quantitative, and mixed methods approaches*. Sage Publications, Inc..

Crick, F. H. C. (1958). On protein synthesis. *Symposia of the Society for Experimental Biology*, *12*, 139–163. PMID:13580867.

Crick, F. H. C. (1970). Central dogma of molecular biology. *Nature*, *227*(5258), 561–563. doi:10.1038/227561a0 PMID:4913914.

Cruz-Neira, C., Sandin, D. J., DeFanti, T. A., Kenyon, R. V., & Hart, J. C. (1992). The CAVE: Audio visual experience automatic virtual environment. *Communications of the ACM, 35*(6), 64–72. doi:10.1145/129888.129892.

Csikszentmihalyi, M. (2011). *The creative personality.* Retrieved October 1, 2012, from http://www.psychologytoday.com/articles/199607/the-creative-personality

Csikszentmihalyi, M. (1996). *Creativity: The work and lives of 91 eminent people.* HarperCollins Publishers.

Csikszentmihalyi, M. (1996). Society, culture, and person: A systems view of creativity. In R. J. Sternberg (Ed.), *The nature of creativity: Contemporary psychological perspectives* (pp. 325–339). New York: Cambridge University Press.

Csikszentmihalyi, M. (1997). *Creativity: Flow and the psychology of discovery and invention.* Harper Perennial.

Csikszentmihalyi, M. (1998). *Finding flow: The psychology of engagement with everyday life.* Basic Books.

Curiosity.com. (2012). *Discovery.* Retrieved January 12, 2012, from http://curiosity.discovery.com/search?query=curiosity

Danto, A. C. (1964). The artworld. *The Journal of Philosophy, 61*, 571–584. doi:10.2307/2022937.

Davies, A., Henrik, C., Dalton, C., & Campbell, N. (2012). Generating 3D morphable model parameters for facial tracking: Factorising identity and expression. In *Proceedings of the International Conference on Computer Graphics Theory and Applications and International Conference on Information Visualization Theory and Applications*, (pp. 309-318). IEEE. ISBN: 978-989-8565-02-0

Davies, S. (1991). *Definitions of art.* London: Cornell University Press.

de Saint-Exupéry, A. (2000). *The little prince.* Mariner Books. (Original work published 1943).

De Saussure, F. (1915). *Ferdinand de Saussure, 1857-1913.* Genève, Switzerland: Kundig.

Dehaene, S., Piazza, M., Pinel, P., & Cohen, L. (2003). Three parietal circuits for number processing. *Cognitive Neuropsychology, 20*(3/4/5/6), 487-506.

Dehaine, S. (1999). *The number sense: How the mind creates mathematics.* Oxford, UK: Oxford University Press.

DeLanda, M. (2000). *A thousand years of nonlinear history.* The MIT Press.

DeLanda, M. (2010). *Deleuze: History and science.* Atropos Press.

Deleuze, G., & Guattari, F. (2009). *Anti-oedipus: Capitalism and schizophrenia.* Penguin Classics.

Deleuze, G., Guattari, F., & Massumi, B. (1987). *A thousand plateaus: Capitalism and schizophrenia.* University of Minnesota Press.

DeLoache, J. S. (2005, August). Mindful of symbols. *Scientific American*, 72–77. doi:10.1038/scientificamerican0805-72 PMID:16053140.

Demazeau, Y., Müller, J., Corchado Rodriguez, J. M., & Bajo, Pérez, J. (Eds.). (2012). *Advances on practical applications of agents and multi-agent systems: 10th international conference on practical applications of agents.* Springer. ISBN 3642287859

Denis, M. (1989). *Image and cognition.* Paris: Presses Universitaires de France.

Derrida, J. (1991). *The truth in painting.* Chicago: The University of Chicago Press.

Desain, P., & Honing, H. (1992). *Music, mind, and machine: Studies in computer music, music cognition, and artificial intelligence (kennistechnologie).* Thesis Pub. ISBN 9051701497

Desain, P., & Honing, H. (1996). Physical motion as a metaphor for timing in music: the final ritard. In *Proceedings of the 1996 International Computer Music Conference*, (pp. 458-460). San Francisco, CA: ICMA.

Dobbs, S. M. (1992). *The DBAE handbook.* Los Angeles, CA: Getty Center for Education in the Arts.

Dorin, A., & Korb, K. (2012). Creativity refined. In *Computers and Creativity.* Berlin: Springer. doi:10.1007/978-3-642-31727-9_13.

Downs, R. M., & Stea, D. (1977). *Maps in minds: Reflections on cognitive mapping.* New York: Harper & Row Publishers.

Duchamp, M. (1917). *The blind man*. Retrieved January 23, 2012, from http://sdrc.lib.uiowa.edu/dada/blind-man/2/05.htm

Duch, W. (2007). Intuition, insight, imagination and creativity. *IEEE Computational Intelligence Magazine*, *2*(3), 40–52. doi:10.1109/MCI.2007.385365.

Dunbar, K. (1997). How scientists think: Online creativity and conceptual change in science. In T. B. Ward, S. M. Smith, & S. Vaid (Eds.), *Conceptual structures and processes: Emergence, discovery and change*. APA Press.

Dunn, J., & Clark, M. A. (1999). Life music: The sonification of proteins. *Leonardo*, *32*(1), 25–32. doi:10.1162/002409499552966.

Dunn, M., Greenhill, S. J., Levinson, S. C., & Gray, R. D. (2011). Evolved structure of language shows lineage-specific trends in word-order universals. *Nature*, *473*, 79–82. doi:10.1038/nature09923 PMID:21490599.

Dweck, C. S. (2006). *Mindset: The new psychology of success*. New York: Random House.

Ebrahimi, A. (2003). *C++ programming easy ways*. Boston: American Press.

Eco, U. (1979). *The open work*. Cambridge, MA: Harvard University Press.

Eco, U. (1990). *The limits of interpretation*. Bloomington, IN: Indiana University Press.

Eco, U. (2002). *Art and beauty in the middle ages* (H. Bredin, Trans.). New Haven, CT: Yale University Press.

Eggebrecht, A. T., White, B. R., Chen, C., Zhan, Y., Snyder, A. Z., Dehlgani, H., & Culver, J. P. (2012, February 10). A quantitative spatial comparison of high-density diffuse optical tomography and fMRI cortical mapping. *NeuroImage*. doi:10.1016/j.neuroimage.2012.01.124 PMID:22330315.

Ehle, R. (2012). *Virtual music*. University of Northern Colorado. Unpublished.

Emmeche, C., Kuli, K., & Stjernfelt, F. (2002). *Reading Hoffmeyer, rethinking biology*. Tartu University Press.

Engelbrecht, A. P. (2003). *Computational intelligence: An introduction*. Wiley.

Eng, H. (1931). *The psychology of children's drawings - Form the first stroke to the coloured drawing*. London: Kegan Paul, Trench, Trübner, & Co.

English Wikipedia. (2012). Retrieved January 19, 2012, from http://en.wikipedia.org/wiki/English_Wikipedia

EON Reality Corporation. (2011). Retrieved June 17, 2012, from http://www.eonreality.com/products_iportal.html

EPA. (2012). Retrieved March 9, 2013, from http://cfpub.epa.gov/npdes/cwa.cfm?program_id=45

Eppler, M. J. (2011). What is an effective knowledge visualization? Insights from a review of seminal concepts. In *Proceedings of the 15th International Conference on Information Visualization*. IEEE. ISBN 978-1-4577-0868-8

Eppler, M. J., & Burkhard, R. A. (2012). *Knowledge communication*. Retrieved June 17, 2012, from http://www.knowledge-communication.org

Eppler, M. J., & Pfister, R. A. (2010). Drawing conclusions: Supporting decision making through collaborative graphic annotations. In *Proceedings of the 14th International Conference on Information Visualization*, (pp. 369-374). IEEE.

Eppler, M. J., Mengis, J., & Bresciani, S. (2008). Seven types of visual ambiguity: On the merits and risks of multiple interpretations of collaborative visualizations. In *Proceedings of iV, 12th International Conference on Information Visualisation*. IEEE.

Eppler, M. J. (2005). Knowledge communication. In *Encyclopedia of Knowledge Management*. Hershey, PA: IGI Global. doi:10.4018/978-1-59140-573-3.ch042.

Eppler, M. J., & Burkhard, R. A. (2007). Visual representations in knowledge management: Framework and cases. *Journal of Knowledge Management*, *4*(11), 112–122. doi:10.1108/13673270710762756.

Erdem, A., & Karaismailoglu, S. (2011). Neurophysiology of emotions. In Affective Computing and Interaction: Psychological, Cognitive, and Neuroscientific Perspectives (pp. 1-24). IGI Global. ISBN 1616928921

Eriksson, Y., Johansson, P., & Björndal, P. (2011). Showing action in pictures. In *Proceedings of the 15th International Conference on Information Visualization*, (pp. 403-408). IEEE. ISBN 978-1-4577-0868-8

Evans, B. (2011). Materials of the data map. *International Journal of Creative Interfaces and Computer Graphics*, *2*(1), 14–26. doi:10.4018/jcicg.2011010102.

Evans, G. W., Marrero, D. G., & Butler, P. A. (1981). Environmental learning and cognitive mapping. *Environment and Behavior*, *13*, 83–104. doi:10.1177/0013916581131005.

Evomusart. (2012). *Proceedings of the 1st international conference and 10th European event on evolutionary and biologically inspired music, sound, art and design*. Retrieved May 31, 2012, from http://evostar.dei.uc.pt/2012/call-for-contributions/evomusart/

Eysel, U. T. (2002) Plasticity of receptive fields in early stages of the adult visual system. In *Perceptual learning*, (pp. 43-66). A Bradford Book. ISBN 0262062216

Fajardo, N., & Vande Moere, A. (2008). ExternalEyes: Evaluating the visual abstraction of human emotion on a public wearable display device. In *Proceedings of the Conference of the Australian Computer-Human Interaction (OZCHI'08)*, (pp. 247-250). Cairns, Australia: OZCHI. Retrieved November 2, 2011, from http://web.arch.usyd.edu.au/~andrew/publications/ozchi08.pdf

Farhi, P. (2008, February 5). CNN hits the wall for the election. *The Washington Post*. Retrieved September 22, 2012, from http://www.washingtonpost.com/wp-dyn/content/article/2008/02/04/AR2008020402796.html

Favareau, D. (2008). Iconic, indexical and symbolic understanding: A commentary on Anna Aragno's the language of empathy. *Journal of the American Psychoanalytic Association*, *56*(3), 783–801. doi:10.1177/0003065108322687 PMID:18802128.

Feigenbaum, L., Herman, I., Hongsermeier, T., Neumann, E., & Stephens, S. (2007). The semantic web in action. *Scientific American*, *297*, 90–97. doi:10.1038/scientificamerican1207-90 PMID:17894177.

Ferguson, H., & Ferguson, C. (1998). *Eightfold way: The sculpture*. Retrieved October 30, 2011, from http://library.msri.org/books/Book35/files/fergall.pdf

Ferreira, M. I. A. (2011). Interactive bodies: The semiosis of architectural forms a case study. *Biosemiotics Online First*. doi 10.1007/s12304-011-9126-0. Retrieved December 20, 2011, from http://www.springerlink.com/content/?Author=Maria+Isabel+Aldinhas+Ferreira

Ferri, F. (2007). *Visual languages for interactive computing: Definitions and formalizations*. Hershey, PA: IGI Global. doi:10.4018/978-1-59904-534-4.

Fisher, R. A. (1958). Cancer and smoking. *Nature*, *182*(596). doi: doi:10.1038/182596a0 PMID:13577916.

Fleur de coin, your online guide to coin collecting. (2012). Retrieved April 27, 2012, from http://www.fleur-de-coin.com/articles/oldestcoin.asp

Floreano, D., & Mattiussi, C. (2008). *Bio-inspired artificial intelligence: Theories, methods, and technologies*. The MIT Press.

Flowing Data. (2012). Retrieved March 24, 2012, from http://flowingdata.com/category/visualization/infographics/

Flynn, J. R. (1980). *Race, IQ, and Jensen*. London: Routledge and Kegan Paul.

Flynn, J. R. (1987). Massive IQ gains in 14 nations: What IQ tests really measure. *Psychological Bulletin*, *101*, 171–191. doi:10.1037/0033-2909.101.2.171.

Flynn, J. R. (1994). IQ gains over time. In R. J. Sternberg (Ed.), *Encyclopedia of human intelligence* (pp. 617–623). New York: Macmillan.

Flynn, J. R. (1999). Searching for justice: The discovery of IQ gains over time. *The American Psychologist*, *54*, 5–20. doi:10.1037/0003-066X.54.1.5.

Flynn, J. R., & Dickens, W. T. (2001). Heritability estimates versus large environmental effects: The IQ paradox resolved. *Psychological Review*, *108*(2), 346–369. doi:10.1037/0033-295X.108.2.346 PMID:11381833.

Forsyth, J. B., & Martin, T., L. (2012). Tools for interdisciplinary design of pervasive computing. *International Journal of Pervasive Computing and Communications*, *8*(2), 112–132. doi:10.1108/17427371211245355.

Fox, P., & Hendler, J. (2011). Changing the equation on scientific data visualization. *Science*, *331*, 705–708. doi:10.1126/science.1197654 PMID:21311008.

Frank, M. C., Everett, D. L., Fedorenko, E., & Gibson, E. (2008). Number as a cognitive technology: Evidence from Piraha language and cognition. *Cognition*, *108*(3), 819–824. doi:10.1016/j.cognition.2008.04.007 PMID:18547557.

Fried, P. A. (2002). Tobacco consumption during pregnancy and its impact on child development. In *Encyclopedia of Early Childhood Development*. Retrieved June 22, 2012, from http://www.child-encyclopedia.com/documents/FriedANGxp.pdf

Friendly, M. (2009). *Milestones in the history of thematic cartography, statistical graphics, and data visualization*. Retrieved March 8, 2012, from http://www.math.yorku.ca/SCS/Gallery/milestone/milestone.pdf

Fuglestad, T., & Tiedemann, S. (2013). *iPads in art education*. Retrieved March 30, 2013, from http://ipadsinart.weebly.com/

Galanter, P. (2010). Complexity, neuroaesthetics, and computational aesthetic evaluation. In *Proceedings of 13th Generative Art Conference GA2010*, (pp. 399-409). GA. Retrieved April 27, 2012, from http://www.generativeart.com/on/cic/GA2010/2010_31.pdf

Galanter, P. (2011). Computational aesthetic evaluation: Past and future. In *Computers and Creativity*. Berlin: Springer.

Gardner, H. (1993a/2011). *Frames of mind: The theory of multiple intelligences*. New York: Basic Books. ISBN 0465024335

Gardner, H. (1993b/2006). *Multiple intelligences: The theory in practice*. New York: Basic Books. ISBN 978-0465047680

Gardner, H. (1978). *Developmental psychology*. Boston: Little, Brown and Co..

Gardner, H. (1983). Artistic intelligences. *Art Education, 36*(2), 47–49. doi:10.2307/3192663.

Gardner, H. (1983/2011). *Frames of mind: The theory of multiple intelligences* (3rd ed.). Basic Books.

Gardner, H. (1988). Toward more effective arts education. *Journal of Aesthetic Education, 22*(1), 157–167. doi:10.2307/3332972.

Gardner, H. (1993). *Art, mind, and brain: A cognitive approach to creativity*. New York: Basic Books, A Division of Harper Collins Publishers.

Gardner, H. (1994). *Creating minds: An anatomy of creativity seen through the lives of Freud, Einstein, Picasso, Stravinsky, Eliot, Graham, and Gandhi*. New York: Basic Books.

Gardner, H. (1997). *Extraordinary minds: Portraits of exceptional individuals and an examination of our extraordinariness*. Basic Books, Harper Collins Publishers.

Gardner, H. (2006). *Multiple intelligences: New horizons in theory and practice*. Basic Books.

Garrigues, A. (2008). *Dressing the horse and rider*. Retrieved December 10, 2012, from http://annisa.garrigues.net/classhandouts/Dressing%20the%20Horse%20and%20Rider.pdf

Gazzaniga, M. S. (1997). The split brain revisited. *Scientific American, 279*(1), 50–55. doi:10.1038/scientificamerican0798-50 PMID:9648298.

Gazzaniga, M. S. (2005). Forty-five years of split-brain research and still going strong. *Nature Reviews. Neuroscience, 6*(8), 653–U651. doi:10.1038/nrn1723 PMID:16062172.

Gelman, R., & Butterworth, B. (2005). Number and language: How are they related? *Trends in Cognitive Sciences, 9*(1), 6–10. doi:10.1016/j.tics.2004.11.004 PMID:15639434.

Gervás, P. (2009). Computational approaches to storytelling and creativity. *AI Magazine, 30*(3), 49–62.

Getzels, J. W., & Csikszentmihalyi, M. (1970). Concern for discovery: An attitudinal component of creative production. *Journal of Personality, 38*, 91–105. doi:10.1111/j.1467-6494.1970.tb00639.x PMID:5435830.

Getzels, J. W., & Csikszentmihalyi, M. (1976). *The creative vision: A longitudinal study of problem finding in art*. New York: Wiley, a Wiley-Interscience Publication.

Getzels, J. W., & Jackson, P. W. (1962). *Creativity and intelligence, explorations with gifted students*. London: Wiley. doi:10.2307/40223437.

Giaccardi, E., & Fischer, G. (2008). Creativity and evolution: A metadesign perspective. *Digital Creativity, 19*(1), 19–32. doi:10.1080/14626260701847456.

Gimeno, J., Morillo, P., Ordu, J. M., & Fernández, M. (2012). An occlusion-aware AR authoring tool for assembly and repair tasks. In *Proceedings of the International Conference on Computer Graphics Theory and Applications and International Conference on Information Visualization Theory and Applications*, (pp. 377-386). ISBN: 978-989-8565-02-0

Girot, C., & Truniger, F. (2006). The walker's perspective: Strategies for conveying landscape perception using audiovisual media. In *Proceedings of the 10th International Conference on Information Visualization*. IEEE Computer Society Press.

Gloor, P. A. (2006). *Swarm creativity: Competitive advantage through collaborative innovation networks*. New York, NY: Oxford University Press.

D. Gökçay, & G. Yildrim (Eds.). (2011). *Affective computing and interaction: Psychological, cognitive, and neuroscientific perspectives*. IGI Global.

Goldsborough, R. (2010/2004). Article. *The Numismatist*. Retrieved May 2, 2012, from http://rg.ancients.info/lion/article.html

Goldstein, E., Saunders, R., Kowalchuk, J. D., & Katz, T. H. (1986). *Understanding and creating art* (Annotated teachers ed.). Dallas, TX: Guard Publishing Company.

Goldstein, S. C., Mowry, T. C., Campbell, J. D., Ashley-Rollman, M. P., De Rosa, M., & Funiak, S. et al. (2009). Beyond audio and video: Using claytronics to enable pario. *AI Magazine, 30*(2).

Goldsworthy, A. (2007). *The biological effect of weak electromagnetic fields*. Retrieved January 6, 2012, from www.electrosense.nl/nl/download/6

Gonçalves, J. P., Madeira, S. C., & Oliveira, A. L. (2009). BiGGEsTS: Integrated environment for biclustering analysis of time series gene expression data. *BMC Research Notes, 2*, 124. doi:10.1186/1756-0500-2-124 PMID:19583847.

Gordon, P. (2004). Numerical cognition without words: Evidence from the Amazonia. *Science, 306*, 496–499. doi:10.1126/science.1094492 PMID:15319490.

Goya, F. (2012). *Grabados de goya caprichos*. Retrieved from http://arte.laguia2000.com/pintura/neoclasicismo-2/las-series-de-grabados-de-francisco-de-goya

Graf, D. (2009). *Point it: Traveller's language kit - The original picture dictionary - Bigger and better*. Graf Ed.

Graham, D. (2005). Information visualization theory and practice. In *Proceedings of iV, International Conference on Information Visualisation,* (pp. 599-603). IEEE.

Grey, W. (2000). Metaphor and meaning. *Minerva – An Internet Journal of Philosophy, 4*. Retrieved February 6, 2012, from http://www.minerva.mic.ul.ie//vol4/metaphor.html

Grigoryev, Y. (2012, March 16). How much information is stored in the human genome? *BitesizeBio*. Retrieved May 9, 2012, from http://bitesizebio.com/articles/how-much-information-is-stored-in-the-human-genome/

Gross, C., Gros, C. P., Abel, P., Loisel, D., Trichaud, N., & Paris, J. P. (2000). Mapping information onto 3D virtual worlds. In *Proceedings of the 4th Conference on Information Visualisation*. IEEE Computer Society Press.

Gruber, H. E. (1973). Courage and cognitive growth in children and scientists. In M. Schwebel, & J. Ralph (Eds.), *Piaget in the classroom*. New York: Basic Books.

Guilford, J. P. (1950). Creativity. *The American Psychologist, 5*, 444–454. doi:10.1037/h0063487 PMID:14771441.

Guilford, J. P. (1959). Traits of creativity. In *Creativity and its cultivation*. New York: Harper and Row.

Guilford, J. P. (1967). *The nature of human intelligence*. New York: McGraw-Hill.

Guilford, J. P. (1968). *Intelligence, creativity and their educational implications*. San Diego, CA: Robert Knapp, Publ..

Halberda, J., Mazzocco, M. M., & Feigenson, L. (2008). Individual differences in non-verbal number acuity correlate with maths achievement. *Nature, 455*, 665-668. doi 10.1038/nature07246. Retrieved January 6, 2012, from http://www.nature.com/nature/journal/v455/n7213/full/nature07246.html

Hammer, E. F. (1984). *Creativity, talent, and personality: An exploratory investigation of the personalities of gifted adolescent artists*. Malabar, FL: R. E. Krieger Publishing Company.

Hardesty, L. (2012, January 24). 'Genetic programming': The mathematics of taste. *PhysOrg*. Retrieved March 4, 2013, from http://phys.org/news/2012-01-genetic-mathematics.html#jCp

Hartman, N. W., & Bertoline, G. R. (2005). Spatial abilities and virtual technologies: Examining the computer graphics learning environment. In *Proceedings of 9th International Conference on Information Visualisation*. IEEE.

Hartman, N. W., & Bertoline, G. R. (2006). *Virtual reality-based spatial skills assessment and its role in computer graphics education*. Retrieved March 19, 2012 from http://www.siggraph.org/s2006/main.php?f=conference&p=edu&s=44

Hartmann, S. A., Pinto, P. C., Runkler, T. A., & Sousa, J. M. C. (2011). Social insect societies for optimization of dynamic NP-hard problems. In *Bio-Inspired Computing and Networking*, (pp. 43-68). CRC Press, Taylor & Francis Group. ISBN 1420080326

Hart, R. A., & Moore, G. T. (1973). The development of spatial cognition: A review. In R. Downs, & D. Stea (Eds.), *Image and Environment*. Chicago: Aldine Press.

Harty, D. (2010). *Drawing//experience: A process of translation*. (Ph.D. Thesis). Loughborough University, Leicestershire, UK.

Harty, H., & Beall, D. (1984). Toward development of a children's science curiosity measure. *Journal of Research in Science Teaching*, *21*(4), 425–436. doi:10.1002/tea.3660210410.

Hayakawa, S. I., & Hayakawa, A. R. (1941/1991). *Language in thought and action* (5th ed.). Harvest Original.

Hayles, K. (2010). How we became posthuman: Ten years on. *Paragraph*, *33*(3), 318–323. doi:10.3366/para.2010.0202.

HBP. (2012). *The human brain project: A report to the european commission*. Retrieved July 21, 2012, from http://www.humanbrainproject.eu/files/HBP_flagship.pdf

Head, B. V. (2012/1886). *Historia numorum: A manual of Greek numismatics*. Retrieved April 27, 2012, from http://www.snible.org/coins/hn/aegina.html

Heaney, S. (2001). *Beowulf*. W. W. Norton & Company.

Heinrich, F. (2010). On the behalf of avatars: What on earth have the aesthetics of Byzantine icons to do with avatar in social technologies? *Digital Creativity, 21*(1), 4-10. Retrieved October 26, 2010 from http://www.informaworld.com/smpp/title~db=all~content=g922551031

Heinrich, F. (2010). On the belief in avatars: What on earth have the aesthetics of the Byzantine icons to do with the avatar in social technologies? *Digital Creativity, 21*(1), 4-10. Retrieved May 3, 2012, from http://www.tandfonline.com/doi/abs/10.1080/14626261003654236

Hennessey, B. A., & Amabile, T. M. (1987). *Creativity and learning: What research says to the teacher*. Washington, DC: National Educational Association.

Henshilwood, C. S., d'Errico, F., Yates, R., Jacobs, Z., Tribolo, C., & Duller, G. A. et al. (2002). Emergence of modern human behavior: Middle stone age engravings from South Africa. *Science*, *295*(5558), 1278–1280. doi:10.1126/science.1067575 PMID:11786608.

Herz, R. S. (2009). Aromatherapy facts and fictions: A scientific analysis of olfactory effects on mood, physiology and behavior. *The International Journal of Neuroscience*, *119*(2), 263–290. doi:10.1080/00207450802333953 PMID:19125379.

Herz, R. S., Schankler, C., & Beland, S. (2004). Olfaction, emotion and associative learning: Effects on motivated behavior. *Motivation and Emotion*, *28*, 363–383. doi:10.1007/s11031-004-2389-x.

History of Printing. (2013). Retrieved March 9, 2013, from http://en.wikipedia.org/wiki/History_of_printing

Hockney, D. (2001). *Secret knowledge: Rediscovering the lost techniques of the old masters*. London: Thames and Hudson.

Hoffman, D. D. (2000). Visual intelligence: How we create what we see. New York: W. W. Norton & Company. ISBN 0-393-04669-9 0393319679

Hoffman, D. H., Greenberg, P., & Fitzner, D. (Eds.). (1980). *Lifelong learning and the visual arts*. Reston, VA: National Art Education Association.

Hoffmeyer, J. (2008). The semiotic body. *Biosemiotics*, *1*(2), 169-190. doi 10.1007/s12304-008-9015-3. Retrieved December 20, 2011, from http://www.springerlink.com/content/03752h27v677377l/

Hoffmeyer, J. (1993). Biosemiotics and ethics. In *Culture and Environment: Interdisciplinary Approaches*. Oslo, Norway: University of Oslo.

Hudson, A. (2011, May 21). Is graphene a miracle material? *BBC News*. Retrieved October 22, 2011, from http://news.bbc.co.uk/2/hi/programmes/click_online/9491789.stm

Huizinga, J. (1950). *Homo ludens: A study of the play element in culture*. New York: Roy Publishers.

IASE. (2012). *Institute for advanced science & engineering*. Retrieved May 31, 2012, from http://iase.info/

ICCC. (2012), *International conference on computational creativity*. Retrieved October 1, 2012, from http://computationalcreativity.net/iccc2012/

iCloud. (2012). Retrieved June 17, 2012 from https://www.icloud.com/

Imaging Research Center & Baltimore Museum of Art. (2005). *Virtual tour: Cone sisters' apartments*. Retrieved July 15, 2012, from www.irc.umbc.edu/2005/10/01/cone-sisters/

Inhelder, B. (1977). Genetic epistemology and developmental psychology. *Annals of the New York Academy of Sciences*, *291*, 332–341. doi:10.1111/j.1749-6632.1977.tb53084.x.

Inhelder, B., & Piaget, J. (1958). *Growth of logical thinking from childhood to adolescence*. New York: Basic Books. doi:10.1037/10034-000.

Inselberg, A. (2009). *Parallel coordinates: Visual multidimensional geometry and its applications*. New York: Springer.

Interactive Storytelling. (2008). *First joint international conference on interactive digital storytelling*. Erfurt, Germany: Springer. ISBN 3540894241

International Visual Literacy Association. (2011). Retrieved September 1, 2011, from http://www.ivla.org/

Isaacson, W. (2011). *Steve Jobs*. Simon & Schuster.

Ittelson, W. H. (2007). The perception of nonmaterial objects and events. *Leonardo*, *40*(3), 279–283. doi:10.1162/leon.2007.40.3.279.

Jackendoff, R. S. (2007). *Language, consciousness, culture: Essays on mental structure*. Cambridge, MA: MIT Press.

Jakobson, R. (1995). *On language*. Boston: Harvard University Press.

Jarvis, E. D. (2004). Learned birdsong and the neurobiology of human language. *Annals of the New York Academy of Sciences*, *1016*, 749–777. doi:10.1196/annals.1298.038 PMID:15313804.

Jenny, H. (2001). *Cymatics: A study of wave phenomena & vibration* (3rd ed.). Macromedia Press.

Jern, M., & Franzen, J. (2006). GeoAnalytics - Exploring spatio-temporal and multivariate data. In *Proceedings of the Information Visualisation (IV) 10th International Conference*, (pp. 25 – 31). IEEE.

Johnson, S. H. (1987). *A cognitive-structural approach to adult creativity*. Paper presented at the Annual Symposium of the Jean Piaget Society. Philadelphia, PA.

Jones, J. (2011). Influences on the formation and evolution of physarum polycephalum inspired emergent transport networks. *Natural Computing*, *10*(4), 1345–1369. doi:10.1007/s11047-010-9223-z.

Jones, O. (2010). *The grammar of ornament*. Deutsch Press. (Original work published 1856).

Jung, C. G. (1976). *Psychological types*. (Original work published 1921).

Juster, N. (2000). *The dot and the line: A romance in lower mathematics*. Chronicle Books. (Original work published 1963).

Jusufi, I., Dingjie, Y., & Kerren, A. (2010). The network lens: Interactive exploration of multivariate networks using visual filtering. In *Proceedings of the Information Visualisation (IV) 14th International Conference*, (pp. 35-42). IEEE.

Kandinsky, W. (2011). *Concerning the spiritual in art*. Empire Books. (Original work published 1911).

Kant, I. (2007). *Critique of judgement* (Oxford World's Classics). Oxford University Press.

Kaptchuk, T. (2000). *The web that has no weaver: Understanding Chinese medicine*. McGraw-Hill.

Karlans, M., Schuerhoff, C., & Kaplan, M. (1969). Some factors related to architectural creativity in graduating architectural students. *The Journal of Genetic Psychology, 81*, 203–215. doi:10.1080/00221309.1969.9711286.

Karr, J. R., Sanghvi, J. C., Macklin, D. N., Gutschow, M. V., Jacobs, J. M., & Bolival, B. ... Covert, M. W. (2012). The dawn of virtual cell biology. *Cell, 150*(2), 389-401. doi 10.1016/j.cell.2012.05.044. Retrieved July 21, 2012, from http://www.cell.com/abstract/S0092-8674%2812%2900776-3

Kashdan, T. B., Gallagher, M. W., Silvia, P. J., Winterstein, B. P., Breen, W. E., Terhar, D., & Steger, M. F. (2009). The curiosity and exploration inventory-II: Development, factor structure, and psychometrics. *Journal of Research in Personality, 43*, 987–998. doi:10.1016/j.jrp.2009.04.011 PMID:20160913.

Katifori, A., Torou, E., Halatsis, C., Lepouras, G., & Vassilakis, C. (2006). A comparative study of four ontology visualization techniques in protégé: Experiment setup and preliminary results. In *Proceedings of the Information Visualisation (IV) 10th International Conference*, (pp. 417-423). IEEE.

Kaufman, J. C., & Sternberg, R. J. (Eds.). (2010). *The Cambridge handbook of creativity*. Cambridge University Press. doi:10.1017/CBO9780511763205.

Kenderdine, S. (2010). Immersive visualization architectures and situated embodiments of culture and heritage. In *Proceedings of the 14th International Conference on Information Visualization*, (pp. 408-414). IEEE.

Keogh, D., Asbeck, P., Chung, T., Dupuis, R. D., & Feng, M. (2006). Digital etching of III-N materials using a two-step Ar/KOH technique. *Journal of Electronic Materials, 35*(4), 772-776. Retrieved June 18, 2012, from http://sigma.ucsd.edu/research/articles/2006/2006_9.pdf

Kepes, G. (1995). *Language of vision*. Dover Publications. (Original work published 1944).

Kernbach, S., & Eppler, M. J. (2010). The use of visualization in the context of business strategies: An experimental evaluation. In *Proceedings of the Information Visualisation (IV) 14th International Conference*. ISBN 978-1-4244-7846-0

Kirby, B., Campbell, J., Aksak, B., Pillai, P., Hoburg, J., Mowry, T., & Goldstein, S. C. (2005). Catoms: Moving robots without moving parts. In *Proceedings of AAAI (Robot Exhibition)*, (pp. 1730-1731). AAAI.

Kite, B. (1994/2000). *Rude awakening at Arles: The aberrant art of Barry Kite: Postcard book by Barry Kite and Katie Burke*. Pomegranate Communications Inc..

Kite, B. (1997). *Sunday afternoon, looking for the car: The aberrant art of Barry Kite*. Pomegranate Communications.

Klee, P. (1948/1979). *On modern art*. New York: Faber & Faber.

Klee, P. (1969). *The thinking eye: Paul Klee notebooks*. G. Wittenborn.

Knipp, T. (2003) Creative performance: Does the computer retard artistic development? In *Proceedings of the International Conference on Information Visualization*, (pp. 621-625). IEEE.

Knödel, S., Hachet, M., & Guitton, P. (2009). Interactive generation and modification of cutaway illustrations for polygonal models. In *Proceedings of the 10th International Symposium on Smart Graphics*, (pp. 140-151). Berlin: Springer-Verlag. ISBN 978-3-642-02114-5

Knowledge-communication.org. (2011). Retrieved September 1, 2011, from http://www.knowledge-communication.org/overview.html

Koestler, A. (1964). *The act of creation*. New York, NY: Penguin Books.

Komar and Melamid, Fineman, M., Schmidt, J., & Eggers, D. (2000). *When elephants paint: The quest of two Russian artists to save the elephants of Thailand*. Harper. ISBN-10: 0060953527

Korzybski, A. (1995). *Science and sanity: An introduction to non-Aristotelian systems and general semantics* (5th ed.). Brooklyn, NY: Institute of General Semantics. (Original work published 1933).

Kosslyn, S. M. (1978). Measuring the visual angle of the mind's eye. *Cognitive Psychology, 10*, 356–389. doi:10.1016/0010-0285(78)90004-X PMID:688748.

Kosslyn, S. M. (1980). *Image and mind*. Cambridge, MA: Harvard.

Kosslyn, S. M. (1991). A cognitive neuroscience of visual cognition: Further developments. In R. H. Logie, & M. Denis (Eds.), *Mental Images in Human Cognition*. Amsterdam: North Holland, Elsevier Science Publishers B.V. doi:10.1016/S0166-4115(08)60523-3.

Kosslyn, S. M., Alpert, N. M., Thompson, W. L., Maljkovic, V., Weise, S. B., & Chabris, C. F. et al. (1993). Visual mental imagery activates topographically organized visual cortex: PET investigations. *Journal of Cognitive Neuroscience*, *5*(3), 263–287. doi:10.1162/jocn.1993.5.3.263.

Kounios, J., & Beeman, M. (2009). The aha! moment: The cognitive neuroscience of insight. *Current Directions in Psychological Science*, *18*(4), 210–216. doi:10.1111/j.1467-8721.2009.01638.x.

Kozin, A. (2008). Translation and semiotic phenomenology: The case of Gilles Deleuze. *Across Language and Culture*, *9*(2), 161-175.

Krippendorff, K. (1990). Product semantics: A triangulation and four design theories. In *Product Semantics '89*. Helsinki: University of Industrial Arts.

Krippendorff, K. (2011). Conversation and its erosion into discourse and computation. In T. Thellefsen, B. Sørensen, & P. Cobley (Eds.), *From First to Third via Cybersemiotics* (pp. 129–176). Frederiksberg, Denmark: SL Forlagene.

Krippendorff, K. (2011). Discourse and the materiality of its artifacts. In *Matters of Communication: Political, Cultural, and Technological Challenges to Communication Theorizing*. New York: Hampton Press.

Kuhn, T. (1970). *The structure of scientific revolutions*. Chicago: University of Chicago Press.

Kuipers, B. (1982). The map in the head metaphor. *Environment and Behavior*, *14*(2), 202–220. doi:10.1177/0013916584142005.

Kull, K. (1999). Biosemiotics in the twentieth century: A view from biology.[from http://www.zbi.ee/~kalevi/bsxxfin.htm]. *Semiotica*, *127*(1/4), 385–414. Retrieved January 22, 2012

Lakoff, G. (1990). The invariance hypothesis: Is abstract reason based on image-schemas? *Cognitive Linguistics*, *1*(1), 39–74. doi:10.1515/cogl.1990.1.1.39.

Lakoff, G., & Núñez, R. E. (2001). *Where mathematics comes from: How the embodied mind brings mathematics into being*. Basic Books.

Lambert, N. (2011). From imaginal to digital: mental imagery and the computer image space. *Leonardo*, *44*(5), 439–443. doi:10.1162/LEON_a_00245.

Lang, S. B. (2006). Merging knowledge from different disciplines in search of potential design axioms. In *Proceedings of the 10th International Conference on Information Visualization*, (pp. 183-188). IEEE.

Lauzzana, R., & Penrose, D. (1987, April 23). A 21st century manifesto. *FINEART Forum*, *1*.

Lauzzana, R., & Penrose, D. (1992). A pre-21st century manifesto. *Languages of Design*, *1*(1), 87.

Lavington, S. (2012, June 19). Alan Turing: Is he really the father of computing? *BBC News Technology*. Retrieved June 23, 2012, from http://www.bbc.co.uk/news/technology-18327261

Lebwohl, B. (2011, July 25). Semir Zeki: Beauty is in the brain of the beholder. *EarthSky*. Retrieved October 19, 2012, from http://earthsky.org/human-world/semir-zeki-beauty-is-in-the-brain-of-the-beholder

LeCorre, M., & Carey, S. (2006). One, two, three, four, nothing more: An investigation of the conceptual sources of the verbal counting principles. *Cognition*, *105*(2), 395–438. doi:10.1016/j.cognition.2006.10.005.

Lehrer, J. (2007). *Proust was a neuroscientist*. Houghton Mifflin.

Lengler, R. (2006). Identifying the competencies of 'visual literacy' – A prerequisite for knowledge visualization. In *Proceedings of 10th International Conference on Information Visualisation*. IEEE.

Lengler, R., & Eppler, M. (2006). *A periodic table of visualization methods*. Retrieved March 26, 2012, from http://www.visual-literacy.org/periodic_table/periodic_table.html

Lengler, R., & Eppler, M. (2007). *Towards a periodic table of visualization methods for management*. Retrieved March 26, 2012, from http://www.visual-literacy.org/periodic_table/periodic_table.pdf

Lengler, R., & Eppler, M. (2011). *A periodic table of visualization methods*. Retrieved November 2, 2011, from http://www.visual-literacy.org/periodic_table/periodic_table.html

Lengler, R., & Vande Moere, A. (2009). Guiding the viewer's imagination: How visual rhetorical figures create meaning in animated infographics. In *Proceedings of the 13th International Conference on Information Visualization*. DOI 10.1109/IV.2009.102

Lewicki, P., Hill, T., & Czyzewska, M. (1992). Nonconscious acquisition of information. *The American Psychologist*, *47*, 796–801. doi:10.1037/0003-066X.47.6.796 PMID:1616179.

Liarokapis, F., White, M., & Lister, P. (2004). Augmented reality interface toolkit. In *Proceedings of 8th International Conference on Information Visualization*, (pp. 761-767). IEEE.

Lifehack. (2012). *4 reasons why curiosity is important and how to develop it*. Retrieved January 12, 2012, from http://www.lifehack.org/articles/productivity/4-reasons-why-curiosity-is-important-and-how-to-develop-it.html

Lima, M. (2013). *Visual complexity*. Retrieved February 20, 2013, from http://www.visualcomplexity.com/vc/

Lima, M. (2011). *Visual complexity: Mapping patterns of information*. New York: Princeton Architectural Press.

Limson, J. (2010). Abstract engravings show modern behavior emerged earlier than previously thought. *Science in Africa*. Retrieved May 9, 2012, from http://www.scienceinafrica.co.za/2002/january/ochre.htm

Lipari, L. (1988). Masterpiece theater: What is discipline based art education and why have so many people learned to distrust it? *Artpaper*, *7*, 14–16.

Liu, L. (2012). *Information systems as a reference discipline for visual design*. Retrieved September 27, 2012, from www.swdsi.org/swdsi2012/proceedings_2012/papers/.../PA137.pdf

Liu, L. (2003). Engineering aesthetics and aesthetic ergonomics: Theoretical foundations and a dual-process research methodology. *Ergononics*, *46*(13/14), 1273–1292. doi:10.1080/00140130310001610829 PMID:14612319.

Liu, L. (2003). The aesthetic and the ethic dimensions of human factors and design. *Ergononics*, *46*(13/14), 1293–1305. doi:10.1080/00140130310001610838 PMID:14612320.

Livemocha. (2012) Retrieved October 1, 2012, from http://www.livemocha.com/

Logothetis, N. K., Pauls, J., Augath, M., Trinath, T., & Oeltermann, A. (2001). Neurophysiological investigation of the basis of the fMRI signal. *Nature*, *412*, 150–157. doi:10.1038/35084005 PMID:11449264.

Lohman, K. J., Lohmann, C. M., & Putman, N. F. (2007). Magnetic maps in animals: Nature's GPS. *The Journal of Experimental Biology*, *210*, 3697–3705. doi:10.1242/jeb.001313 PMID:17951410.

Loizides, A. (2012). *Andreas Loizides research home page*. Retrieved February 12, 2012, from http://www.cs.ucl.ac.uk/staff/a.loizides/research.html

Loizides, A., & Slater, M. (2002). The empathic visualisation algorithm (EVA) - An automatic mapping from abstract data to naturalistic visual structure. In *Proceedings of iV02, 6th International Conference on Information Visualisation*. IEEE.

Loizides, A., & Slater, M. (2001). The empathic visualisation algorithm (EVA), Chernoff faces re-visited. In *Conference Abstracts and Applications, Technical Sketch SIGGRAPH 2001*. ACM Press.

LoPiccolo, P. (2003). A virtual exhibit transports museum-goers back in time to view a famous art collection in its original setting. *Computer Graphics World, 26*(3). Retrieved July 14, 2012, from http://www.cgw.com/Publications/CGW/2003/Volume-26-Issue-3-March-2003-/Backdrop-3-03.aspx

Lowenfeld, V. (1947). *Creative and mental growth*. New York: Macmillan Publishing Company.

Lowenfeld, V., & Brittain, W. L. (1987). *Creative and mental growth*. New York: Macmillan Publishing Company.

Luger, G. F. (2004). *Artificial intelligence: Structures and strategies for complex problem solving* (5th ed.). Addison Wesley.

Luo, J., Knoblich, G, & Lin, C. (2009). Neural correlates of insight phenomena. *On Thinking, 1*(3), 253-267. doi: 10.1007/978-3-540-68044-4_15

Macdonald, S. (1970). *The history and philosophy of art education.* New York: American Elsevier Publishing Company, Inc..

Mackworth, N. H. (1965). Originality. *The American Psychologist, 20,* 51–66. doi:10.1037/h0021900 PMID:14251990.

Madden, P. A. F., Heath, A. C., Pedersen, N. L., Kaprio, J., Koskenvuo, M. J., & Martin, N. G. (1999). The genetics of smoking persistence in men and women: A multicultural study. *Behavior Genetics, 29*(6), 423-431. doi 001-8244/00/1100-0423

Maeda, J., & Bermont, R. J. (2011). *Redesigning leadership (simplicity: design, technology, business, life).* Boston: The MIT Press.

Makeuseof. (2013). Retrieved March 30, 2013, from http://www.makeuseof.com/tag/7-ways-ipad-students-excel-school/

Malchiodi, C. A. (2006). *Expressive therapies.* The Guilford Press.

Mangels, J. A., Butterfield, B., Lamb, J., Good, C. D., & Dweck, C. S. (2006). Why do beliefs about intelligence influence learning success? A social-cognitive-neuroscience model. *Social Cognitive and Affective Neuroscience, 1,* 75–86. doi:10.1093/scan/nsl013 PMID:17392928.

Mann, S. (2001). *Intelligent image processing.* John Wiley & Sons, Inc. doi:10.1002/0471221635.

Marchese, F. T. (2011). Exploring the origins of tables for information visualization. In *Proceedings of the Information Visualisation 15th International Conference,* (pp. 395-402). London. ISBN 978-1-4577-0868-8

Marchese, F. M., & Banissi, E. (Eds.). (2013). *Knowledge visualization currents: From text to art to culture.* London: Springer-Verlag. doi:10.1007/978-1-4471-4303-1.

March, S., & Storey, V. (2008). Design science in the information systems discipline: An introduction to the special issue on design science research. *Management Information Systems Quarterly, 32*(4), 725–730. Retrieved from misq.org/misq/downloads/download/editorial/152/.

Markham, A. (2011). Adaptive social hierarchies: From nature to networks. In *Bio-Inspired Computing and Networking,* (pp. 305-350). CRC Press, Taylor & Francis Group. ISBN 1420080326

Markoff, J. (2012, July 20). In first, software emulates lifespan of entire organism. *The New York Times.* Retrieved July 21, 2012, from http://www.nytimes.com/2012/07/21/science/in-a-first-an-entire-organism-is-simulated-by-software.html?_r=1&partner=rss&emc=rss

Marr, D. (1982/2010). *Vision: A computational investigation into the human representation and processing of visual information.* The MIT Press.

Marshall, J., Chamberlain, A., & Benford, S. (2011). I seek the nerves under your skin: A fast interactive artwork. *Leonardo Journal, 44*(5), 401–404. doi:10.1162/LEON_a_00239.

Martinet, A., & Palmer, E. (1982). *Elements of general linguistics.* Chicago: University of Chicago Press.

Maslow, A. H. (1943). A theory of human motivation. *Psychological Review, 50,* 370–396. doi:10.1037/h0054346.

Maslow, A. H. (1950). Self-actualizing people: a study of psychological health. *Personality Symposia, 1.*

Maslow, A. H. (1954/1987). *Motivation and personality.* New York: Harper.

Maslow, A. H. (1968/1998). *Towards a psychology of being* (3rd ed.). New York: Wiley.

Masud, L., Valsecchi, F., Ciuccarelli, P., Ricci, D., & Caviglia, G. (2010), From data to knowledge - Visualizations as transformation processes within the data-information-knowledge continuum. In *Proceedings of the Information Visualisation 14th International Conference* (pp. 445-449). IEEE.

Mateo, J. L., & Sauter, F. (Eds.). (2007). *Natural metaphor: Architectural papers III.* Zurich: ACTAR & ETH Zurich.

Mäthger, L. M., Barbosa, A., Miner, S., & Hanlon, R. T. (2005). Color blindness and color perception in cuttlefish (Sepia officinalis) determined by a visual sensorimotor assay. *Vision Research, 46,* 1746-1753. Retrieved October 15, 2012, from http://hermes.mbl.edu/mrc/hanlon/pdfs/mathger_et_al_visres_2006.pdf

Matthews, L., & Perin, G. (2011). Digital images: Interaction and production. *International Journal of Creative Interfaces and Computer Graphics*, *2*(1), 27–41. doi:10.4018/jcicg.2011010103.

Mattison, C. (2007). *300 frogs: A visual reference to frogs and toads from around the world*. Firefly Books.

Maturana, H. (1970). *Biology of cognition*. Urbana, IL: University of Illinois.

Maturana, H., & Varela, F. (1987/1992). *The tree of knowledge: The biological roots of human understanding*. Boston: D. Reidel.

Maya Mathematical System. (2012). *Maya world studies center*. Retrieved November 1, 2011, from http://www.mayacalendar.com/f-mayamath.html

Mayrhofer, R., Sommer, A., & Saral, S. (2012). Airwriting: A platform for scalable, privacy-preserving, spatial group messaging. *International Journal of Pervasive Computing and Communications*, *8*(1), 53–78. doi:10.1108/17427371211221081.

McCall Smith, A. (2010). *The unbearable lightness of scones*. New York: Anchor.

McDonald, Q. (2011). *Raw art journaling*. Cincinnati, OH: North Light Books.

McLaughlin, J. (1998). Good vibrations. *American Scientist*, *86*(4), 342. doi: doi:10.1511/1998.4.342.

McWhinnie, H. J. (1989). *The computer & the right side of the brain*. Journal Announcement: RIESEP90. (ERIC Document Reproduction Service No. ED 318461).

Medical Research News. (2009). Molecular link between intelligence and curiosity discovered. *Medical Research News*. Retrieved January 11, 2012, from http://www.news-medical.net/news/20090917/Molecular-link-between-intelligence-and-curiosity-discovered.aspx

Miller, P. (2010). *The smart swarm: How understanding flocks, schools, and colonies can make us better at communicating, decision making, and getting things done*. Avery.

Milne, A. A., & Sheppard, E. H. (2009). *Winnie the pooh*. Dutton Juvenile.

Mirsky, S. (2011, November). Article. *Scientific American*, 96.

Mitchell, L. (2005). *A posthuman methodology for nondual world*. (Unpublished Thesis). York University, Toronto, Canada.

Mitsunaga, N., & Asada, M. (2004). Visual attention control by sensor space segmentation for a small quadruped robot based on information criterion. *Lecture Notes in Computer Science*, 2377.

MOMA. (2012). *What is a print*? Retrieved June 18, 2012, from http://www.moma.org/interactives/projects/2001/whatisaprint/flash.html

Moore, D. M. (1985). *Field independence-dependence: Multiple and linear imagery in a visual location task*. Paper presented at the Annual Convention of the Association for Educational Communications and Technology. Anaheim, CA.

Murakami, H. (2011). *1Q84*. New York: Knopf.

Muralikrishnan, T. R. (2010). Translator as reader: Phenomenology and text reception: An investigation of indulekha. *Language in India*, *10*(1), 255–266.

Mythology. (2008). *National Geographic essential visual history of world mythology*. Berlin: Peter Delius Verlag GmbH & Co KG. ISBN 978-1-4262-0373-2

NASA. (2012). *Mars science laboratory*. Retrieved January 11, 2012, from http://www.nasa.gov/mission_pages/msl/index.html

National Cancer Institute. (2012). *What you need to know about™ lung cancer*. Retrieved June 22, 2012, from http://cancer.gov/cancertopics/wyntk/lung

National Research Council of the National Academies. (2008). *Inspired by biology: From molecules to materials to machines*. Washington, DC: The National Academies Press.

Needham, J. (1962). *Science and civilisation in China*. Cambridge University Press.

Nguyen, L. (2010, April 27). Teen obesity linked to prebirth tobacco exposure: Study. *Gazette*.

Nöth, W. (2000). *Handbook of semiotics*. Stuttgard, Germany: Metzler.

O'Brien, J., & Marakas, G. (2009). *Introduction to information systems*. McGraw Hill Higher Education.

O'Brien, J., & Marakas, G. (2010). *Management information systems*. McGraw-Hill/Irwin.

O'Reilly, T. (2005). *Design patterns and business models for the next generation of software*. Retrieved February 2, 2013, from http://oreilly.com/web2/archive/what-is-web-20.html

Ogden, C. K., & Richards, I. A. (1923). *The meaning of meaning: A study of the influence of language upon thought*. New York: Harcourt Brace and Co.

Ogden, C. K., & Richards, I. A. (1989). *Meaning of meaning*. Mariner Books. (Original work published 1923).

Olson, D. R., & Bialystok, E. (1983). *Spatial cognition*. Hillsdale, NJ: Lawrence Erlbaum.

Osawa, N. (2004). Visual and sound glyphs for representing constraints. *YLEM, Artists Using Science & Technology, 24*(2).

Othman, A., El Ghoul, O., & Jemni, M. (2012). An automatic approach for facial feature points extraction from 3d head. In *Proceedings of the International Conference on Computer Graphics Theory and Applications and International Conference on Information Visualization Theory and Applications* (pp. 369-372). IEEE. ISBN: 978-989-8565-02-0

Owen, O. (2001/1856). *The grammar of ornament*. DK Adult. ISBN 10789476460

Ox, J. (1999). Synesthesia. *Leonardo, 32*(5), 391. doi:10.1162/002409499553622.

Ozansoy, M., & Denizhan, Y. (2009). The endomembrane system: A representation of the extracellular medium? *Biosemiotics, 2*, 255–267. doi:10.1007/s12304-009-9063-3.

Pacific Northwest National Laboratory. (2008). *Cognitive informatics*. Retrieved April 2, 2013, from http://www.pnl.gov/coginformatics/

Paivio, A. (1970). On the functional significance of imagery. *Psychological Bulletin, 73*, 385–392. doi:10.1037/h0029180.

Paivio, A. (1971). *Imagery and verbal processes*. New York: Holt, Rinehart, and Winston.

Paivio, A. (1986/1990). *Mental representations: A dual coding approach*. Oxford University Press.

Paivio, A. (1991). Dual coding theory: Retrospect and current status. *Canadian Journal of Psychology, 45*(3), 255–287. doi:10.1037/h0084295.

Panas, T., Berrigan, R., & Grundy, J. C. (2003). A 3D metaphor for software production visualization. In *Proceedings of the 7th International Conference on Information Visualization* (pp. 314-319). IEEE.

ParkWise. (2012). Migration basics. *ParkWise*. Retrieved January 11, 2012, from http://www.nps.gov/akso/parkwise/students/referencelibrary/general/migrationbasics.htm

Paton, R., Nwana, H. S., Shave, M. J. R., Bench-Sapon, T. J. M., & Hughes, S. (1990). Transfer of natural metaphors to parallel problem solving applications. *Lecture Notes in Computer Science*, 496.

Peak, D. A., Prybutok, V., Gibson, M., & Xu, C. (2012). Information systems as a reference discipline for visual design. In *Proceedings of the 2012 SWDSI Conference*. Retrieved September 27, 2012, from http://www.swdsi.org/swdsi2012/proceedings_2012/papers/Papers/PA137.pdf

Peak, D. A., Prybutok, V., Gibson, M., & Xu, C. (2012). Information systems as a reference discipline for visual design. *International Journal of Art, Culture and Design Technologies, 2*(2), 57–71. doi:10.4018/ijacdt.2012070105.

Pease, R. (2012, June 19). Alan Turing: Inquest's suicide verdict not supportable. *BBC News Technology*. Retrieved June 23, 2012, from http://www.bbc.co.uk/news/science-environment-18561092

Peirce, C. S. (1958). *Collected papers of Charles Sanders Peirce*. Cambridge, MA: Harvard University Press.

Pelletier, T. (2011). *Ask a naturalist*. Retrieved January 8, 2012, from http://askanaturalist.com/why-don't-ducks'-feet-freeze/

Pérez, J. B., Corchado Rodríguez, J. M., Ortega, A., & María, N. Moreno, M. N., Navarro E., Mathieu, P. (Ed.). (2012). Highlights on practical applications of agents and multi-agent systems. In *Proceedings of the 10th International Conference on Practical Applications of Agents*. Berlin: Springer. ISBN-10: 3642287611

Pessoa Forte, J. A., Arruda, D., Gomes, C. A., Nogueira, G., & Cavalcante de Almeida, C. F. (2011). Educational services in second life: A study based on flow theory. *International Journal of Web-Based Learning and Teaching Technologies*, *6*(2), 1–17. doi:10.4018/jwltt.2011040101.

Piaget, J., & Inhelder, B. (1956). *The child's conception of space*. London: Routledge & Kegan Paul.

Piaget, J., & Inhelder, B. (1971). *Mental imagery in the child: A study of development of imaginal representation*. New York: Basic Books Inc., Publishers.

Piestrup, A. M. (1982). *Young children use computer graphics*. Cambridge, MA: Harvard Univ.

Pink, D. (2006). *A whole new mind: Why right-brainers will rule the future*. Riverhead Trade.

Plato (2000). *The Republic*. Cambridge, UK: Cambridge University Press.

Poslad, S. (2009). *Ubiquitous computing: Smart devices, environments and interactions*. Wiley. doi:10.1002/9780470779446.

Posner, R. (2009). *Eight historical paradigms of cultural studies and their semiotic explication*. Retrieved June 3, 2012, from http://www.ut.ee/CECT/docs/CECT_II_PPT/I_session/Posner_CECT09.pdf

Posner, M. I. (1993). *Foundations of cognitive science*. Cambridge, MA: The MIT Press.

Posner, R. (1992). Origins and development of contemporary syntactics. *Languages of Design*, *1*(1), 37–54.

Potter, B. (2006). *The complete tales*. Frederick Warne & Co.(Original work published 1902).

Pouivet, R. (2000). On the cognitive functioning of aesthetic emotions. *Leonardo*, *33*(1), 49–53. doi:10.1162/002409400552234.

Powell, B. A. (2003, October 27). Framing the issues: UC Berkeley professor George Lakoff tells how conservatives use language to dominate politics. *UC Berkeley News*. Retrieved February 6, 2012, from http://berkeley.edu/news/media/releases/2003/10/27_lakoff.shtml

Price, A. (2011). *The sundagger explorer*. Retrieved July 14, 2012, from http://accad.osu.edu/%7Eaprice/works/sundagger/index.html

Proust, M. (2011). A fragment of the *"Remembrance of Things Past: Swann Way."* Retrieved January 17, 2012, from http://www.authorama.com/remembrance-of-things-past-3.html

Pumpa, M., & Wyeld, T. G. (2006). Database and narratological representation of Australian aboriginal knowledge as information visualisation using a game engine. In *Proceedings of the iV, 10th International Conference on Information Visualization*. IEEE Computer Society Press.

Purchase, H. C., Plimmer, B., Baker, R., & Pilcher, C. (2010). Graph drawing aesthetics in user-sketched graph layouts. In *Conferences in Research and Practice in Information Technology (CRPIT)*. Brisbane, Australia: CRPIT.

QR Code. (2012). Retrieved June 22, 2012, from http://en.wikipedia.org/wiki/QR_code

Rasch, M. (1988). Computer-based instructional strategies to improve creativity. *Computers in Human Behavior*, *4*, 23–28. doi:10.1016/0747-5632(88)90029-5.

Raven, J. (1981). *Manual for Raven's progressive matrices and vocabulary scales*. San Antonio, TX: Harcourt Assessment.

Redström, J., Skog, T., & Hallnäs, L. (2000). Informative art: Using amplified artworks as information displays. In *Proceedings of DARE 2000, On Designing Augmented Reality Environments*. Elsinore, Denmark: ACM. Retrieved January 20, 2012, from http://www.johan.redstrom.se/thesis/pdf/infoart.pdf

Reynolds, C. (2011). Interactive evolution of camouflage. *Artificial Life*, *17*(2), 123-136. doi:10.1162/artl_a_00023. Retrieved May 13, 2012, from http://www.mitpressjournals.org/doi/abs/10.1162/artl_a_00023?journalCode=artl

Reynolds, C. (2011). Interactive evolution of camouflage. *Artificial Life*, *17*(2), 123–126. doi:10.1162/artl_a_00023 PMID:21370960.

Rezaei, A. R., & Katz, L. (2004). Evaluation of the reliability and validity of the cognitive style analysis. *Personality and Individual Differences*, *36*(6), 1317–1327. doi:10.1016/S0191-8869(03)00219-8.

Riley, B. (2012). *Brigdet Riley*. Retrieved July 13, 2012, from http://www.op-art.co.uk/bridget-riley/

Rind, A., Wang, T., Aigner, W., Miksch, S., Wongsuphasawat, K., Plaisant, C., & Shneiderman, B. (2010). Interactive information visualization for exploring and querying electronic health records: A systematic review. *Human-Computer Interaction Lab*. Tech Report HCIL-2010-19.

Ritchie, G. (2007). Some empirical criteria for attributing creativity to a computer program. *Minds and Machines*, *17*(1), 67–99. doi:10.1007/s11023-007-9066-2.

Roam, D. (2008). *The back of the napkin: Solving problems and selling ideas with pictures*. New York: The Penguin Group.

Rockley, A., & Cooper, C. (2012). *Managing enterprise content: A unified content strategy* (2nd ed.). New Riders.

Rogers, K. G. (1985). The museum and gifted child. *Roeper Review*, *7*(4), 239–241. doi:10.1080/02783198509552906.

Rooney, P. (2002). *Microsoft's CEO: 80-20 rule applies to bugs, not just features*. Retrieved March 27, 2011 from http://www.crn.com/

Runco, M. A. (1986). Divergent thinking and creative performance in gifted and nongifted children. *Educational and Psychological Measurement*, *46*, 375–384. doi:10.1177/001316448604600211.

Russell, S., & Norvig, P. (2009). *Artificial intelligence: A modern approach* (3rd ed.). Prentice Hall.

Russo Dos Santos, C., & Gros, C. P. (2002). Multiple views in 3D metaphoric information visualization. In *Proceedings of iV02, 6th International Conference on Information Visualisation*, (pp. 468-473). IEEE.

Rusu, A., Fabian, A. J., Jianu, R., & Rusu, A. (2011). Using the gestalt principle of closure to alleviate the edge crossing problem in graph drawings. In *Proceedings of the Information Visualisation 15th International Conference*, (pp. 329-336). ISBN 978-1-4577-0868-8

Ryder, A. P. (n.d.). *Artcyclopedia*. Retrieved May 12, 2012, from http://www.artcyclopedia.com/artists/ryder_albert_pinkham.html

Saffran, J. R. (2003). Absolute pitch in infancy and adulthood: The role of tonal structure. *Developmental Science*, *6*, 35–47. doi:10.1111/1467-7687.00250.

Saloman, G. (1979). *Interaction of media, cognition, and learning*. San Francisco, CA: Jossey-Bass.

Samuels, R. (2010). The double articulation of Schubert: Reflections on der doppelgänger. *The Musical Quarterly*, *93*(2), 192–233. doi:10.1093/musqtl/gdq008.

Sanders, L. (2011, December). He's not a rat, he's my brother: Rodents exhibit empathy by setting trapped friends free. *Science News*, 16. doi:10.1002/scin.5591801414.

Sandkühler, S., & Bhattacharya, J. (2008). Deconstructing insight: EEG correlates of insightful problem solving. *PLoS ONE*, *3*(1), e1459. doi:10.1371/journal.pone.0001459 PMID:18213368.

Saneyoshi, A., Niimi, R., Suetsugu, T., Kaminaga, T., & Yokosawa, K. (2011). Iconic memory and parietofrontal network: fMRI study using temporal integration. *Neuroreport*, *22*(11), 515–519. doi:10.1097/WNR.0b013e328348aa0c PMID:21673607.

Santamaría, R., Therón, R., & Quintales, L. (2008). A visual analytics approach for understanding biclustering results from microarray data. *BMC Bioinformatics*, *9*, 247. doi:10.1186/1471-2105-9-247 PMID:18505552.

Satyanarayanan, M. (2010). The role of cloudlets in mobile computing. *Microsoft Networking Research Summit*. Retrieved July 13, 2012, from research.microsoft.com/en-us/events/mcs2010/satya.ppt

Satyanarayanan, M., Bahl, V., Caceres, R., & Davies, N. (2009). The case for VM-based cloudlets in mobile computing. *IEEE Pervasive Computing*, *99*(1). doi:10.1109/MPRV.2009.64.

Sawyer, R. K. (2007). *Group genius: The creative power of collaboration*. Basic Books.

Sawyer, R. K. (2012). *Explaining creativity: The science of human innovation*. Oxford, UK: Oxford University Press.

Schumacher, M. (2011). *Zodiac calendar & lore*. Retrieved November 1, 2012, from http://www.onmarkproductions.com/html/12-zodiac.shtml

Schwartz, L. F. (1985). The computer and creativity. *Transactions of the American Philosophical Society*, *75*, 30–49. doi:10.2307/20486639.

Schwartz, L. F., & Schwartz, L. R. (1992). *The computer artist's handbook: Concepts, techniques, and applications*. New York: W.W. Norton & Company.

Sebeok, T. A. (1991). *A sign is just a sign*. Bloomington, IN: Indiana University Press.

Seifert, C., Kump, B., Kienreich, W., Granitzer, G., & Granitzer, M. (2008). On the beauty and usability of tag clouds. In *Proceedings of 12th International Conference Information Visualisation*, (pp. 17-25). IEEE.

Shepard, R. N. (1978). The mental image. *The American Psychologist*, *17*, 179–188.

Shneiderman, B. (1996). The eyes have it: A task by data type taxonomy for information visualizations. In *Proceedings of the 1996 IEEE Conference on Visual Languages*, (pp. 336-343). IEEE.

Shreve, J. (2010). Drawing art into the equation: Aesthetic computing gives math a clarifying visual dimension. *Edutopia*. Retrieved April 21, 2012, from http://www.edutopia.org/drawing-art-equation

Simon, H. A. (1996). *The sciences of the artificial* (3rd ed.). Cambridge, MA: The MIT Press.

Simon, J., & Wegman, W. (2006). *Funney/strange*. New Haven, CT: Yale University Press.

Skogen, M. G. R. (2006). An investigation into the subjective experience of icons: A pilot study. In *Proceedings of the 10th International Conference on Information Visualization*, (pp. 368-373). IEEE.

Slingsby, A., Dykes, J., Wood, J., & Clarke, K. (2007). Interactive tag maps and tag clouds for the multiscale exploration of large spatio-temporal datasets. In *Proceedings of iV07, 11th International Conference Information Visualisation*, (pp. 497-504). IEEE.

Slobodchikoff, C. N. (2002). Cognition and communication in prairie dogs. In M. Beckoff, C. Allen, & G. M. Burghardt (Eds.), *The Cognitive Animal*, (pp. 257-264). Cambridge, MA: A Bradford Book.

Slotnick, S. D., Thompson, W. L., & Kosslyn, S. M. (2012). Visual memory and visual mental imagery recruit common control and sensory regions of the brain. *Cognitive Neuroscience, 3*(1), 14-20. Retrieved May 6, 2012, from http://www.tandfonline.com/doi/abs/10.1080/17588928.2011.578210

Slusallek, P. (2009). *Computer graphics, ray tracing III*. Retrieved March 23, 2012 from http://graphics.cs.uni-saarland.de/fileadmin/cguds/courses/ws0809/cg/slides/CG04-RT-III.pdf

Solé, R. (2008). On networks and monsters: The possible and the actual in complex systems. *Leonardo*, *41*(3), 253–258. doi:10.1162/leon.2008.41.3.253.

Sookhanaphibarn, K., & Thawonmas, R. (2010). Digital museums in 3D virtual environment. In *Handbook of Research on Methods and Techniques for Studying Virtual Communities: Paradigms and Phenomena*, (pp. 713-730). Academic Press.

Sperling, G. (1960). The information available in brief visual presentations. *Psychological Monographs*, *74*, 1–29. doi:10.1037/h0093759.

Stam, R. (2000). *Film theory*. Oxford, UK: Blackwell.

Stanford University. (2012). *Symbolic systems*. Retrieved January 20, 2012, from http://www.stanford.edu/dept/registrar/bulletin/6141.htm

Stanovich, K. E., West, R. F., & Toplak, M. E. (2011). Intelligence and rationality. In *The Cambridge Handbook of Intelligence*. Cambridge, UK: Cambridge University Press. doi:10.1017/CBO9780511977244.040.

Stapput, K., Güntürkün, O., Hoffmann, K.-P., Wiltschko, R., & Wiltschko, W. (2010). Magnetoreception of directional information in birds requires nondegraded vision. *Current Biology, 20*(14), 1259-1262. Retrieved October 17, 2012, from http://dx.doi.org/10.1016/j.cub.2010.05.070

Sternberg, R. (2011). *Cognitive psychology*. Wadsworth Publishing.

Sternberg, R. J. (1985). *Beyond IQ: A triarchic theory of human intelligence*. New York: Cambridge University Press.

Sternberg, R. J. (1999). The theory of successful intelligence. *Review of General Psychology, 3*, 292–316. doi:10.1037/1089-2680.3.4.292.

Sternberg, R. J. (2006). The nature of creativity. *Creativity Research Journal, 18*(1), 87–98. doi:10.1207/s15326934crj1801_10.

Sternberg, R. J. (2007). *Wisdom, intelligence, and creativity synthesized*. New York: Cambridge University Press.

Sternberg, R. J. (2008). *Cognitive psychology*. Wadsworth Publishing.

Sternberg, R. J. (Ed.). (1998/2011). *The nature of creativity: Contemporary psychological perspectives*. Cambridge University Press.

Sternberg, R. J., & Kaufman, S. B. (Eds.). (2011). *The Cambridge handbook of intelligence*. Cambridge University Press. doi:10.1017/CBO9780511977244.

Sternberg, R. J., & Lubart, T. I. (1995). An investment perspective on creative insight. In R. J. Sternberg, & J. E. Davidson (Eds.), *The nature of insight* (pp. 386–426). Cambridge, MA: MIT Press.

Sternberg, R. J., & Lubart, T. I. (1999). The concept of creativity. In R. J. Sternberg (Ed.), *Handbook of creativity* (pp. 3–15). New York: Cambridge University Press.

Sternberg, R. J., & Lubart, T. I. (1999). The concept of creativity: Prospects and paradigms. In R. J. Sternberg (Ed.), *Handbook of creativity* (pp. 137–152). Cambridge, UK: Cambridge University Press.

Stillwaggon, L., & Goldberg, S. L. (2010). How is meaning grounded in the organism? *Biosemiotics, 3*(2), 131–146. doi:10.1007/s12304-010-9072-2.

Sturm, B. L. (2005). Pulse of an ocean: Sonification of ocean buoy data. *Leonardo, 38*(2), 143–149. doi:10.1162/0024094053722453.

Sullivan, K. (2000). Between analogue and digital. *Computer Graphics, 34*(3), 5.

Tanaka-Ishii, K. (2010). *Semiotics of programming*. Cambridge University Press.

Targowski, A. (2011). *Cognitive informatics and wisdom development: Interdisciplinary approaches*. Hershey, PA: IGI Global.

Tatarkiewicz, W. (1976). *Dzieje szesciu pojec*. Warsaw, Poland: PWN.

Tatarkiewicz, W. (1999). *History of aesthetics*. Thoemmes Press.

Taylor, C. W. (1959). The identification of creative scientific talent. *The American Psychologist, 14*, 100–102. doi:10.1037/h0046057.

Tennyson, R. D., Thurlow, R., & Breuer, K. (1987). Problem-oriented simulations to improve higher-level thinking strategies. *Computers in Human Behavior, 3*, 151–165. doi:10.1016/0747-5632(87)90020-3.

Thbz. (2000). The double articulation of language: Computer languages. *Everything2*. Retrieved June 2, 2012, from http://everything2.com/title/The+double+articulation+of+language

The Carnegie Library of Pittsburg. (2011). *The handy science answer book*. Visible Ink Press.

The New York Times. (2012, June 5). Tattoos. *The New York Times*. Retrieved June 5, 2012, from http://topics.nytimes.com/top/reference/timestopics/subjects/t/tattoos/index.html

The Teaching Palette. (2013). Retrieved March 30, 2013, from http://theteachingpalette.com/2012/02/24/theres-an-app-for-that-ipads-in-the-art-room/

The United States Department of Justice. (1999). *Office of consumer protection litigation against tobacco companies*. Retrieved June 22, 2012, from http://www.justice.gov/civil/cases/tobacco2/index.htm

Thomas, N. J. T. (2010). *Stanford encyclopedia of philosophy*. Retrieved November 18, 2012, from http://plato.stanford.edu/entries/mental-imagery/mental-rotation.html

Thomas, J. J., & Cook, K. A. (2006). A visual analytics agenda. *IEEE Computer Graphics and Applications, 26*(1), 10–13. doi:10.1109/MCG.2006.5 PMID:16463473.

Tominski, C., & Schumann, H. (2008). Visualization of gene combinations. In *Proceedings of iV, 12th International Conference on Information Visualisation*. IEEE.

Tomitsch, M., Grechenig, T., Vande Moere, A., & Renan, S. (2008). Information sky: Exploring the visualization of information on architectural ceilings. In *Proceedings of iV, 12th International Conference on Information Visualisation*. IEEE.

Tong, E. K., & Glantz, S. A. (2007). Tobacco industry efforts undermining evidence linking second hand smoke with cardiovascular disease. *Circulation*, *116*, 1845–1854. doi:10.1161/CIRCULATIONAHA.107.715888 PMID:17938301.

Torrance, E. P. (1962). *Guiding creative talent*. Englewood Cliffs, NJ: Prentice-Hall, Inc. doi:10.1037/13134-000.

Torrance, E. P. (1974). *Torrance tests of creative thinking: Norms and technical manual*. Lexington, MA: Personnel Press.

Torrance, E. P. (1990). *Torrance test of creative thinking*. Benseville, IL: Scholastic Testing Service, Inc..

Trajkovski, G. (2010). *Developments in intelligent agent technologies and multi-agent systems: Concepts and applications*. Hershey, PA: IGI Global. doi:10.4018/978-1-60960-171-3.

Trajkovski, G., & Collins, S. G. (2009). *Handbook of research on agent-based societies: Social and cultural interactions*. Hershey, PA: IGI Global. doi:10.4018/978-1-60566-236-7.

Trapp, M., Glander, T., Buchholz, H., & Döllner, J. (2008). 3D generalization lenses for interactive focus + context visualization of virtual city models. In *Proceedings of iV, 12th International Conference Information Visualisation*, (pp. 356-361). IEEE Computer Society Press. ISBN: 9780769532684. DOI: 10.1109/IV.2008.18

Trochim, W. M. (2006). *Concept mapping*. Retrieved March 24, 2012, from http://www.socialresearchmethods.net/kb/conmap.htm

Tufte, E. R. (1983/2001). *The visual display of quantitative information*. Cheshire, CT: Graphics Press.

Tufte, E. R. (1992/2005). *Envisioning information*. Cheshire, CT: Graphics Press.

Tufte, E. R. (1997). *Visual explanations: Images and quantities, evidence and narrative*. Graphics Press.

Tufte, E. R. (2001). *The visual display of quantitative information*. Cheshire, CT: Graphics Press. (Original work published 1983).

Tufte, E. R. (2003). *The cognitive style of powerpoint*. Cheshire, CT: Graphics Press.

UbiComp. (2012). *14th ACM international conference on ubiquitous computing*. Retrieved August 10, 2012, from http://www.ubicomp.org/ubicomp2012/

Ubirobots. (2012). *ACM international workshop, smart gadgets meet ubiquitous and social robots on the web*. Retrieved August 10, 2012, from https://sites.google.com/site/ubirobots2012/

UCLA Academic Technology Services. (2012). *Stat computing*. Retrieved June 17, 2012, from http://www.ats.ucla.edu/stat/mult_pkg/whatstat/nominal_ordinal_interval.htm

UCLA. (2011). *Instructional research lab: Chladni plate*. Retrieved December 28, 2011, from http://www.physics.ucla.edu/demoweb/demomanual/acoustics/effects_of_sound/chladni_plate.html

Ursyn, A., & Banissi, E. (Eds.). (2003). Visualizing data sets. *The YLEM Journal: Artists Using Science & Technology*, 23(10).

Ursyn, A., & Scott, T. (2007). Web with art and computer science. In *Proceedings of the ACM SIGGRAPH Conference on Computer Graphics and Interactive Technics*. ACM. ISBN 978-1-59593-648-6

Ursyn, A., & Lohr, L. (2010). Pretenders and misleaders in product design. *Design Principles and Practices: An International Journal*, *4*(3), 99–108.

Vallortigara, G., & Rogers, L. J. (2005). Survival with an asymmetrical brain: advantages and disadvantages of cerebral lateralization. *The Behavioral and Brain Sciences*, *28*(4), 575–633. doi:10.1017/S0140525X05000105 PMID:16209828.

Van Scoy, F. L., & Gifu, U. (2004). Sonification of remote sensing data: Initial experiment. In *Proceedings of the International Conference on Information Visualization*. IEEE.

Van Tonder, B., & Wesson, J. (2008). Visualization of personal communication patterns using mobile phones. In *Engineering Interactive Systems: EIS 2007 Joint Working Conferences EHCI*. Springer.

Vanacker, B., & Heider, D. (2012). Ethical harm in virtual communities. *Convergence*, *18*(1), 71–84. doi: doi:10.1177/1354856511419916.

Vande Moere, A. (2008). Beyond the tyranny of the pixel: Exploring the physicality of information visualization. In *Proceedings of 12th International Conference on Information Visualisation*. IEEE.

Vande Moere, A., & Boltzmann, L. (2009). Beyond ambient display: A contextual taxonomy of alternative information display. *International Journal of Ambient Computing and Intelligence*, *1*(2), 39–46. doi:10.4018/jaci.2009040105.

Various Artists. (2000). *Double articulation*. Audio CD album. ASIN B000024ENV

Varley, R. A., Klessinger, N. J. C., Romanowski, C. A. J., & Siegal, M. (2005). Agrammatic but Numerate. *Proceedings of the National Academy of Sciences of the United States of America*, *102*, 3519–3524. doi:10.1073/pnas.0407470102 PMID:15713804.

Victorri, B. (2007). Analogy between language and biology: A functional approach. *Cognitive Process, 8*(1), 11-9. Retrieved July 9, 2012, from http://www.ncbi.nlm.nih.gov/pubmed/17171371

Vihma, S. (Ed.). (1992). *Objects and images: Studies in design and advertising*. Helsinki: University of Industrial Arts.

Vinciarelli, A. (2011). Towards a technology of nonverbal communication: Vocal behavior in social and affective phenomena. In *Affective Computing and Interaction: Psychological, Cognitive, and Neuroscientific Perspectives* (pp. 133–156). Hershey, PA: IGI Global.

Viola, I., & Gröller, M. E. (2005). Smart visibility in visualization. In L. Neumann, M. Sbert, B. Gooch, & W. Purgathofer (Eds.), *Computational Aesthetics in Graphics, Visualization, and Imaging*. Retrieved June 18, 2012, from http://www.cg.tuwien.ac.at/research/publications/2005/Viola-05-Smart/Viola-05-Smart-Paper.pdf

Visual Literacy.org. (2011). Retrieved September 1, 2011 from http://www.visual-literacy.org/index.html

Vivekjivandas, S., & Dave, J. (2011). *Hinduism: An introduction*. Ahmedabad, India: Swaminarayan Aksharpith.

Voigt, R. (2002). *An extended scatterplot matrix and case studies in information visualization*. (Unpublished Masters thesis). Virtual Reality and Visualization Research Center, Vienna, Austria.

von Foerster, H. (2002). *Understanding understanding: Essays on cybernetics and cognition*. New York: Springer-Verlag.

von Glasersfeld, E. (1989). Constructivism in education. In T. Husen, & T. N. Postlethwaite (Eds.), *International encyclopedia of education*. Oxford, UK: Pergamon Press.

von Glasersfeld, E. (1991). Knowing without metaphysics: Aspects of the radical constructivist position. In F. Steier (Ed.), *Research and reflexivity*. London: Sage.

von Glasersfeld, E. (2007). The constructivist view of communication. In A. Müller, & K. Müller (Eds.), *An unfinished revolution?* (pp. 351–360). Vienna: Echoraum.

Vygotsky, L. S. (1972). *Thought and language*. Cambridge, MA: MIT Press.

Vygotsky, L. S. (1986). *Thought and language* (A. Kozulin, Ed.). The MIT Press.

W3C Semantic Web Activity. (2011). Retrieved June 17, 2012, from http://www.w3.org/2001/sw/

W3C. (2012). *Graphics*. Retrieved July 13, 2012, from http://www.w3.org/standards/webdesign/graphics#uses

Wahl, H., Winiwarter, W., & Quirchmayr, G. (2011). Towards an intelligent integrated language learning environment. *International Journal of Pervasive Computing and Communications, 7*(3), 220 – 239. doi 10.1108/17427371111173013. Retrieved July 15, 2012, from http://www.emeraldinsight.com/journals.htm?issn=1742-7371&volume=7&issue=3&

Walberg, H. J. (1969). A portrait of the artist and scientist as young man. *Exceptional Children, 36*(1), 5–11.

Wallach, M. A., & Kogan, N. (1965). *Modes of thinking in young children*. New York: Holt, Rinehart, & Winston.

Wands, B. (2001). *Digital creativity: Techniques for digital media and the internet*. New York: John Wiley & Sons, Inc.

Wang, T. D. (2010). *Interactive visualization techniques for searching temporal categorical data*. (PhD dissertation). University of Maryland, University Park, MD. Retrieved July 12, 2012, from http://hcil2.cs.umd.edu/trs/2010-15/2010-15.pdf

Wang, T., Deshpande, A., & Shneiderman, B. (2010). A temporal pattern search algorithm for personal history event visualization. *IEEE Transactions on Knowledge and Data Engineering, 99*. doi: 10.1109/TKDE.2010.257. HCIL-2009-14

Wang, T., Wongsuphasawat, K., Plaisant, C., & Shneiderman, B. (2010). Visual information seeking in multiple electronic health records: Design recommendations and a process model. In *Proceedings of the 1st ACM International Informatics Symposium* (IHI '10), (pp. 46-55). ACM.

Wang, Y. (2011). Towards the synergy of cognitive informatics, neural informatics, brain informatics, and cognitive computing. *International Journal of Cognitive Informatics and Natural Intelligence, 5*(1), 75–93. doi:10.4018/jcini.2011010105.

Ward, M., Grinstein, G. G., & Keim, D. (2010). Interactive data visualization: Foundations, techniques, and applications. Natick, MA: A K Peters Ltd. ISBN 1568814739

Ware, C. (2012). Visual thinking algorithms. In *Proceedings of the International Conference on Computer Graphics Theory and Applications and International Conference on Information Visualization Theory and Applications*. ISBN: 978-989-8565-02-0

Ware, C. (2000). *Information visualization: Perception for design (interactive technologies)*. Morgan Kaufmann.

Wattenberg, M. (2005). *From data to pictures to insight*. Retrieved March 18, 2006, from http://www.alphaworks.ibm.com/contentnr/introvisualization

Web Science 2012 Conference. (2012). Retrieved July 15, 2012, from http://www.websci12.org/program

Wegrzyn, W. (2012). *Plug-in book offers digital content when the page is turned*. Retrieved November 22, 2012, from http://www.springwise.com/media_publishing/plug-in-book-offers-digital-content-page-turned/

Wei, H., Liu, E., Zhao, X., McFarlane, N. J. B., & Clapworthy, G. J. (2011). Article. In *Proceedings of the 15th International Conference on Information Visualization*, (pp. 632-637). IEEE.

Weiser, M. (1991). The computer for the 21st century. *Scientific American*. Retrieved January 17, 2012, from http://classes.dma.ucla.edu/Winter06/256/text/Weiser-21stCentury.pdf

Weitz, M. (1956). The role of theory in aesthetics. *The Journal of Aesthetics and Art Criticism, 15*, 27–35. doi:10.2307/427491.

Weng, J. (2012). *Natural and artificial intelligence: Introduction to computational brain-mind*. BMI Press.

Whorf, B. L. (1956). *Language, thought, and reality*. Cambridge, MA: MIT Press.

Wiggins, G. (2006). Searching for computational creativity: New generation computing, computational paradigms and computational intelligence. *Computational Creativity, 24*(3), 209–222.

Wikiversity. (2012). *Level of measurement*. Retrieved March 18, 2012, from http://en.wikiversity.org/wiki/Level_of_measurement

Wilkinson, W. K., & Schwartz, N. H. (1991, January). A factor-analytic study of epistomological orientation and related variables. *The Journal of Psychology*, 91–101. doi:10.1080/00223980.1991.10543274.

Wisegeek. (2012). *What is double articulation?* Retrieved June 2, 2012, from http://www.wisegeek.com/what-is-double-articulation.htm

Witkin, H. A. (1954/1972). *Personality through perception*. London: Greenwood Press.

Witkin, H. A., Dyk, R. B., Faterson, G. E., Goodenough, D. R., & Karp, S. A. (1974). *Psychological differentiation: Studies of development*. Potomac, MD: Lawrence Erlbaum Associates.

Wohlwill, J. F. (1988). Artistic imagination during the latency period revealed through computer graphics. In G. Forman, & P. B. Pufall (Eds.), *Constructivism in the Computer Age* (pp. 129–150). Hillsdale, NJ: Lawrence Erlbaum Associates, Publishers.

Wolfram MathWorld. (2012). Retrieved August 28, 2012, from http://mathworld.wolfram.com/VoronoiDiagram.html

Wong, K. (2002, January 11). Ancient engravings push back origin of abstract thought. *Scientific American*. Retrieved April 21, 2012, from http://www.scientificamerican.com/article.cfm?id=ancient-engravings-push-b

Wong, P. C., & Thomas, J. (2004). Visual analytics. *IEEE Computer Graphics and Applications*, *24*(5), 20–21. doi:10.1109/MCG.2004.39 PMID:15628096.

Wood, G. (1930). *American gothic*. Retrieved May 30, 2012, from http://upload.wikimedia.org/wikipedia/commons/7/71/Grant_DeVolson_Wood_-_American_Gothic.jpg

World Question Center. (2010). Retrieved October 27, 2012, from www.edge.org/questioncenter.html, http://www.edge.org/q2010/q10_3.html#kosslyn

Wueringer, B. E., Squire, L., Kajiura, S. M., Hart, N. S., & Collin, S. P. (2012). The function of the sawfish's saw. [from http://www.cell.com/current-biology/current]. *Current Biology*, *22*(5), R150–R151. Retrieved March 6, 2012. doi:10.1016/j.cub.2012.01.055 PMID:22401891.

Wyeld, T. (2011). The implications of David Hockney's thesis for 3D computer graphics. In *Proceedings of the Information Visualisation 15th International Conference*, (pp. 409-413). London: IEEE. ISBN 978-1-4577-0868-8

Wynn, K. (1992). Children's acquisition of the number words and the counting system. *Cognitive Psychology*, *24*(2), 220–251. doi:10.1016/0010-0285(92)90008-P.

Xu, Z., & Taylor, D. (2000). Using motion platform as a haptic display for virtual inertia simulation. In *Proceedings of the Information Visualization Conference*, (pp. 498-504). IEEE.

Yang, X. (2011). *Bio-inspired computing and networking*. CRC Press.

Yau, N. (2011). *Visualize this: The flowing guide to design, visualization, and statistics*. Wiley.

Yau, N. (2013). *Data points: Visualization that means something*. Wiley.

Yotsumoto, Y., & Watanabe, T. (2008). Defining a link between perceptual learning and attention. *PLoS Biology*, *6*(8), e221. doi:10.1371/journal.pbio.0060221 PMID:18752357.

Young, J. G. (1985). What is creativity? *The Journal of Creative Behavior*, *19*(2), 77–87. doi:10.1002/j.2162-6057.1985.tb00640.x.

Yu, D., Dong, Y., Qin, Z., & Wan, T. (2011). Exploring market behaviors with evolutionary mixed-games learning model. In *Proceedings of the Computational Collective Intelligence: Technologies and Applications*. ICCCI. doi:10.1007/978-3-642-23935-9_24

Yu, L. (2011). *A developer's guide to the semantic web*. Berlin: Springer-Verlag. doi:10.1007/978-3-642-15970-1.

Zbikowski, L. M. (1998). Metaphor and music theory: Reflections from cognitive science. *Music Theory Online*, *4*(1).

Zbikowski, L. M. (2005). *Conceptualizing music: Cognitive structure, theory, and analysis*. Oxford University Press.

Zeit Online. (2012). Retrieved April 6, 2012, from http://twitpic.com/3pesre

Zeki, S. (2009). *Splendors and miseries of the brain: Love, creativity, and the quest for human happiness*. Wiley-Blackwell. ISBN 1405185570

Zeki, S. (1993). Vision of the brain. *Wiley-Blackwell.*, *ISBN-10*, 0632030542.

Zeki, S. (1999). Art and the brain. *Journal of Conscious Studies: Controversies in Science and the Humanities*, *6*(6/7), 76–96.

Zeki, S. (2001, July 6). Artistic creativity and the brain. *Scienc*, *293*(5527), 51–52. doi:10.1126/science.1062331 PMID:11441167.

Zuanon, R., & Lima, G. C., Jr. (2008). *BioBodyGame*. Retrieved March 09, 2011, from http://www.rachelzuanon.com/biobodygame/

Zuanon, R., & Lima, G. C., Jr. (2010). *NeuroBodyGame*. Retrieved March 09, 2011, from http://www.rachelzuanon.com/neurobodygame/

Zuanon, R. (2012). Bio-interfaces: Designing wearable devices to organic interactions. In *Biologically-Inspired Computing for the Arts: Scientific Data through Graphics*. Hershey, PA: IGI Global Publishing. doi:10.4018/978-1-4666-0942-6.ch001.

About the Author

Anna Ursyn is a Professor in the School of Art and Design at University of Northern Colorado, where she is also the Computer Graphics Area Head. Her research-based art and pedagogy interests include an integrated approach to art, science, and computer art graphics. Her publications include: Editor of the book *Biologically-Inspired Computing for the Arts: Scientific Data through Graphics* (2012) (ISBN 1466609427), various book chapters, over 30 articles published in professional journals, as well as published poetry and artwork. Her professional work includes: Chair of the Symposium and Digital Art Gallery, D-ART iV, International Conference on Information Visualization, IEEE (Institute of Electrical and Electronics Engineers) Computing Society Press, Los Alamitos, CA 1997-2013 London; and Computer Graphics, Image & Visualization Asia; co-editor for the Proceedings of the iV, Int'l Conference on Information Visualization, IEEE (Institute of Electrical and Electronics Engineers) Computing Society Press, Los Alamitos, CA, 2004-2013; Associate Editor, *Int'l Journal of Creative Interfaces & Computer Graphics* (IJCICG), IGI Global, www.igi-global.com/IJCICG, 2009-2013; Associate Editor, *Design Principles and Practices: An International Journal*, http://designprinciplesandpractices.com/journal/http://ijg.cgpublisher.com; online art gallery organizer and keynote speaker for the Special Year on Art and Mathematics, University of Colorado at Boulder 2005. She has had over 30 single art shows and participated in over 100 juried and invitational fine art exhibitions.

Index

A

aesthetic-communicative features 188
allusionary base 108
Amazonian culture 40
arteriogram 135
art history 2, 9, 13, 64, 94, 105, 155, 162-163, 291
artificial intelligence (AI) 10, 99, 166
augmented reality 24, 332-333, 342, 349, 351, 354

B

basic art concept 61, 65-66, 94
biclustering 277, 303, 328, 330
bio-derivation 14-15
bio-inspiration 14-15
bio-interface 47, 53, 60, 116
biological constructivism 175
biological simulations 332, 337
biology-inspired art 16
biology-inspired technology 1, 16
bio-mimicry 14-15
biosemiotics 174, 178, 188-191
brain-computer interface 47

C

calligraphy 247-249, 271, 275
camouflage 36-38, 172, 181, 193, 205-207
canonical object 48-51, 193-194, 196
clustering 40, 48, 75, 163, 277, 284, 303-304, 319, 336, 338, 344
cognitive computation 230
cognitive informatics 132, 134, 140, 150, 167-168, 171-173
cognitive psychologists 134, 138, 322
cognitive science 22-23, 26, 98, 123, 131-134, 150, 166-168, 211-212, 214, 224
cognitive structure 49, 131-132, 137, 148, 152, 155, 165, 224
cognitive study 3
collaborative visualization 277, 319, 326
computational creativity 98-99, 126, 129
computational intelligence 124, 129, 332-333, 353
computational solution 25-26, 35, 42, 47, 96, 174, 299
computed axial tomography (CAT) 136
computed tomography (CT) 136
computer-assisted instruction (CAI) 232
computer languages 2, 9-10, 24, 178, 231
computer versus pencil 226
concept mapping 137, 142, 209, 277, 298, 301, 303-304, 322, 330
conceptual metaphor 209-210, 214, 301
constructivist epistemology 175
coordinate system 226, 234, 238, 279
creative machine 98
creative potential 104, 108, 121
creative thinking 96, 103-105, 122-123, 128, 131
critical thinking 96, 104, 122
cultural evolution 40

D

data graphics 74, 226, 234, 236, 238-239, 241, 293, 323, 333
data map 124, 233, 238-239, 295
data mining 97, 134, 167, 208-209, 221-222, 241, 277-278, 285, 287, 297, 301, 303-305, 309, 318, 322-323
degenerative disease 135
design science 332, 340, 355
developing imagination 108, 120
developmental psychologists 151
diffuse optical imaging (DOI) 136
diffuse optical tomography (DOT) 136

digital art 64-65, 96, 118, 162-163, 337
digital creativity 96, 118, 122-123, 125, 129, 354
digital illustration 247, 252, 257, 266, 268, 271, 275
discipline based art education (DBAE) 103, 155
divergent thinking 96, 98, 101, 104-105, 122, 127, 142
double articulation 1, 3-8, 24
double-duty gadget 193, 197, 203, 206

E

electroencephalography (EEG) 109, 137
electromagnetic wave 25, 27, 29, 79
elements of design 2, 77, 91, 112
engraving 247, 249, 275
etching 247-249, 275
ethologist 184
European Particle Physics Laboratory (CERN) 240
evolutionary computing 116, 120, 226, 244, 333

F

Facebook 115, 240-241, 285
fifth generation computer systems (FGCS) 232
fine arts 2, 9, 13, 44
first-generation computer 97
flexibility of mind 105
folk art 61, 74, 77, 89
functional magnetic resonance imaging (FMRI) 109, 113, 136

G

genetics translation 14
gestalt psychology 101, 160-161
graphic metaphors 208-209, 220-222

H

high-density diffuse optical tomography (HD-DOT) 136
higher order thinking 54, 131, 138, 152
house-tree-person drawing test (HTP) 107
human memory 138, 151
hypertext markup language (HTML) 240

I

iconicity 195
industrial design 175, 188
information aesthetics 285
informer 18, 193, 197, 200-201, 203, 205-206
iPhone 14, 62, 228, 310

K

kinesthesia 115
knowledge map 222, 277, 298-301, 304, 320, 327
knowledge representation 3, 134, 166, 287, 307-309

L

language skills 74
large-scale integrated (LSI) circuits 231
linocut 250, 275
lithography 248-249, 251, 275
living organism 27, 36, 117, 135, 150, 174, 188-189, 209, 255

M

magnetic resonance imaging (MRI) 136
magnetoencephalography (MEG) 137
media study 5, 177
methods of inquiry 236, 245
misleaders 18, 38, 193, 197, 199-200, 206-207
modal phenomena 29-30
molecular biology 2, 9, 14, 22, 255
morphemes 1, 3, 6
Morse Code 27
motion video capture capability 42
multi agent systems (MAS) 39, 166-167, 244, 291, 332, 340
multifunctional tools 197, 203, 206
multiple intelligences theory 131
music visualization 277, 311
MySQL 241

N

NASA 110, 127
natural language 2-3, 10, 42, 166, 174, 178, 342
nature derived metaphors 209, 219, 222
network visualization 10, 241, 277, 301, 304, 306
neuroimaging 41, 109, 136, 334

O

open source intelligence 277, 301, 307

P

parental care 39
pen-and-ink illustration 248, 275
pheromones 26-27, 30, 33, 38, 44
phonemes 1, 3, 30

photolithograph 214, 345
photosilkscreen 214, 252, 275, 345
pitch recognition 30
pitch response 25
playing cards design 272, 275
positron emission tomography (PET) 136
preoperational stage 152
pretender 18, 38, 193, 196-201, 203, 205-207
principles of design 20, 61, 75, 77, 80, 87, 92-94, 112
printmaking 249, 252, 275
proceduralist 65
product design 18, 38, 146, 174, 182, 187, 193, 195-198, 206-207, 211, 270, 275
product semantics 188, 191, 193, 196, 199-200, 270, 307

Q

qualitative research design 236

R

Rorschach test 107

S

screenprinting 251, 275
semiotic content of product design 174, 187
semiotic phenomenology 13, 23
sensorimotor stage 152
sensory data 30, 160
sensory experience 34, 64, 111, 115, 137, 212
sensory information 25-27, 44-45, 49, 132, 135
signage 18, 195-196, 323
single photon emission computed tomography (SPECT) 136
Skype 115, 285, 342
smartphones 180, 344, 352
spatial perception 48-49, 294
spatial visualization abilities test (SVAT) 55
stage of formal operational thinking and conceptual thought 152
street art 268, 275
symbolic thinking 153
symbolist theory 39, 139

T

tag cloud visualization 277, 301, 310-311
taxonomy 100, 296-297, 307, 321-322
telecarving 215
telepresence 336, 349, 351
the internet of things 351
thematic apperception test (TAT) 107
traditional illustration 247-248, 275

U

ubiquitous computing 232, 320, 346-349, 355-356
U.S. Psychotronics Association 45

V

venogram 135
verbal communication 2, 9-10, 211
Vimeo 11, 115, 264
virtual music 334, 353
virtual reality immersive simulations 335, 352
visual analytics 277, 287, 301, 303, 309-310, 330-331
visual and verbal metaphor 208-209
visual guidance 301
visual intelligence 62, 106, 126, 131-132, 161-163
visualization in education 277, 317
visualization of the semantic web 277, 301, 307
visual learning 149, 162, 221-222, 277, 291, 314, 318, 326
visual literacy 61-64, 94-95, 115, 161-163, 171, 220-221, 223-224, 261, 295, 310
visual playfulness 301
visual reasoning 61, 131-132, 137, 145, 147, 167
visual unfreezing 301
visual variety 301
vocal learning 30, 38

W

web graphics 241, 285
web-search result visualization 277, 287, 304, 306
woodcut 249-250, 275

Y

Youtube 78, 115, 285, 310, 351